About Island Press

Island Press is the only nonprofit organization in the United States whose principal purpose is the publication of books on environmental issues and natural resource management. We provide solutions-oriented information to professionals, public officials, business and community leaders, and concerned citizens who are shaping responses to environmental problems.

In 2000, Island Press celebrates its sixteenth anniversary as the leading provider of timely and practical books that take a multidisciplinary approach to critical environmental concerns. Our growing list of titles reflects our commitment to bringing the best of an expanding body of literature to the environmental community throughout North America and the world.

Support for Island Press is provided by The Jenifer Altman Foundation, The Bullitt Foundation, The Mary Flagler Cary Charitable Trust, The Nathan Cummings Foundation, The Geraldine R. Dodge Foundation, The Charles Engelhard Foundation, The Ford Foundation, The German Marshall Fund of the United States, The George Gund Foundation, The Vira I. Heinz Endowment, The William and Flora Hewlett Foundation, The W. Alton Jones Foundation, The John D. and Catherine T. MacArthur Foundation, The Andrew W. Mellon Foundation, The Charles Stewart Mott Foundation, The Curtis and Edith Munson Foundation, The National Fish and Wildlife Foundation, The New-Land Foundation, The Oak Foundation, The Overbrook Foundation, The David and Lucile Packard Foundation, The Pew Charitable Trusts, The Rockefeller Brothers Fund, Rockefeller Financial Services, The Winslow Foundation, and individual donors.

Whitebark Pine Communities

Whitebark Pine Communities

Ecology and Restoration

Edited by
Diana F. Tomback
Stephen F. Arno
Robert E. Keane

ISLAND PRESS
Washington ♦ Covelo ♦ London

Library of Congress Cataloging-in-Publication Data

Whitebark pine communities : ecology and restoration / Diana F. Tomback, editor ; Stephen F. Arno, Robert E. Keane, consulting editors.
p. cm.
Includes bibliographical references and index.
ISBN 1-55963-717-X (cloth : alk. paper) — ISBN 1-55963-718-8 (pbk. : alk. paper)
1. Whitebark pine. 2. Whitebark pine—Ecology. 3. Restoration ecology. I. Tomback, Diana F. II. Arno, Stephen F. III Keane, Robert E.
QK494.5.P66 W55 2001
585'.2—dc21

00-011161

British Library Cataloging-in-Publication Data available.

Printed on recycled, acid-free paper 

Printed in Canada
10 9 8 7 6 5 4 3 2 1

This book is dedicated
to all those who know whitebark pine communities and
find special joy, pleasure, and appreciation for nature in their midst;
and to the memories of Seth Diamond, a pioneer in whitebark pine
restoration, and Bob Ogilvie, who provided invaluable information
about whitebark pine distributions in Canada.

Contents

Preface

In the northwestern United States and southwestern Canada, much of the whitebark pine (*Pinus albicaulis*) is disappearing. Fire exclusion at the upper elevations has resulted in advancing succession and a major decline in whitebark pine, which was historically an important seral component of many forests. The introduced disease white pine blister rust (*Cronartium ribicola*) has spread throughout the whitebark pine range with telltale signs of red-brown foliage on dying upper branches; dead bare branches; and cankers on branches and stems. On healthy trees, the upper branches bear the seed cones.

Those of us who have studied whitebark pine communities cannot help but take the losses and the imminent rangewide decline of whitebark pine personally. Several of us have devoted careers to understanding the ecological dynamics of these special communities. From our studies, we researchers realize that the current and future losses have negative implications for regional and local biodiversity, both plant and animal, as well as possible impacts on water quality and stream flows in subalpine watersheds. The losses of whitebark pine are inescapable to those of us who hike, hunt, and ski in the high country. This conspicuous mortality of an important subalpine species implies to us some fundamental failure in our current approach to the management of natural lands.

But there is hope. As researchers, we now understand the ecological dynamics of whitebark pine communities and the destructive processes currently at work. We have more than fifteen years of collective studies focused on whitebark pine. Based on detailed knowledge of blister rust infection and natural resistance in western white pine (*Pinus monticola*) and preliminary work in whitebark pine, we have devised strategies for restoring whitebark pine on both local and regional scales. Implementation will require institutional commitment and the hard work of dedicated people.

Another story that is implicit in the pages of this book is how a loosely organized group of researchers interested in whitebark pine coordinated their work and focused on filling in holes in the knowledge, without the benefit of a major funding source. Perhaps this story can serve as a model for other threatened communities and ecosystems as well. One instigator of this effort was Stephen Arno, who had worked extensively in whitebark pine communities in the course of studying subalpine forest communities and the natural role of fire. In 1986, he published "Whitebark Pine Cone Crops—A Diminishing Source of Wildlife Food," a short paper that described the synergistic destructive forces impacting whitebark pine in the northern part of its range; this paper was a call to action. Arno also began collaborating with agency and university researchers who had studied or were interested in whitebark pine communities. Under the informal sponsorship of the USDA Forest Service's Intermountain Research Station (now the Rocky Mountain Research Station) and especially the station's Fire Sciences Laboratory in Missoula, Montana, workshops were held to determine what was known about whitebark pine and what new knowledge was needed, and directed research gradually began filling gaps. Thus a fledgling scientific network was established. In 1989, a major symposium devoted to whitebark pine communities was held at Montana State University in Bozeman. Hosted by research silviculturists Ward McCaughey and Wyman Schmidt, this symposium attracted researchers, managers, educators, naturalists, and the lay public.

The timing of our work and interest in whitebark pine meshed well with the developing field of "ecosystem management" and "new perspectives in forestry." Forest geneticist Ray Hoff synthesized the wealth of information on blister rust resistance in western white pine and assessed its potential application to whitebark pine; he began blister rust surveys and studies of rust resistance in whitebark pine. Robert Keane undertook a detailed, multifaceted study of whitebark pine communities in the Bob Marshall Wilderness Complex and developed ecological process models to predict the future of whitebark pine communities under different fire regimes and blister rust epidemic scenarios. Several scientists—including Diana Tomback, Ronald Lanner, Katherine Kendall, and David Mattson—studied the use and importance of whitebark pine seeds to the animal community, notably pioneering studies of whitebark pine seed dispersal by Clark's nutcracker (*Nucifraga columbiana*) and the use of whitebark pine seeds by grizzly bears (*Ursus arctos*). The other research collaborators are too numerous to list, but many are authors of the chapters in this book. Several forest biologists, silviculturists, and fire managers also joined this group, which was recognized by the USDA Forest Service Intermountain Research Station with a Centennial Conservation Award in 1991.

This book has been a labor of love for the editors and contributors, and an important opportunity to communicate whitebark pine's precarious status and summarize the state of our knowledge to a broader audience. The editors wish to point out that the opinions expressed and scenarios described by several authors on controversial subjects such as global warming and management in protected areas are those of the authors and not necessarily shared by all contributors. We believe that the differing viewpoints offer readers useful perspectives and challenge all of us to reexamine our own assumptions.

The botanically astute will notice that we continue to refer to "subalpine fir" (*Abies lasiocarpa*) instead of "Rocky Mountain subalpine fir" (*Abies bifolia*), as listed in the *Flora of North America* (1993, Vol. 2, Oxford University Press, New York), recognizing the need for additional studies.

The editors wish to acknowledge those individuals who served as reviewers, sometimes on short notice, and the helpful staff at Island Press, particularly Barbara Dean for guidance in developing and writing the book, and Barbara Youngblood, for help with the technical aspects. We thank the USDA Forest Service Rocky Mountain Research Station for permission to use the cover illustration drawn by Gina Gahagan, which should be familiar to attendees of the 1989 Symposium on Whitebark Pine Ecosystems: Ecology and Management of a High-Mountain Resource. A number of authors kindly contributed photographs to illustrate the book. Also, Norma Williamson, editor of *The Dubois Frontier,* Dubois, Wyoming, graciously permitted us to use her photos of whitebark pine cone collecting operations. We also wish to acknowledge the University of Colorado at Denver for institutional support during the process of editing and assembling this book, and the Fire Sciences Laboratory of the USDA Forest Service Rocky Mountain Research Station, for staff and institutional support.

As this book goes to press, several of the contributors have joined together to establish a Whitebark Pine Ecosystem Foundation (WPEF), whose mission is to "promote the conservation of whitebark pine ecosystems by supporting educational, management, and research projects that enhance knowledge and stewardship." For more information about this effort, please write to WPEF at P.O. Box 16503, Missoula, Montana 59808, or contact one of the three editors of this volume.

DIANA F. TOMBACK
STEPHEN F. ARNO
ROBERT E. KEANE
February 6, 2000

Whitebark Pine Communities

Part I

Statement of the Problem

This first part of *Whitebark Pine Communities: Ecology and Restoration* contains a single chapter written by the editors, "The Compelling Case for Management Intervention," which serves both as an introduction to the book and as an overview of the "whitebark pine story." We crafted this chapter as a précis for the reader to communicate what is unique about whitebark pine (*Pinus albicaulis*), the multiple ways in which whitebark pine is a keystone species in subalpine communities, the major problems threatening whitebark pine, and finally, the knowledge and tools available to restore whitebark pine communities. With this overview, the rationale of dividing the book into Parts II, III, and IV becomes apparent. Chapters in each subsequent part of the book provide detailed information on topics briefly treated by this first chapter, and the reader can better integrate the information presented in the subsequent chapters within the context of the larger picture.

In this first chapter, we make one additional contribution: We raise the question and trace the history of how a tree species that grows at high elevations and in remote places in "protected areas"—wilderness, national parks, and national forests—can be so profoundly affected by anthropogenic events. What has happened to whitebark pine underscores the fact that human influence is everywhere, and no ecosystem will escape perturbation. With anthropogenic effects ranging in scale from climate change to long-distance transport of pollutants, to regional fire suppression and the local introduction of exotic organisms, this pervasiveness of human influence requires that we rethink our strategies and goals for both wildland management and preservation.

Chapter 1

The Compelling Case for Management Intervention

Diana F. Tomback, Stephen F. Arno, and Robert E. Keane

Why Is Whitebark Pine in Trouble?

Whitebark pine (*Pinus albicaulis*) is familiar to all western high-mountain recreationists for its distinctive spreading canopies and majestic, wind-battered growth forms in the upper subalpine zone (Figures 1-1 and 1-2). It occurs in small stands on steep, windswept slopes and ridges where other conifers can scarcely grow—often the last erect trees before the forest is reduced to krummholz (shrublike) patches and then the tundra above. Whitebark pine is also known to naturalists for its bird-dispersed seeds and its designation as critical grizzly bear habitat in the northern Rocky Mountains of the United States. Many of us cannot imagine backcountry and wilderness without whitebark pine. In our minds we picture our destinations, the rugged ridges and peaks and the alpine lakes; among their guardians, the windswept whitebark pine.

That raises the paradoxical question: Why is a high-mountain species, a widespread conifer that is usually found in places rarely disturbed by humans, in trouble? Why aren't whitebark pine communities pristine and thriving?

Whitebark pine could be a twenty-first-century environmental symbol—a sort of ecological "poster child"—for the combined consequences of altered fire regimes and the introduction of exotics to western North America. It is also symbolic in the larger sense—for the awareness that "preservation" alone is not the answer to saving biodiversity. In about half of its range in the northern Rocky Mountains of the United States and Canada, this once abundant pine is rapidly declining from the combination of successional replacement and the exotic fungal disease white pine blister rust (*Cronartium ribicola*). Although

Figure 1-1. Whitebark pine (*Pinus albicaulis*) in a mixed upper subalpine forest community on windswept ridgelines in the Greater Yellowstone Area. The spreading canopies of whitebark pine are highly distinctive. Photo credit: Diana F. Tomback.

Figure 1-2. Whitebark pine with multiple stems at Tioga Pass, Yosemite National Park, California. Photo credit: Diana F. Tomback.

whitebark pine is less fire-dependent elsewhere, mass mortality caused by blister rust is now a rangewide threat.

The Pervasive Human Influence

People in general perceive the landscape of wilderness areas, national forests, national parks, and other public forestlands as "natural" and thus relatively free from human influence. Beyond the concessions, campgrounds, hotels, and roads, they see a vast area that has been shaped by timeless, natural processes. They assume that what they see has existed for millennia and will continue to exist in the same state if we provide the proper safeguards. What many do not understand is that important changes occur across these forest landscapes as both direct and indirect results of human influence.

These changes are not a recent phenomenon. Historical records show that in presettlement times Native Americans set fires to improve forage for game and horses and to prevent the establishment of trees (Gruell 1985; Boyd 1999). In areas such as the Bitterroot Valley, Montana, Indian-ignited fires nearly doubled fire frequencies for montane and lower subalpine zone communities, and this pattern appears to have existed for several centuries prior to Euro-American settlement in the mid-1800s (Arno 1976; Barrett and Arno 1982). Postsettlement and recent landscape alterations result from a variety of activities, including logging, grazing, mining, air and water pollution, recreational uses, and urban expansion. The scales of these disturbances are usually local but cumulative. In contrast, we are faced with changes on massive scales—regional or global disturbances of uncertain consequences from climate change, and from alteration of biogeochemical cycling from pollutants such as nitrogen and sulfur.

Two additional consequences of the human presence have been both large-scale and pervasive and only recently recognized for their damaging effects—the exclusion of fire from many historically fire-dependent communities throughout the United States for much of the twentieth century (e.g., Weaver 1943; Arno 1996a), and the introduction of exotic species, particularly plants but also animals and diseases, into terrestrial and aquatic ecosystems (e.g., Coblentz 1990). Both change the composition and even the ecosystem processes of native landscapes. For example, the spread of cheatgrass (*Bromus tectorum*)—a grass introduced from Europe to improve cattle forage—over much of the Great Basin has increased local fire frequencies, replaced native grasses (West 1988), and reduced the diversity of small mammal communities (L. Carpenter, Denver Museum of Natural History, personal communication).

People excluded fire from our natural lands for most of the twentieth century to protect the land from what they regarded as a destructive force. They did not understand that periodic fire is essential to maintaining many plant communities and their biodiversity (Botkin 1990; Agee 1993). Now that we understand the historical importance of fire, modern society has imposed additional constraints to returning natural fire regimes: The spread of residential development to higher elevations and wilder places prohibits natural fire and even prescribed burns from many areas. The Clean Air Act and other environmental legislation place barriers to allowing fire to resume its natural role. In the montane west, fires often begin at lower elevations and burn into higher elevations—now incorporated into our national forests and wilderness areas (Czech 1996). Because these low-elevation fires must be quickly suppressed, our high-elevation forest communities are buffered in many areas from their historical source of fire at lower elevations and, consequently, are experiencing changes in their composition and function.

Our Goals for This Book

We selected the topics and contributors and determined the organization of this book with several goals in mind: (1) to convey what is known about whitebark pine communities and their ecological value, (2) to make known the precarious status of whitebark pine communities, (3) to present the state of knowledge of restoration alternatives, and (4) to encourage and facilitate the restoration process.

Much of the information presented by the contributors to this volume results from more than two decades of research, with the last fifteen years of work coordinated to address gaps in our understanding of the basic biology of whitebark pine and the effects of fire suppression and blister rust on whitebark pine communities. In fact, we believe that the research and modeling strategies that were developed for whitebark pine communities could be applied to other fire-dependent forest communities that are widespread throughout the world (Pyne 1997), but also for disturbed ecosystems in general.

This first chapter presents an overview of the ecology and status of whitebark pine communities and the case for management intervention. It is followed by Part II, "The Biology of Whitebark Pine," which contains a series of chapters presenting basic understanding of whitebark pine taxonomy, distribution, and ecology, including environmental tolerances, community disturbance processes, regeneration processes, species interactions, and genetic population structure. The chapters under Part III, "Whitebark Pine Communities: Threats and Consequences," identify the threats and perturbations to whitebark pine communities and survey the extent of impact and losses to date. Part IV, "Restoring Whitebark Pine Communities," demonstrates that we have the basic knowledge and management tools to begin restoring these communities locally and on a significant scale regionally. Obviously, monitoring and refining restoration methods and goals are a critical part of this process. What we do *not* have, however, is the luxury of taking plenty of time to implement these techniques; implementation must begin now.

About Whitebark Pine

Whitebark pine is a hardy subalpine conifer that tolerates poor soils, steep slopes, windy exposures, and tree-line environments (Arno and Hoff 1990). It ranges from the coastal mountains of central British Columbia south along the crest of the Cascade Range and Sierra Nevada, and from the Canadian Rocky Mountains south through the northern Rocky Mountains of Idaho and Montana to the edge of the Wyoming Basin (Chapter 2). In the northwestern United States and southwestern Canada, many whitebark pine communities that occur in the upper subalpine zone are successional and thus subject to replacement by shade-tolerant conifers. In the northern Rocky Mountains of the United States, whitebark pine communities historically accounted for 10 to 15 percent of the total forest cover (Arno 1986). More than half of the whitebark pine communities in this region occur on relatively moist, productive sites and are successionally replaced through time (Chapter 4). These successional communities depend on periodic fire for renewal. In the more droughty regions, such as the Sierra Nevada, whitebark pine occurs primarily as a self-perpetuating climax species, but its communities cover a more limited area of the high country than in the Rocky Mountains (Tomback 1986; Barbour 1988).

From the perspectives of taxonomy, evolution, and ecology, whitebark pine is particularly interesting: It is the only New World species in the pine (*Pinus*) subsection *Cembrae,* which otherwise includes the European and Asian species Swiss stone pine (*Pinus cembra*), Korean stone pine (*P. koraiensis*), Japanese stone pine (*P. pumila*), and Siberian stone pine (*P. sibirica*) (Critchfield and Little 1966). *Cembrae* pines have large, wingless seeds, and cones that do not open when seeds are ripe. Consequently, the ripe seeds are not released; and even if they were, their large size and lack of wings would otherwise preclude effective dispersal by wind.

For all *Cembrae* pines, seed dispersal depends primarily on the seed-harvesting and caching behavior of birds of the genus *Nucifraga,* the "nutcrackers" (Family Corvidae) (Tomback and Linhart 1990 and references therein; Lanner 1996). Whitebark pine seeds are dispersed primarily by Clark's nutcracker (*Nucifraga columbiana*) (Tomback 1978, 1982; Hutchins and Lanner 1982), which ranges throughout the higher mountains of western North America (Tomback 1998). In fact, tree establishment from seed dissemination by nutcrackers profoundly affects several aspects of the population biology and population genetic structure of whitebark pine and its relatives (e.g., Rogers et al. 1999; Feldman et al. 1999; also Chapters 5 and 8). Nutcrackers also commonly feed on the seeds of other relatively large-seeded pines, especially limber pine (*Pinus flexilis*), the piñon pines (*P. monophylla, P. edulis*), and ponderosa pine (*P. ponderosa*) (Tomback and Linhart 1990; Lanner 1996).

Why Restore Whitebark Pine Communities?

Whitebark pine serves several important ecological functions throughout its range, such as protecting watersheds, promoting post-fire forest regeneration, and providing a food source for wildlife. In fact, with respect to subalpine zone biodiversity, whitebark pine is considered a keystone species.

Whitebark Pine as a Keystone Species

Whitebark pine is a *keystone species* of upper subalpine ecosystems. The keystone species concept, developed from empirical ecological studies, is now important to the field of conservation biology for its management implications. R. B. Primack (1998, 44) defines keystone species as follows: "Within biological communities, certain species may determine the ability of large numbers of other species to persist in the community." In other words, a keystone species performs a role or function that increases the biodiversity of a community. Most keystone species play a single important role, such as fig trees (*Ficus* spp.) in the tropics providing fruit for frugivorous birds, primates, and bats while other tree species are not fruiting (Terbourgh 1986; Lambert 1991), or the predator starfish *Pisaster ochraceus* maintaining population balance among prey species (Paine 1966). In contrast, whitebark pine maintains biodiversity in multiple ways.

Whitebark Pine Seeds as a Wildlife Food Source

A well-known keystone function of whitebark pine is its role as a food source in late summer for a number of birds and mammals, including nutcrackers, squirrels, and bears.

Food source for small birds and mammals. Whitebark pine has the largest seeds of all conifers at subalpine elevations throughout its range, with a mean of 175 milligrams per seed. For comparison, subalpine zone pines that may co-occur with whitebark pine and their mean seed weight include limber (93 mg), foxtail (*P. balfouriana,* 27 mg), bristlecone (*P. aristata,* 25 mg), western white (*P. monticola,* 16 mg), and lodgepole (*P. contorta,* 4 mg) (seed weights from Table 3 in Tomback and Linhart 1990).

Shelled whitebark pine seeds contain about 21 percent carbohydrates, 21 percent protein, and about 52 percent fat by weight (Lanner and Gilbert 1994). The seed protein comprises sixteen amino acids, including relatively large quantities of lysine, which is considered important for the diet of birds (Lanner and Gilbert 1994 and references therein). Also, the fairly high fat content of whitebark pine seeds provides a high-energy food, which is particularly important for animals during cold weather. Whitebark pine seeds are therefore a good dietary fit for nutcrackers and other avian consumers.

When whitebark seeds are ripe in mid- to late August, nutcrackers and a number of other birds and small mammals begin taking whole seeds from cones, both for eating and for storage as winter food, and pine squirrels (*Tamiasciurus* spp.) cut down cones for storage in middens (Tomback 1978; Hutchins and Lanner 1982; Kendall 1983). Wildlife species that eat whitebark pine seeds vary geographically, but include woodpeckers, jays, ravens, chickadees, nuthatches, finches, chipmunks, ground squirrels, and probably mice (Tomback 1978; Hutchins and Lanner 1982; Chapter 12).

Food source for grizzly bears and black bears. Whitebark pine seeds are a major seasonal food source for grizzly bears (*Ursus arctos*) and black bears (*U. americanus*) throughout the range of whitebark pine, but particularly in the Greater Yellowstone Area (Kendall 1983; Mattson et al. 1991; Mattson and Reinhart 1994), the east front of the Montana Rocky Mountains, and, until recent declines, the greater Bob Marshall ecosystem (Craighead et al. 1982; Chapter 7). Based on twenty years of scat data from the Greater Yellowstone Area, Mattson et al. (1991) concluded that the major food types consumed by grizzly bears are ungulates in spring, grasses in summer, and whitebark pine seeds in fall. When whitebark pine seeds are available, Yellowstone grizzly bears will feed on the seeds nearly exclusively. Whitebark pine seeds were recorded in the grizzly bear diet from May through October, but particularly in September and October.

Bears in more maritime (coastal) climates eat whitebark pine seeds to a lesser extent, because of lower availability, and rely primarily on fleshy fruits for fattening in the months before hibernation. In regions of continental climate, fleshy fruits are scarce, and whitebark pine seeds are the preferred food of grizzly bears and black bears (Mattson et al. 1991; Mattson and Reinhart 1994).

Bears usually obtain whitebark pine seeds from cones hoarded by pine squirrels (*Tamiasciurus hudsonicus* and *T. douglasii*) in middens, which are piles of cones and cone debris that have accumulated over years in a squirrel territory (Chapter 7). During good cone production years, bears take freshly harvested cones that lie on the surface of the middens; in fall, spring, or summer they also dig out middens in search of cones (Kendall 1983; K. C. Kendall, personal communication). Pine squirrels are usually present in mixed whitebark pine communities—for example, in several subalpine fir (*Abies lasiocarpa*) habitat types in the Greater Yellowstone Area (Chapter 4). In pure whitebark pine (climax) communities, pine squirrels are rarer (Mattson and Reinhart 1994).

Whitebark pine communities as critical grizzly bear habitat. In the Greater Yellowstone Region, whitebark pine seeds and ungulates are considered the two most important foods of grizzly bears. Mattson et al. (1992, 433) state that the "availability of whitebark pine seeds has the greatest potential of any food-related factor to impact behavior and demography of the Yellowstone grizzly bear population." In years when large whitebark pine cone crops are produced in the Greater Yellowstone Area, grizzly bears forage for seeds at high elevations, away from major roads, development, and most tourists. In poor whitebark pine cone production years, bears wander at lower montane elevations in search of alternative food sources, often interacting with humans. This results in six times more management actions (e.g., trapping problem bears), twice the mortality of mature females, and three times the mortality of subadult males, compared with good cone crop years (Mattson et al. 1992). Because of the importance of whitebark pine seeds to grizzly bear ecology in the Greater Yellowstone Area, whitebark pine communities are designated as critical habitat in grizzly bear recovery plans for that area and others (e.g., U.S. Fish and Wildlife Service 1997).

Within the last fifty or sixty years, grizzly bears were still numerous in the wetter mountain ranges of northwestern Montana and northern Idaho. Whitebark pine seeds were considered an important food for grizzly bears in these areas. However, mountain pine beetles (*Dendroctonus ponderosae*), blister rust, and advancing succession have since killed most of the trees in these regions, and grizzly bears have disappeared or declined in many areas (Kendall and Arno 1990). Today, bears still feed on whitebark pine seeds in the greater Bob Marshall ecosystem (Craighead et al. 1982), but with recent declines in whitebark pine from blister rust and fire, this food source is rapidly diminishing as well (Keane et al. 1994).

Whitebark Pine Communities Are Diverse

The structure and composition of whitebark pine communities vary greatly over their broad geographic range and locally with differences in elevation (Chapter 4) and community successional status (Chapter 9). Whitebark pine communities taken as a whole harbor a wide diversity of understory plants (e.g., see Pfister et al. 1977; Steele et al. 1983; Chapter 12). Also, whitebark pine communities often form ecotones with other community types, especially wet or dry meadows, wetlands, and alpine tundra, producing additional structural complexity and greater plant diversity (Arno and Hammerly 1984; Chapter 12). To this basic plant diversity must be added diversity in mycorrhizal fungi (symbiotic root fungi), other multicelled and single-celled fungi, soil and leaf microorganisms, lichens, mosses, and bryophytes.

Throughout their range, whitebark pine communities provide food, shelter, nesting sites, tree holes, and burrows, territories, and home ranges to a variety of animals—both vertebrate and invertebrate. The animals vary both seasonally and geographically, and with elevation, community type, and successional status of the community (Chapter 12). Although few species appear restricted in distribution to whitebark pine communities, the particular associations of species in whitebark pine communities and the number of species summed over all community types represent important contributions to western North American forest biodiversity.

Whitebark Pine Communities Regulate Runoff and Reduce Soil Erosion

Classified as a stress-tolerant pine (McCune 1988), whitebark pine pioneers after fire or other disturbance, and forms climax stands on dry, windswept sites throughout the subalpine zone up to tree line throughout its range (Arno and Hoff 1990). The species assumes a matlike, or krummholz, growth form at tree line, occurring in dense tree islands (Arno and Hammerly 1984). Its hardiness enables it to grow where other conifers cannot.

The following information, specific to the hydrological effects of whitebark pine communities, comes from Farnes (1990): Snowpack melts off completely from whitebark pine communities as early as mid-May to as late as mid-July. Because the soils at these elevations are poorly developed with little water-holding capacity in general, 35 to 60 percent of the annual precipitation becomes runoff in these watersheds. The presence of trees at higher elevations, and of whitebark pine in particular, slows the progression of snowmelt, resulting in later melt-off and higher stream flows in summer months, and reduced spring flooding at lower elevations. Also, whitebark pine communities accumulate more snow than annual precipitation measurements would indicate because of local drifting and snow redistribution. The broad, spreading, open canopies of whitebark pine, compared to the narrow, dense canopies of other conifers, shade larger surface areas and further reduce the rates of snowmelt.

The fact that whitebark pine is present at high elevations on poor sites not tolerated by other conifers both regulates runoff and reduces soil erosion. Slower, more protracted runoff is less damaging to unstable soils. Tree roots physically stabilize soils and take up water, further slowing runoff rates and reducing erosion potential. This results in high-quality water for human use and for low-elevation aquatic ecosystems.

Whitebark Pine Regenerates After Fire

As a climax species, whitebark pine replaces itself through time on exposed sites in the upper subalpine zone and at tree line. As a "pioneer," or early seral species, whitebark pine may be the first conifer to become established after disturbance, initiating the successional process in upper subalpine forests (Lanner 1980; Tomback 1986; Tomback et al. 1993; Tomback 1994). How long after disturbance before the first whitebark pine seedlings appear depends on both seed availability and seed dormancy requirements (McCaughey 1993; Tomback 1994; Chapter 6).

Two factors contribute to the early establishment of whitebark pine following a disturbance, such as fire: First, whitebark pine seedlings are exceptionally hardy, and more tolerant of exposed sites and drought than are the seedlings of associated conifers (e.g., Arno 1986; Tomback 1986, 1994). Newly germinated whitebark pine seedlings have thick stems and long, sturdy cotyledons, and develop deep taproots quickly (Arno and Hoff 1990; Chapter 6). Second, Clark's nutcrackers readily cache whitebark pine seeds in recently disturbed open areas, such as burns and clear-cuts (Tomback 1986, 1994; Chapter 5). They transport whitebark pine seeds 8 kilometers or more from seed sources to caching sites in large burns, thereby initiating whitebark pine regeneration over large areas (Tomback et al. 1990; Tomback et al. 1993; Tomback et al. 1995). Thus, between the hardiness of its seedlings and its highly effective dispersal system, whitebark pine becomes established early in succession.

Whitebark Pine Facilitates Succession

The environmental conditions after stand-replacing fire are extremely harsh, particularly on south- and west-facing slopes, which receive direct solar radiation for much of the day. Loss of canopy cover and understory reduces shade and increases wind exposure at ground level. In general, soil that is directly exposed to solar radiation in summer attains surface temperatures over 60°C, but soil that is either charred by fire or covered by a layer of conifer needle litter (duff) can achieve temperatures of 72°C and 68°C, respectively, or higher (Kimmins 1997). These high soil surface temperatures can kill many seedlings in their first growing season.

Furthermore, stand-replacing fires and severe surface fires, which remove all vegetation and destroy the organic soil layers, change soil water relations. In whitebark pine ecosystems, soils are generally coarse-textured with minimal organic matter (Arno and Hoff 1990); but, for these soil types most water storage occurs within the organic layers. With the loss of organic layers to fire, and the structural and chemical changes to the mineral layer, soils become drier with less water infiltration and greater surface evaporation (Kimmins 1997).

Seedlings of whitebark pine and other conifers such as subalpine fir, Engelmann spruce (*Picea engelmannii*), and lodgepole pine may appear nearly synchronously after fire (Tomback et al. 1993). The hardiness of whitebark pine seedlings, however, enables them to tolerate harsh postfire conditions on all slope aspects and exposures (Tomback et al. 1993; Tomback 1994). The presence of whitebark pine on harsh sites provides microenvironments with shade, moisture, and shelter from wind that facilitates establishment of conifer associates and understory vegetation. Thus, whitebark pine as a pioneer facilitates the successional process by creating favorable habitats for establishment and growth (e.g., Lanner 1980).

In addition, at upper subalpine elevations in the northern Rocky Mountains, whitebark pine may act as a "nurse" tree, facilitating the survival and growth of its forest competitors subalpine fir and Engelmann spruce on burned or open sites. These sites at high elevations experience lower soil water availability, lower snow depth, and higher wind speeds; and often many subalpine fir trees are aggregated around larger whitebark pine trees. Callaway (1998) found that subalpine fir trees without a neighboring whitebark pine experience significantly lower growth rates than fir trees growing near whitebark pine; the death of the "nurse" whitebark pine resulted in a 24 percent decrease in subalpine fir growth rates. As subalpine fir saplings grow above the snowpack, their needles are wind-blasted by particles of ice and snow, but those trees growing in proximity to a large whitebark pine are sheltered to some extent.

On a variety of sites in the upper subalpine zone (e.g., on moraines that rise above meadows, on rocky cliffs, and in other exposed locations), whitebark pine alone is the pioneer. Lanner (1980, 1996, and references therein) describes the following scenario for Squaw Basin, Bridger-Teton National Forest, Wyoming: Small groves of whitebark pine first become established on the tops of moraines, initially no more than a few seedlings. As more trees become established and grow larger, and the canopy closes, producing a shadier, protected microenvironment in winter and summer, Engelmann spruce invades—about 120 years later. In the meantime, the sheltered environment develops a forest understory of woody and herbaceous plants, which replaces the meadow vegetation. Similar observations for extremely dry, wind-exposed sites elsewhere indicate that whitebark pine is the pioneer, but stands are later colonized and even replaced by shade-tolerant species, such as subalpine fir and mountain hemlock (*Tsuga mertensiana*) (Franklin and Dyrness 1973) or form a coclimax with lodgepole pine (Steele et al. 1983).

Whitebark Pine Is the "Quintessential" High-Mountain Conifer

Whitebark pine defines the upper subalpine and tree-line experience for the western wilderness and backcountry explorer. The range of whitebark pine includes every high-elevation national park but one (Rocky Mountain National Park, Colorado) in the western United States. In the northern Rocky Mountains alone, whitebark pine is found in twenty-five national forests. Within the three largest wilderness complexes in the western United States—the Bob Marshall, the Selway-Bitterroot–Frank Church, and the Greater Yellowstone—about 49 percent, 23 percent, and 47 percent, respectively, of their areas are potential whitebark pine habitat. In addition, about 98 percent of all whitebark pine communities occur on public lands, with 48 percent in national forests and 49 percent in wilderness areas and national parks (R. E. Keane, unpublished data).

Reaching the highest whitebark pine communities, the gnarled and windswept forms on the last forested cliffs and ridgetops, signals achievement to the hard-driving backpacker, rock climber, or backcountry skier—the "extreme" sports enthusiast—and spiritual satisfaction to the nature lover and more casual hiker. Alpine or backcountry skiing among the whitebark pine stands conveys a special thrill; you are indeed in the highest place, otherwise inaccessible, where survival in winter is a challenge to every plant and animal. Mountain recreationists associate their high-elevation experiences with whitebark pine's spreading, upswept canopies silhouetted against the darkening western sky. These stands are often open and parklike, creating a pastoral high-mountain setting. In summer and in the early days of fall, visitors to this high country camp, picnic, or relax within the sun-drenched whitebark pine stands, which may be teeming with nutcrackers and other birds and squirrels harvesting pine seeds, activities that are accompanied by a cacophony of communication calls and scolding.

Why Is Whitebark Pine Declining?

The decline of whitebark pine comes from a synergism of natural and human-driven causes. Periodic, massive outbreaks of mountain pine beetle, killing mature whitebark pines, have been exacerbated by suppression of natural fires. A major reduction in high-elevation fires since the early 1900s has led to successional replacement of whitebark pine on more productive sites in the part of its range where it otherwise should be abundant. Finally, white pine blister rust is killing whitebark pine trees in the intermountain region, coastal ranges, and Canadian Rocky Mountains, and rangewide mortality is expected within one to several decades.

Fire Exclusion

In regions where whitebark pine communities are successional, fires occurred historically every fifty to four hundred years (Arno 1980, 1986; Romme 1982; Barrett 1994). These fires were of two severities—stand-replacement fires and mixed-severity fires (Chapter 9). Fires were ignited at subalpine elevations by lightning strikes, or fires burned upslope from lower elevations.

The slow-moving surface and ground fires spare more whitebark pine than its main competitors Engelmann spruce and subalpine fir, which are more fire-sensitive and flammable (Arno 1986). The newly open forest understory favors the surviving whitebark pine, thereby maintaining their dominance. Stand-replacement fires provide whitebark pine a successional advantage over other conifers because its seeds are planted by nutcrackers throughout the burned area. Nutcrackers disperse seeds farther and faster than wind disperses the seeds of its forest associates. Fire is particularly important on moist sites, where successional replacement by subalpine fir occurs rapidly (Arno 1986).

Since about 1929, fire exclusion—the result of active fire suppression—has reduced the frequency of fires in subalpine communities, resulting in the successional replacement of whitebark pine by shade-tolerant species on many landscapes (Arno 1986). Although some lightning-ignited fires were allowed to burn in national parks and wilderness areas beginning in the early 1970s, the area burned per year since still falls considerably short of the rate prior to human settlement (Brown et al. 1994). Arno (1986) examined the extent of area inhabited by successional whitebark pine with respect to area recently burned by prescribed fires and concluded that the contemporary fire return interval is now about three thousand years, at least ten times the historical average. The major consequence is widespread senescence and mortality of whitebark pine among successional community types. Instead of the historic mosaic of whitebark pine communities at different seral stages—from open stands of whitebark pine and fir to predominantly fir and spruce associations—the landscape is becoming homogeneous with late seral forests of subalpine fir and spruce (Chapter 9).

Exclusion of fire has similarly advanced succession in lower-elevation communities, resulting in abnormal fuel accumulation and denser understory (Arno 1996b). However, residential development now extends well outside cities and towns, up hillsides and canyons to middle elevations (Hill 1998), and in the case of ski facilities, even to high elevations. With the combination of dense forest, heavy fuel loadings, and nearby development, natural and prescribed fires are no longer a viable option in many areas. Furthermore, because fires ignited at the lower elevations often burn upslope, fire exclusion at lower elevations greatly reduces the occurrence of fire in wilderness areas and at subalpine elevations in general.

Mountain Pine Beetles and Mistletoe

Mountain pine beetles attack large, mature whitebark pine, killing healthy trees with inner bark thick enough to support larvae. Mountain pine beetles are commonly found in lodgepole pine forests, but there are often low levels of endemic infestation in whitebark pine. Dispersing mountain pine beetles move from the lodgepole pine zone up into adjacent whitebark pine stands, especially in epidemic peaks (Arno 1986; Bartos and Gibson 1990). Periodically, mountain pine beetle outbreaks cause widespread mortality in both lodgepole and whitebark pine. During the twentieth century, large-scale infestations of mountain pine beetle killed entire forests of whitebark pine in Idaho and Montana, including Glacier National Park, and the epidemic remains active (Bartos and Gibson 1990; Kendall and Arno 1990; Chapter 11).

Where whitebark pine is fire-dependent, fire suppression results in a greater abundance of late successional forests, which increases the scale and frequency of beetle infestations. In these areas, whitebark pine is less apt to survive a beetle infestation, because it is already stressed from competition with subalpine fir. In the absence of fire, both lodgepole pine and whitebark pine forests increase in age, and thus suitability for mountain pine beetles (Peterman 1978; Marsden 1983; McGregor and Cole 1985). With persistent infestations in older lodgepole pine, whitebark pine populations become highly vulnerable.

Another synergistic effect may occur between mistletoe and fire exclusion in whitebark pine communities. Several dwarf mistletoe species (*Arceuthobium* spp.) parasitize whitebark pine, but the most prevalent mistletoe is limber pine dwarf mistletoe (*A. cyanocarpum*) (Arno and Hoff 1990). Although it is not a rangewide problem, dwarf mistletoe can cause major local mortality in whitebark pine and can lower the vigor of parasitized trees (Mathiasen and Hawksworth 1988). Prolonged fire return intervals provide the time and opportunity for dispersal of mistletoe among host trees.

White Pine Blister Rust

Native to Eurasia, white pine blister rust is a stem rust that infects only five-needled white pines. The rust initially enters white pines through needle stomata, grows into the branches and stems, and erupts as spore-producing cankers that kill the branches—thus ending cone production—and finally kills the tree itself. Because the rust is slow-growing, infected mature trees may live more than a decade. Blister rust requires the widely distributed genus *Ribes* (the currants or gooseberries) as an alternate host for its complex life cycle. White pine blister was inadvertently introduced to western North America in 1910 near Vancouver, British Columbia, and the first blister rust infections in stands of western white pine and whitebark pine were observed in the region in the northwest in 1921 or shortly thereafter (Hoff and Hagle 1990; Chapter 10). From there, blister rust has relentlessly spread throughout the ranges of western white pine, an important timber species, whitebark pine, and more recently to sugar pine (*Pinus lambertiana*), limber pine, and southwestern white pine (*P. strobiformis*).

The northwestern Rocky Mountains, and the Olympic and western Cascade Ranges of the United States, and the Coastal and Rocky Mountains of southwestern Canada have the highest whitebark pine infection rates and the greatest whitebark pine mortality to date—more than 90 percent in some areas (Kendall and Arno 1990; Keane et al. 1994; Chapter 11, Figure 11-1). In areas such as Glacier National Park and the Selkirk and Cabinet Mountains, whitebark pine losses are so great that seed production is sparse and regeneration is unlikely. For example, twenty-five years after the 1967 Sundance Burn in the Selkirk Range of northern Idaho, whitebark pine regeneration density in the burn was significantly lower than in burns of similar age and size elsewhere; 29 percent of the whitebark pine seedlings in the burn were infected by blister rust (Tomback et al. 1995). High numbers of blister rust–infected trees are now present in the Pacific Northwest coastal ranges (Kendall and Arno 1990) and Canadian Rocky Mountains (Stuart-Smith 1998), and infected trees occur virtually everywhere throughout the range of whitebark pine (Chapters 10 and 11).

Thus, the combination of advancing succession, which leads to the replacement of whitebark pine by shade-tolerant conifers; mountain pine beetle infestations, which kill trees; and the blister rust epidemic, which kills cone-bearing branches years before the entire tree is killed, has devastated whitebark pine populations throughout the Pacific Northwest and northern Rocky Mountains of the United States and Canada. Further declines in these and other areas are imminent from all factors, but particularly the spread and intensification of blister rust infection. With the loss of trees, we also lose seed source for new regeneration. The situation is now so dire that land stewards are faced with a major dilemma: Even if they used fire or cutting techniques to open up stands for whitebark pine regeneration, the few remaining whitebark pine trees may be an inadequate source of seeds, and nutcrackers may consume the limited supply of unripe seeds long before cone ripening. Ideally, major tracts of forest should be disturbed to enable the regeneration of large numbers of whitebark pine seedlings, so that the small percentage with some resistance to blister rust have an opportunity to survive (Chapter 17). This works only with sufficient seed availability. Land stewards now must consider a more costly alternative, but one that is already being implemented on a limited scale: planting whitebark pine seedlings grown in nurseries. In addition, with selective cone harvesting from trees that survived blister rust infection, a proportion of seedlings may be rust-resistant.

Genetic evidence suggests that white pine blister rust mutates into different strains, and a more drought-tolerant strain has recently been documented (Chapter 17). The slower invasion of blister rust into the more arid regions of the whitebark pine range and throughout the range of the more drought-tolerant white pines, particularly limber and southwestern white pine, could be accelerated at any time.

What Are the Consequences of Losing Whitebark Pine?

The consequences of losing whitebark pine as a major subalpine forest component together with other western five-needled white pines vary in ecological impact from local to regional scales, from watershed to rangewide, and ultimately to western North America.

- *Altered watershed hydrology.* Throughout much of its range, whitebark pine is an important presence in the upper subalpine, growing where other conifers cannot. The loss of whitebark pine will alter local patterns of snow accumulation and snowmelt, with watershed effects on the timing, levels, and quality of stream flow.
- *Altered successional processes.* The hardiness of whitebark pine seedlings leads to timely regeneration after disturbance, and particularly after fire, on steep slopes and all aspects; to pioneering on moraines and rocky cliffs and other exposed locations; and to facilitation of the establishment of other conifers. Whitebark pine roots hold the loose, rocky substrate together, which allows other woody plants to colonize in association with it. That the loss of such woody plant cover could result in accelerated erosion is readily apparent from studies of comparable cover in the high mountains of Eurasia (Lowdermilk 1953; Holtmeier 1973; Arno and Hammerly 1984). Consequently, the loss of whitebark pine will slow the successional process in many kinds of sites after disturbance, particularly in the upper subalpine on poor soils and wind-blown exposures, and also result in the prolonged absence of trees from the more stressful sites in subalpine. The time frame of occurrence and composition of early successional communities will be greatly altered.
- *Homogenization of the subalpine zone landscape.* From a landscape perspective, as whitebark pine disappears, subalpine zone forests will lose ecological and structural diversity. In the absence of whitebark pine, forest area will shrink at high elevations. The composition of forest communities early in succession will

also be the composition of near-climax communities. As subalpine fir, Engelmann spruce, or mountain hemlock come to dominate the landscape, the forests will be more vulnerable to large, severe, stand-replacing fires as well as insect and disease epidemics, with declines in landscape structural and compositional diversity. Isolated stands of whitebark pine will no longer occupy exposed sites.

- *Impacts on grizzly bear populations.* In the midcontinental range of grizzly and black bears (that is from the Bob Marshall Wilderness Complex south through the mountains of western Wyoming, a distance of 725 kilometers), as whitebark pine disappears, a major food source for bears is also lost. This area includes the Greater Yellowstone Ecosystem and the even larger proposed reintroduction area in north-central Idaho in the Clearwater and Salmon River drainages. Given the stresses imposed on the grizzly bear population in this region—including reduction of historic range, development and urbanization, and human intrusion—the loss of whitebark pine could be the final blow to an already precarious situation. One outcome is certain: The absence of whitebark pine seeds in the subalpine zone will send bears wandering far and wide for food in late summer and fall, thereby increasing the incidence of encounters with humans. Inevitably, this will lead to many bears being destroyed.
- *Reduced nutcracker populations and accelerated whitebark pine losses.* Whitebark pine seeds are a primary food source for Clark's nutcrackers, and nutcrackers are the primary dispersers of whitebark pine seeds. As whitebark pine declines, nutcrackers will rely on other seed sources. Blister rust is spreading throughout the range of limber pine, so this food source will also decline with time. Eventually, nutcracker populations will shrink in proportion to their regional food base; fewer birds will spend time at subalpine zone elevations, and fewer birds will be available as seed dispersers (Chapter 12). The point may be moot anyhow: Wandering nutcrackers in midsummer may feed on unripe seeds from the few remaining whitebark pine cones, leaving no seeds for dispersal in fall, resulting in no regeneration. At this advanced stage, only intensive, costly management intervention would prevent total loss of whitebark pine, and it would be impossible to restore whitebark pine to any semblance of its historic range.
- *Impaired aesthetic and recreational values.* The loss of whitebark pine will have a major aesthetic impact on the backcountry experience. This is beginning to happen already. Year-round recreationists will be exposed to dead and dying trees throughout subalpine zone forests and stands of tree skeletons instead of groves of majestic, old-growth whitebark pine in the high mountains. Even now these depressing scenes are widespread in the northern Rocky Mountains. For example, in the upper subalpine zone on the Piegan Pass Trail in Glacier National Park, whitebark pine tree skeletons intermix with half-dead trees—with breathtaking snowcapped peaks beyond. Other examples include the upper subalpine zone of the Selkirk Range in the Idaho panhandle, where, looking west from the crest, the dead whitebark pine trees are silhouetted against Priest Lake in the distance; and all around the slopes of ski areas, such as Big Mountain and Snow Bowl in western Montana, whitebark pine trees are dying. The informed recreationists who know the reasons for the dead and dying trees will be particularly disturbed by this apparent disruption of a high-mountain ecosystem, and the introduced disease that is a harsh reminder of detrimental human influence. Some may wonder if there is a lack of stewardship in these public forest lands.

Restoring Whitebark Pine Communities

Management strategies and tools currently exist to restore whitebark pine communities on local and landscape scales. Appropriate treatments for a stand or area depend on the area's management direction (e.g., wilderness versus road-accessible, commercial forest). Important site-specific conditions include tree composition, community structure, and the extent of blister rust infection (Keane and Arno 1996). The different treatments, briefly described below, are discussed in greater detail in Chapter 18; they do not generally involve complex burning or cutting techniques and are designed to be implemented by local resource management personnel (Keane and Arno 1996).

Restoration Techniques

Successful restoration must address the two major factors causing the decline of whitebark pine—advanced succession and blister rust damage—which are synergistic. Where blister rust has destroyed cone production capacity and thus the ability to initiate regeneration, it may be advisable to use fire (often along with some felling of competing trees) followed by planting of rust-resistant seedlings (Table 1-1). Recent genetically based seed transfer guidelines for whitebark pine allow broader seed transfer zones than for most western conifers, probably because of avian seed dispersal, so that seeds may be moved about at a regional or rangewide scale (USDA Forest Service 1999; Chapters 15 and 16).

Basic restoration methods are summarized from Keane and Arno (1996; Chapter 18) and generally include the use of fire and cutting competing trees

Table 1-1. Techniques for local or larger-scale restoration of whitebark pine communities, based on Keane and Arno (1996; Chapter 18).

SILVICULTURAL TECHNIQUES
- Planting with rust-resistant whitebark pine seedlings
- Release cuttings
- Thinning
- Tree understory removal
- Selective tree removal
- Cutting small openings (50 m diameter) for caching by Clark's nutcracker
- Cutting for fuel enhancement

PRESCRIBED FIRE TECHNIQUES

High intensity
- Natural stand-replacement
- Prescribed operational stand-replacement

Mixed severity to low severity
- Broadcast burn with varying intensity in natural fuel
- Broadcast burn with fuel enhancement
- Underburn

Figure 1-3. The Smith Creek research area, Bitterroot Mountains, western Montana, was used to compare different whitebark pine restoration techniques. In one treatment unit, trees had been harvested to create nutcracker openings for seed caching, and some slash was left for fuel enhancement for prescribed burning. See also Chapter 18. Photo credit: Stephen F. Arno.

(Table 1-1, Figure 1-3). Tree cutting is possible only on a local scale, but can mimic the effects of mixed-severity fire by killing primarily the competing fir and spruce trees. It can also provide excellent fuel (cured slash) that allows prescription fires to be applied successfully under low to moderate wildfire hazard conditions. Cutting techniques may be fine-tuned to specific stand conditions. For example, tree thinning or selective tree removal targets whitebark pine competitors, particularly subalpine fir. Historically, low-intensity and mixed-severity fires killed these competing species, which are less fire-resistant than whitebark pine. Understory removal simulates the effects of low-intensity fire. Where there is still significant cone production, removing trees to make openings in the forest (at least 50 meters in diameter) encourages Clark's nutcrackers to cache seeds in these clearings (Chapter 5).

Prescribed fire is possible at both a stand-level and a landscape scale. High-intensity fire from lightning ignitions, burning more than 1,000 hectares, occurred historically and should be allowed to burn wherever feasible. Without this natural fire regime, whitebark pine in its lower elevational range is outcompeted by a closed forest of shade-tolerant species. Prescribed fires of mixed intensity may be used within stands with natural fuel buildup or with enhanced fuel as a result of thinning. Alternatively, a low-intensity fire simulates another historical fire pattern that thinned out whitebark pine's competitors.

Restoration Efforts: Planning and Management

Implementing whitebark pine restoration efforts either locally or regionally should not be done haphazardly or spontaneously. First, restoration projects should be part of a local or regional strategic plan; and second, there are many constituents who, for whatever their agendas, will misunderstand restoration actions and need to be educated. Demonstration projects are an excellent device for getting started, learning techniques, evaluating initial results, and educating the public (Chapter 18).

The restoration process has been organized into seven steps by Keane and Arno (Chapter 18), which include, in sequence, the initial inventory of landscape and stand characteristics, a description of ecosystem processes at landscape and stand scale, the prioritization of landscapes and stands for restoration, the selection of stands and landscapes for treatment, the design of treatments for each selected unit, the implementation of treatments, and monitoring the results for evaluation. In addition, we recommend that a campaign to inform and educate the public about the value of whitebark pine and whitebark pine restoration be included in this process. Finding the funds for restoration may also require planning, depending on the agency or organization. The importance of monitoring the aftereffects of treatment to provide feedback for future restoration efforts cannot be overemphasized. This information must be shared among forest managers and forest scientists.

Related, Pressing Issues

The circumstances of whitebark pine decline and the challenges of whitebark pine restoration raise a number of pressing and confounding issues. First of all, restoration itself is difficult in a human-dominated landscape. The proximity of towns, mountain homes, ski areas, and ranches to whitebark pine communities—often just a few kilometers away—limits the implementation of natural fire and places restrictions on all prescribed fire, including factors such as the effects of smoke and public fears. Ironically, prescription fires, like cutting, are management actions that draw attention and require good public relations, whereas fire exclusion is taken for granted despite damaging ecological effects.

Another issue in restoration, particularly in wilderness settings, is whether we can determine what the historical, "natural" landscape patterns were, and whether such a landscape is possible today even after restoration. Fire exclusion has had a large effect on many wilderness areas (Brown et al. 1994). The Wilderness Act of 1964 states that these areas should remain "untrammeled" by man; but by restricting fires and thus the important ecological effects of fire, we trammel (restrict or constrain) entire natural ecosystems (Chapter 13).

Ecological processes may now be impacted by regional and worldwide climate change and by airborne pollutants. Blister rust will evidently have a prolonged, major effect on whitebark pine communities, and the best we can hope for is a slow decline in infection level as natural and augmented rust resistance spreads.

A third issue is the cost of restoration and who will bear it. The cost ultimately depends on how many restoration projects are proposed and the logistics of accomplishing them. Given the extensive area impacted by advanced succession and whitebark pine blister rust, restoration could be very costly over time in both money and manpower. Even with a burgeoning economy, we are confronted with downsizing institutions and making do with less. Alternatively, costly fire suppression practices (Hill 1998) could be altered to allow more fuel treatment and prescription fire to be folded into a fire management program that ultimately might be no more expensive than the present efforts. Whitebark pine restoration must be elevated to a high priority among scientists, managers, and the public in order to compete successfully for limited institutional and private funds.

Finally, as we move ahead with restoration projects, we do not yet know all the institutional, cultural, and societal challenges that we will encounter. Are there stumbling blocks out there that could seriously delay or diminish our efforts (e.g., Chapter 19)? Most certainly this is the case, and overcoming each challenge will require attention and planning

On a positive note, restoration of whitebark pine communities fits neatly within the concept of ecosystem-based management, which has been embraced since the early 1990s by the USDA Forest Service, most other public agencies, and even nonprofit organizations like The Nature Conservancy (Salwasser and Pfister 1994; Reid 1998). But the urgency of ecosystem management implementation is higher in whitebark pine ecosystems than in most other major forest types. Unfortunately, we do not have the luxury of time to investigate all issues and problems. Blister rust is spreading, seed sources are being lost, and whitebark pine is disappearing over a large portion of its range. We also do not have the time to make restoration of whitebark pine communities part of a political agenda or to devise clever economic or commercial arguments for their preservation. Instead, we must restore whitebark pine communities now for their crucial watershed and ecosystem services, their wildlife relationships, and their aesthetic and symbolic value in the high mountains of the West.

Acknowledgments

The authors thank Harry E. Hutchins, Itasca Community College, Grand Rapids, Minnesota, and Katherine C. Kendall for their comments and suggestions for this chapter.

Literature Cited

Agee, J. K. 1993. Fire ecology of Pacific Northwest forests. Island Press, Washington, D.C.

Arno, S. F. 1976. The historical role of fire on the Bitterroot National Forest. USDA Forest Service Intermountain, Research Station Research INT-187, Ogden, Utah.

———. 1980. Forest fire history in the Northern Rockies. Journal of Forestry 78:460–465.

———. 1986. Whitebark pine cone crops—a diminishing source of wildlife food? Western Journal of Applied Forestry 1:92–94.

———. 1996a. The seminal importance of fire in ecosystem management—impetus for this publication. Pages 3–5 *in* C. C. Hardy and S. F. Arno, editors. The use of fire in forest restoration. USDA Forest Service Intermountain Research Station, General Technical Report INT-GTR-341, Ogden, Utah.

———. 1996b. The concept: restoring ecological structure and process in ponderosa pine forests. Pages 37–38 *in* C. C. Hardy and S. F. Arno, editors. The use of fire in forest restoration. USDA Forest Service Intermountain Research Station, General Technical Report INT-GTR-341, Ogden, Utah.

Arno, S. F., and R. P. Hammerly. 1984. Timberline: Mountain and arctic forest frontiers. The Mountaineers, Seattle, Washington.

Arno, S. F., and R. J. Hoff. 1990. *Pinus albicaulis* Engelm. Whitebark pine. Pages 268–279 *in* R. M. Burns and B. H. Honkala, technical coordinators. Silvics of North America. USDA Forest Service, Agriculture Handbook 654, Washington, D.C.

Barbour, M. G. 1988. California upland forests and woodlands. Pages 131–164 *in* Barbour, M. G., and W. D. Billings, editors. North American terrestrial vegetation. Cambridge University Press, New York.

Barrett, S. W. 1994. Fire regimes on andesitic mountain terrain in northeastern Yellowstone National Park, Wyoming. International Journal of Wildland Fire 4:65–76.

Barrett, S. W., and S. F. Arno. 1982. Indian fires as an ecological influence in the Northern Rockies. Journal of Forestry 80:647–651.

Bartos, D. L., and K. E. Gibson. 1990. Insects of whitebark pine with emphasis on mountain pine beetle. Pages 171–178 *in* W. C. Schmidt and K. J. McDonald, compilers. Proceedings—Symposium on whitebark pine ecosystems: Ecology and management of a high-mountain resource. USDA Forest Service Intermountain Research Station, General Technical Report INT-270, Ogden, Utah.

Botkin, D. B. 1990. Discordant harmonies—a new ecology for the twenty-first century. Oxford University Press, New York.

Boyd, R. (ed.). 1999. Indians, fire, and the land in the Pacific Northwest. Oregon State University, Corvallis.

Brown, J. K., S. F. Arno, S. W. Barrett, and J. P. Menakis. 1994. Comparing the prescribed natural fire program with presettlement fires in the Selway-Bitterroot wilderness. International Journal of Wildland Fire 4:157–168.

Callaway, R. M. 1998. Competition and facilitation on elevation gradients in subalpine forests of the northern Rocky Mountains, USA. Oikos 82:561–573.

Coblentz, B. E. 1990. Exotic organisms: A dilemma for conservation biology. Conservation Biology 4:261–265.

Craighead, J. J., J. S. Sumner, and G. B. Scaggs. 1982. A definitive system for analysis of grizzly bear habitat and other wilderness resources. Wildlife-Wildlands Institute Monograph 1. University of Montana, Missoula.

Critchfield, W. B., and Little, E. L. Jr. 1966. Geographic distribution of the pines of the world. USDA Forest Service, Miscellaneous Publication 991, Washington, D.C.

Czech, B. 1996. Challenges to establishing and implementing sound natural fire policy. Renewable Resources Journal 14:14–19.

Farnes, P. E. 1990. SNOTEL and snow course data describing the hydrology of whitebark pine ecosystems. Pages 302–304 *in* W. C. Schmidt and K. J. McDonald, compilers. Proceedings—Symposium on whitebark pine ecosystems: Ecology and management of a high-mountain resource. USDA Forest Service Intermountain Research Station, General Technical Report INT-270, Ogden, Utah.

Feldman, R., D. F. Tomback, and J. Koehler. 1999. Cost of mutualism: Competition, tree morphology, and pollen production in limber pine clusters. Ecology 80:324–329.

Franklin, J. F., and C. T. Dyrness. 1973. Natural vegetation of Oregon and Washington. USDA Forest Service, Pacific Northwest Forest and Range Experiment Station, General Technical Report PNW-8, Portland, Oregon.

Gruell, G. E. 1985. Fire on the early western landscape: An annotated record of wildland fires 1776–1900. Northwest Science 59:97–107.

Hill, B. T. 1998. Western national forests: Catastrophic wildfires threaten resources and communities. U.S. General Accounting Office, GAO/T—RCED-98-273, Washington, D.C.

Hoff, R., and S. Hagle. 1990. Diseases of whitebark pine with special emphasis on white pine blister rust. Pages 179–190 *in* W. C. Schmidt and K. J. McDonald, compilers. Proceedings—Symposium on whitebark pine ecosystems: Ecology and management of a high-mountain resource. USDA Forest Service Intermountain Research Station, General Technical Report INT-270, Ogden, Utah.

Holtmeier, F.-K. 1973. Geoecological aspects of timberline in northern and central Europe. Arctic and Alpine Research 5:A45–A54.

Hutchins, H. E., and R. M. Lanner. 1982. The central role of Clark's nutcracker in the dispersal and establishment of whitebark pine. Oecologia 55:192–201.

Keane, R. E., and S. F. Arno. 1996. Whitebark pine ecosystem restoration in Montana. Pages 51–53 *in* C. C. Hardy and S. F. Arno, editors. The use of fire in forest restoration. USDA Forest Service Intermountain Research Station, General Technical Report INT-GTR-341, Ogden, Utah.

Keane, R. E., P. Morgan, and J. P. Manakis. 1994. Landscape assessment of the decline of whitebark pine (*Pinus albicaulis*) in the Bob Marshall Wilderness Complex, Montana, USA. Northwest Science 68:213–229.

Kendall, K. C. 1983. Use of pine nuts by grizzly and black bears in the Yellowstone area. International Conference on Bear Research and Management 5:166–173.

Kendall, K. C., and S. F. Arno. 1990. Whitebark pine—an important but endangered wildlife resource. Pages 264–273 *in* W. C. Schmidt and K. J. McDonald, compilers. Proceedings—Symposium on whitebark pine ecosystems: Ecology and management of a high-mountain resource. USDA Forest Service, Intermountain Research Station, General Technical Report INT-270, Ogden, Utah.

Kimmins, J. P. 1997. Forest ecology: A foundation for sustainable management, 2d edition. Prentice-Hall, Upper Saddle River, New Jersey.

Lambert, F. 1991. The conservation of fig-eating birds in Malaysia. Biological Conservation 58:31–40.

Lanner, R. M. 1980. Avian seed dispersal as a factor in the ecology and evolution of limber and whitebark pine. Pages 15–48 *in* B. P. Dancik and K. O. Higginbotham, editors. Proceedings of sixth North American forest biology workshop. University of Alberta, Edmonton, Alberta, Canada.

———. 1996. Made for each other: A symbiosis of birds and pines. Oxford University Press, New York.

Lanner, R. M., and B. K. Gilbert. 1994. Nutritive value of whitebark pine seeds and the question of their variable dormancy. Pages 206–211 *in* W. C. Schmidt and F.-K. Holtmeier, compilers. Proceedings—International workshop on subalpine stone pines and their environment: The status of our knowledge. USDA Forest Service, Intermountain Research Station, General Technical Report INT-GTR-309, Ogden, Utah.

Lowdermilk, W. C. 1953. Conquest of the land through 7,000 years. USDA Soil Conservation Service, Agricultural Bulletin 99.

Marsden, M. A. 1983. Modeling the effect of wildfire frequency on forest structure and succession in the northern Rocky Mountains. Journal of Environmental Management 16:45–62.

Mathiasen, R. I., and F. G. Hawksworth. 1988. Dwarf mistletoes on western white pine and whitebark pine in northern California and southern Oregon. Forest Science 34:429–440.

Mattson, D. J., and D. P. Reinhart. 1994. Bear use of whitebark pine seeds in North America. Pages 212–220 *in* W. C. Schmidt and F.-K. Holtmeier, compilers. Proceedings—International workshop on subalpine stone pines and their environment: The status of our knowledge. USDA Forest Service Intermountain Research Station, General Technical Report INT-GTR-309, Ogden, Utah.

Mattson, D. J., B. M. Blanchard, and R. R. Knight. 1991. Food habits of Yellowstone grizzly bears, 1977–1987. Canadian Journal of Zoology 69:1619–1629.

———. 1992. Yellowstone grizzly bear mortality, human habituation, and whitebark pine seed crops. Journal of Wildlife Management 56:432–442.
McCaughey, W. W. 1993. Delayed germination and seedling emergence of *Pinus albicaulis* in a high elevation clearcut in Montana, U.S.A. Pages 67–72 *in* D. G. W. Edwards, compiler and editor. Dormancy and barriers to germination. Proceedings of an International Symposium, IUFRO Project Group (Seed Problems). Forestry Canada, Pacific Forestry Centre, Victoria, British Columbia, Canada.
McCune, B. 1988. Ecological diversity in North American pines. American Journal of Botany 75:353–368.
McGregor, M. D., and D. M. Cole. 1985. Integrating management strategies for the mountain pine beetle with multiple-resource management of lodgepole pine forests. USDA Forest Service Intermountain Research Station, General Technical Report INT-174, Ogden, Utah.
Paine, R. T. 1966. Food web complexity and species diversity. American Naturalist 100:65–75.
Peterman, R. M. 1978. The ecological role of mountain pine beetle in lodgepole pine forests. Pages 16–26 *in* A. A. Berryman, G. D. Amman, and R. W. Stark, editors. Proceedings—Symposium on theory and practice of mountain pine beetle management in lodgepole pine forests. Washington State University, Pullman.
Pfister, R. D., B. L. Kovalchik, S. F. Arno, and R. C. Presby. 1977. Forest habitat types of Montana. USDA Forest Service Intermountain Forest and Range Experiment Station, General Technical Report INT-34, Ogden, Utah.
Primack, R. B. 1998. Essentials of conservation biology, 2d edition. Sinauer Associates, Sunderland, Massachusetts.
Pyne, S. J. 1997. World fire: The culture of fire on earth. University of Washington Press, Seattle.
Reid, B. 1998. A clearing in the forest. The Nature Conservancy 48:18–24.
Rogers, D. L., C. I. Millar, and R. D. Westfall. 1999. Fine-scale genetic structure of whitebark pine (*Pinus albicaulis*): Associations with watershed and growth form. Evolution 53:74–90.
Romme, W. H. 1982. Fire and landscape diversity in subalpine forests of Yellowstone National Park. Ecological Mongraphs 52:199–221.
Salwasser, H., and R. D. Pfister. 1994. Ecosystem management: From theory to practice. Pages 150–160 *in* W. W. Covington and L. F. DeBano, compilers. Sustainable ecological systems: Implementing an ecological approach to land management. USDA Forest Service Rocky Mountain Research Station, General Technical Report RM-247, Fort Collins, Colorado.
Steele, R., S. V. Cooper, D. M. Ondov, D. W. Roberts, and R. D. Pfister. 1983. Forest habitat types of eastern Idaho–western Wyoming. USDA Forest Service Intermountain Forest and Range Experiment Station, General Technical Report INT-144, Ogden, Utah.
Stuart-Smith, G. J. 1998. Conservation of whitebark pine in the Canadian Rockies: Blister rust and population genetics. M.S. thesis. University of Alberta, Edmonton, Alberta, Canada.
Terbourgh, J. 1986. Keystone plant resources in the tropical forest. Pages 330–344 *in* M. E. Soule, editor. Conservation biology: The science of scarcity and diversity. Sinauer Associates, Sunderland, Massachusetts.
Tomback, D. F. 1978. Foraging strategies of Clark's nutcracker. Living Bird 16:123–161.
———. 1982. Dispersal of whitebark pine seeds by Clark's nutcracker: A mutualism hypothesis. Journal of Animal Ecology 51:451–467.
———. 1986. Post-fire regeneration of krummholz whitebark pine: A consequence of nutcracker seed caching. Madroño 33:100–110.

———. 1994. Effects of seed dispersal by Clark's nutcracker on early postfire regeneration of whitebark pine. Pages 193–198 *in* W. C. Schmidt and F.-K. Holtmeier, compilers. Proceedings—International workshop on subalpine stone pines and their environment: The status of our knowledge. USDA Forest Service, Intermountain Research Station, General Technical Report INT-GTR-309, Ogden, Utah.

———. 1998. Clark's nutcracker (*Nucifraga columbiana*), No. 331. *In* A. Poole and F. Gill, editors. The birds of North America. The Birds of North America, Inc., Philadelphia, Pennsylvania.

Tomback, D. F., and Y. B. Linhart. 1990. The evolution of bird-dispersed pines. Evolutionary Ecology 4:185–219.

Tomback, D. F., L. A. Hoffmann, and S. K. Sund. 1990. Coevolution of whitebark pine and nutcrackers: Implications for forest regeneration. Pages 118–129 *in* W. C. Schmidt and K. J. McDonald, compilers. Proceedings—Symposium on whitebark pine ecosystems: Ecology and management of a high-mountain resource. USDA Forest Service, Intermountain Research Station, General Technical Report INT-270, Ogden, Utah.

Tomback, D. F., J. K. Clary, J. Koehler, R. J. Hoff, and S. F. Arno. 1995. The effects of blister rust on post-fire regeneration of whitebark pine: The Sundance Burn of northern Idaho (U.S.A.). Conservation Biology 9:654–664.

Tomback, D. F., S. K. Sund, and L. A. Hoffman. 1993. Post-fire regeneration of *Pinus albicaulis:* Height-age relationships, age structure, and microsite characteristics. Canadian Journal of Forest Research 23:113–119.

USDA Forest Service. 1999. Transfer rules. Chapter 4 *in* Seed handbook. Draft Forest Service Handbook 2409.26f, Missoula, Montana.

U.S. Fish and Wildlife Service. 1997. Grizzly bear recovery in the Bitterroot Ecosystem. Draft Environmental Impact Statement. U.S. Fish and Wildlife Service. Bitterroot Grizzly Bear EIS, P.O. Box 5127, Missoula, MT 59806.

Weaver, H. 1943. Fires as an ecological and silvicultural factor in the ponderosa pine region of the Pacific Slope. Journal of Forestry 41:7–14.

West, N. E. 1988. Intermountain deserts, shrub steppes, and woodlands, chapter 7. Pages 209–230 *in* M. G. Barbour and W. D. Billings, editors. North American terrestrial vegetation. Cambridge University Press, New York.

Part II

The Biology of Whitebark Pine

Whitebark pine has many unique life history attributes. It is the only North American pine of the pine subsection *Cembrae,* a taxon characterized by large, wingless seeds and cones that do not open when ripe. Whitebark pine, like the other *Cembrae* pines, has seeds dispersed by birds—the nutcrackers (*Nucifraga* spp., family Corvidae)—and not by wind, which disperses seeds for most other conifers. The evolution and population biology of the *Cembrae* pines has been profoundly affected by interaction with nutcrackers. Whitebark pine is found only at tree line and subalpine elevations throughout its range; in its "center of abundance" in the northern Rocky Mountains of the United States and southern Canada, whitebark pine forms successional communities in the upper subalpine, which depend on fire for renewal. The seven chapters in Part II provide important background information on the taxonomic relationships, distribution zone, evolution, ecology, and population biology of whitebark pine. Much of this information serves as a foundation for the chapters in Parts III and IV. The contents of Part II and the logic of presentation are described below.

Chapter 2, "Taxonomy, Distribution, and History," by Ward W. McCaughey and Wyman C. Schmidt, provides an historic context and taxonomic perspective for whitebark pine, which is essential for subsequent chapters. The chapter also describes the geographic range, fossil record, and dendrochronology of the species. Chapter 3, "Whitebark Pine and Its Environment" by T. Weaver, is a logical extension of Chapter 2. It describes in detail the physical environment and apparent physiological tolerances of whitebark pine, often drawing comparisons with other *Cembrae* pines. It also provides insights into the latitudinal and elevational limits to whitebark pine's geographic distribution. In Chapter 4, "Community Types and Natural Disturbance Processes," Stephen F. Arno describes how community structure and the nature and frequency of disturbance, and particularly fire, vary across the range of whitebark pine, providing an important review of key literature. With Weaver's chapter as background, the reasons for latitudinal and elevational differences in community structure are more apparent. Chapter 4 provides the background information necessary for readers to understand why fire exclusion has had such damaging effects on whitebark pine communities in the inland northwestern United States and adjacent areas of Canada.

With Chapter 5, "Clark's Nutcracker: Agent of Regeneration," by Diana F. Tomback, Part II focuses on both the unique seed dispersal mode of whitebark pine and the important interactions between the pine and wildlife. Tomback describes the coevolved mutualistic interaction between whitebark pine and nutcrackers (*Nucifraga* spp.). Most important, Chapter 5 explains how seed dispersal by nutcrackers affects the ecology and population biology of whitebark pine: successional status, local distribution, tree growth form, and population genetic structure. The significant role of nutcrackers in regeneration of whitebark pine, and particularly after fire, becomes increasingly apparent in subsequent chapters. The theme of tree regeneration is continued by Ward W. McCaughey and Diana F. Tomback in Chapter 6, "The Natural Regeneration Process." They discuss the development of whitebark pine cones and seeds: the timing, or phenology, of reproductive events, cone production, seed dispersal, seed germination, and early seedling growth. They also discuss the phenomenon of delayed germination in whitebark pine seeds, which makes growing seedlings a challenge for nursery operations—the basis for Chapter 16 in Part IV.

The theme of wildlife interactions is continued by David J. Mattson, Katherine C. Kendall, and Daniel P. Reinhart in Chapter 7, "Whitebark Pine, Grizzly Bears, and Red Squirrels," which discusses the importance of whitebark pine seeds as a food resource for grizzly bears (*Ursus arctos*) in the northern Rocky Mountains and particularly in the Greater Yellowstone Area. Grizzly bear use of whitebark pine seeds is a fascinating story of wildlife interactions, with the red squirrel (*Tamiasciurus hudsonicus*) as the intermediary. It is clear from this chapter that the viability of the Yellowstone population of grizzly bears rests on the health of whitebark pine communities.

"Population Genetics and Evolutionary Implications" is Chapter 8, the final chapter in Part II. Authors Leo P. Bruederle, Deborah L. Rogers, Konstantin V. Krutovskii, and Dmitri V. Politov provide basic background information on the genetic diversity and population genetic structure in whitebark pine. This chapter examines the taxonomic relationships of whitebark pine using recent molecular evidence, and also illustrates how seed dispersal by nutcrackers has affected fine-scale population structure and possibly the genetic differentiation of populations, building on discussion in Tomback's Chapter 5. Studies of genetic differentiation discussed in Chapter 8 were used by the USDA Forest Service to construct seed transfer guidelines for whitebark pine, which are essential for restoration efforts, as discussed in Part IV.

Chapter 2

Taxonomy, Distribution, and History

Ward W. McCaughey and Wyman C. Schmidt

Whitebark pine (*Pinus albicaulis*) is an important upper subalpine species in western North America and particularly in the northern Rocky Mountains of the western United States where its seeds are an essential food source for the endangered grizzly bear (*Ursus arctos horribilis*) and other mammals and birds (Knight et al. 1987; Hutchins and Lanner 1982; Chapter 12). Whitebark pine was first described by Engelmann in 1863 as occurring at high elevations from central British Columbia, Canada, south into the United States to the southern Sierra Nevada range, the Ruby Mountains of Nevada, and the mountains of northwestern Wyoming (Critchfield and Little 1966). Whitebark pine is facing serious declines throughout its range as the result of introduced disease, native insect infestations, and fire suppression (Chapter 1).

Whitebark pine is one of several North American pine species with large seeds that were used historically by indigenous peoples as an important food source. Its seeds were eaten by the Interior Salish tribes from Montana to British Columbia, the Lilooet, Nlaka' pamux, Shuswap, Okanagan-Colville, Chilcotin, Kootenay, and Flathead (Lanner 1996). However, rarely has the wood of whitebark pine been used, except when it happened to be growing near human settlements or when found in a stand with other valued conifers (Losensky 1990). Its wood has been burned for domestic firewood; used for construction associated with mining, such as mine supports, railroad tracks, and mill structures; or burned for ore processing.

Until twenty years ago, little was known about the basic ecology and biology of whitebark pine. Its evolutionary history prior to the Pleistocene, however, remains incompletely known. Our inability to reconstruct historical events is reflected in controversies surrounding the taxonomy of whitebark pine and, to some extent, pines in general. This chapter describes whitebark pine's taxonomy and classification, present distribution throughout North America, and evolutionary history.

Taxonomy and Classification

Whitebark pine is one of more than a hundred species of pines found in the family Pinaceae within the kingdom Plantae, division Spermatophyta, subdivision Gymnospermae and order Coniferales (Harlow et al. 1979; Price et al. 1998). This North American conifer is classified in the family Pinaceae, genus *Pinus,* subgenus *Strobus,* section *Strobus* and subsection *Cembrae* (Critchfield and Little 1966; Price et al. 1998). Whitebark pine is generally considered to be one of five stone pines worldwide, comprising subsection *Cembrae* within the section *Strobus;* the *Cembrae* pines include whitebark pine, Swiss stone pine (*P. cembra*), Korean stone pine (*P. koraiensis*), Siberian stone pine (*P. sibirica*), and Japanese stone pine (*P. pumila*) (Table 2-1). These are characteristics shared by the five stone pines: five needles per fascicle; indehiscent cones that, at maturity, remain essentially closed; and wingless seeds that are dispersed by birds of the genus *Nucifraga* (family Corvidae), the nutcrackers (Chapter 5).

Historically, Shaw (1914) was the first taxonomist to use cone and seed characteristics as a basis for pine taxonomy. Shaw classified whitebark pine with the Swiss, Korean, Siberian, and Japanese stone pines. Since that time, taxonomists have disagreed on whether whitebark pine belongs in subsection *Cembrae* or in subsection *Strobi* of section *Strobus* (Table 2-1). Pilger (1926) recognized subsection *Cembrae* to consist of whitebark pine, Swiss stone pine, Japanese stone pine Korean stone pine, Armand pine (*Pinus armandii*), and limber pine (*P. flexilis*). Critchfield (1986) did not believe that whitebark pine belonged within Pilger's (1926) subsection *Cembrae,* because its cones were not completely indehiscent. Critchfield also believed that whitebark pine cone scales only sometimes separate, but they almost always separate in Korean stone pine and Japanese stone pine. In reply to Critchfield's (1986) assertion that whitebark pine cones may not be indehiscent, Lanner (1990) explained that although the cone scales become loose enough to open slightly and expose the seeds, the seeds do not fall out when the cone is turned or shaken. Lanner argued to maintain whitebark pine within subsection *Cembrae.* Price et al. (1998) agreed with Lanner that subsection *Cembrae* includes the five stone

Table 2-1. Commonly accepted stone pine species of the world in subsection *Cembrae.*

Common Name	Species	Geographical Region
Whitebark pine	*Pinus albicaulis*	Western North America
Swiss stone pine	*Pinus cembra*	Central Europe
Korean stone pine	*Pinus koraiensis*	Eastern Siberia, Korea, northeast China, Japan
Siberian stone pine	*Pinus sibirica*	Siberia, Mongolia
Japanese stone pine	*Pinus pumila*	Eastern Siberia, Japan

Table 2-2. Currently recognized *Pinus* species within sections and subsections of subgenus *Strobus* (from Price et al. 1998).

Subgenus *Strobus*	
Section *Strobus*	**Section *Parrya***
Subsection *Strobi*	Subsection *Balfourianae*
P. armandii	*P. aristata*
P. ayacahuite	*P. balfouriana*
P. bhutanica	*P. longaeva*
P. chiapensis	Subsection *Krempfianae*
P. dabeshanensis	*P. krempfii*
P. dalatensis	Subsection *Cembroides*
P. fenzeliana	*P. cembroides*
P. flexilis	*P. culminicola*
P. lambertiana	*P. discolor*
P. monticola	*P. edulis*
P. morrisonicola	*P. johannis*
P. parviflora	*P. juarezensis*
P. peuce	*P. maximartinezii*
P. strobus	*P. monophylla*
P. wallichiana	*P. nelsonii*
P. wangii	*P. pinceana*
Subsection *Cembrae*	*P. remota*
P. albicaulis	Subsection *Rzedowskianae*
P. cembra	*P. rzedowskii*
P. koraiensis	Subsection *Gerardianae*
P. pumila	*P. bungeana*
P. sibirica	*P. gerardiana*
	P. squamata

pines (Table 2-2), and relegated Armand pine and limber pine to subsection *Strobi*.

Critchfield (1986) also noted interspecific variation among the *Cembrae* species in oleoresin composition and heartwood phenolics. He combined whitebark pine with limber pine, Mexican white pine (*P. ayacahuite*), and the *Cembrae* core species Korean stone pine and Swiss stone pine to form this subsection. Other taxonomists believe it would be a mistake to place whitebark pine and limber pine together in the same subsection based on oleoresin composition and heartwood phenolics, primarily because of the great variation in these compounds among all pines (Lanner 1990). Cone differences used in Critchfield's (1986) treatment can also be disputed. For example, whitebark pine and limber pine often grow in the same general area, have similar crown configurations, and can easily be misidentified, with the exception of decidedly different cone characteristics. Mature whitebark pine cones are oblong, 5 to 8 centimeters long (2 to 3 inches), and dark brown to dark purple in color; cones of limber pine are cylindrical, 8 to 15 centimeters long (3 to 6 inches), and green in color (Harlow et al. 1979). Pollen cones of whitebark pine are red in color, while those of limber pine are green (Chapter 6). Price et al. (1998) considered the seed wing on limber pine to be adnate (attached) to the body of the seed or very reduced, while whitebark seeds are truly wingless. Cones of limber pine are dehiscent and release their seeds, as do the cones of other pines in subsection *Strobi* (Young and Young 1992).

Site preference often separates whitebark pine and limber pine. Whitebark pine usually occurs on cold, moist, high-elevation sites and limber pine on warm, dry sites at lower and middle elevations (Pfister et al. 1977). Where their ranges overlap, the two species sometimes grow in the same sites on droughty soils. Occasionally, limber pine grows above whitebark pine in isolated areas such as in the Bridger Mountains in southwestern Montana (McCaughey and Schmidt 1990). In California and Colorado, beyond the southern limits of whitebark pine, limber pine often ascends to the alpine tree line. Taxonomically, the closed-cone persistence of whitebark pine and the open-cone characteristic of limber pine definitely separate the two, placing whitebark pine in subsection *Cembrae* and limber pine in subsection *Strobi,* even though both are five-needled pines with wingless seeds.

Critchfield (1986) grouped whitebark pine with Mexican white pine because they have nearly identical heartwood constituents. However, some morphological characteristics of Mexican white pine tend to place it within subsection *Strobi.* The characteristics of Mexican white pine that are different from whitebark pine but similar to western white pine (*Pinus monticola*) and sugar pine (*P. lambertiana*) include longer needles, shorter needle longevity (the number of years needles remain on the tree), longer cones, and taller tree height (Chapter 8). Furthermore, the cones of Mexican white pine open upon ripening—that is, they are dehiscent—and the seeds exhibit subspecific variation with regard to seed wings. The seeds of *P. ayacahuite* var. *brachyptera* are wingless, whereas the seeds of *P. a.* var. *veitchii* and *P. a.* var. *ayacahuite* are winged (Ledig 1998).

Current molecular and statistical approaches to phylogenetics are promising to resolve the controversy surrounding the pines of subsections *Cembrae* and *Strobi,* as well as the placement of whitebark pine. DNA sequences of several genes, both chloroplast and nuclear, indicate that the pines traditionally classified in subsection *Cembrae* (Critchfield and Little 1966), including whitebark pine, are closely related. However, several white pines currently in subsection *Strobi* are equally closely related to the *Cembrae* pines, such as Armand pine, suggesting that subsection *Cembrae* should be included in subsection *Strobi* taxonomically (R. Price, personal communication; Wang et al. 1999; Chapter 8). In other words, the *Cembrae* pines appear to be *Strobi* pines that are morphologically specialized for seed dispersal by nutcrackers, which may provide the rationale for keeping them taxonomically separate.

Distribution

Whitebark pine is limited in distribution to the high mountains of western North America (Figure 2-1), where it has been present for the past 8,000 years through most of the Holocene. Because whitebark pine occurs along high-elevation ridges, many stands are geographically isolated, with large intervening valleys between them (Arno and Hoff 1990). Although not contiguous, whitebark pine forests extend longitudinally between 107 and 128 degrees west and latitudinally between 37 and 55 degrees north (McCaughey and Schmidt 1990; Ogilvie 1990). In Canada, whitebark pine grows along the coastal range from northwestern British Columbia to the southern end of the province. In the United States, the species' range extends from the Canadian border of the Cascade Mountains of Washington, through central Oregon to the southern Sierra Nevada range of California and western Nevada. It also grows in areas of eastern Oregon, western Idaho, and eastern Nevada, and is found extensively in the Rocky Mountains of Montana, Idaho, and western Wyoming (Schmidt 1994; Chapter 4).

The distribution of whitebark pine is divided into western and eastern ranges, whose only tenuous connection is an area in southern British Columbia and northeastern Washington where a few isolated stands occur (Figure 2-1). The western range extends from the coastal ranges of the Bulkley Mountains in northern British Columbia through the Blue and Wallowa Mountains of northeastern Oregon and ending in southeastern California, as far south as the Kern River, with outliers in the Warner and Siskiyou Mountains in California (Critchfield and Little 1966; Ogilvie 1990). Small isolated stands occur in mountain ranges in northeastern California, south-central Oregon, and northern Nevada (Arno and Hoff 1990).

The eastern range of whitebark pine consists of the northern Rocky Mountains of British Columbia, Alberta, Idaho, Montana, Wyoming, and Nevada (Figure 2-1). The northern extent of the eastern range occurs along the Continental Divide of the Rocky Mountains just below the 54th latitude, northeast of McBride, British Columbia. The southern extent of the eastern range consists of isolated occurrences in small mountain ranges of northeastern Nevada. Whitebark pine forms extensive stands in the Yellowstone ecosystem of Wyoming, Idaho, and Montana, with contiguous stands extending into the eastern ranges of the Salt River and Wind River ranges of Wyoming (Thompson and Kuijt 1976). The Wind River range represents the eastern-most extension of the species (Schmidt 1994).

Small outlier stands occur throughout the distributional range of whitebark pine. For example, small stands occur in the Sweetgrass Hills in north-central Montana, 145 kilometers (90 miles) from the nearest stand in the Rocky Mountains. Outlier stands also occur in the Blue and Wallowa Mountains of northeastern Oregon and in small isolated ranges in northeastern California, south-central Oregon, and northern Nevada (Arno and Hoff 1990).

Within this geographic range, the upper elevational limits of whitebark pine, and of alpine timberlines in general, decrease with increasing latitude. Whitebark pine occurs as high as 3,050 to 3,660 meters (10,000 to 12,000 feet) in the Sierra Nevada and 2,590 to 3,200 meters (8,500 to 10,500 ft.) in western Wyoming; and as low as 900 meters (2,950 ft.) in the northern limits of its range in British Columbia (Arno and Hoff 1990; Ogilvie 1990).

Whitebark pine is typically found just below the alpine timberline and extends downward in elevation into associations with several other conifers, most commonly Rocky Mountain lodgepole pine (*Pinus contorta* var. *latifolia*), Engelmann spruce (*Picea engelmannii*), subalpine fir (*Abies lasiocarpa*), and

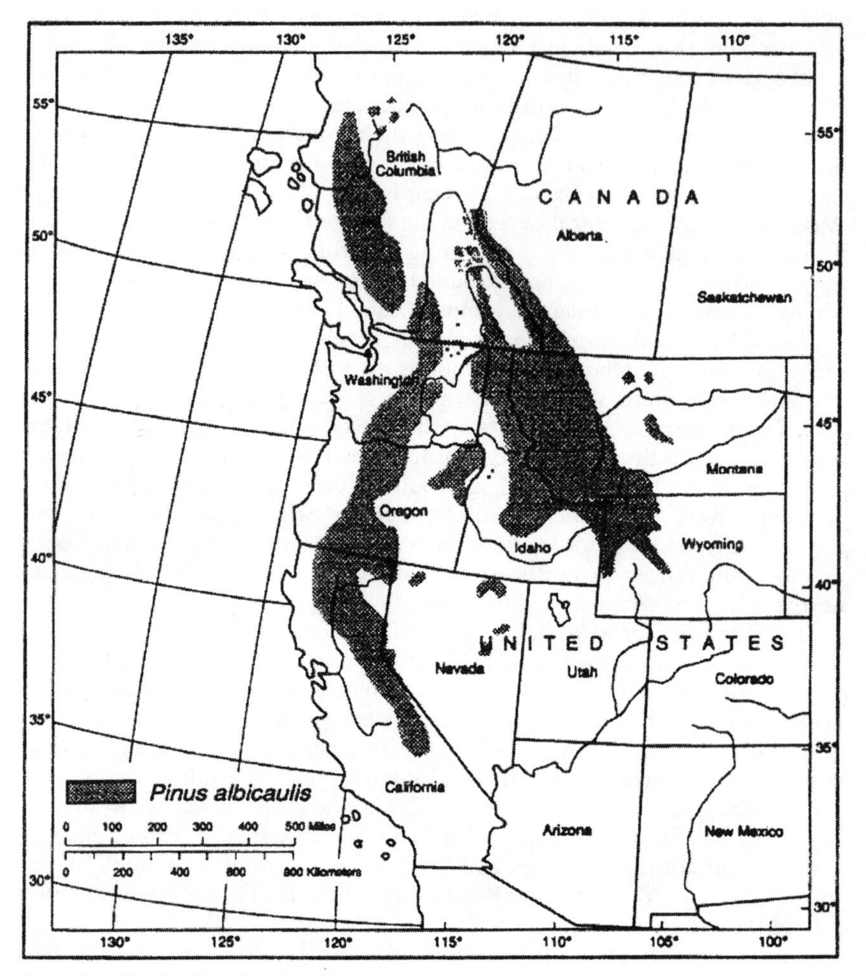

Figure 2-1. Distribution of whitebark pine (*Pinus albicaulis*) in North America (modified from Schmidt 1994).

mountain hemlock (*Tsuga mertensiana*) in the Rocky Mountains, and *Pinus contorta* var. *murrayana* in the Sierra Nevada and the Cascade and Blue Mountains. Other associated tree species include interior Douglas-fir (*Pseudotsuga menziesii* var. *glauca*), limber pine, alpine larch (*Larix lyallii*), Rocky Mountain juniper (*Juniperus scopulorum*), California red fir (*Abies magnifica*), western white pine, and foxtail pine (*Pinus balfouriana*) (McCaughey and Schmidt 1990).

Historical Perspectives

The evolutionary history of whitebark pine is poorly known. Many methods have been used to determine the origin of the first conifers, the first pines, the initial radiation of pines, and the first evidence of whitebark pine. These methods include evaluations of wood, needle, and cone fossils; pollen analyses; and cladistic analyses of molecular genetic data.

Early interpretations of Mesozoic fossil flora suggest that the first conifers originated in Asia in the Permian period of the Paleozoic era, over 250 million years ago (Smith et al. 1981). Mirov (1967) thought that the subsequent evolution of pines unfolded from the northern latitudes in a progressive southward migration during the Mesozoic and Cenozoic eras, culminating during the Pleistocene epoch. However, at the beginning of the Mesozoic, there appeared to be one landmass—Pangaea (Smith et al. 1981). By the early Jurassic, the northern supercontinent Laurasia had separated, disrupting continuity of the land masses during the radiation of the pines (Wolfe 1985; Millar 1998). This separation of the land masses, along with no fossil evidence for a high-latitude Mesozoic center of origin, indicates that pines were more likely to originate in the middle latitudes (Millar 1998). Paleobotanists and taxonomists have reanalyzed many of the fossils previously identified to *Pinus* and have reclassified them as the extinct Pinaceous genera *Pityostrobus* and *Pseudoaraucaria,* genera considered most closely related to *Pinus* and most likely to have been the ancestral gene pool to *Pinus* (Miller 1976, 1988; Millar 1998). As a result of this reclassification, coupled with fossil discoveries during the last two decades, Millar (1998) proposed a middle latitude Mesozoic origin for the genus, in contrast to the Palaeozoic origin as suggested by Mirov (1967).

Current thought places the center of origin of *Pinus* in the northeastern United States, western Europe, or eastern Asia (Miller 1976; Equiluz Piedra 1985; Axelrod 1986; Millar and Kinloch 1991). Many pine species became extinct during the late Mesozoic era approximately 90 million to 120 million years ago (Millar 1998). The tectonic events and consequent migrations of the early Tertiary period appear to have created several new secondary centers of pine diversity (refugia). There were centers in northern North America, Mexico, and Central America. In Eurasia, centers of pine diversity appear to have been divided into northern and southern (midlatitude) refugia (Millar 1998). Pine fossils or pollen have never been found in geologic formations of the Southern Hemisphere. However, when pines are planted there today, including in New Zealand, Chile, and South Africa, they often grow at exceptionally rapid rates (Mirov 1967).

The first conifers may have appeared during the late Paleozoic (Mirov 1967); however, the earliest known pine pollen, from *Pinus belgica,* did not appear until about 130 million years ago (Harland et al. 1990; Gradstein et al. 1995; Millar 1998) (Table 2-3). From the time that pines were first documented throughout the Northern Hemisphere, including North America, they already had split into two readily discriminated subgenera: *Strobus* and *Pinus* (Mirov 1967). The pines comprising subgenus *Strobus* (haploxylon pines) have needles with one fibrovascular bundle, while those comprising subgenus *Pinus* (diploxylon pines) have needles with two bundles.

The late Mesozoic period was a time of continued expansion of pines into additional geographic areas with subsequent differentiation. There are two schools of thought regarding the main paths of migration, with Millar (1998) suggesting an east and west pathway in North America, and not predominantly

Table 2-3. History of whitebark pine and its progenitors in relation to geologic time (from Harland et al. 1990; Gradstein et al. 1995).

Geologic Time				
Era	Period	Epoch	Age	Pine Events
Cenozoic	Quaternary	Holocene	Present	Oldest known living whitebark pine germinated >1,270 years ago.
			4,000 yrs	Slightly cooler period but whitebark pine forests have changed little.
			4,000 to 8,000 yrs	Warming period with whitebark pine largely restricted to high elevations. Lodgepole pine became strong competitor.
			8,000 to 10,000 yrs	Abundant whitebark pine, spruce, and poplar occupied previously glaciated sites.
		Pleistocene	1.64 M*	Whitebark pine extends back at least 100,000 years, maybe as early as 600,000 to 1.3 million years ago. Whitebark pine and associates advanced and retreated with glaciers of this era.
	Tertiary	Pliocene	5.2 M	Some cooling with pine migrations south via cordilleras.
		Miocene	23.3 M	Many pine species in western North America.
		Oligocene	35.4 M	Numerous pine species.
		Eocene	56.5 M	Pines widely distributed in North America.
		Paleocene	65 M	
Mesozoic	Cretaceous	Senonian	89.9 M	Pines occur throughout Northern Hemisphere. Have split into *Strobus* and *Pinus*.
		Gallic	127 M	First pines, possibly in north eastern North America, western Europe, or eastern Asia (Millar 1998).
	Jurassic			
		Neocomian	144.2 M	
		Malm	159.4 M	
		Dogger	180.1 M	
		Lias	205.7 M	
	Triassic	Tr3	227.4 M	
		Tr2	241.7 M	
		Scythian	248.2 M	
Paleozoic	Permian	Zechstein	256.1 M	First conifers? (Mirov 1967)
		Rotliegendes	290 M	

* M=millions of years

southward, as Mirov (1967) suggested. According to Axelrod (1986), the gradual cooling during the Pliocene resulted in increased migration southward via the cordilleras. This cooling was a prelude to the Pleistocene, when extensive glaciation in North America and Eurasia resulted in the loss of many pine species, particularly in Europe where southward movement was restricted by the Alps.

The Pleistocene was a time of fluctuating climates (Millar 1998). Apparently, whitebark pine along with other pines advanced and retreated northward and southward with the glaciers during this era, with refugia likely occurring in the middle latitudes in areas not covered by glacial ice (Van Devender and Spaulding 1979; Critchfield 1985). Nutcrackers dispersed the seeds, enabling whitebark pine to move rapidly both latitudinally and elevationally (Chapter 5).

Exactly when whitebark pine first appeared is still unclear. Molecular genetic data suggest that it may have originated only 600,000 to 1,300,000 years ago, presumably diverging from the Eurasian stone pines (Krutovskii et al. 1990, 1994). Lanner (1996) suggested that a pine similar to today's Siberian stone pine crossed from northeast Asia to Alaska over the Bering Strait land bridge. The bridge disappeared sometime prior to 1.8 million to 3.5 million years ago, so this migration most likely occurred in the Pliocene (Krutovskii et al. 1994). After the land bridge connection was terminated, the pine became isolated from Asia and subsequently differentiated into whitebark pine. The fact that there are four species of stone pines in Europe and Asia with at least six named subspecies, nine cultivars, and several varieties, but only one commonly recognized stone pine in North America supports Lanner's hypothesis. However, molecular genetic data do not preclude a North American origin for whitebark and the stone pines.

Two fossil sites in the Rocky Mountains, both older than the last glacial advance, contain whitebark pine pollen and macrofossils dating back to the Pleistocene epoch (Baker 1990). They reveal whitebark pine to be present in the Yellowstone National Park region for more than 100,000 years. Following the retreat of the glaciers in the Holocene, about 15,000 to 10,000 years ago, abundant whitebark pine, spruce (*Picea* spp.), and poplars (*Populus* spp.) occupied the harsh conditions found on the previously glaciated sites. As warming continued into the period from 8,000 to 4,000 years ago, whitebark pine became largely restricted to high-elevation sites (MacDonald et al. 1989; Richardson and Rundel 1998). A slight cooling period about 4,000 years ago essentially created the conditions that we find today for whitebark pine forests. Whitebark pine is able to survive at high elevations where soils are young, coarse and shallow; where fires are fairly frequent, reducing or eliminating competitors; and where climatic disturbances, such as damaging winds, ice storms, snow, summer frost, and winter desiccation prevent stand closure by competitors (Arno and Weaver 1990; Weaver 1990).

Chronologies

Tree-ring chronologies of old whitebark pine are a source of recent history as well as climatic information on processes affecting its current distribution. For example, tree-ring studies in British Columbia show ages of more than 700 years for whitebark pine (Luckman et al. 1984). Studies by Perkins and Swetnam (1996) in the Sawtooth and adjacent mountain ranges of central Idaho show that whitebark pine is a promising species for dendroclimatic studies. Because it occupies harsh, cold sites, interactions with associated species are often very limited, particularly on rocky, shallow soils and windswept exposures, enhancing the possibilities of cross-dating back into millennium-length chronologies. Perkins and Swetnam constructed tree-ring chronologies of 700 to more than 1,000 years for four stands of whitebark pine. Using climate variable analysis, they determined that over 50 percent of the variance in ring width was explained by climate variables, such as precipitation and monthly average temperatures, although their sample size was inadequate for full climate reconstruction. They concluded that additional sampling of open-grown stands of whitebark pine and further time series analysis were needed to clarify relationships between climate and growth variables.

At this time we are unaware of any other dendroclimatic studies that permit correlation with long-term climates of the Holocene. However, Perkins and Swetnam (1996) have helped set the stage; and their discoveries of ancient trees exceeding 1,270 years old and abundant well-preserved logs and snags on their central Idaho study sites may make it possible to reach further back into time.

Acknowledgments

We acknowledge Stephen F. Arno, Robert E. Keane, Diana F. Tomback, and Leo P. Bruederle for their review of this chapter. We also thank Harold Ziolkowski with the USDA Animal and Plant Health Inspection Service, Bozeman, Montana, for help with developing our whitebark pine distribution map, and Judy O'Dwyer with the USDA Forest Service, Rocky Mountain Research Station in Bozeman for her editorial review throughout all phases of chapter preparation.

Literature Cited

Arno, S. F., and R. J. Hoff. 1990. *Pinus albicaulis* Engelm. Whitebark pine. Pages 268–279 *in* R. M. Burns and B. H. Honkala, technical coordinators. Silvics of North America, Vol. 1, Conifers. USDA Forest Service, Agriculture Handbook 654, Washington, D.C.

Arno, S. F., and T. Weaver. 1990. Whitebark pine community types and their patterns on the landscape. Pages 97–105 *in* W. C. Schmidt and K. J. McDonald, compilers. Proceedings—Symposium on whitebark pine ecosystems: Ecology and management of a high-mountain resource. USDA Forest Service, Intermountain Research Station, General Technical Report INT-270, Ogden, Utah.

Axelrod, D. I. 1986. Cenozoic history of some western American pines. Annals of the Missouri Botanical Garden 73:565–641.

Baker, R. G. 1990. Late Quaternary history of whitebark pine in the Rocky Mountains. Pages 40–48 *in* W. C. Schmidt and K. J. McDonald, compilers. Proceedings—Symposium on whitebark pine ecosystems: Ecology and management of a high-mountain resource. USDA Forest Service, Intermountain Research Station, General Technical Report INT-270, Ogden, Utah.

Critchfield, W. B. 1985. The late Quaternary history of lodgepole and jack pine. Canadian Journal of Forest Research 15:749–772.

———. 1986. Hybridization and classification of the white pines (*Pinus* section *Strobus*). Taxonomy 35:647–656.

Critchfield, W. B., and E. L. Little. 1966. Geographic distribution of the pines of the world. USDA Forest Service, Miscellaneous Publication 991, Washington, D.C.

Equiluz Piedra, T. 1985. Origen y evolucion del genero *Pinus*. Dasonomia Mexicana 3:5–31.

Gradstein, F. M., F. P. Agterberg, J. G. Ogg, J. Hardenbol, P. van Veen, J. Thierry, and Z. Haung. 1995. A *Triassic, Jurassic* and *Cretaceous* timescale. Pages 95–126 *in* W. A. Berggren, D. V. Kent, M.-P. Aubry, and J. Hardenbol, editors. Geochronology, timescales, and global stratigraphic correlation. Society for Sedimentary Geology Special Publication No. 54.

Harland, W. B., R. Armstrong, A. Cox, C. Lorraine, A. Smith, and D. Smith. 1990. A geologic timescale 1989. Cambridge University Press, New York.

Harlow, W. M., E. S. Harrar, and F. M. White. 1979. Textbook of dendrology, 6th edition. McGraw-Hill, New York.

Hutchins, H. E., and R. M. Lanner. 1982. The central role of Clark's nutcracker in the dispersal and establishment of whitebark pine. Oecologia 55:192–201.

Knight, R. R., B. M. Blanchard, and D. J. Mattson. 1987. Yellowstone grizzly bear investigations. Report of the Interagency Study Team. USDI, National Park Service, Bozeman, Montana.

Krutovskii, K. V., D. V. Politov, and Y. P. Altukov. 1990. Interspecific genetic differentiation of Eurasian stone pines for isoenzyme loci. Soviet Genetics 26:440–451.

———. 1994. Genetic differentiation and phylogeny of stone pine species based on isozyme loci. Pages 19–30 *in* W. C. Schmidt, and F.-K. Holtmeier, compilers. Proceedings—International workshop on subalpine stone pines and their environment: The status of our knowledge. USDA Forest Service, Intermountain Research Station, General Technical Report INT-309, Ogden, Utah.

Lanner, R. M. 1990. Biology, taxonomy, evolution, and geography of stone pines of the world. Pages 14–23 *in* W. C. Schmidt and K. J. McDonald, compilers. Proceedings—Symposium on whitebark pine ecosystems: Ecology and management of a high-mountain resource. USDA Forest Service, Intermountain Research Station, General Technical Report INT-270, Ogden, Utah.

———. 1996. Made for each other: A symbiosis of birds and pines. Oxford University Press, New York.

Ledig, F. T. 1998. Genetic variation in pines. Pages 251–280 *in* D. M. Richardson, editor. Ecology and biogeography of *Pinus*. Cambridge University Press, New York.

Losensky, B. J. 1990. Historical uses of whitebark pine. Pages 191–197 *in* W. C. Schmidt and K. J. McDonald, compilers. Proceedings—Symposium on whitebark pine ecosystems: Ecology and management of a high-mountain resource. USDA Forest Service, Intermountain Research Station, General Technical Report INT-270, Ogden, Utah.

Luckman, B. H., L. A. Jozsa, and P. J. Murphy. 1984. Living seven-hundred-year-old *Picea engelmannii* and *Pinus albicaulis* in the Canadian Rockies. Arctic and Alpine Research 16: 419–422.

MacDonald, G. M., L. C. Cwynar, and C. Whitlock. 1989. Late Quaternary dynamics of pines: Northern North America. Pages 122–136 *in* D. M. Richardson, editor. Ecology and biogeography of *Pinus*. Cambridge University Press, New York.

McCaughey, W. W., and W. C. Schmidt. 1990. Autecology of whitebark pine. Pages 85–96 *in* W. C. Schmidt, and K. J. McDonald, compilers. Proceedings—Symposium on whitebark pine ecosystems: Ecology and management of a high-mountain resource. USDA Forest Service, Intermountain Research Station, General Technical Report INT-270, Ogden, Utah.

Millar, C. I. 1998. Early evolution of pines. Pages 69–91 *in* D. M. Richardson, editor. Ecology and biogeography of *Pinus*. Cambridge University Press, New York.

Millar, C. I., and B. K. Kinloch. 1991. Taxonomy, phylogeny, and coevolution of pines and their stem rusts. Pages 1–38 *in* Y. Hiratsuka, J. Samoil, P. Blenis, P. Crane, and B.

Laishley, editors. Rusts of pines. Proceedings—IUFRO Rusts of Pine. Working Party Conference. Forestry Canada, Edmonton, Alberta, Canada.

Miller, C. N. 1976. Early evolution in the *Pinaceae*. Reviews in Palaeontology and Palynology 21:101–117.

———. 1988. The origin of modern conifer families. Pages 448–487 *in* C. B. Beck, editor. Origin and evolution of gymnosperms. Columbia University Press, New York.

Mirov, N. T. 1967. The genus *Pinus*. Ronald Press, New York.

Ogilvie, R. T. 1990. Distribution and ecology of whitebark pine in western Canada. Pages 40–48 *in* W. C. Schmidt and K. J. McDonald, compilers. Proceedings—Symposium on whitebark pine ecosystems: Ecology and management of a high-mountain resource. USDA Forest Service, Intermountain Research Station, General Technical Report INT-270, Ogden, Utah.

Perkins, D. L., and T. W. Swetnam. 1996. A dendrochronological assessment of whitebark pine in the Sawtooth-Salmon River region, Idaho. Canadian Journal of Forest Research 26:2123–2133.

Pfister, R. D., B. L. Kovalchik, S. F. Arno, and R. C. Presby. 1977. Forest habitat types of Montana. USDA Forest Service, Intermountain Forest and Range Experiment Station, General Technical Report INT-34, Ogden, Utah.

Pilger, R. 1926. Genus *Pinus. In* A. Engler and K. Prantl, editors. Die natürlichen Ppflanzenfamilien. Vol. XIII. Gymnospermae. Wilhelm Engelmann, Leipzig, Germany.

Price, R. A., A. Liston, and S. H. Strauss. 1998. Phylogeny and systematics of *Pinus*. Pages 49–68 *in* D. M. Richardson, editor. Ecology and biogeography of *Pinus*. Cambridge University Press, New York.

Richardson, D. M., and P. W. Rundel. 1998. Pine ecology and biogeography—An introduction. Pages 3–46 *in* D. M. Richardson, editor. Ecology and biogeography of *Pinus*. Cambridge University Press, New York.

Schmidt, W. C. 1994. Distribution of stone pines. Pages 1–6 *in* W. C. Schmidt and F.-K. Holtmeier, compilers. Proceedings—International workshop on subalpine stone pines and their environment: The status of our knowledge. USDA Forest Service, Intermountain Research Station, General Technical Report INT-309, Ogden, Utah.

Shaw, G. R. 1914. The genus *Pinus*. Arnold Arboretum Publication No. 5, Houghton Mifflin Co., Boston.

Smith, A. G., A. M. Hurley, and J. C. Briden. 1981. Phanerozoic paleocontinental world maps. Cambridge University Press, Cambridge, United Kingdom.

Thompson, L. S., and J. Kuijt. 1976. Montane and subalpine plants of the Sweetgrass Hills, Montana, and their relationship to early post-glacial environments of the Northern Great Plains. Canadian Field-Naturalist 90:432–448.

Van Devender, T. R., and W. G. Spaulding. 1979. Development of vegetation and climate in the southwestern United States. Science 204:701–710.

Wang, X. R., Y. Tsumura, H. Yoshimaru, K. Nagasaka, and A. E. Szmidt. 1999. Phylogenetic relationships of Eurasian pines (*Pinus, Pinaceae*) based on chloroplast rbcL, matK, rp120-rps18 spacer, and trnV intron sequences. American Journal of Botany 86:1742–1753.

Weaver, T. 1990. Climate of subalpine pine woodlands. Pages 72–79 *in* W. C. Schmidt and K. J. McDonald, compilers. Proceedings—Symposium on whitebark pine ecosystems: Ecology and management of a high-mountain resource. USDA Forest Service, Intermountain Research Station, General Technical Report INT-270, Ogden, Utah.

Wolfe, J. A. 1985. Distributions of major vegetational types during the Tertiary. Geophysical Monograph 32:357–375.

Young, J. A., and C. G. Young. 1992. Seeds of woody plants in North America. Dioscoridies Press, Portland, Oregon.

Chapter 3

Whitebark Pine and Its Environment

T. Weaver

Whitebark pine (*Pinus albicaulis*) is a member of a small group of circumboreal species known as stone pines (*Pinus* subsection *Cembrae,* Critchfield and Little 1966) that occur at high elevations or high latitudes in the Northern Hemisphere (Chapter 2). Whitebark pine occupies timberline environments from 53° N latitude southward to 37° N in the Sierra Nevada and 42° N in the Rocky Mountains (Little 1971).

Understanding the relationship between whitebark pine and its environment is essential to all management plans for the species. Thus, this chapter explores four aspects of this relationship. The first section describes the environment of whitebark with respect to factors most likely to influence its performance. Comparisons of environments will show both the similarity of whitebark pine's environment to those of other stone pines and dissimilarities with environments above and below it on the altitudinal gradient. The second section estimates whitebark's physiological tolerances to these environmental factors from measurements made on all five stone pine species. A master tolerance can be integrated across single-factor tolerances to outline the tree's potential ecological niche (Hutchinson 1958). The niche outlined is the potential niche, the sum of environments currently occupied by whitebark pine (the realized niche) and environments from which it is excluded by competitors (the unrealized niche). The third section uses the description of whitebark's physiological niche (potential niche) to understand the behavior of whitebark pine in the entire environmental field available to it. Specifically, I address whitebark pine's geographic range; altitudinal range; responses to climate change; productivity; responses to substrate, nutrients, and pollution; phenology; and responses to humans. The last section lists the reciprocal influences of whitebark on its ecosystem.

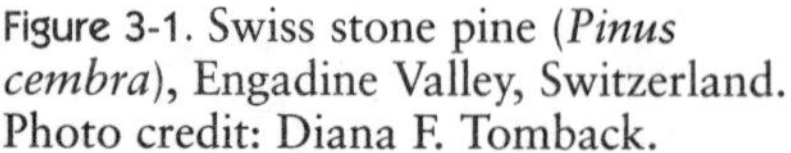

Figure 3-1. Swiss stone pine (*Pinus cembra*), Engadine Valley, Switzerland. Photo credit: Diana F. Tomback.

Information about other stone pines of the world is included in this review, either to tentatively fill gaps in our knowledge about whitebark pine's interaction with its environment or to corroborate existing, but scanty, information about *P. albicaulis*. This is reasonable, because the stone pines are closely related both taxonomically and ecologically (Figure 3-1). Stone pines are characterized morphologically by five-needled fascicles, large wingless seeds, and cones that do not open on their own (Mirov 1967; Lanner 1990; Chapter 2). Recent DNA gene sequence data indicate that all the stone pines, *P. cembra* of the Alps, *P. sibirica* of western Siberia, *P. pumila* of eastern Siberia, and *P. koraiensis* of eastern China, are closely related to one another (Chapter 8). The ecological equivalence of these species is demonstrated by their occurrence in similar environments, all located at the southern boundary of Pleistocene glaciation (Weaver 1994a). Common names and scientific names for all tree species mentioned in this chapter are listed in Table 3-1.

The Environment of Whitebark Pine Communities

The environment of whitebark pine is discussed under the headings of climate, soils, and substrates.

Climate

The presence of whitebark pine on the various sites it occupies indicates a more or less equivalent climate, described below by an average condition and condition limits.

1. The average climate is described using data from all the long-term weather stations located in whitebark pine forests between central California and central Alberta (five stations, Table 3-2). Average climates of stone pine ecosystems around the world are remarkably similar to the average whitebark pine climate (Table 3-3). Note, however, that whereas temperature at tree line is nearly constant over whitebark pine's north-south range, there may be considerable variation in other factors such as photoperiod, wind, ultraviolet radiation, and snow depth and persistance (Billings 2000; Koerner 1989).
2. The limits of the whitebark climate are specified by contrasting the average whitebark pine climate with climates of contrasting ecosystems located above and below it on the altitudinal gradient, that is, tundra above and warmer forests and grasslands below (Table 3-2).
3. The variation around both geographic and altitudinal means is discussed by Weaver (1994a, 1994b).
4. Except where noted, the climate data come from Weaver (1980, 1990, 1994a, 1994b).

Temperature. Winters in the whitebark pine ecosystem seem cold to humans but are mild for subnivian (under the snow) occupants (Weaver 1994a). Organisms functioning above the snow, such as trees and large vertebrates, experience

Table 3-1. Synonomy of tree and grass species.[1,2]

Latin Name	Common Name
Abies lasiocarpa	Subalpine fir
Picea engelmannii	Engelmann spruce
Pinus albicaulis	Whitebark (stone) pine
Pinus aristata	Bristlecone pine
Pinus balfouriana	Foxtail pine
Pinus cembra	Swiss stone pine
Pinus contorta	Lodgepole pine
Pinus flexilis	Limber pine
Pinus koraiensis	Korean stone pine
Pinus monticola	Western white pine
Pinus ponderosa	Ponderosa pine
Pinus pumila	Japanese stone pine
Pinus sibirica	Siberian stone pine
Pseudotsuga menziesii	Douglas-fir
Tsuga mertensiana	Mountain hemlock
Agropyron spicatum	Bluebunch wheat grass
Bouteloua gracilis	Blue grama grass
Festuca idahoensis	Idaho fescue

[1]Latin names are often coded with the first two letters of the generic and specific epithets; for example, PIAL indicates *Pinus albicaulis.*
[2]The species BOGR, AGSP, FEID, PSME, ABLA, PIAL indicate environmental zones from the plains to the timberline (cf. Tables 3-2 and 3-4).

Table 3-2. Comparison of climates in the whitebark pine ecosystems[1] and other Rocky Mountain forest ecosystems. Standard errors, related data, and methods are available in Weaver (1980, 1990, 1994b).

	Temperatures[2]					Water[3]			GS[4]
Parameter	Abs min °C	Jan min °C	T_{gs} mean °C	July max °C	Abs max °C	Pptn ann mm	Drt mo mo	H_2O deficit mm	mo
Alpine tundra[1]	–36	–16	6	12	20	778	0	0	3.6
Pinus albicaulis	**–36**	**–13**	**9**	**19**	**29**	**705**	**0**	**3**	**4.3**
Abies lasiocarpa	–47	–18	12	22	36	820	0	0	4.5
Pseudotsuga menziesii	–43	–16	12	27	36	58	0	1	3.6
Festuca idahoensis	–42	–12	12	27	39	38	1	6	5.1
Agropyron spicatum	–42	–13	12	28	39	38	1	17	4.9
Pinus flexilis	–46	–13	12	28	41	—	2	20	5.2
Pinus ponderosa	–41	–14	13	32	43	39	2	20	4.7
Bouteloua gracilis	–43	–15	14	31	42	35	2	25	4.4

[1]The ecosystems compared range from alpine down through high forests (*P. albicaulis, A. lasiocarpa*), low forests (*P. menziesii*), savanna (*P. ponderosa, P. flexilis*), and grasslands (*F. idahoensis, A. spicatum,* and *B. gracilis*). They are named for their climax vegetation.

Pinus albicaulis data were recorded at Ellery Lake, California; Crater Lake, Oregon; Kings Hill, Montana; Old Glory, British Columbia; and Banff, Alberta (Weaver 1990, 1994). Climate data for other ecosystems were gathered in the alpine (California, Colorado), *Abies lasiocarpa* (Montana, Idaho), *Pseudotsuga menziesii* (Montana, Idaho, Wyoming), *Pinus flexilis* (Montana), *Pinus ponderosa* (Montana), *Festuca idahoensis* (Montana), wheat grass (Montana), *Bouteloa gracilis* (Montana), and *Atriplex sarcobatus* (Wyoming, Weaver 1980, 1990). *Pinus flexilis* data represent sites below the *Pseudotsuga menziesii* zone.

[2]Temperatures (Weaver 1980, 1990) include the absolute minimum, January minimum, growing season average, July maximum, and absolute maximum.

[3]Water data (Weaver 1980, 1990) are annual precipitation, drought months, and annual water deficit (mm).

[4]Growing season months are defined as those with moist soils and average air temperature above 0°C (Weaver 1990, 1994).

Table 3-3. Comparison of climates in the whitebark pine ecosystems and other stone pine ecosystems. These and similar data were compiled with standard errors by Weaver (1990, 1994a).

Ecosystem[1] Location[2]	PIAL RM	PICE Alps	PISI C Asia	PIPU E Asia	PIKO China
WINTER TEMPERATURES, °C					
Absolute min.	–34	–23	–55	–52	–42
Jan. min.	–14	–8	–27	–30	–27
max.	–5	–1	–19	–24	–11
SUMMER TEMPERATURES, °C					
Absolute max.	29	27	37	33	36
July max.	18	14	21	15	26
min.	4	5	12	8	14
Growing season (T_{gs})[3]	9	8	13	9	11
PRECIPITATION, MM[4]					
Avg Annual	931	939	432	407	636
October–June	829	616	245	264	242
July–September	102	323	187	143	394
Driest summer mo.	4	45	8	4	32
Drought months	0	0	0	0	0
Water deficit	0	0	0	0	0
Rain days in gro-season	8	16	14	1	18

[1]Pine species are *P. albicaulis* (*PIAL*), *P. cembra* (*PICE*), *P. sibirica* (*PISI*), *P. pumila* (*PIPU*), and *P. koraiensis* (*PIKO*).
[2]Locations are northern Rocky Mountains USA, European Alps, north-central Asia, northeast Asia, and China and Korea.
[3]Growing season includes those months when soils are moist and average air temperature exceeds 0°C (Weaver 1994b).
[4]Annual precipitation, October–June precipitation, July–September precipitation, precipitation in the driest summer month, months with drought as defined by Walter (1984), "water deficit" is the excess of evapotranspiration over rainfall, and "rain days" are the number of days with any precipitation.

average maximum/minimum temperatures of –5°/–14°C in January (Table 3-3). Absolute minimum temperatures are around –34°C (Table 3-3). In contrast, organisms under deep snow—including microbes, insects, mammals, and roots—on all but windswept sites function at near 0°C (Sirucek et al. 1999), because snow insulates against the loss of heat stored by the soil in the previous summer (Seligman 1939; Sirucek et al. 1999). Note that winter air temperatures are similar or colder in ecosystems at lower altitudes (Table 3-2). Soil temperatures are identical under lower forests, but drop sharply under grasslands, because snow cover becomes intermittent in space or time (Sirucek et al. 1999). Air temperatures are slightly warmer in *Pinus cembra* (–1°/–8°C) and cooler in *P. pumila* (–24°/–30°C), *P. sibirica* (–19°/–27°C), and *P. koraiensis* (–11°/–27°C) ecosystems (Table 3-3).

Summer temperatures are cool in whitebark pine environments. Average maximum/minimum July temperatures are 18°/4°C (Table 3-3). Absolute maximums are about 30°C (Table 3-2). Soil temperatures at 50 centimeters in whitebark pine and subalpine fir (*Abies lasiocarpa*) stands rose from 1°C in mid-May to 9°C in mid-August and fell again to 4°C in October (Sirucek et al. 1999). Soil temperatures may be higher near the soil surface (5°C at 7 centimeters, Sirucek 1996) and on steep south-facing slopes (60°C at 0 centimeters, Turner et al. 1975). Summer temperatures rise as one moves downslope from an average July maximum of 19°C in the whitebark pine ecosystem to 30°C in the foothills and plains below (Table 3-2). The average July maximum in the *P. albicaulis* ecosystem (18°C) is higher than in *P. cembra* (14°C) or *P. pumila* (15°C), and lower than in *P. sibirica* (21°C) or *P. koraiensis* (26°C) ecosystems (Table 3-3).

Water. Precipitation in whitebark pine environments ranges, at the least, from 600 to 1,600 millimeters/year (Weaver 1990). Over 85 percent of this falls in winter and accumulates as 170 to 300 centimeters of snow (Weaver 1990), which, when it melts, contributes to runoff important to streams, rivers, and irrigation. The 70 to 300 millimeters falling in summer is deposited in 5 to 11 showers. The number of showers falling in July–August falls from 37 in grasslands to 31 in montane forests to 29 in subalpine forests. Over half of the showers are less than 3 millimeters (Weaver 1985). In the Rocky Mountains, total precipitation in the alpine/subalpine zone is 20 times that in grasslands below (Table 3-2); in the Cascades, the ratio is higher (Gerlach 1970).

Across stone pine environments, the average number of summer showers is one for *P. pumila,* eight for *P. albicaulis,* and fourteen to eighteen for the remaining species (Table 3-3). Summer precipitation is less for whitebark pine (102 mm) than for any other stone pine—*P. pumila* (143 mm), *P. sibirica* (187 mm), *P. cembra* (323 mm), or *P. koraiensis* (394 mm). The number of rain days in *P. albicaulis* (2/month) is greater than in *P. pumila* (0.3/month) and similar to *P. sibirica* (3.5/month), *P. cembra* (4/month), and *P. koraiensis* (4.5/month, Table 3-3).

Stone pine ecosystems experience soil drought rarely, if ever. That is, soils are saturated in spring, evaporation is probably less than 2 millimeters per month (Walter 1985; Nielson 1992), and precipitation exceeds that evaporation. Drought months rise from 0 in whitebark pine forests, to 1 in subalpine fir forests, to 2 in foothills grasslands (Table 3-2). Exceptional microsites could be droughty due to absence of soil buffering (low water-holding capacity), high evaporation rates (especially windy or sunny), or high runoff/low absorption. *P. albicaulis* forests are more droughty for organisms not using soil water—for example, lichens and insects—than for trees (Weaver 1994a).

Wind. Wind flow across mountain ridges is greater than at lower altitudes, especially in winter. Winds high above plains flow faster than at low altitudes, because they are not slowed by friction against the ground. When winds meet mountains, a complex of accelerating and slowing forces usually results in some slowing, but with maintenance of the vertical profile (Barry 1992). This is illustrated by data from Colorado (Barry 1973) where January/July flows decline from exposed ridges to subridge alpine sites (13.9/5.8 meter/second) to the subalpine fir (5.8/1.9 m/sec) and Douglas-fir (3.6/1.3 m/sec) zones. Gusts decline from alpine (27.8/1.6 m/sec) to subalpine fir (17.7/0 m/sec) and Douglas-fir (5.4/0 m/sec) zones in July/August. The gust speeds are underestimated. (For reference, 45 m/sec [meters/second] equals 100 miles per hour).

Radiation. Average daily solar radiation in the whitebark pine region is 0.2–0.5 kW/m^2 [=100–220 cal cm^{-2} day^{-1}] in January and 1.1–1.3 kW/m^2 [=550–650 cal cm^{-2} day^{-1}] in July (Gerlach 1970). Summer radiation varies little over the altitudinal gradient (Barry 1973). Whereas the fall-winter values are among the lowest in the United States, summer values approach those over deserts of the Southwest.

Soils

Soils of the whitebark pine ecosystem are most often classified as cryochrepts, that is, cold-climate soils with light-colored (leached) surface horizons and minimally developed subsurface horizons (Hansen-Bristow et al. 1990). They are sometimes less well developed (cryorthents) and sometimes better developed (cryoboralfs or cryoborolls).

Rather than emphasizing soil taxonomy, I instead emphasize the capacity of these soils to store water and minerals. Thus, I compare conditions in *P. albicaulis* stands (Weaver and Dale 1974) with conditions in other ecosystems of the northern Rocky Mountains (Weaver 1978). Similar data appear in Hansen-Bristow et al. (1990).

The water-holding capacity of a soil is its tendency to buffer against drought, that is, to absorb water and release it in the dry period between rainstorms. The water-holding capacity of soil increases as its clay- and organic matter–content increase (Decker et al. 1975). Clay contents are low at high altitude, both because little clay forms there and much of what does form is exported to lower ecosystems in running water; thus, the clays and silts that do exist in the alpine and *P. albicaulis* zones (Table 3-4) were probably deposited as loess. Organic carbon (OC × 1.7 = organic matter) contents are high in tundra, decline through the *P. albicaulis* zone to closed conifer forests, and rise again in the grasslands (Table 3-4). Integrating this, the water storage capacity of whitebark pine soils, 2.1 centimeter in the top decimeter, is higher than most other high-altitude ecosystems of the northern Rocky Mountains (Table 3-4).

The high organic carbon (OC) of alpine soils, noted above, results from both the high production and the slow decomposition of roots in the cool soils of this environment (Weaver 1978). The carbon content of adjacent *P. albicaulis* soils is also high—as are the carbon contents of *P. cembra* forests (Blaser 1980). I suggest three possible explanations: (1) High OC under whitebark pine could be due to high production and slow decomposition, as in tundra. (2) Some of the soil organic matter of the whitebark pine ecosystem could be residual from colder eras, when tundra reached downslope to altitudes currently occupied by *P. albicaulis* or from warmer eras when mountain meadows reached further upslope; cryoborolls of the whitebark pine zone (Hansen-Bristow et al. 1990) may represent such remnant organic matter. (3) Or systematic differences in soil sampling methods could be responsible.

Soils also serve as a reservoir for nutrients essential to plant growth. Because most soil nitrogen is covalently bound to carbon, quantities present

Table 3-4. Soils of whitebark pine and associated ecosystem types[1] compared (Weaver 1978).

	Soil property[2]									
	WHC cm/dm	WHC cm/rt	Clay %	OC %	N %	C/N	K meq[2]	Ca meq[2]	Mg meq[2]	P meq[2]
ALPN[1]	0.9	3.8	10	5.7	0.50	0.08	0.3	6.7	1.6	0.07
PIAL	2.1	—	8.5	3.6	0.22[3]	0.06	0.4	1.3	0.3	—[3]
ABLA	1.5	3.0	4	1.6	0.07	0.04	0.5	1.8	0.6	0.09
PSME	3.1	10.3	12	1.2	0.07	0.05	5.0	5.4	1.7	0.16
FEID	2.0	10.1	27	2.8	0.30	0.01	1.0	14.0	3.4	0.03
AGSP	2.4	11.7	32	3.1	0.28	0.09	1.7	17.9	3.1	0.03
BOGR	2.5	11.7	28	1.5	0.15	0.10	1.2	14.6	5.2	0.02

[1]Ecosystems, from low to high, are *Bouteloua gracilis* (BOGR), *Agropyron spicatum* (AGSP), *Festuca idahoensis* (FEID), *Pseudotsuga menziesii* (PSME), *Abies lasiocarpa* (ABLA), *Pinus albicaulis* (PIAL), and alpine tundra (ALPN).
[2]Soil properties are water-holding capacity (WHC, cm in the topmost 10 cm of the soil); water-holding capacity (WHC, cm in the rooted zone); clay content (%) and organic carbon content (%), which control water-holding capacity; nitrogen content (%); carbon/nitrogen ratio; and quantities of potassium, calcium, magnesium, and phosphorus in meq/100 gm of soil. Meq/100 gm is translated into ppm by multiplying by 391, 200, 122, and 320 for K, Ca, Mg, and P, respectively.
[3]N is estimated for PIAL from OC and the average of C/N ratios from ABLA and ALPN. Phosphorus contents for PIAL (0.49 meq/100 gm, Bray extraction) are not comparable to phosphorus status measured in other ecosystems with the Olson extraction.

are well correlated with organic carbon. Thus, nitrogen is plentiful in the tundra (and grasslands below) but scarce in the forest zone (Table 3-4). Though nitrogen has not been measured in whitebark pine stands, I estimate it at 0.22 percent by multiplying the soil's organic carbon content by the average of the N/C ratios in tundra (0.08) and in the subalpine fir (*Abies lasiocarpa*) (0.04) zones. Relevant conversion factors are organic carbon × 1.7 = organic matter (Allison 1965), and approximately 5 percent of OM (organic matter) is nitrogen (Jackson 1958).

Elements appearing in ionic form are leached from soils with low ion-binding capacities, such as those low in clays and organic carbon. For this reason potassium is relatively scarce in the alpine and high forests (Table 3-4). Calcium and magnesium are also scarcer in high forests than in grasslands below (Table 3-4). Calcium and magnesium are, however, surprisingly plentiful in tundra (Table 3-4), perhaps due to external sources (loess) or low leaching losses associated with large quantities of materials that bind well, that is, clay and organic carbon.

Substrates

Harlow and Harrar (1958) reported that whitebark pine avoids limestone. This is true for most of Montana (Weaver and Dale 1974) and apparently holds to the south. On the other hand, in wetter areas near and north of the Canadian border, the tree appears mostly on limestone (Arno and Weaver 1990; P. Acuff, personal communication). Other stone pines, including *P. cembra* (Sauermoser 1994) and *P. sibirica* (Gorchakovsky 1994), also avoid limestone in the core of their range, but favor it in the north. A rationalization for this inconstant behavior is outlined in "Substrate Effects on Performance" on page 61.

Physiological Niche of Whitebark Pine

The performance of whitebark pine varies with several factors, including light, temperature, water, and nutrients. Its physiological niche includes the hyperspace defined by plotting its tolerances on each of these orthagonal factor axes (Hutchinson 1958). Its optimum performance (i.e., best growth) may occur in a smaller hypervolume where conditions are optimal for all factors. Its realized niche includes all of the physiological niche minus those parts of its physiological niche from which whitebark is excluded by competition, predation, or destructive forces, including fire, strong winds, or avalanches. Because data on tolerances are scarce, measurements made on ecologically and taxonomically related stone pines, especially *P. cembra,* are reported as surrogate or supportive data; the application of such data is, of course, tentative.

Light

For two-year-old whitebark pine seedlings, photosynthesis first exceeds respiration (light compensation point) at 100–200 u moles m^{-2} sec^{-1} of photosynthetically active photons (5–10 percent of full sun = FS). Net photosynthesis increases with increasing light intensity. The benefits of increasing light ceases (saturation) above 500–1,000 u moles m^{-1} sec^{-1} (30–40 percent FS, Jacobs and Weaver 1990). Similarly, for young *P. cembra,* light compensation is reached at 2 percent of full sun, light limits through 20 percent of full sun, light is optimal at 20 to 40 percent of full sun, and above 40 percent excesses limit photosynthesis (Tranquillini 1979). Although the data are qualitative and opinions differ, whitebark may be moderately shade tolerant (McCaughey and Schmidt 1990) and tolerates full sun well.

Temperature

Temperature tolerances vary with season. Summer high temperature tolerances equal or exceed the 35° to 45°C maxima observed for germination, photosynthesis, and growth of two-year stock (Jacobs and Weaver 1990). Midwinter killing temperatures are –70°C in *Pinus pumila* (and *Picea glauca,* Sakai 1978) to –80°C in *Pinus cembra* (Bauer et al. 1994). Air temperatures in the field probably never reach either extreme—absolute highs and lows measured in stone pine environments are 37°C and –55°C (Table 3-3). Similarly, in the field, absolute May minima of –10° to –19°C are rare and probably come early in the month; thus at bud break (24 May, Schmidt and Lotan 1980) frost must be rare or nonexistent. High summer soil surface temperatures on south-facing slopes in the Alps kill *P. cembra* seedlings (60°C, Aulitzky and Turner 1982). The death of whitebark pine seedlings on southwest-facing slopes may also be due to heat (D. F. Tomback, personal communication).

The range of temperatures suitable for growing season activity may be indexed by physiological processes that require several enzymes, such as photosynthesis, respiration, and growth. Two-year-old whitebark pine seedlings photosynthesize between 0° to 35°C with an optimum at 20°C. Photosynthesis is considerably reduced below 10°C in *P. albicaulis* (Jacobs and Weaver 1990), *P. pumila* (Kajimoto 1990), and *P. cembra* (Bauer et al. 1994). At the lower extreme, photosynthesis may extend to –10°C in *P. cembra* (Bauer et al. 1994). At the upper extreme, photosynthesis rates decrease above 30°C in *P. albicaulis* (Jacobs and Weaver 1990), above 35°C in *P. cembra* (Tranquillini 1955), and above 15°C in *P. pumila* (Kajimoto 1990). Well-watered whitebark pine trees planted in a grassland climate grow well despite an average July maximum of 27° to 28°C and an absolute high of 39°C (Weaver 1994b). The rate of photosynthesis apparently increases with increases in soil as well as air temperature (Koike et al. 1994)

Apical growth of *P. cembra* is most rapid at air temperatures near 10°C, a temperature warmer than plants at this phenologic stage normally experience (Kronfuss 1994). Root growth in whitebark pine seedlings occurred at 0° to 45°C, with an optimum near 30°C (Jacobs and Weaver 1990). Diameter growth increased 7 to 15 percent per degree C above average in *P. sibirica* (Kokorian and Nazarod 1995). Length of mature needles increased with increasing growing season temperature (Armstrong et al. 1988; Kajimoto 1993). Needle longevity increases with declines in temperature (Kajimoto 1993) and increasing altitude (Nebel and Matile 1992; Reich et al. 1996).

Water

The generalization that pines are intolerant of flooding, and especially as seedlings (Kozlowski et al. 1991), probably applies to stone pines, plants of well-drained sites. At the other extreme, whitebark pine trees probably never experience soil drought greater than the –0.2 to –0.5 MPa (megapascals) range (Table 3-2, Daubenmire 1968; Weaver 1977; Sirucek et al. 1999) and probably could not tolerate them. In contrast, water potentials in dry grasslands commonly fall below –1.5 MPa (Weaver 1977; Sirucek et al. 1999). In *P. cembra* stomata begin to close at osmotic pressures of < –0.4 MPa, and photosynthesis ceases near –1.5 MPa (Havranek and Benecke 1978; Baig and Tranquillini 1980). These plant water potentials depend on the balance between water supply from the soil and losses from the needles.

Dry air, even with moist soils, will reduce the rate of photosynthesis of *P. cembra*. For example, photosynthesis at 25 percent relative humidity is 43 percent of that at 80 percent relative humidity (Tranquillini 1979).

Wind

Growing season winds have little effect on photosynthesis. Wind speeds less than 10 meters/second had no effect on photosynthesic rates of *P. cembra* (Tranquillini 1979), but higher winds can reduce photosynthesis by 20 to 40 percent (Caldwell 1970). The reduction was not due to closing of stomata, but to compression of needles and thus reduced light supply (Caldwell 1970). Wind has little effect on transpiration rates of *P. cembra* and Norway spruce (*Picea abies*) (Tranquillini 1979).

Growing season winds affect the growth form of whitebark pine. Wind reduces shoot growth and increases fascicle density in *P. pumila* (Kajimoto 1993). Chronic wind stimulates the formation of reaction wood and thus contributes much to the differences in form between trees on windy and sheltered sites (Kozlowski et al. 1991; Wooldridge et al. 1996).

Winter winds stunt or eliminate trees from severe sites. In high wind, water potentials of *P. albicaulis* twigs fall critically, and especially on warmer days (Marshall and Zhang 1994). The drying is attributed to abrasion of cuticle from needles by winds, which increase by ten times as one climbs 200 meters through the krummholz zone from low sites with flagged trees 8 meters tall to sites high in the "kampf-zone" where the tree forms a mat with no protruding stems (Hadley and Smith 1989). The heightened drying observed on warm days may indicate some stomatal control, as well.

Nutrients

Essential nutrients comprise the bulk of a pine's dry weight. Data from Scots pine (*Pinus sylvestris*) (Rodin and Bazelevich 1965) are representative. Young needles, old needles, fine roots, and wood are 97.2, 98.2, 98.6, and 99.6 percent carbohydrate and lignin, respectively. This material is approximately 40 percent carbon, 7 percent hydrogen, and 53 percent oxygen, all absorbed as CO_2 or water. In young needles 47 percent of the remainder is nitrogen, 44 percent is mesonutrients, and 9 percent is inert. The mesonutrients are potassium, calcium, phosphorus, magnesium, sulfur, manganese, and iron, in order of weight. The important nonnutrient components are aluminum, silicon, and sodium, in that order. Micronutrients were not measured and are thus included with the carbohydrates. The possibility that nutrients limit production is discussed in the behavior section, on page 61.

Nutrients for new growth, except carbon and oxygen, are either absorbed by roots and mycorrhizae or resorbed for recycling from senescing needles. Mycorrhizae associated with *P. cembra* may absorb nitrogen from inorganic or organic substrates: ammonia is absorbed by most mycorrhizae, nitrate is absorbed by nitrate specialists, and even amino acids and proteins (representing less decomposed organic materials) are used by some mycorrhizal species (Keller 1996). Roots absorb ammonia, nitrate, and other nutrients in ionic forms. As needles age and abscission approaches, nitrogen, phosphorus, potassium, and sulfur are withdrawn; calcium is not withdrawn (Nebel and Matile 1992).

Pollution

The scant literature on the tolerances of other stone pines may suggest the pollution tolerance ranges of unstudied whitebark pine. *Pinus cembra* and *P. koraiensis* are resistant to ozone. *Pinus cembra* is resistant to SO_2, but *P. koraiensis* is sensitive to it (Genys and Heggestad 1978). *Pinus cembra* pollination may be inhibited by acid rain (Paolett and Bellani 1990). Of thirteen pines tested, *P. cembra* was simultaneously most susceptible to sodium chloride damage and second in ability to exclude it (Townsend and Kwolek 1987). I attribute its sensitivity to the rarity of salt in its environment, and its resistance to the possession of an unusually dense cuticle—evolved not to resist salt damage but to resist wind damage. Airborne dust may reduce photosynthesis by filling stomata (Minarcic and Kubicek 1991).

Mechanical Stresses

Whitebark pine is abraded by windborne snow, sand, and silt. Resultant loss of protective cuticle and bark allows desiccation; and, due to the drying, growth is reduced on windy sites and especially on the more exposed sides of the affected trees (Hadley and Smith 1989; Wooldridge et al. 1996). Girdling of trunks and branches by bark beetles (*Dendroctonus ponderosae*) can be fatal (Bartos and Gibson 1990), as can removal of bark by porcupines *(Erethizon dorsatum)* and bears (*Ursus* spp.*)*. Browsing of twigs by deer (*Odocoileus hemionus*) and elk (*Cervus elaphus*) causes minimal damage (Lonner and Pac 1990). Woody plants, like young whitebark pine and associated shrubs, are much less tolerant of trampling by large animals and machines than are grasses and prostrate-leaved forbs (Weaver and Dale 1978).

Seasonality

The niche of whitebark pine has a time dimension in which growth is concentrated in the spring, fruiting follows in summer and fall, and inactivity occurs in winter. Both the requirements for resources and the tolerance of stress vary among these stages in the annual cycle. The tree's coordination with the season (phenology) is cued by day length (Kozlowski et al. 1991), as well as seasonal changes in other factors, including temperature (e.g., Tranquillini 1979). The phenology of whitebark pine (Schmidt and Lotan 1980) is described on pages 63–65, as a response to environment.

Whitebark Pine Behavior

The interaction of whitebark pine's genetic constitution and its environment determines its behavior. The following sections consider whitebark pine's interaction with climate to determine its range, yield, and form; its interaction with substrate to determine details of its within-range distribution; its interaction with the physical environment and competitors to determine its form; its interaction with the seasons to determine its phenology; and the effects of some, mostly indirect, interactions with humans.

Climate, Competition, and the Geographic Range

Whitebark pine forms pure stands in high-forest ecosystems of central Idaho, western Montana and northern Wyoming (Weaver and Dale 1974; Arno and Weaver 1990; Chapter 4), and northern California (Barbour 1988). From these core areas its dominance declines in all directions (Franklin 1988; Peet 1988), including downslope. The following paragraphs consider the ecological basis of declines away from the core in each cardinal direction.

Near the Canadian border (50° N) subalpine fir (*Abies lasiocarpa*), white spruce (*Picea glauca*), Engelmann spruce (*P. engelmannii),* and lodgepole pine (*Pinus contorta*) invade sites occupied by whitebark pine and share dominance with it. Near 53° N, whitebark is entirely displaced by these competing species (Little 1971). The increased success of competitors near the border is probably due to increased water availability (Peet 1988). This hypothesis is supported by the fact that near the border, the tree is most dominant on excessively drained limestone sites (Peter Acuff, personal communication). To explain whitebark pine's final disappearance in the far north, I favor the last two of the following four hypotheses.

1. That cold eliminates whitebark pine is doubtful, because trees that seem less cold tolerant than *P. albicaulis* (*Abies lasiocarpa* and especially *P. contorta*) in the U.S. Rocky Mountains range much further to the north (Little 1971).
2. Whereas lack of Clark's nutcrackers (*Nucifraga columbiana*) to the north (Peterson 1990; Tomback 1998) could have inhibited seed dispersal farther north, it is more likely that the lack of whitebark pine, that is, unavailable large pine seeds, ultimately excludes the bird.
3. Wetter sites to the north favor *Pinus contorta, Abies lasiocarpa, Picea glauca,* and *Larix occidentalis* (western larch) in competition with whitebark pine, a tree of well-drained soils. Increasing wetness is deduced from cooling northward, which results in less evaporation per unit of precipitation and is demonstrated by muskegs (swamps) of tiaga in the region (Elliott-Fisk 1988).
4. Fire-adapted species, especially *P. contorta* and *Populus tremuloides* (aspen), are favored over whitebark pine in the gentler topography of northern forests. Forest topography is gentle, because forests of the north are displaced from broken peaks to gentler foothills by expanding glaciers, tundra, and permafrost (Harris 1986). The gentle topography favors fire, and fire-resistant species, because it supports larger, more frequent burns and provides fewer refuges (Whitlock 1993). The fire factor is maximized where tiaga abuts grassland in the plains of northern Alberta and Manitoba (Wells 1965; Elliott-Fisk 1988).

To the west in the Cascades, *Pinus albicaulis* yields dominance to *Tsuga mertensiana* (mountain hemlock) and other conifers, including *P. contorta, P. monticola* (western white pine), *Picea engelmannii,* and *Abies lasiocarpa* (Franklin 1988). The strong competitive position of alternate species is probably also due to greater precipitation levels (Gerlach 1970; Arno and Weaver 1990; Weaver 1990; Chapter 4). This hypothesis is supported by whitebark pine's ability to dominate—despite the wet climate—both on excessively drained pumice soils near Crater Lake and on locally dry sites in the Olympic rain-shadow. As whitebark pine yields to other species in the wetter north and west, so does its sister species *P. sibirica* decline from the eastern Urals to the moister western Urals (Gorchakovsky 1994).

Southward, whitebark's range is apparently restricted by excessive drought. In both the Rocky and Cascade Mountains, its southern limit is coincident with decreased cloudiness, decreased spring and fall precipitation, decreased humidity, and increased evaporation (Gerlach 1970; Mitchell 1976). On drier sites to the south, whitebark pine is replaced by the more drought resistant *P. flexilis* (limber pine) in both southern Wyoming (42° N) and central California/Nevada (37° N, Little 1971; Barbour 1988). Further south in both the Rockies and the Great Basin, *P. flexilis* is replaced by *P. aristata* (bristlecone pine) (Peet 1988).

The nature of the competition between whitebark pine and its associates differs between stands from the northern (moist) and southern (dry) ends of its range in three ways:

1. In the north and west, whitebark—maturing in gradually closing stands—is excluded by associates, such as *Abies lasiocarpa* and *Pinus contorta,* which eventually capture most of the resources. In the core and southward, on the other hand, *P. albicaulis* or *P. flexilis* seedlings establish in (and outcompete other species in) rare "safe sites" in stressful high- and low-altitude zones. These stands develop into open woodlands apparently with little direct competition between species or among individuals of the same species.
2. In savanna environments of the core and southward, whitebark pine and limber pine probably do compete for the safe sites, however. Where whitebark is present, it shows its superior cold tolerance by dominating high savanna forests. But where whitebark pine is absent, limber pine occupies high sites (Colorado, Peet 1988), thus showing a considerable tolerance of cold. Where both species are present, limber pine shows its superior drought tolerance by occupying dry, low-altitude sites. But where limber pine is absent, whitebark ranges much lower than usual (northern Nevada, Critchfield and Allenbaugh 1969), indicating some tolerance for heat and drought. Thus, the competitors require similar safe sites in a woodland environment and differ either in their methods for dispersing to them or in their ability to dominate them.
3. The nutcracker may create pure stands by giving "the planting advantage" to the tree that is most numerous locally, that is, to whitebark at high altitude and limber pine at low altitude. Stands with both species are found occasionally, presumably where both seed sources are equally convenient to the birds.

Whitebark pine disappears in the lowlands east of both the Rockies and the Sierras. South of the Canadian border and east of both ranges, its disappearance may be due to the dryness of the grassland and desert environments. A presence on the northern plains in moister post-Pleistocene times is suggested, however, by fossil pollen present in the Rocky Mountain foothills (Whitlock 1993) and relict stands ranging eastward at least to 110° W longitude in the Great Plains (Sweetgrass, Highwood, and Crazy Mountains, Little 1971; S. F. Arno, personal communication). The reasons for whitebark's absence in the coniferous forests of northern Alberta and Manitoba are less clear. It may be unable to survive there, because in contrast to the mountain and foothill sites where relief provides firebreaks (Whitlock 1993), fires sweeping over smooth plains are so frequent that species better adapted to fire (e.g., lodgepole pine and aspen) exclude whitebark pine. It is also possible that whitebark pine never followed retreating glaciers to tiaga latitudes, because fires on the plains to the south prevented the production of sufficient seed to supply either propagules or food to support a dispersing nutcracker population. In contast, the rougher topography of northern Russia has allowed *P. sibirica* to follow retreating glaciers northward by dispersing between fire-sheltered sites (Smolonogov 1994).

Climate and the Whitebark Pine Altitudinal Range

In mountains, the whitebark zone has both upper and lower altitudinal limits whose determination is discussed separately below.

The upper boundary of whitebark pine. We consider nine factors, numbered below, as possible controls of whitebark's upper limit. Growing season length is probably the primary factor, and high winds, acting through the supply of soil water, is the principal modifying factor.

Low temperatures of winter and spring do not set the upper limit. (1) Absolute minimum temperatures in the alpine above whitebark pine are –35° to –45°C. Since this is far warmer than temperatures tolerated by stone pines (–70° to –80°C, Sakai 1978; Bauer et al. 1994), winter extremes must not determine the upper limits of the tree. (2) Absolute minimum temperatures of May (–10° to –19°C, Weaver 1990) probably occur before buds break (Schmidt and Lotan 1980), but suggest possible rare damage by spring frosting.

Growing season conditions are more likely controlling. I consider four growing season temperature parameters. Since plant growth depends on accumulation of photosynthate, one expects the product of growing season temperature and growing season length to set the upper limits of plant growth (Weaver 1994b). (3) Thus, one might expect low temperature to set the upper altitudinal limit for any species (Daubenmire 1954; Holtmeier 1996; Koerner 1998). On this altitudinal gradient, trees fail sooner than herb/shrub associates, both because their growing tips extend beyond the "climate near the ground" and because they shade and cool the ground in which their roots grow (Koerner 1998); in support of his conclusion, trees reaching the highest altitudes appear in the stunted krummholz form. (4) We also expect accumulation of photosynthate to depend on the length of the warm season, which also declines with increasing altitude. The relative importance of growing season temperature and season length for Rocky Mountain vegetation was tested with regression analysis against productivity (Weaver 1994b). When growing season length falls below about three months, whether due to cold or to drought, wood production falls to zero, and an upper or lower timberline is reached. For Rocky Mountain vegetation, season length predicted annual productivity significantly better than either temperature or the product of season length and temperature (Weaver 1994b). Thus, season length appears to be the primary control for whitebark pine production. *Pinus pumila* production is also limited primarily by the length of the growing season (Kajimoto 1994). The *P. cembra* response could be explained either by temperature or by season length (Turner and Streule 1988).

On a more local scale, two other growing season temperature factors may contribute significantly to control of whitebark pine's upper limit. (5) Temperature (climatic) timberline is often depressed where wind shortens the growing season by scouring snow away. Snow removal allows freezing of the soil, which does not occur under snow cover (Sirucek et al. 1999), lowers root temperatures, postpones bud break in the spring (Frey 1983), and thus shortens the already short growing season. (6) In addition, seedlings may be excluded from steep south-facing sites at middle altitudes in the whitebark pine zone if soil surface temperatures rise too high in the summer (*P. cembra,* Aulitzky and Turner 1982). Lowered establishment rates of whitebark pine on south slopes (Tomback et al. 1990) may be due to high temperature.

I doubt that there is any general restriction of whitebark upward due to lack of soil water. (7) Alpine soils in California were not drier than high-forest soils (Klikoff 1965). In 1979–1987 soils in representative subalpine fir and whitebark pine stands never dried below –0.2 to –0.4 MPa (Weaver 1977; Sirucek et al. 1999); and, despite late summer browning of herbaceous vegetation, alpine soils in central Montana did not dry in the three years observed (T. Weaver, unpublished data).

Wind-induced drought may lower whitebark timberline locally, and perhaps extensively. Drought can limit photosynthesis, in spite of moist soils, if transpiration exceeds water uptake (Zweifel and Haesler 1999). Transpiration can be increased by drying winds in either summer or winter. (8) Wind scouring of a site in summer may directly induce transpiration high enough to cause drought. (9) More likely, wind scouring of a site in winter will indirectly increase drought by simultaneously increasing transpiration (Turner et al. 1975; Tranquillini 1979; Hadley and Smith 1989) and decreasing uptake—that is, removing snow, allowing soil to freeze, and thus preventing delivery of soil water. Tranquillini (1979) calls this "frost drought."

The lower boundary of whitebark pine. Whitebark pine's lower boundary is currently set by competition with *Abies lasiocarpa, Picea engelmannii,* and/or *Pinus contorta* for light, water, or nutrients. That is, the factor setting the upper limit of its competitors sets the lower boundary for *P. albicaulis*. The following discussion concentrates, therefore, on the effects of physical factors on the upward limits of these competing species. Note, however, that where disturbance such as fire or insect kill eliminates competition, whitebark pine ranges far below its normal lower limits, at least to the upper limit of the Douglas-fir (*Pseudotsuga menziesii*) zone (Arno and Weaver 1990), and, with irrigation, even down to *Agropyron* and *Festuca* grassland habitats (Table 1, Weaver 1994b). Arno (Chapter 4) shows that *P. albicaulis* occupies more of its physiologic niche when competitors are removed and argues that the larger range represents its "true" ecological niche.

While the factor setting the upper boundary of competing trees and the lower boundary of whitebark pine is often assumed to be temperature (e.g., Romme and Turner 1991), this appears to be species-dependent. I offer two arguments against temperature limitation of *Abies* (subalpine fir) and *Picea*. These trees often reach the same upper limit as whitebark pine. In addition, on a warmth index including elevation, aspect, and slope, spruce is insensitive to cold, and *Abies* is sensitive at timberline only (Mattson and Reinhart 1990). On the other hand, the upper limit of lodgepole pine probably is set by temperature (season length), since it rarely reaches to whitebark tree line in either the Rocky Mountains (Mattson and Reinhart 1990) or the Sierra Nevada (Barbour 1988), and its productivity falls well below timberline (Mattson and Reinhart 1990).

Alternatively, wind-induced drought might set the upper boundary of trees competing with whitebark, especially *Abies* and *Picea*.

1. Soil drought rarely limits upland species. Climate data indicate no lack of soil water in alpine, timberline, or subalpine zones (Weaver 1990) or in the ranges of other stone pines (Weaver 1990, 1994a). Soil water measurements in northwest Montana support this conclusion (Daubenmire 1968; Weaver 1977; Sirucek et al. 1999).
2. Wind-induced transpiration might, nevertheless, cause summer or winter drought stress. Such transpiration stress is suggested by the short- (Armstrong et al. 1988) and thin-cuticled (Sowell et al. 1982) needles of high-altitude trees. Similar responses are observed in *Pinus cembra* (Tranquillini 1979) and *P. pumila* (Kajimoto 1993).
3. On windswept sites, transpiration stress is surely aggravated by wind erosion of needles (Hadley and Smith 1989).
4. The relative tolerances of the competing trees have been measured by plotting relative productivity against summer wind flows. *Abies* and *Picea* are far more sensitive to wind than either *Pinus albicaulis* or *P. contorta* (Mattson and Reinhart 1990). This conclusion is supported by two additional observations. At timberline, the degree of flagging increases from whitebark pine to *Picea* to *Abies*. And, flagging on *Abies,* the most sensitive, can be significant in windy gaps even near the bottom of the subalpine fir zone, that is, where wind flow is high and low temperature is not a factor. Thus, while the temperature limits of whitebark pine, *Abies,* and *Picea* may be similar, depression of the upward ranges of *Abies* and *Picea* by drying wind opens a niche for whitebark pine.

Effects of Climate Change on the Range of Whitebark Pine

Anthropogenic deposition of greenhouse gases in the atmosphere is expected to modify climatic factors that determine the geographic and altitudinal distribution of whitebark pine. Temperature increases of 1° to 5°C (Romme and Turner 1991) or even 10°C (Bartlein et al. 1997) may result from a doubling of atmospheric CO_2. Precipitation may decrease in the south and increase in the north (Romme and Turner 1991), perhaps with a doubling of winter precipitation and a slight decrease of summer precipitation in the whitebark core area (Bartlein et al. 1997). Neither paper considers season length, a factor correlated with temperature, but, as indicated above, likely a better predictor of climate effects.

Ecologists estimate changes in the geographic distribution of vegetation types by mapping the new climate expected and overlaying the vegetation associated with environments like those expected. Thus, in a simple model deducible from the preceding discussion of geographic and altitudinal controls (Romme and Turner 1991), cooling might convert whitebark pine sites to tundra, whereas considerable warming might allow the replacement of *Pinus albicaulis* by *Pseudotsuga menziesii* or *P. contorta.* Similarly, with drying of the climate, whitebark pine may be replaced by *P. flexilis* (Colorado/Utah, Peet 1988), and on wetting, it might be replaced by *Abies* or *Tsuga mertensiana* (Franklin 1988). The model of Bartlein et al. (1997) uses conceptually similar correlation methods, but it incorporates more parameters and more quantitative measures of them to seek additional detail. Conclusions drawn from the two models are similar for high altitude (whitebark pine), but the Bartlein et al. (1997) model is more provocative for low altitudes.

Thus, with simple warming, we expect the southern boundaries of whitebark's range to retreat northward, whereas the northern boundary is unchanged, if it is substrate- or fire-determined. Climate maps (e.g., Gerlach 1970) provide some basis for speculative quantification of northward shifts in the range due to any proposed degree of warming; perhaps 50 to 300 kilometers with a 1 to 5 degrees C rise in temperature. Within its projected range, the tree will assume greater dominance if the climate dries, and will become less dominant if the climate becomes wetter.

Similarly, warming at one location in whitebark pine's range is expected to shift vegetation bands upward altitudinally. Romme and Turner (1991) predict that, with a 1 to 5 degrees C increase in temperature and with neither increased precipitation nor CO_2-induced increases in water-use efficiency, the zone of *P. albicaulis* in the Yellowstone ecosystem will rise at least 500 meters. Since the area of a band decreases when it is pushed up a cone, the area of whitebark forest will shrink by 90 percent. Predicted effects of 10°C warming are similar (Bartlein et al. 1997). Reduction in area will proportionally reduce both the whitebark pine gene pool and food supplied to consumers of whitebark pine seeds, including squirrels, bears, and—with possible negative feedback—nutcrackers, which plant the trees (Chapter 5). The precipitation changes predicted by Bartlein et al. (1997) may have little impact on *P. albicaulis*, because most of the water added as snow will run off, and slight reductions in summer rain may retard *Abies* competition. If warming were accompanied instead by increased summer precipitation, water-requiring species, especially *Abies*, would increase in the whitebark zone and, through competition, further reduce its population size and production.

Suitable environments may move too rapidly, both latitudinally and altitudinally, for some plants to follow. Seed dispersal rates, finite and variable among species, are rapid for whitebark pine when they are carried by nutcrackers (Tomback et al.1990). Range margins advancing into forests of other trees may be slowed by competition with the original inhabitants. Vertical movements may require hundreds of years as in *Pinus balfouriana* (foxtail pine) (Lloyd and Graumlich 1997). Horizontal movements may require thousands of years (Barnosky et al. 1987; Tallis 1991; Bartlein et al. 1997). But since whitebark occupies a range that is shrinking inward, it may require little human assistance to reach newly opening adjacent habitat.

The "temperature band" model presented above is partially validated by the response of vegetation to warming as the climate warmed from full glacial times through conditions similar to the present, to the warmest postglacial times (hypsithermal, Whitlock 1993). Thawing of glaciers in Yellowstone National Park left unvegetated sites, which were revegetated, in turn, with *Artemisia*-rich tunda (>11,500 years before present), a spruce parkland (>10,500 years before present), a *Picea/Abies/Pinus albicaulis* forest approximating present vegetation (>9,500 years before present) and a *Pinus contorta/Pseudotsuga* forest of the dry hypsithermal (>5,000 years before present). With cooling since the hypsithermal, the *Picea/Abies/Pinus albicaulis* forest has regained its dominance. Interaction of temperature with precipitation may modify this response.

The preceding predictions of the effects of climate change are uncertain, because so many factors are involved and the magnitudes of the likely changes are unknown (Koerner 1994; Holtmeier 1994). Thus, projections of the temperature band model are simplistic and may err in at least five ways:

1. Wind may exclude the vegetation type forecast for ridges and plateaus. Specifically, water-requiring species—including whitebark pine, *Picea,* and *Abies*—may fail to invade windswept sites. Exclusion of trees from windy sites could allow tundra to survive in a temperature zone where the temperature model forecasts its elimination. For example, the unexplained survival of tundra during the hypsithermal, when mountain peaks and ridges were probably warm enough for forests (Romme and Turner 1991), may have been due to the availability of sites from which wind excluded forests. Downslope, on slightly less exposed sites, drying wind may exclude the less tolerant *Abies/Picea* from timberline/whitebark sites. Thus, near a mountain ridge, wind may defend both tundra from upward-migrating whitebark pine and whitebark from upward-migrating *Abies/Picea.* If so, under climate warming, the *Abies/Picea* zone may be pinched between upward-migrating *Pseudotsuga* and unbudging whitebark pine.
2. Soil conditions can also prevent migration. Rhyolitic substrates of the central plateau in Yellowstone supported *Artemisia*/tundra while *Picea/Abies*/whitebark pine forest occupied surrounding substrates (10,500 to 9,500 years before present). Today, Yellowstone's rhyolite supports *Pinus contorta,* in contrast to the *Abies/Picea*/whitebark pine forests of adjacent substrates. Similarly, the dominance of whitebark on loess-free parts of the Pitchstone Plateau may be due to edaphic (soil) exclusion of *Abies/Picea.*
3. Fire favors fire-resistant species, except where its effects are minimized by barriers such as lakes (Whitlock 1993) or rivers and ridges (Wells 1965). Where it occurs, fire thus favors *Pinus albicaulis* somewhat at high altitude and *Pseudotsuga/Pinus contorta* considerably at low altitude, both over *Abies/Picea.*
4. After a climate shift, some species invade suitable sites faster than others, leaving the impression of a unique vegetation type. Thus, the postglacial *Picea* parkland of Yellowstone National Park (Whitlock 1993) may have been a simplified forest waiting for "normal" associates. I know of no place in the world where such a *Picea* parkland exists as a stable type.
5. The alternate explanation for vegetation types without known analogues is that large areas had climates with factor combinations that do not currently exist. The *Artemisia*/tundra and the *Picea* parkland of postglacial Yellowsone (Whitlock 1993; Bartlein et al. 1997) might, for example, have depended on a unique distribution of temperature and precipitation. In this case, summer rainfall might have favored grassland or parkland over forests that depend on winter precipitation (Romme and Turner 1991). Similarly, Lloyd and Graumlich (1997) suggest that, in analogous sites in the Sierra Nevada, vegetation bands will rise with warming, but only if precipitation equals or exceeds current levels.

Climate Effects on Production and Standing Crop

We expect whitebark pine to produce best where it experiences no water stress and where July temperatures are above 20°/4°C (maximum/minimum in the whitebark zone) or even 27°/6°C (maximum/minimum in the Douglas-fir zone) or higher (Table 3-2). Growth form changes from stately trees in high forests to clumped trees of woodlands to krummholz over short altitudinal distances are probably more wind-driven than temperature-driven, because wind acts on plant water balance both by drying and by preventing snowdrift supplied recharge.

Production in woodland and forest whitebark pine is 0.6 to 0.8 $kg.m^{-2}yr^{-1}$ (Forcella and Weaver 1977). Production in *Pinus pumila,* which always assumes the low or bent krummholz growth form, is 0.37 to 0.41 $kg.m^{-2}yr^{-1}$; 80 percent of this goes to roots and most of the remainder is incorporated into needles (Kajimoto 1994).

Three lines of evidence suggest that production in all stone pines would increase with increases in season length, temperature, or moisture:

1. Production is proportional to season length (Weaver 1994b; Kajimoto 1994).
2. Production of *Pinus sibirica* in the northern tiaga increases 15 percent for each degree C above the average temperature (Kokorian and Nazarod 1995). In the southern tiaga, production increases 7 percent for each degree C above the average. Whitebark pine's response would probably be similar, that is, with smaller increases in yield per degree in warmer portions of its range.
3. In Idaho, diameter growth was greatest in years with high winter or spring precipitation, low May or June temperatures, and high August temperatures (Perkins and Swetnam 1996). The first four factors increase the availability of water for growth, and these increases may raise production. The high August temperature extends the growing season and increases production, as discussed on page 55.

The biomass of trees in a forest (its "standing crop") depends on the rate of production and the length of the productive period. Thus, in a given environment whitebark pine standing crop tends to increase, due to continued tree growth, for more than four centuries (Forcella and Weaver 1977). One expects standing crops of a given age to decline with growth form, from whitebark pine forests to open whitebark woodlands to krummholz patches. The comparison of 150- to 400-year-old *Pinus albicaulis* woodlands (300 tons/hectare equals 30 kg/m^2, Forcella and Weaver 1977) and *P. pumila* krummholz (120 to 180 t/ha, Kajimoto 1989, 1994) supports this hypothesis. The comparison is weak, however, because the trees are different species, the climates are different, and *P. pumila* krummholz is denser than Rocky Mountain whitebark pine krummholz. In any case, both production and standing crop tend to asymptote as stand age increases (Forcella and Weaver 1977).

Climate and Cone Production

Cone production is important to whitebark pine reproduction, bears, squirrels, nutcrackers, and, in the past, human hunter-gathers (Forcella and Weaver 1980; Chapters 7 and 12). Cone crops measured across eight years and for twenty-eight whitebark pine stands ranged from 0 to 6.2 cones/m^2, and stand-to-stand averages ranged from 0.4 to 3.6 cones/m^2 (Weaver and Forcella 1986). Cone production in *Pinus pumila* over twenty years ranged from 1 to 8 cones/m^2 and differed little between low and high stress sites (Nakashinden 1994). Both data sets were acquired by analysis of cone scars (Weaver and Forcella 1986; Nakashinden 1994; Vorobjev et al. 1994).

The considerable year-to-year variation in whitebark pine cone yield (Weaver and Forcella 1986; Morgan and Bunting 1992) showed little relationship to variation in climate. Comparable variation was observed in *Pinus pumila;* yields were about twice the mean in "mast years" and half the mean in "fail years" (Nakashinden 1994). Smith and Balda (1979) and Forcella (1981) hypothesized that avoidance of predators rather than current production or carbohydrate reserves controls annual yields. Thus, they believe that intermittent, unpredictable masting is induced by a rare factor (or factor level) that makes bumper crops hard for predators or researchers to predict. Whenever large seed crops coincide with low to average predator densities, the predators are over-provisioned and a good seedling year, based on unconsumed seeds, follows.

Substrate Effects on Performance

Changes in substrate cause discontinuities in the range of whitebark pine: (1) Well-drained substrates exclude whitebark in dry areas such as the central Rocky Mountains (e.g., limestone, Harlow and Harrar 1958; Weaver and Dale 1974; Parker 1989). On the other hand, in wetter areas, limestone (Canadian Rocky Mountains) and pumice (Cascade Range) exclude competing trees by allowing "excess" water to drain off and so support *Pinus albicaulis.* Other stone pines, including *P. cembra* (Sauermoser 1994) and *P. sibirica* (Gorchakovsky 1994), also avoid limestone in the core of their range and favor it in the north. While I attribute the substrate relationships primarily to water, these authors describe limestone sites as both warmer and drier. (2) Nutrients supplied by a substrate could also be influential, but I doubt that the tree's response to limestone is due to calcium, because whitebark pine favors or avoids it according to water availability in the part of its range considered.

Nutrient Effects on Production

Nutrient limitation of whitebark pine's range or production is unstudied. The following paragraphs consider major nutrients one at a time.

Increases in atmospheric CO_2 should increase the production of whitebark. (1) Production of *Pinus cembra* is reduced by 10 to 20 percent from a sea level "standard," because CO_2 (air) pressure at timberline is only 75 percent of that in lowlands (Tranquillini 1979; Koerner 1994). The CO_2 effect on whitebark pine is surely similar. Thus, human-caused increases in atmospheric CO_2 concentrations may increase high-altitude production, and such increases may have already been observed (Koerner 1994). (2) Increases in CO_2, if they are accompanied by increases in photosynthesis, may also support upward migration of whitebark. Thus, if growing-season length sets the upper limit of *P. albi-*

caulis, this could be due either to inadequate time to complete growth and hardening before fall (Tranquillini 1979) or to inadequate time to accumulate sufficient material for growth (Sowell et al. 1982; Kajimoto 1993). If photosynthate is limiting, CO_2 fertilization will support some upward migration. Note, however, that the CO_2 fertilization effects discussed could alternatively be due to other changes in the environment, such as changes in precipitation, nitrogen pollution, or warming (Holtmeier 1994).

Calculations suggest that supplies of nitrogen, potassium, and calcium in the soil are sufficient for whitebark pine growth. (1) Soil nutrient pools were estimated from soil data (Table 3-5d). For comparison, tree requirements (Table 3-5b) were estimated by multiplying average composition of conifer species (gm nutrient gm^{-1} plant, Weaver and Forcella 1977) by production (gm growth $tree^{-1}$ yr^{-1}, Table 3-5a). The nutrient pools exceeded requirements for

Table 3-5. Estimating macronutrient requirements of WBP in stands aged 100–600 years.

	Organ produced				
	Leaf	Branch	Bole	Roots	Total
A. Stand production (gm m^{-1} yr^{-1})					
	430	90	120	120	706
B. Nutrients consumed[2] (gm m^{-1} yr^{-1}) = production (gm m^{-1} yr^{-1}) x concentration (g g^{-1})					
Nitrogen	4.5	0.4	0.6	0.5	6.0
Phosphorus	0.4	0.1	0.1	0.1	0.7
Potassium	2.4	0.2	0.2	0.2	3.0
Calcium	2.4	0.6	0.5	0.8	4.3
C. Senescence allows recycling to provide most nutrients required for needle and bark growth.[3]					
D. Soil nutrient pools available for new (unrecycled) growth (gm m^{-2})[4]					
Nitrogen					26–96
Potassium					3–8
Calcium					24–26

Assumptions:
[1]Production estimates (3-5B) were made in 100–600-year-old forests (Forcella and Weaver 1977).
[2]For each organ, the nutrients consumed (gm m^{-1} yr^{-1}) in tree production equals production (gm m^{-1} yr^{-1}, 3-5B) × nutrient concentration (gm/gm, Weaver and Forcella 1977) and total consumption is the sum across the component organs. It is also assumed that roots have compositions similar to branches.
[3]Recycling of nutrients from dying organs supplies part of the nutrient needs. Recycling from needle and bark loss may nearly compensate for their losses. Thus, nutrient uptake needs only to provide for nutrients associated with increases in tree size.
[4]Soil nutrient availability in both the *Abies lasiocarpa* and alpine ecosystems (3-5D, Weaver 1978) is presented. Supply (3-5D) exceeds demands (3-5C) even if no recycling occurs. Extractions were Kjeldahl for nitrogen, and ammonium acetate for potassium and calcium. Thus, the nitrogen pool could be overestimated, if cool soils prevent decomposition.

total production, whether average conifer composition or the richest conifer compositions were considered (Rodin and Bazelevich 1965). Mineral nutrient availability seems even more adequate when one considers that nutrients are released from falling needles. Nutrient-rich needles are constantly lost at rates similar to new needle production. Thus, most of the nutrients required for needle growth are supplied by resorption of nutrients from dying needles before abscission, resorption of nutrients from fallen needles via mycorrhizae, or resorption of nutrients from decomposed needles via roots. (2) The presence of nutrients may overstate their availability in the upper subalpine zone. Nitrogen excesses may be overestimated, if cool soils slow decomposition and make organic nitrogen (Kjeldahl-N) less available. Organic material on high, north-facing *Pinus cembra* sites decomposes poorly, and trees on these exposures may be nutrient-limited (Blaser 1980). In addition, cool soils may minimize absorption of free nitrogen, phosphorus, and other elements (Chapin and Shaver 1985; Bliss 2000).

If other essential nutrients do not limit production in similarly leached alpine sites (Koerner 1989), one doubts that they generally limit production in whitebark forests. However, I cite two reports of phosphorus and calcium influences: Holtmeier (1994) suggested that phosphorus limits growth of conifers at the Colorado timberline. Whitebark pine's tendency to avoid limestone (Harlow and Harrar 1958; Weaver and Dale 1974; Parker 1989) is more likely due to water relations than to high pH, which could conceivably reduce tree establishment (Jacobs and Weaver 1990).

Growth Form and Environment

Three aspects of the growth form of whitebark are notable:

1. In open forests, older whitebark pine trees have a diffuse (lyrate), flat-topped crown in contrast to the conical form of their associates. This is due to weak apical dominance (Weaver and Jacobs 1990). The crown is narrower in closed forests because of shading by adjacent trees, and in young trees. The apical dominance of *Pinus cembra* also weakens with age.
2. The tree is often multi-stemmed. This could be due to branching at the base, simultaneous germination of cached seeds, and even seeds with multiple embryos (Weaver and Jacobs 1990; Tomback et al. 1993; Chapter 6). Multi-stemmed trees are more common in open woodlands than in closed forests, presumably because allocation of photosynthesis to multiple shoots conflicts with height growth. Reduced height puts the multi-stemmed trees at a competitive disadvantage in forests (Weaver and Jacobs 1990).
3. The tree varies from stately, in sheltered sites, to shorn and flagged on windy sites (Wooldridge et al. 1996). Stately trees on the Canadian border near Glacier National Park, now mostly blister-rust killed, had single stems and reached heights of 30 meters (S. F. Arno, personal communication).

Phenology and Environment

Activity varies with season in *Pinus albicaulis* (Schmidt and Lotan 1980), *P. cembra* (Tranquillini 1979; Turner and Streule 1988), and other stone pines. The observations reported below were made on the east slope of the Rocky Mountains in Montana.

Table 3-6. Phenology of Rocky Mountain pines from high, middle, and low altitude[1] east of the Rocky Mountains in Montana (Schmidt and Lotan 1980).

	PHENOLOGICAL STAGE									
	Bark slips	Shoot growth	Bud break	Pollen starts	Pollen ends	Shoot ceases	Buds winter	Bark set	Cone max	Cone "opens"
PIAL	24MA[2]	17JN	14JN	21JL	5AU	27JL	24AU	12SE	13AU	5SE
PICO	6MA	16MA	30MA	17JN	3JL	22JL	3AU	18AU	19AU	3SE
PIPO	28AP	6MA	24MA	4JN	17JN	21JL	30AU	21SE	22AU	20SE

[1]Trees are *Pinus albicaulis* (PIAL) at timberline, *Pinus contorta* (PICO) at middle altitude, and *Pinus ponderosa* (PIPO) at low altitude.
[2]Months and their abbreviations are April (AP), May (MA), June (JN), July (JL), August (AU), September (SE).

The active season for whitebark pine opens with cambial growth (24 May), bud break (14 June), and shoot growth (17 June) (Table 3-6). Because spring comes to lower altitudes first, these events occur increasingly earlier for trees of middle altitudes (*P. contorta*), and trees of the lowest altitudes (*P. ponderosa*). Shoot growth ceases almost simultaneously for the three species (21–27 July). If most wood production occurs between bud break and winter bud formation, the wood-growing period lengthens downslope from *P. albicaulis* (sixty-one days) to *P. contorta* (seventy-five days) and *P. ponderosa* (ninety-eight days). Cambial growth and bud break in *P. cembra* are cued by reduced frosting (Tranquillini 1979) and increasing day length. Height growth in *P. cembra* begins earlier and ends earlier (1 April–15 July, Kronfuss 1994) than for *P. albicaulis*.

Cone formation begins shortly before pollination and is completed about fourteen months later in whitebark pine (early August). Pollination progresses from low to high altitude and is well separated among the pine species, perhaps as a device evolved for avoiding cross-pollination. Although whitebark is last to flower, its cone expansion is complete (13 August) a week earlier than *P. contorta* and *P. ponderosa*. The shorter, more dynamic growing season of whitebark pine ensures seed maturity before the return of winter. Cones of *P. albicaulis* and *P. contorta* mature in early September, and those of *P. ponderosa* open in late September.

The active season for whitebark pine closes as winter buds form (24 August), cambial growth ceases (12 September), and cones ripen (5 September, Table 3-6). Plant activity might be expected to cease from higher to lower sites, because winter comes first to the highlands. As expected, low-altitude ponderosa pine activity ceases in all parameters later than mid-altitude lodgepole pine. Contrary to expectation, whitebark activity also ceases after lodgepole pine in all stages. It is as if whitebark pine cannot complete growth and hardening before lower-altitude *P. contorta* and may thus enter the cold season unmatured and unhardened in winters that begin early. While this has not been reported for whitebark pine, twigs of *P. cembra* (Tranquillini 1979) and *P.*

pumila (Kajimoto 1990) near timberline suffer frost dessication when they fail to complete development.

Human Effects and Whitebark Pine

People may influence whitebark pine directly by harvest, regulating fire (Chapter 9), or introducing/regulating disease (Chapter 10). While grand specimens may be cut for lumber, most whitebark pine trees are reserved for watershed cover and wildlife use. People may also influence whitebark indirectly by influencing climate (with the results discussed earlier in the range, altitude, production, and phenology sections), pollution, or possibly Clarks's nutcracker populations (reducing seedling establishment, Chapter 12).

Reciprocal Effects

Besides being affected by its environment, whitebark pine should have effects on its environment. Although largely unstudied, there are several possibilities.

During early post-glacial times, when vast, previously glaciated areas were unvegetated, trees probably trapped blowing soil to create loess deposits. Trapping of loess by subalpine trees, including whitebark pine, continues; for example, current deposition rates in the Bridger Range, Montana, are about 0.5 millimeter/year (T. Weaver, unpublished data). A deeper, nutrient-rich rooting medium results.

Growing in highlands either as ribbon forests or woodlands (Billings 1969; Arno and Weaver 1990), *Pinus albicaulis* also traps snow. Melting of the resulting high-altitude snowdrifts generates late-season stream flow (Anderson 1963; Berndt 1964; Martinelli 1964). The drift sites are loess-rich since soil blows with the snow, and their soils are leached (Tursich and Weaver 1998) and compacted (McNeal and Weaver 1982). Soils under the trees may be even more leached than those under the drift sites (Buchanan 1972).

While trees probably reduce erosion overall, this may depend on the site. Widely spaced trees on relatively level soil-rich sites could conceivably increase wind/water erosion by excluding soil-binding grasses; trees on these sites may compensate both by production of a litter carpet and by reducing winds near ground level to non-erosive speeds. On steep alluvial sites, roots reduce mass wasting (Burroughs and Thomas 1977). *Pinus cembra* trees on steep slopes bind snow and reduce avalanching (Frey 1994). On the other hand, when whitebark pine roots penetrate cliffs, breakdown probably increases.

When whitebark colonizes alpine and mountain meadow vegetation, it creates microsites that are colonized by *Abies lasiocarpa* (subalpine fir) and forest understory plants, such as *Vaccinium scoparium* and *Arnica cordifolia*. The vegetation of forest islands contrasts with the meadow/alpine turf on sites between them. Effects of whitebark pine on other vegetation are discussed in Chapter 1.

Whitebark pine trees are used by various animals. Its pea-sized seeds are eaten by bears (*Ursus arctos* and *U. americanus*), squirrels (*Tamiasciurus* spp.), and nutcrackers (Hutchins 1994; Lanner and Gilbert 1994; Mattson and Reinhardt 1994; Chapters 1, 7, and 12). In the past, the seeds were probably eaten by humans (Forcella and Weaver 1980), and the seeds of other stone pines are still eaten in Europe and Asia (Lanner 1996). The tree furnishes winter browse for deer and elk (Lonner and Pac 1990) and bark for porcupines and bears. It furnishes hiding places, as well as wind/radiation cover to all of these animals.

Acknowledgments

I gratefully acknowledge enlightening discussion with many colleagues; reviews of S. Arno, R. Calloway, D. Despain, R. Haesler, R. Keane, D. Perry, and D. Tomback; and the patient editing of D. Tomback. J. O'Dwyer kindly checked the manuscript for formatting details. Flaws that undoubtedly remain are surely mine.

Literature Cited

Allison, L. E. 1965. Organic matter. Pages 1367–1378 *in* C. A. Black, editor. Methods of soil analysis. Agronomy monograph #9. American Society of Agronomy, Madison, Wisconsin.

Anderson, H. W. 1963. Managing California's snow zone lands for water. USDA Forest Service, Pacific Southwest Forest and Range Experimental Station, Research Paper PSW-6, Berkeley, California.

Armstrong, J. K., K. Williams, L. F. Hunneke, and H. A. Mooney. 1988. Topographic position effects on growth depression of Sierra Nevada (California) pines during the 1982–1983 El Nino. Arctic and Alpine Research 20:252–257.

Arno, S. F., and T. Weaver. 1990. Whitebark pine community types and their patterns on the landscape. Pages 97–105 *in* W. C. Schmidt and K. J. McDonald, compilers. Whitebark pine ecosystems—Ecology and management of a high mountain resource. USDA Forest Service, Intermountain Research Station, General Technical Report INT-270, Ogden, Utah.

Aulitzky, H., and H. Turner. 1982. Bioklimatische Grundlagen einer standortsgemaessen Bewirtschaftung des subalpinen Laerchen-Zibenwaldes. Mitteilungen, Eidgenössische Anstalt für das forstliche Versuchswesen 8:325–580.

Baig, M. N., and W. Tranquillini. 1980. The effects of wind and temperature on cuticular transpiration of *Picea–Abies* and *Pinus cembra* and their significance in desiccation damage at the alpine treeline. Oecologia 47:252–256.

Baker, R. G. 1990. Late quaternary history of whitebark pine in the Rocky Mountains. Pages 40–48 *in* W. C. Schmidt and K. J. McDonald, compilers. Proceedings—Symposium on whitebark pine ecosystems: Ecology and management of a high mountain resource. USDA Forest Service, Intermountain Research Station, General Technical Report INT-270, Ogden, Utah.

Barbour, M. G. 1988. Californian upland forests and woodlands. Pages 131–164 *in* M. G. Barbour and W. D. Billings, editors. North American terrestrial vegetation. Cambridge University Press, New York.

Barnosky, C. W., P. M. Anderson, and P. J. Bartlein. 1987. The northeastern United States during deglaciation; vegetational history and paleoclimatic implications. Pages 289–231 *in* W. Ruddiman and H. Wright, editors. North America and adjacent oceans during the last deglaciation. Geology of North America, V K-3. Geological Society America, Boulder, Colorado.

Barry, R. G. 1973. A climatological transect on the east slope of the front range, Colorado. Arctic and Alpine Research 5:89–110.

———. 1992. Mountain weather and climate. Routledge, New York.

Bartlein, P. J., C. Whitlock, and S. L. Shafer. 1997. Future climate in the Yellowstone

National Park region and its potential impact on vegetation. Conservation Biology 11:782–792.

Bartos, D. L., and K. E. Gibson. 1990. Insects of whitebark pine with emphasis on mountain pine beetle. Pages 171–178 *in* W. C. Schmidt and K. J. McDonald, compilers. Whitebark pine ecosystems—Ecology and management of a high mountain resource. USDA Forest Service, Intermountain Research Station, General Technical Report INT-270, Ogden, Utah.

Bauer, H., M. Nagele, M. Comploj, V. Galler, M. Mair, and E. Unterpertinger. 1994. Photosynthesis in cold acclimated leaves of plants with various degrees of freezing tolerance. Physiologia Plantarum 9:403–412.

Berndt, H. W. 1964. Inducing snow accumulation on mountain grasslands watersheds. J. Soil and Water Conservation 19:196–198.

Berndt, H. W., and G. W. Swank. 1970. Forest land use and streamflow in central Oregon. USDA Forest Service, Pacific Northwest Forest and Range Experimental Station, Research Paper PNW-93, Portland, Oregon.

Billings, W. D. 1969. Vegetational pattern near alpine timberline as affected by fire snowdrift interaction. Vegetatio 19:192–207.

———. 2000. Alpine vegetation. Pages 537–572 *in* M. G. Barbour and W. D. Billings, editors. North American terrestrial vegetation. Cambridge University Press, New York.

Blaser, P. 1980. Der Boden als Standortsfactor bei Aufforstungen in der subalpinen Stufe (Stillberg, Davos). Mitteilungen, Eidgenössische Anstalt für das forstliche Versuchswesen 56:527–611.

Bliss, L. C. 2000. Arctic tundra and polar desert biome. Pages 1–40 *in* North American terrestrial vegetation. Cambridge University Press, New York.

Bruun, B. 1982. Der Kosmos-Vogel fuehrer (Deutschland u. Europas) Claus Koenig, Stuttgart, Germany.

Buchanan, B. A. 1972. Ecological effects of weather modification: Relationships of soil, vegetation and microclimate. Doctoral Dissertation, Montana State University, Bozeman.

Burroughs, E. R., Jr., and B. R. Thomas. 1977. Declining root strength in Douglas-fir after felling as a factor in slope stability. USDA Forest Service, Intermountain Forest and Range Experimental Station, Research Paper INT-190, Ogden, Utah.

Caldwell, M. M. 1970. The effect of wind on stomatal aperature, photosynthesis, and transpiration of *Rhododendron ferrugineum* L. and *Pinus cembra* L. Zentralblatt Gesamte Forstwes 87:193–201.

Chapin, F. S., and G. R. Shaver. 1985. Arctic. Pages 16–40 *in* B. F. Chabot and H. A. Mooney, editors. Physiological ecology of North American plant communities. Chapman and Hall, New York.

Critchfield, W. B., and G. L. Allenbaugh. 1969. The distribution of Pinaceae in and near northern Nevada. Madrono 20:12–25.

Critchfield, W. B., and E. L. Little. 1966. Geographic distribution of the pines of the world. USDA Forest Service, Miscellaneous Publication #991. Washington, D.C.

Dale, D., and T. Weaver. 1974. Trampling effects on vegetation of the trail corridors of the northern Rocky Mountain forests. Journal of Applied Ecology 11:767–772.

Daubenmire, R. 1954. Alpine timberlines in the Americas and their interpretation. Butler University, Botanical Studies 11:119–136.

———. 1968. Soil moisture in relation to vegetation distribution in the mountains of northern Idaho. Ecology 49:431–438.

Decker, G. L., G. A. Nielsen, and J. W. Rodgers. 1975. The Montana automated data processing system for soil inventories. Montana Agricultural Experiment Station, Research Report 89, Bozeman.

Elliott-Fisk, D. L. 1988. The boreal forest. Pages 33–62 *in* M. G. Barbour and W. D.

Billings, editors. North American terrestrial vegetation. Cambridge University Press, New York.

Forcella, F. 1981. Ovulate cone production in pinyon: Negative exponential relationship with late summer temperature. Ecology 62:488–491.

Forcella, F., and T. Weaver. 1977. Biomass and productivity of the subalpine *Pinus albicaulis–Vaccinium scoparium* association in Montana, USA. Vegetatio 35:95–105.

———. 1980. Food production in the *Pinus albicaulis–Vaccinium scoparium* association. Montana Academy of Science Proceedings 39:73–80.

Franklin, J. F. 1988. Pacific northwest forests. Pages 103–130 *in* M. G. Barbour and W. D. Billings, editors. 1988. North American terrestrial vegetation. Cambridge University Press, New York.

Frey, W. 1983. The influence of snow on growth and survival of planted trees. Arctic and Alpine Research 15:241–242.

———. 1994. Silvicultural treatment and avalanche protection of Swiss stone pine forests. Pages 290–293 *in* W. Schmidt and F. -K. Holtmeier, compilers. Proceedings—International workshop on subalpine stone pines and their environment: The status of our knowledge. USDA Forest Service, Intermountain Research Station, General Technical Report INT-GTR-309, Ogden, Utah.

Genys, J. B., and H. E. Heggestad. 1978. Susceptibility of different species clones and strains of pines to acute injury caused by ozone and sulfur dioxide. Plant Disease Reporter 62:687–691.

Gerlach, A. C., editor. 1970. The national atlas. USDI Geological Survey, Washington, D.C.

Gorchakovsky, P. L. 1994. Distribution and ecology of the Siberian stone pine in the Urals. Pages 56–60 *in* W. C. Schmidt and F. -K. Holtmeier, compilers. Proceedings—International workshop on subalpine stone pines and their environment: The status of our knowledge. USDA Forest Service, Intermountain Research Station, General Technical Report INT-GTR-309, Ogden, Utah.

Hadley, J. L., and W. K. Smith. 1989. Wind erosion of leaf surface wax in alpine timberline conifers. Arctic and Alpine Research 21:392–398.

Hansen-Bristow, K., C. Montagne, and G. Schmid. 1990. Geology, geomorphology, and soils within whitebark pine ecosystems. Pages 62–71 *in* W. C. Schmidt and K. J. McDonald, compilers. Whitebark pine ecosystems—Ecology and management of a high mountain resource. USDA Forest Service, Intermountain Research Station, General Technical Report INT-270, Ogden, Utah.

Harlow, W. M., and E. G. Harrar. 1958. Textbook of dendrology. McGraw-Hill, New York.

Harris, H. A. 1986. Permafrost distribution, zonation and stability along the eastern ranges of the cordillera of North America. Arctic 39:29–38.

Havranek W. M., and U. Benecke. 1978. The influence of soil moisture on water potential, transpiration, and photosynthesis of conifer seedlings. Plant and Soil 49:91–104.

Heumader, J. 1994. Cultivation of cembra pine plants for high elevation afforestation. Pages 298–301 *in* W. C. Schmidt and F. -K. Holtmeier, compilers. Proceedings—International workshop on stone pines and their environment: The status of our knowledge. USDA Forest Service, Intermountain Research Station, General Technical Report INT-GTR-309, Ogden, Utah.

Holtmeier, F. -K. 1994. Ecological aspects of climatically caused timberline fluctuations. Pages 220–233 *in* M. Beniston, editor. Mountain environments in changing climates. Routledge Press, New York.

———. 1996. Timberline research in Europe and North America. Pages 23–36 *in* L. Loven and S. Salmela, editors. Finnish Forest Research Institute, Research Paper #623, Rovaniemi, Finland.

Hutchins, H. E. 1994. Role of various animals in dispersal of whitebark pine in the

Rocky Mountains, USA. Pages 163–171 *in* W. C. Schmidt and F. -K. Holtmeier, compilers. Proceedings—International workshop on subalpine stone pines and their environment: The status of our knowledge. USDA Forest Service, Intermountain Research Station, General Technical Report INT-GTR-309, Ogden, Utah.

Hutchinson, G. E. 1958. Concluding remarks. Cold Spring Harbor Symposium on Quantitative Biology 22:415–427.

Jackson, M. L. 1958. Soil chemical analysis. Prentice-Hall, Englewood Cliffs, New Jersey.

Jacobs, J., and T. Weaver. 1990. Effects of temperature and temperature preconditioning on seeding performance of whitebark pine. Pages 134–139 *in* W. C. Schmidt and K. J. McDonald, compilers. Whitebark pine ecosystems—Ecology and management of a high mountain resource. USDA Forest Service, Intermountain Research Station, General Technical Report INT-270, Ogden, Utah.

Kajimoto, T. 1989. Aboveground biomass and litterfall of *Pinus pumila* shrubs growing on the Kiso mountain range, Japan. Ecological Research 4:55–70.

———. 1990. Photosynthesis and respiration of *Pinus pumila* needles in relation to needle age and season. Ecological Research 5:333–340.

———. 1993. Shoot dynamics of *Pinus pumila* in relation to exposure gradients on the Kiso mountain range, central Japan. Tree Physiology 13:41–53.

———. 1994. Aboveground net production and dry matter allocation of *Pinus pumila* forests in the Kiso mountain range, central Japan. Ecological Research 9:193–204.

Keller, G. 1996. Utilization of inorganic and organic nitrogen sources by high subalpine ectomycorrhizal fungi of *Pinus cembra* in pure culture. Mycological Research 100:989–998.

Klikoff, L. 1965. Micro-environmental influence on vegetational pattern near timberline in the central Sierra Nevada. Ecological Monographs 35:187–211.

Koerner, C. 1989. The nutritional status of plants from high altitudes. Oecologia 81:379–391.

———. 1994. Impact of atmospheric changes on high mountain vegetation. Pages 155–166 *in* M. Beniston, editor. Mountain environments in changing climates. Routledge Press, New York.

———. 1998. A re-assessment of high elevation treeline positions and their explanation. Oecologia 115:445–459

Koike, T., R. Haessler, and H. Item. 1994. Needle longevity and photosynthetic performance in Cembran pine and Norway spruce growing on north and east slopes at timberline. Pages 78–80 *in* W. C. Schmidt and F. -K. Holtmeier, compilers. Proceedings—International workshop on subalpine stone pines and their environment: The status of our knowledge. USDA Forest Service, Intermountain Research Station, General Technical Report INT-GTR-309, Ogden, Utah.

Kokorian, A. O., and I. M. Nazarod. 1995. The analysis of growth parameters of Russian boreal forests in warming and its use in a carbon budget model. Ecological Modeling 82:139–150.

Kozlowski, T. T., P. J. Kramer, and S. G. Pallardy. 1991. Physiological ecology of woody plants. Academic Press, New York.

Kronfuss, H. 1994. Height growth in cembran pine as a factor of air temperature. Pages 99–104 *in* W. C. Schmidt and F. -K. Holtmeier, compilers. Proceedings—International workshop on subalpine stone pines and their environment: The status of our knowledge. USDA Forest Service, Intermountain Research Station, General Technical Report INT-GTR-309, Ogden, Utah.

Krutovskii, K. V., D. V. Politov, and Yu P. Altukhov. 1994. Genetic differentiation and phylogeny of stone pine species based on isozyme loci. Pages 19–30 *in* W. C. Schmidt and F. -K. Holtmeier, compilers. Proceedings—International workshop on subalpine stone pines and their environment: The status of our knowledge. USDA Forest Service, Intermountain Research Station, General Technical Report INT-GTR-309, Ogden, Utah.

Lanner, R. M. 1990. Biology, taxonomy, evolution, and geography of stone pines of the world. Pages 14–24 *in* W. C. Schmidt and K. J. McDonald, compilers. Whitebark pine ecosystems—Ecology and management of a high mountain resource. USDA Forest Service, Intermountain Research Station, General Technical Report INT-270, Ogden, Utah.

———. 1996. Made for each other: A symbiosis of birds and pines. Oxford University Press, New York.

Lanner, R. M., and B. K. Gilbert. 1994. Nutritive value of whitebark pine seeds and the question of their variable dormancy. Pages 206–211 *in* W. C. Schmidt and F. -K. Holtmeier, compilers. Proceedings—International workshop on subalpine stone pines and their environment: The status of our knowledge. USDA Forest Service, Intermountain Research Station, General Technical Report INT-GTR-309, Ogden, Utah.

Little, E. L. 1971. Atlas of United States trees. Vol I: Conifers and important hardwoods. USDA Forest Service, United States Government Printing Office, Miscellaneous Publication 1146, Washington, D.C.

Lloyd, A. H., and L. J. Graumlich. 1997. Holocene dynamics of treeline forests in the Sierra Nevada. Ecology 74:1199–1210.

Lonner, T. N., and D. F. Pac. 1990. Elk and mule deer use of whitebark pine forests in southwestern Montana: An ecological perspective. Pages 237–244 *in* W. C. Schmidt and K. J. McDonald, compilers. Whitebark pine ecosystems—Ecology and management of a high mountain resource. USDA Forest Service, Intermountain Research Station, General Technical Report INT-270, Ogden, Utah.

Marshall, J. D., and J. Zhang. 1994. Carbon isotope discrimination and water use efficiency in native plants of the north central Rockies. Ecology 75:1887–1895.

Martinelli, M. 1964. Watershed management in the Rocky Mountain alpine and subalpine zones. USDA Forest Service, Rocky Mountain Forest and Range Experiment Station, Research Note RM-36, Fort Collins, Colorado.

Mattson, D. J., and D. P. Reinhart. 1990. Whitebark pine on the Mount Washburn massif, Yellowstone National Park. Pages 106–117 *in* W. C. Schmidt and K. J. McDonald, compilers. Whitebark pine ecosystems—Ecology and management of a high mountain resource. USDA Forest Service, Intermountain Research Station, General Technical Report INT-270, Ogden, Utah.

———. 1994. Bear use of whitebark pine seed in North America. Pages 212–220. *in* W. C. Schmidt and F. -K. Holtmeier, compilers. Proceedings—International workshop on subalpine stone pines and their environment: The status of our knowledge. USDA Forest Service, Intermountain Research Station, General Technical Report INT-GTR-309, Ogden, Utah.

McCaughey, W. W., and W. C. Schmidt. 1990. Autecology of whitebark pine. Pages 85–96 *in* W. C. Schmidt and K. J. McDonald, compilers. Whitebark pine ecosystems—Ecology and management of a high mountain resource. USDA Forest Service, Intermountain Research Station, General Technical Report INT-270, Ogden, Utah.

McNeal, A., and T. Weaver. 1982. Soil compaction as a determinant of subalpine snowdrift vegetation. Montana Academy Science Proceedings 41:46–50.

Minarcic, P., and F. Kubicek. 1991. Localization of emissions on the needle surface. I. Morphological characterization of the stomata of *Picea abies* and *Pinus cembra.* Ekologia (CSFR) 10:405–413.

Mirov, N. T. 1967. The genus *Pinus*. Ronald Press, New York.

Mitchell, V. L. 1976. The regionalization of climate in the western United States. Journal of Applied Meterology 15:920–927.

Morgan, P., and S. C. Bunting. 1992. Using cone scars to estimate past cone crops of whitebark pine. Western Journal of Applied Forestry 7:71–73.

Nakashinden, I. 1994. Japanese stone pine production estimated from cone scars, Mount Kisokomagatake, central Japanese Alps. Pages 199–192 *in* W. C. Schmidt and F. -K. Holtmeier, compilers. Proceedings—International Workshop on subalpine stone pines and their environment: The status of our knowledge. USDA Forest Service, Intermountain Research Station, General Technical Report INT-GTR-309, Ogden, Utah.

Nebel, B., and P. Matile. 1992. Longevity and senescence of needles of *Pinus cembra* L. Trees 6:156–161.

Nielson, R. P. 1992. Toward a rule-based biome model. Landscape Ecology 7:27–43.

Paolett, E., and L. M. Bellani. 1990. The in-vitro response of pollen germination and tube length to different types of acidity. Environmental Pollution 67:279–286.

Parker, A. J. 1989. Forest environment relationships in Yosemite National Park, California, USA. Vegetatio 82:41–54.

Peet, R. K. 1988. Forests of the Rocky Mountains. Pages 63–102 *in* M. G. Barbour and W. D. Billings, editors. North American terrestrial vegetation. Cambridge Press, New York.

Perkins, D. L., and T. Swetnam. 1996. A dendroecological assessment of whitebark pine in the Sawtooth–Salmon River region, Idaho. Canadian Journal of Forest Research 26:2123–2133.

Peterson, D. L. 1994. Recent changes in the growth and establishment of subalpine conifers in western North America. Pages 234–243 *in* M. Beniston, editor. Mountain environments in changing climates. Routledge Press, New York.

Peterson, R. T. 1990. Field guide to western birds. Houghton Mifflin, Boston, Massachusetts.

Reich, P. B., J. Oleksyn, J. Modrzynski, and M. G. Tjoelker. 1996. Evidence that longer needle retention of spruce and pine populations at high elevations and high latitudes is largely a phenotypic response. Tree Physiology 16:643–647.

Rodin, L. E., and N. I. Bazelevich. 1965. Production and mineral cycling in terrestrial vegetation. Oliver and Boyd, Edinburgh, Scotland.

Romme, W. H., and M. G. Turner. 1991. Implications of global change for biogeographic patterns in the greater Yellowstone ecosystem. Conservation Biology 5:373–386.

Sakai, A. 1978. Low temperature exotherms of winter buds of hardy conifers. Plant and Cell Physiology 19:1439–1446.

Sauermoser, S. 1994. Current distribution of cembran pine in the Lechtal Alps. Pages 269–274 *in* W. C. Schmidt and F. -K. Holtmeier, compilers. Proceedings—International workshop on subalpine stone pines and their environment: The status of our knowledge. USDA Forest Service, Intermountain Research Station, General Technical Report INT-GTR-309, Ogden, Utah.

Schmidt, W. C., and J. Lotan. 1980. Phenology of common forest flora of the northern Rockies: 1928–1937. USDA Forest Service, Intermountain Research Station, Research Paper INT-259, Ogden, Utah.

Seligman, G. 1939. Snow structure and snow fields. Foister & Jagg LTD., Cambridge University Press, New York.

Sirucek, D. A. 1996. Soil temperature and moisture characteristics of forest habitat types of Montana and Idaho. Master's thesis, Montana State University, Bozeman.

Sirucek, D., J. Wraith, and T. Weaver. 1999. Soil water, soil temperature, and elevation as predictors of site quality (HT). In review.

Smith, C. C. 1970. The co-evolution of pine squirrels (*Tamiasciurus*) and conifers. Ecological Monographs 40:349–371.

Smith, C. C., and R. P. Balda. 1979. Competition among insects, birds, and mammals for conifer seeds. American Zoologist 19:1065–1083.

Smolonogov, E. Y. 1994. Geographical differentiation and dynamics of Siberian stone pine forests in Eurasia. Pages 275–279 *in* W. C. Schmidt and F. -K. Holtmeier, compilers. Proceedings—International workshop on subalpine stone pines and their environments: The status of our knowledge. USDA Forest Service, Intermountain Research Station, General Technical Report INT-GTR-309, Ogden, Utah.

Sowell, J. B., D. L. Koutnik, and A. J. Lansing. 1982. Cuticular transpiration of whitebark pine (*Pinus albicaulis*) within a Sierra Nevadan timberline ecotone, USA. Arctic and Alpine Research 14:97–104.

Tallis, J. H. 1991. Plant community history: Long-term changes in plant distribution and diversity. Chapman and Hall, New York.

Tomback, D. F. 1998. Clark's nutcracker (*Nucifraga columbiana*), No. 331. *In* A. Poole and F. Gill, editors. The birds of North America. The Birds of North America, Inc., Philadelphia.

Tomback, D. F., L. A. Hoffman, and S. K. Sund. 1990. Co-evolution of whitebark pine and nutcrackers: Implications for forest regeneration. Pages 118–129 *in* W. C. Schmidt and K. J. McDonald, compilers. Proceedings—Symposium on whitebark pine ecosystems: Ecology and management of a high-mountain resource. USDA Forest Service, Intermountain Research Station, General Technical Report INT-270, Ogden, Utah.

Tomback, D. F., F. -K. Holtmeier, H. Mattes, K. S. Casey, and M. L. Powell. 1993. Tree clusters and growth form distribution in *Pinus cembra,* a bird-dispersed pine. Arctic and Alpine Research 25:374–381.

Townsend, A. M., and W. F. Kwolek. 1987. Relative susceptibility of thirteen pine species to sodium chloride spray. Journal Arboriculture 13:225–228.

Tranquillini, W. 1955. Die Bedeutung des Lichtes und der Temperatur fur die Kohlensaureassimilation von *Pinus cembran*-Jungwuchs an einem hochalpinen Standort. Planta 46:14–178.

———. 1979. Physiological ecology of the alpine timberline. Springer, New York.

Turner, H., and A. Streule. 1988. Wurzelwachstum und Sprossentwicklung juger Koniferen im Klimastress der alpinen Waldgrenze. Schweizerische Zeitschrift für Forstwesen 139:785–789.

Turner, H., P. Rochat, and A. Streule. 1975. Thermische Charakteristik von Hauptstandortstypen im Bereich der oberen Waldgrenze (Stillberg, Dischmatal bei Davos). Mitteilungen, Eidgenössische Anstalt für das forstliche Versuchswesen 51:95–119.

Tursich, N., and T. Weaver. 1998. Plant growth in soils from snowdrift versus adjacent driftless meadow sites. Intermountain Journal Science 4:22–26.

Vorobjev, N. N., S. N. Goroshkevich, and D. A. Savchuk. 1994. New trend in dendrochronology: A method for retrospective study of seminiferous dynamics in Pinaceae. Pages 201–204 *in* W. C. Schmidt and F. -K. Holtmeier, compilers. Proceedings—International workshop on subalpine stone pines and their environment: The status of our knowledge. USDA Forest Service, Intermountain Research Station, General Technical Report INT-GTR-309, Ogden, Utah.

Walter, H. 1985. Vegetation of the earth. Springer-Verlag, NewYork.

Weaver, T. 1977. Root distribution and soil water regimes in nine habitat types on the northern Rocky Mountains. Pages 239–244 *in* J. K. Marshall, editor. The belowground ecosystem. Colorado State University, Range Science Series #26, Fort Collins.

———. 1978. Change in soils along a vegetation-altitudinal gradient of the northern Rocky Mountains. Pages 14–29 *in* C. T. Youngberg, editor. Forest soils and land use. Soil Science Society America and Forest and Wood Science Department, Colorado State University, Fort Collins.

———. 1980. Climates of vegetation types of the northern Rocky Mountains and adjacent plains. American Midland Naturalist 103:392–398.

———. 1985. Summer showers: Their sizes and interception by surface soils. American Midland Naturalist 114:409–413.

———. 1990. Climates of subalpine pine woodlands. Pages 72–79 *in* W. C. Schmidt and K. J. McDonald, compilers. Whitebark pine ecosystems—Ecology and management of a high mountain resource. USDA Forest Service, Intermountain Research Station, General Technical Report INT-270, Ogden, Utah.

———. 1994a. Climates where stone pines grow, a comparison. Pages 85–89 *in* W. C. Schmidt and F. -K. Holtmeier, compilers. Proceedings—International workshop on subalpine stone pines and their environment: The status of our knowledge. USDA Forest Service, Intermountain Research Station, General Technical Report INT-GTR-309, Ogden, Utah.

———. 1994b. Vegetation distribution and production in Rocky Mountain climates—with emphasis on whitebark pine. Pages 142–152 *in* W. C. Schmidt, and F.-K. Holtmeier, compilers. Proceedings—International workshop on subalpine stone pines and their environment: The status of our knowledge. USDA Forest Service, Intermountain Research Station, General Technical Report INT-GTR-309, Ogden, Utah.

Weaver, T., and D. Dale. 1974. *Pinus albicaulis* in central Montana: Environment, vegetation and production. American Midland Naturalist 92:222–230.

———. 1978. Trampling effects of hikers, motorcycles and horses in meadows and forests. Journal Applied Ecology 5:451–457.

Weaver, T., and F. Forcella. 1977. Biomass of fifty conifer forests and nutrient exports associated with their harvest. Great Basin Naturalist 37:395–400

———. 1986. Cone production in *Pinus albicaulis* forests. Pages 68–76 *in* R. C. Shearer, compiler. Conifer tree seed in the inland mountain West. USDA Forest Service, Intermountain Research Station, General Technical Report INT-203, Ogden, Utah.

Weaver, T. and J. Jacobs. 1990. Occurrence of multiple stems in whitebark pine. Pages 156–159 *in* W. C. Schmidt and K. J. McDonald, compilers. Whitebark pine ecosystems—Ecology and management of a high mountain resource. USDA Forest Service, Intermountain Research Station, General Technical Report INT-270, Ogden, Utah.

Wells, P. V. 1965. Scarp woodlands, transported grassland soils, and the concept of grassland climate in the Great Plains region. Science 148:246–249.

Whitlock, C. 1993. Stability of holocene climate regimes in the Yellowstone region. Quaternary Research 43:433–436.

Whittaker, R. E. 1975. Communities and ecosystems. Macmillan, Inc., New York.

Wooldridge, G. L., R. C. Musselman, R. A. Sommerfield, D. G. Fox, and B. H. Connell. 1996. Mean wind patterns and snow depths as measures of damage to coniferous trees. Journal of Applied Ecology 33:100–108.

Zweifel, R., and R. Haesler. 1999. Diurnal and seasonal limits of root water uptake affecting the water balance of mature subalpine Norway spruce trees. Pages 61–79 *in* R. Zweifel, editor. The rhythm of trees: Watorage dynamics in subalpine Norway spruce. Swiss Federal Institute of Technology, Dissertation # Eth-13391, Zurich.

Chapter 4

Community Types and Natural Disturbance Processes

Stephen F. Arno

For a species that is narrowly restricted to the upper subalpine and timberline zones, whitebark pine (*Pinus albicaulis*) occurs in a surprisingly diverse array of community types. For example, at the highest elevations it forms stands of shrublike trees (krummholz) growing in association with a variety of alpine cushion plants. At lower elevations in the subalpine zone it may be associated with grassland species on semiarid sites or, conversely, with shrubs and herbs characteristic of wet and snowy sites. It even occupies subalpine fens, where its principal undergrowth associates are Labrador tea (*Ledum* spp.), bog laurel (*Kalmia* spp.), and bog huckleberry (*Vaccinium occidentale*).

In much of its geographic range whitebark pine is primarily a timberline tree, confined to sites that are so cold, snowbound, rocky, or wind-scoured that its competitors are suppressed by the harsh environment (Arno and Hoff 1990). This is generally the case in the Cascade Range, the Sierra Nevada, and in the Canadian Rocky Mountains north of about 50° N latitude. In contrast, in the northern Rocky Mountains of Montana, Idaho, and northwestern Wyoming, historically whitebark pine extended downslope from the timberline to occupy much of the subalpine forest zone. This historical distribution is evidenced by an abundance of whitebark pine snags, remnants of large trees that died during the last hundred years (Arno and Hoff 1990). This broader historical distribution appears linked to fire ecology—large fires occurred frequently enough to allow periodic recolonization by whitebark pine (Arno 1986). In contrast, in wetter mountain regions like the Cascades, large fires were uncommon in the subalpine zone, and highly competitive conifers precluded the establishment of whitebark pine except near the uppermost limit of forest growth (Arno and Hoff 1990).

Successional Relationships

Whitebark pine is a relatively slow growing tree that is outcompeted for growing space by more shade-tolerant species—subalpine fir (*Abies lasiocarpa*), Engelmann spruce (*Picea engelmannii*), and mountain hemlock (*Tsuga mertensiana*) (Minore 1979). However, whitebark pine has superior hardiness on dry, cold, and windy sites, and thus is able to become established and persist in harsher environments than the competitors can tolerate (Chapter 3).

Conversely, whitebark pine appears able to coexist indefinitely with shade-intolerant associates lodgepole pine (*Pinus contorta*) and alpine larch (*Larix lyallii*) as well as with the midtolerant inland Douglas-fir (*Pseudotsuga menziesii* var. *glauca*). The successional role of whitebark pine on different sites can be characterized as follows (Figure 4-1):

- *Climax:* It persists in pure stands on the coldest (highest elevation) and driest sites, where competing species are scarce because of their inferior hardiness. Examples are the whitebark pine habitat types (a site classification based on potential vegetation) described by Pfister et al. (1977) and other publications listed in Table 4-1.
- *Co-climax:* This applies to sites capable of supporting shade-tolerant tree species,

Table 4-1. Principal publications and theses describing whitebark pine communities, listed by state and province. Complete citations appear in the Literature Cited section.

State or Province	Citation(s)
Alberta	Achuff 1989; Baig 1972; Ogilvie 1990
British Columbia	Achuff 1989; Brink 1959; McAvoy 1931; Ogilvie 1990; Selby and Pitt 1984
California	Barbour 1988; Cooke 1940, 1955; Klikoff 1965; Riegel et al. 1990; Sawyer and Thornburgh 1977; Taylor 1976; Vale 1977
Idaho	Cooper et al. 1991; Daubenmire and Daubenmire 1968; Steele et al. 1981, 1983
Montana	Arno 1970; Craighead et al. 1982; Forcella 1977, 1978; Pfister et al. 1976, 1977; Weaver and Dale 1974
Nevada	Loope 1969
Oregon	Cole 1982; Franklin and Dyrness 1973; Hall 1973; Hopkins 1979; Jackson and Faller 1973; Johnson and Clausnitzer 1992; Lueck 1980; Topik et al. 1995
Washington	Agee and Kertis 1987; Arno 1970; Daubenmire and Daubenmire 1968; del Moral 1979; Franklin and Dyrness 1973; Lillybridge et al. 1995; Topik et al. 1995; Williams and Lillybridge 1983; Williams et al. 1995
Wyoming	Forcella 1978; Gruell 1980; Steele et al. 1983

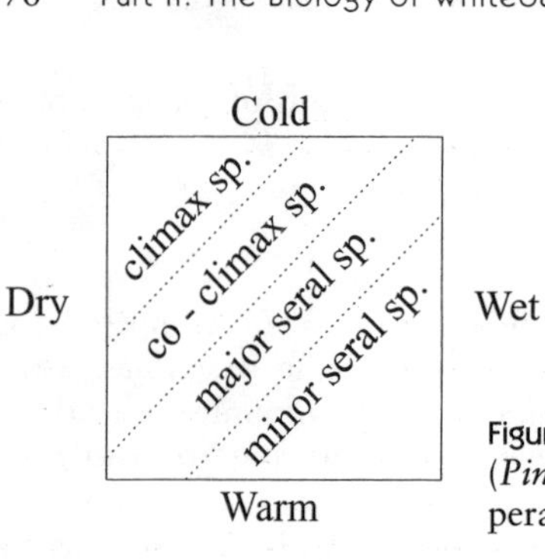

Figure 4-1. Successional status of whitebark pine (*Pinus albicaulis*) arrayed on moisture and temperature gradients.

but where they are unable to grow vigorously enough to replace whitebark pine. Examples are the whitebark pine–subalpine fir habitat types.

- *Major seral:* Whitebark pine was maintained historically as a major stand component in the inland northwestern United States, northern Rocky Mountains of the United States, and adjacent regions of Canada, as a result of periodic fires. Examples are the "upper subalpine" habitat types of Pfister et al. (1977). After long periods without fire, shade-tolerant conifers become dominant.
- *Minor seral:* Whitebark pine was maintained historically as a minor stand component in the inland northwestern United States, northern Rocky Mountains of the United States, and adjacent regions of Canada as a result of periodic fires, with other shade-intolerant species, especially lodgepole pine, being more abundant. This condition applies in the "lower subalpine" habitat types of Pfister et al. (1977).

Community Types

A variety of reports describe whitebark pine communities in almost every state and province occupied by this tree (Table 4-1). Whitebark pine habitats are abundant, diverse, and best-documented in the northern Rocky Mountains of the United States. Here, two remarkably consistent community complexes linked to site moisture gradients appear repeatedly:

1. In the continental climatic region of east-central Idaho, western Wyoming, and central Montana, whitebark forest communities (both seral and potential climax) are abundant (Arno and Hammerly 1984). Here, on north-facing slopes and other cool sites, the undergrowth is often dominated by *Vaccinium scoparium.* On progressively warmer and drier sites, the dominant undergrowth changes to *Carex geyeri, Juncus parryi, Arnica cordifolia,* and *Festuca idahoensis* or other bunchgrasses.
2. In the inland-maritime mountain climates of northeastern Oregon, northern and west-central Idaho, and western Montana, whitebark pine stands are codominated by subalpine fir and, lower in the subalpine zone, by Rocky Mountain lodgepole pine (*Pinus contorta* var. *latifolia*). On cool exposures, the undergrowth is commonly *Luzula hitchcockii, Phyllodoce empetriformis,* and *Vaccinium scoparium.* As sites become warmer and drier, the dominants are usually *Xerophyllum tenax, Vaccinium globulare, Festuca viridula, Carex geyeri,* and *Juncus parryi.* The wettest upland forest sites often have whitebark pine as a seral component mixed with Engelmann spruce, subalpine fir, and

sometimes mountain hemlock or alpine larch. Characteristic undergrowth includes numerous wet-meadow forbs and sedges, and the shrubs *Ledum glandulosum, Phyllodoce empetriformis, Menziesia ferruginea,* and *Rhododendron albiflorum.*

In the continental climatic regions of the northern Rockies, whitebark pine is the potential climax tree in several subalpine forest habitat types as well as being a seral associate in several others (Table 4-2, Figure 4-2). In the inland-maritime mountains further west, whitebark pine is seral except in the timberline zone (Table 4-3).

Heading west from whitebark pine's abundant habitat in the northern Rockies, there is a transition toward drier-appearing whitebark pine communities (Table 4-4). In west-central Idaho, the Seven Devils Mountains have whitebark pine communities often with undergrowth of *Luzula hitchcockii, Polemonium pulcherrimum,* and *Lewisia columbiana* (C. G. Johnson Jr., Wallowa-Whitman National Forest, personal communication). Farther west across the Snake River Canyon, the Wallowa Mountains form a cluster of high peaks rising from Oregon's Blue Mountain region. The Wallowas support extensive whitebark pine communities between about 2,300 meters and 2,800 meters (7,500 and 9,200 feet) in elevation, including krummholz growth forms on windswept ridges, groves of large spreading trees interspersed with subalpine grasslands, and dense forests of whitebark pine, lodgepole pine, subalpine fir, and Engelmann spruce. These communities occupy diverse substrates, including granodiorite, basalt, greenstone, and limestone, which influence their composition (C. G. Johnson Jr., personal communication). Whitebark pine is the principal tree at timberline on dry, cold sites with undergrowth of *Juncus parryi* and *Stipa lemmonii.* To the west, a small subalpine area called the Elkhorn Mountains has whitebark pine communities in which the undergrowth dominants are commonly *Vaccinium scoparium, Carex geyeri,* and *Polemonium pulcherrimum* (C. G. Johnson Jr., personal communication). Still farther west, in the Blue Mountain region, another small subalpine area, the Strawberry Mountains, contains some of the apparently warmest and driest sites occupied by whitebark pine, where undergrowth is characterized by *Arenaria aculeata, Carex geyeri,* and *Stipa occidentalis* (C. G. Johnson Jr., personal communication). Serpentine geologic substrates here have climax communities of whitebark pine with undergrowth of *Juniperus communis* and *Arctostaphylos nevadensis.*

Northward in the Rocky Mountains of Alberta and British Columbia, whitebark pine remains widespread but is less often a major component of the upper subalpine forest or timberline communities. Achuff (1989) and Ogilvie (1990) provide detailed descriptions of whitebark pine communities in Canada. In Alberta, whitebark pine is most common in the timberline zone as a codominant with subalpine fir, Engelmann spruce, and sometimes with alpine larch (Baig 1972). Characteristic undergrowth includes *Phyllodoce empetriformis* in moist sites along or near the Continental Divide (inland-maritime climate of Arno and Hammerly [1984]), *Phyllodoce glanduliflora* in comparable sites in mountains farther inland (continental climate), *Vaccinium scoparium* on well-drained sites, and *Juniperus communis* on the driest south-facing slopes.

Table 4-2. Composition of whitebark pine communities of western Wyoming arranged in a moisture gradient (modified from Steele et al. 1983, and Arno and Weaver 1990).

Site Moisture	Habitat Type[1]	Tree Species[2]						
		Pinus albicaulis	*Abies lasiocarpa*	*Picea engelmannii*	*Pinus contorta* var. *latifolia*	*Pinus flexilis*	*Pseudotsuga menziesii* var. *glauca*	*Populus tremuloides*
Wet ↑								
	PIEN/VASC	(S)	c	C	(S)	(s)		
	PIEN/CALE	s	C	C	S			
	ABLA/VAGL, VASC	s	C	S	S			
	ABLA/VASC, PIAL	C	C	S	S		(s)	
	ABLA/VASC, VASC	s	C	S	S			
	ABLA/ARLA	(S)	C	S	(S)		(S)	(S)
	ABLA/THOC	s	C	S	S		S	(S)
	ABLA/JUCO	s	C	S	S	s	(S)	
	ABLA/RIMO,RIMO	s	C	S	(s)			
	ABLA/RIMO,PIAL	C	C	C				
	ABLA/ARCO,SHCA	s	C	S	S	s	s	s
	PIAL/VASC	C	C	C	C			
	PIAL/CAGE	C			(C)			
	PIAL/JUCO	C			C	S		
	PIAL/CARO	C	(c)	(c)	(C)	(s)		
	PIAL/FEID	C						
Dry ↓								

[1]Abbreviations consist of the first two letters of the genus and species name. Tree species are: PIEN = *Pinus engelmannii;* ABLA = *Abies lasiocarpa;* PIAL = *Pinus albicaulis.* Undergrowth species are ARCO = *Arnica cordifolia;* ARLA = *A. latifolia;* CAGE = *Carex geyeri;* CALE = *Caltha leptospeala;* CARO = *Carex rossii;* FEID = *Festuca idahoensis;* JUCO = *Juniperus communis;* RIMO = *Ribes montigenum;* SHCA = *Shepherdia canadensis;* THOC = *Thalictrum occidentale;* VAGL = *Vaccinium globulare;* VASC = *V. scoparium.*

[2]C = climax dominant; S = seral dominant; c = minor climax species; s = minor seral species; () = in part of the habitat type only.

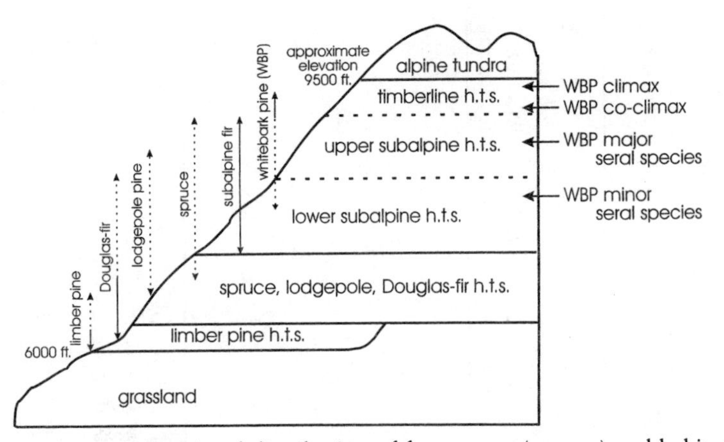

Figure 4-2. General elevational distribution of forest trees (arrows) and habitat types (potential climax) on noncalcareous geologic substrates in south-central Montana. Solid portion of arrow indicates where a species is the potential climax, and dotted portion shows where it is seral (modified from Pfister et al. 1977 and Arno and Weaver 1990).

In the Cascade Range, whitebark pine is most common on dry sites. In the rain shadow of the Cascade Crest in the Stuart Range in central Washington, whitebark pine is abundant in the upper subalpine forest and the timberline zone (Arno and Hammerly 1984). Timberline communities are dominated by whitebark pine on dry sites and warm aspects and by alpine larch in moist-cool situations (del Moral 1979), as they are in some mountain ranges of western Montana (Arno and Habeck 1972). In the granitic Stuart Range, relatively moist whitebark pine–alpine larch communities have an undergrowth of *Vaccinium myrtillus,* which is ecologically similar to *V. scoparium* (del Moral 1979). With increasing dryness, whitebark pine communities have undergrowths characterized by *Lewisia columbiana, Phlox diffusa, Juniperus communis,* and *Penstemon davidsonii.*

In the excessively well-drained pumice substrate of the Oregon Cascades, whitebark pine communities are characterized by sparse undergrowth. On moist microsites, *Vaccinium scoparium* and *Luzula hitchcockii* are characteristic; on sites with average moisture conditions, undergrowth is typically *Penstemon davidsonii* (Jackson and Faller 1973; Lueck 1980). On coarse volcanic substrates in the Warner Mountains of south-central Oregon and northeastern California, whitebark pine and Sierra Nevada lodgepole pine (*P. contorta* var. *murrayana*) are the climax codominants in community types where the principal undergrowth consists of dry-site herbs, *Carex pensylvanica, Poa nervosa, Stipa californica, Arenaria aculeata,* and *Penstemon* spp. (Hopkins 1979; Riegel et al. 1990).

Southward on the Cascade–Sierra Nevada axis, whitebark pine is common but largely confined to timberline communities in northern and central California. Subalpine fir and Engelmann spruce are absent or rare, and the timberline communities tend to be open, with only sparse, scattered undergrowth

Table 4-3. Composition of whitebark pine communities in west-central Montana arranged on an elevational gradient (modified from Pfister et al. 1977, and Arno and Weaver 1990).

			STAND COMPONENTS[2]					
Elevational Zone	Site Moisture	Habitat Types[1]	*Pinus albicaulis*	*Abies lasiocarpa*	*Picea engelmannii*	*Larix lyallii*	*Pinus contorta* var. *latifolia*	*Pseudotsuga menziesii* var. *glauca*
Timberline	Dry	PIAL	C					
	Intermediate	PIAL-ABLA	C	C	c			
	Moist	LALY-ABLA	C	C	c	C		
Upper subalpine forest	Dry	ABLA-PIAL/VASC	S^{300y}	C	s			
	Intermediate	ABLA/LUHI	S^{250y}	C	s		S^{200y}	
	Moist		S^{200y}	C	S^{400y}	s	s	
Lower subalpine forest	Dry	ABLA/XETE, VASC	s	C	s		S	s

[1]Abbreviations consist of the first two letters of the genus and species names. Tree species are PIAL = *Pinus albicaulis;* ABLA = *Abies lasiocarpa;* LALY = *Larix lyallii.* Undergrowth species are LUHI = *Luzula hitchcokii;* VASC = *Vaccinium scoparium;* XETE = *Xerophyllum tenax.*

[2]C = climax dominant; S = seral dominant for number of years (as indicated) after fire or other disturbance; c = minor climax species; s = minor seral species.

Table 4-4. Composition of whitebark pine communities of the Seven Devils Mountains, Idaho, and the Blue Mountains region, Oregon, arranged on a moisture gradient (C. G. Johnson Jr., personal communication).

Site Moisture	Plant Associations[1]		Tree Species[2]			
	PIAL-ABLA Co-climax	PIAL Climax	*Pinus albicaulis*	*Abies lasiocarpa*	*Picea engelmannii*	*Pinus contorta*
Wet			C	C		
	ABLA-PIAL/PHEM		C	C		
	ABLA-PIAL/RIMO/POPU		C	C	S	
	ABLA-PIAL/VASC/ARCO		C	C	S	
	ABLA-PIAL/VASC/CARO		C	C	S	
	ABLA-PIAL/VASC/ARAC2		C	C		
	ABLA-PIAL/POPU		C	C		
	ABLA-PIAL/FEVI		C	C		
	ABLA-PIAL/CAGE		C	C		S
	ABLA-PIAL/JUPA-STLE2		C	C		
		PIAL/RIMO/POPU	C	c		
		PIAL/VASC/LUHI	C	c	S	
		PIAL/FEVI	C			
		PIAL/CAGE	C	C		s
		PIAL/JUCO4-ARNE	C			S
		PIAL/ARAC2	C			
Dry						

[1]Tree species abbreviations: ABLA = *Abies lasiocarpa;* PIAL = *Pinus albicaulis.* Undergrowth species abbreviations: ARAC2 = *Arenaria aculeate;* ARCO = *Arnica cordifolia;* ARNE = *Arctostaphylos nevadensis;* CAGE = *Carex geyeri;* CARO = *Carex rossii;* FEVI = *Festuca viridula;* JUCO4 = *Juniperus communis;* JUPA = *Juncus parryi;* LUHI = *Luzula hitchockii;* PHEM = *Phyllodoce empetriformis;* POPU = *Polemonium pulcherimum;* RIMO = *Ribes montigenum;* STLE2 = *Stipa lemmonii;* VASC = *Vaccinium scoparium.*
[2]C = climax dominant; c = minor climax species; S = seral dominant; s = minor seral species.

(Barbour 1988). These communities consist of mixtures of whitebark pine with mountain hemlock on moist sites and with Sierra lodgepole pine, western white pine (*Pinus monticola*), and foxtail pine (*Pinus balfouriana*) on drier sites.

Across much of its distribution, whitebark pine occurs in an intricate pattern of community types dominated by tall or dwarf trees of various species, and by shrubs, subalpine herbs, or alpine tundra plants. These community mosaics and their microenvironmental controls are not well documented, with the exception of del Moral's (1979) work in central Washington's Stuart Range and Pfister et al.'s (1976) quantitative description and map of an extensive whitebark pine community mosaic in northwestern Montana. Pfister et al.'s (1976) map (published as Figure 17 in Craighead et al. 1982) differentiated six habitat types and phases with a major representation of whitebark pine and several habitat types in which whitebark pine is a minor component (data published as Tables 5 and 6 in Arno and Weaver 1990).

Role of Disturbance

Natural perturbations are vital to the perpetuation of whitebark pine in the habitat types where it is seral, and disturbances are also important in shaping the structure of whitebark pine communities. In the timberline zone, the climate is so harsh that climatic factors such as damaging winds, ice storms, snowloads, summer frost, and winter desiccation prevent stand closure and often result in parklands where groves of trees form a mosaic with grasslike or heath vegetation. This rigorous microclimate also allows competition-intolerant species like whitebark pine to coexist indefinitely with their tolerant competitors (Franklin and Dyrness 1973; Pfister et al. 1977; Arno and Hammerly 1984). Conversely, in subalpine forest habitats, whitebark pine's perpetuation depends upon occasional disturbances, most notably fires. Without these disturbances, succession leads to dominance by subalpine fir, spruce, or mountain hemlock (Chapters 9 and 18). Habitat types where whitebark pine is seral probably account for 75 percent or more of the species' historic distribution in the interior Columbia River Basin (Keane et al. 1996).

In large portions of the inland northwestern United States, the area covered by seral whitebark pine communities has diminished in recent decades. This decline is due to successional replacement linked to fire suppression and is aggravated by epidemics of the native mountain pine beetle (*Dendroctonus ponderosae*) and especially by the introduced white pine blister rust (*Cronartium ribicola*) (Arno 1986; Kendall and Arno 1990; Chapters 10 and 11). For example, in a detailed study of landscapes on the Bitterroot National Forest, Montana, Hartwell (1997) found that in the subalpine forest between elevations of 2,100 meters and 2,300 meters (6,900 and 7,500 feet), historically (circa 1900) whitebark pine made up about 39 percent of the basal area, but by 1995 it amounted to only 11 percent. The sample areas experienced no logging. Apparently, these losses of whitebark pine are attributable to the mountain pine beetle epidemic of the early 1900s that killed many large whitebark and lodgepole pines, coupled with successional replacement related to fire exclusion. Substantial mortality from blister rust has also occurred in the last twenty to thirty years (Keane and Arno 1993). Hartwell (1997) found that subalpine fir had the largest increase in abundance, expanding from 13 percent of the basal area in 1900 to 32 percent in 1995.

Whitebark pine snags, standing or fallen, can frequently be identified based on their distinctively spreading upper-crown branches. Many snags of whitebark pine trees that died during the last hundred years are still identifiable. My observations from numerous areas in western Montana suggest that the whitebark pine–dominated forests historically covered 610 to 670 meters (2,000 to 2,200 feet) in elevation, extending down the mountainsides from the limit of erect trees (Figure 4-1). Today, the lower 240 meters (800 feet) of this elevation zone has lost most of its whitebark pine to subalpine fir and other species by successional replacement. Because only a small amount of the mountain terrain extends as high as timberline, the lower 240 meters probably represents at least half of the area historically occupied by this species.

Prior to 1900, fires at intervals averaging between thirty and three hundred years were widespread and important in perpetuating seral whitebark pine communities (Table 4-5). These often burned in a patchy pattern with different

Table 4-5. Pre-1900 mean fire intervals for whitebark pine forests. Data are mean fire intervals in years in small sample areas; range of interval lengths are given in parentheses.

Fire Interval, Years (range)	Fire Regime (Brown 1995)	Geographical Area and Reference
144 (55 to 304)	mixed-severity	Bob Marshall Wilderness, NW Montana (Keane et al. 1994)
57 to 94	mixed-severity	W side of Bitterroot Valley, W Montana (Arno and Petersen 1983)
56 to 180	mixed-severity	Selway-Bitterroot Wilderness, Idaho and Montana (Brown et al. 1994)
29 (13 to 46)	mixed-severity	Russell Peak area, NW of Cody, Wyoming (Morgan and Bunting 1990)
66 to >400	mixed-severity	NE Yellowstone National Park; Wyoming (Barrett 1994)
300 to 400	stand replacement	Central Plateau, Yellowstone National Park (Romme 1982)

severities, suggestive of a "mixed-severity fire regime" (Brown 1995). The mixed-severity fire regime may have been prevalent in rugged topography where there is pronounced spatial variation in microsites, stand composition, and fuels. In some areas—particularly where there is an extensive, continuous subalpine forest, as in Yellowstone National Park—large, stand-replacement burns were characteristic, corresponding to the "stand-replacement fire regimes" (Brown 1995). Both light surface fires and stand-replacing fires favor whitebark pine in relation to its shade-tolerant competitors. High-intensity, stand-replacing fires in dense subalpine fir-spruce forests often allow whitebark pine to become established as a result of seed caching by the Clark's nutcracker (*Nucifraga columbiana*) (Chapter 5). After establishment, some of these seral whitebark pine communities have been perpetuated by low-intensity fires that killed understory fir and spruce (Arno and Hoff 1990).

As a result of fire exclusion during the 1900s, natural fire cycles in seral whitebark pine communities have been postponed and, as a result, this species is being replaced by its competitors (Arno 1986). Management programs that allow some natural fires to burn are probably insufficient for mimicking whitebark pine fire cycles of the past (Brown et al. 1994). The most effective fires in the highly discontinuous whitebark pine habitats, which occur atop isolated high ridges, are those that spread over large expanses—tens of thousands of hectares. However, most whitebark pine habitats lie near developed or commercially utilized lands where such massive fires are not tolerable politically, even in wilderness areas or national parks. In other words, today, few areas supporting whitebark pine are far enough from development or agricultural communities that natural, large-scale fires are possible.

Fire suppression during this century has no doubt resulted in a decrease in the establishment of new whitebark pine communities (Brown et al. 1994).

These young stands are needed to compensate for aging stands in which whitebark pine is being replaced successionally by shade-tolerant species. This replacement is being speeded up by the effects of blister rust on whitebark pine. Avalanches and severe blowdowns also create open microenvironments that allow whitebark pine to enter as a pioneer species. These disturbances are no substitute for fire, however, because they produce only small areas suitable for seral whitebark pine forests. Also, they generally do not kill small firs or prepare the site for regeneration as well as fire, which kills the undergrowth, reduces the duff layer, and releases nitrogen and other nutrients (Hungerford et al. 1991).

Concluding Remarks

It is apparent that whitebark pine communities historically represented a significant proportion of the biological diversity associated with high-mountain landscapes in the American West (Chapters 1 and 12). This diversity has already declined with the major loss of whitebark pine from subalpine forest habitats where it is being replaced successionally. Exclusion of most fires is the management approach used in the majority of whitebark pine habitats. As a result, the few early-successional whitebark pine communities found today are largely a result of old clear-cuts or wildfires that escaped suppression. Today, clear-cutting at such high elevations is generally not considered an appropriate management option.

Whitebark pine is one of several fire-dependent forest types in North America that are experiencing major losses in natural biodiversity. Notable progress in restoration is being made in the historic ponderosa pine (*Pinus ponderosa*) forest, the most extensive forest type in the western United States (Covington and Moore 1994; Arno et al. 1995). Here, major efforts are underway on national forests and other federal lands to implement management that simulates the historical fire process, using uneven-aged silviculture, fuel reduction treatments, and prescribed fire. Conversely, another large fire-dependent forest type—coastal Douglas-fir (*Pseudotsuga menziesii* var. *menziesii*) (Pinchot 1899; Morrison and Swanson 1990)—has received minimal interest in returning fire as a component of ecosystem-based management (Agee 1993, Means et al. 1996). In the case of both ponderosa pine and coastal Douglas-fir forests, ecologists have helped to produce ecologically based management strategies. Similarly, ecologists will have to assume an important share of the responsibility for restoration and maintenance of biodiversity in whitebark pine ecosystems. The question is, will ecologists accept this challenge or evade it by continuing to subscribe to a simplistic view that fire exclusion is the best approach to recommend (Botkin 1990; Dickmann and Rollinger 1998).

Acknowledgments

Steve Cooper, Montana Natural Heritage Program, Helena, Montana, and Charles Johnson Jr., Wallowa-Whitman National Forest, Baker City, Oregon, provided helpful reviews and information that contributed to this chapter.

Literature Cited

Achuff, Peter L. 1989. Old-growth forests of the Canadian Rocky Mountain national parks. Natural Areas Journal 9:12–26.

Agee, J. K. 1993. Fire ecology of Pacific Northwest forests. Island Press, Washington, D.C.

Agee, J. K., and J. Kertis. 1987. Forest types of the North Cascades National Park Service complex. Canadian Journal of Botany 65:1520–1530.

Arno, S. F. 1970. Ecology of alpine larch (*Larix lyallii* Parl.) in the Pacific Northwest. Ph.D. Dissertation, University of Montana, Missoula.

Arno, S. F. 1986. Whitebark pine cone crops: A diminishing source of wildlife food. Western Journal of Applied Forestry 1:92–94.

Arno, S. F, and J. R. Habeck. 1972. Ecology of alpine larch (*Larix lyallii* Parl.) in the Pacific Northwest. Ecological Monographs 42:417–450.

Arno, S. F., and R. P. Hammerly. 1984. Timberline—Mountain and arctic forest frontiers. The Mountaineers, Seattle, Washington.

Arno, S. F., and R. Hoff. 1990. *Pinus albicaulis* Engelm. Whitebark pine. Pages 268–279 *in* R. M. Burns and B. H. Honkala, technical coordinators. Silvics of North America. United States Department of Agriculture, Forest Service, Agriculture Handbook 654, Washington, D.C.

Arno, S. F., and T. D. Petersen. 1983. Variation in estimates of fire intervals: A closer look at fire history on the Bitterroot National Forest. USDA Forest Service, Intermountain Forest and Range Experiment Station, Research Paper INT-301, Ogden, Utah.

Arno, S. F., and T. Weaver. 1990. Whitebark pine community types and their patterns on the landscape. Pages 97–105 *in* W. C. Schmidt and K. J. McDonald, compilers. Proceedings—Symposium on whitebark pine ecosystems: Ecology and management of a high-mountain resource. USDA Forest Service, Intermountain Research Station, General Technical Report INT-270, Ogden, Utah.

Arno, S. F, J. H. Scott, and M. G. Hartwell. 1995. Age-class structure of old-growth ponderosa pine/Douglas-fir stands and its relationship to fire history. USDA Forest Service, Intermountain Research Station, Research Paper INT-RP-481. Ogden, Utah.

Baig, M. N. 1972. Ecology of timberline vegetation in the Rocky Mountains of Alberta. Ph.D. Dissertation, University of Calgary, Calgary, Alberta, Canada.

Barbour, M. G. 1988. Californian upland forests and woodlands. Pages 131–164 *in* M. G. Barbour and W. D. Billings, editors. North American terrestrial vegetation. Cambridge University Press, New York.

Barrett, S. W. 1994. Fire regimes on andesitic mountain terrain in northeastern Yellowstone National Park, Wyoming. International Journal of Wildland Fire 4:65–76.

Botkin, D. B. 1990. Discordant harmonies—A new ecology for the twenty-first century. Oxford University Press, New York.

Brink, V. C. 1959. A directional change in the subalpine forest-heath ecotone in Garibaldi Park, British Columbia. Ecology 40:10–16.

Brown, J. K. 1995. Fire regimes and their relevance to ecosystem management. Pages 171–178 *in* Proceedings of Society of American Forestry National Convention, Society of American Foresters, Bethesda, Maryland.

Brown, J. K., S. F. Arno, S. W. Barrett, and J. P. Menakis. 1994. Comparing the prescribed natural fire program with presettlement fires in the Selway-Bitterroot wilderness. International Journal of Wildland Fire 4:157–168.

Cole, D. N. 1982. Vegetation of two drainages in Eagle Cap wilderness, Wallowa Mountains, Oregon. Research. USDA Forest Service, Intermountain Forest and Range Experiment Station, Paper INT-288. Ogden, Utah.

Cooke, W. B. 1940. Flora of Mount Shasta. America Midland Naturalist 23:497–572.

———. 1955. Fungi of Mount Shasta (1936–1951). Sydowia, Annales Mycologici 9:94–215.

Cooper, S. V., K. E. Neiman, and D. W. Roberts. Revised 1991. Forest habitat types of northern Idaho: A second approximation. USDA Forest Service, Intermountain Research Station, General Technical Report, INT-236, Ogden, Utah.

Covington, W. W., and M. M. Moore. 1994. Southwestern ponderosa forest structure: Changes since Euro-American settlement. Journal of Forestry 92:39–47.

Craighead, J. J., J. S. Sumner, and G. B. Scaggs. 1982. A definitive system for analysis of grizzly bear habitat and other wilderness resources. University of Montana, Wildlife-Wildlands Institute Monograph 1, Missoula, Montana.

Daubenmire, R., and J. B. Daubenmire. 1968. Forest vegetation of eastern Washington and northern Idaho. Washington State University, Washington Agricultural Experiment Station, Technical Bulletin 60, Pullman, Washington.

del Moral, R. 1979. High-elevation vegetation of the Enchantment Lakes Basin, Washington. Canadian Journal of Botany 57:1111–1130.

Dickmann, D. I., and J. L. Rollinger. 1998. Fire for restoration of communities and ecosystems. Bulletin of the Ecological Society of America 79:157–160.

Forcella, F. 1977. Flora, chorology, biomass and productivity of the *Pinus albicaulis–Vaccinium scoparium* association. Master's Thesis, Montana State University, Bozeman.

———. 1978. Flora and chorology of the *Pinus albicaulis–Vaccinium scoparium* association. Madrono 25:131–150.

Franklin, J. F., and C. T. Dyrness. 1973. Natural vegetation of Oregon and Washington. USDA Forest Service, Pacific Northwest Forest and Range Experiment Station, General Technical Report PNW-8, Portland, Oregon.

Gruell, G. E. 1980. Fire's influence on wildlife habitat on the Bridger-Teton National Forest, Wyoming. Vol. I—Photographic record and analysis. USDA Forest Service, Intermountain Forest and Range Experiment Station Research, Paper INT-235, Ogden, Utah.

Hall, F. C. 1973. Plant communities of the Blue Mountains in eastern Oregon and southeastern Washington. USDA Forest Service, Pacific Northwest Region, R6 Area Guide 3-1, Portland, Oregon.

Hartwell, M. G. 1997. Comparing historic and present conifer species compositions and structures on forested landscapes of the Bitterroot Front. USDA Forest Service, Rocky Mountain Research Station, Intermountain Fire Sciences Laboratory, Contract Report RJVA-94928, Missoula, Montana.

Hopkins, W. E. 1979. Plant associations of the Fremont National Forest. USDA Forest Service, Pacific Northwest Region, R6-Ecol.-79-005. Portland, Oregon.

Hungerford, R. D., M. G. Harrington, W. H. Frandsen, K. C. Ryan, and G. J. Niehoff. 1991. Influence of fire on factors that affect site productivity. Pages 32–50 *in* A. E. Harvey and L. F. Neuenschwander, compilers. Proceedings—Symposium on management and productivity of western-montane forest soils. USDA Forest Service, Intermountain Research Station, General Technical Report INT-280, Ogden, Utah.

Jackson, M. T., and A. Faller. 1973. Structural analysis and dynamics of the plant communities of Wizard Island, Crater Lake National Park. Ecological Monograph 43:441–461.

Johnson, C. G. Jr., and R. R. Clausnitzer. 1992. Plant associations of the Blue and Ochoco Mountains. USDA Forest Service, Pacific Northwest Region, R6 ERW TP 036-92, Portland, Oregon.

Keane, R. E., and S. F. Arno. 1993. Rapid decline of whitebark pine in western Montana: Evidence from 20-year remeasurements. Western Journal of Applied Forestry 8:44–47.

Keane, R. E., P. Morgan, and J. P. Menakis. 1994. Landscape assessment of the decline of whitebark pine (*Pinus albicaulis*) in the Bob Marshall Wilderness Complex, Montana, USA. Northwest Science 68:213–229.

Keane, R. E., P. Morgan, and S. W. Running. 1996. FIRE-BGC—A mechanistic ecological process model for simulating fire succession on coniferous forest landscapes of the northern Rocky Mountains. USDA Forest Service, Intermountain Research Station, Research Paper INT-RP-484. Ogden, Utah.

Kendall, K. C., and S. F. Arno. 1990. Whitebark pine—An important but endangered wildlife resource. Pages 264–273 *in* W. C. Schmidt and K. M. McDonald, compilers. Proceedings—Symposium on whitebark pine ecosystems: Ecology and management of a high-mountain resource. USDA Forest Service, Intermountain Research Station, General Technical Report INT-270, Ogden, Utah.

Klikoff, L. G. 1965. Microenvironmental influence on vegetational pattern near timberline in the central Sierra Nevada. Ecological Monographs 35:187–211.

Lillybridge, T. R., B. L. Kovalchik, C. K. Williams, and B. G. Smith. 1995. Field guide for forested plant associations of the Wenatchee National Forest. USDA Forest Service, Pacific Northwest Research Station, General Technical Report PNW-GTR-359, Portland, Oregon.

Loope, L. L. 1969. Subalpine and alpine vegetation of northeastern Nevada. Ph.D. Dissertation, Duke University, Durham, North Carolina.

Lueck, D. 1980. Ecology of *Pinus albicaulis* on Bachelor Butte. Master's Thesis, Oregon State University, Corvallis.

McAvoy, B. 1931. Ecological survey of the Bella Coola region. Botanical Gazette 92:141–171.

Means, J. E., J. H. Cissel, and F. J. Swanson. 1996. Fire history and landscape restoration in Douglas-fir ecosystems of western Oregon. Pages 61–67 *in* C. C. Hardy and S. F. Arno. The use of fire in forest restoration. Proceedings, Society for Ecological Restoration Annual Meeting, Seattle, Washington, USA, Sept. 14–16, 1995. USDA Forest Service, Intermountain Research Station, General Technical Report INT-341, Ogden, Utah.

Minore, D. 1979. Comparative autecological characteristics of northwestern tree species—A literature review. USDA Forest Service, Pacific Northwest Forest and Range Experiment Station, General Technical Report PNW-87, Portland, Oregon.

Morgan, P., and S. C. Bunting. 1990. Fire effects in whitebark pine forests. Pages 97–105 *in* W. C. Schmidt and K. J. McDonald, compilers. Proceedings—Symposium on whitebark pine ecosystems: Ecology and management of a high-mountain resource. USDA Forest Service, Intermountain Research Station, General Technical Report INT-270, Ogden, Utah.

Morrison, P. H., and F. J. Swanson. 1990. Fire history and pattern in a Cascade Range landscape. USDA Forest Service, Pacific Northwest Forest and Range Experiment Station, General Technical Report PNW-GTR-254, Portland, Oregon.

Ogilvie, R. T. 1990. Distribution and ecology of whitebark pine in western Canada. Pages 54–60 *in* W. C. Schmidt and K. M. McDonald, compilers. Proceedings—Symposium on whitebark pine ecosystems: Ecology and management of a high-mountain resource. USDA Forest Service, Intermountain Research Station, General Technical Report INT-270, Ogden, Utah.

Pfister, R. P., B. L. Kovalchik, and R. Ringleb. 1976. Scapegoat nonalpine vegetation. Unpublished report on file at School of Forestry, University of Montana, Missoula.

Pfister, R. P., B. L. Kovalchik, S. F. Arno, and R. Presby. 1977. Forest habitat types of

Montana. USDA Forest Service, Intermountain Forest and Range Experiment Station, General Technical Report INT-34, Ogden, Utah.

Pinchot, G. 1899. The relation of forests and forest fires. National Geographic 10:393–403.

Riegel, G. M., D. A. Thornburgh, and J. O. Sawyer. 1990. Forest habitat types of the South Warner Mountains, Modoc County, California USA. Madroño 37:88–112.

Romme, W. H. 1982. Fire and landscape diversity in subalpine forests of Yellowstone National Park. Ecological Monographs 52:199–221.

Sawyer, J. O., and Thornburgh, D. A. 1977. Montane and subalpine vegetation of the Klamath Mountains. Pages 699–732 *in* M. Barbour and J. Major, editors. Terrestrial vegetation of California. John Wiley and Sons, New York.

Selby, C. J., and M. D. Pitt. 1984. Classification and distribution of alpine and subalpine vegetation in the Chilcotin Mountains of southern British Columbia. Syesis 17:13–41.

Steele, R., R. Pfister, R. Ryker, and J. Kittams. 1981. Forest habitat types of central Idaho. USDA Forest Service, Intermountain Forest and Range Experiment Station, General Technical Report INT-114, Ogden, Utah.

Steele, R., S. V. Cooper, D. Ondov, D. Roberts, and R. Pfister. 1983. Forest habitat types of eastern Idaho–western Wyoming. USDA Forest Service, Intermountain Forest and Range Experiment Station, General Technical Report INT-144, Ogden, Utah.

Taylor, D. W. 1976. Vegetation near timberline at Carson Pass, central Sierra Nevada, California. Ph.D. Dissertation, University of California, Davis.

Topik, C. D, D. Smith, N. Diaz, and T. High. 1995. Plant association and management guide for the mountain hemlock zone. Mt. Hood and Gifford Pinchot National Forests. USDA Forest Service, Pacific Northwest Region, MTH-GP-TP-08-95, Portland, Oregon.

Vale, T. R. 1977. Forest changes in the Warner Mountains, California. Annals of the Association of American Geographers 67:28–45.

Weaver, T., and D. Dale. 1974. *Pinus albicaulis* in central Montana: Environment, vegetation, and production. American Midland Naturalist 92:222–230.

Williams, C. K., and T. Lillybridge. 1983. Forested plant associations of the Okanogan National Forest. USDA Forest Service, Pacific Northwest Region, R6-Ecol-132b-1983, Portland, Oregon.

Williams, C. K., B. F. Kelly, B. G. Smith, and T. R. Lillybridge. 1995. Forested plant associations of the Colville National Forest. USDA Forest Service, Pacific Northwest Research Station, General Technical Report PNW-GTR-360, Portland, Oregon.

Chapter 5

Clark's Nutcracker: Agent of Regeneration

Diana F. Tomback

Although whitebark pine (*Pinus albicaulis,* Family Pinaceae) was first described in 1863, only within the last quarter century have we learned of its dependence for seed dispersal on a bird, the Clark's nutcracker (*Nucifraga columbiana,* Family Corvidae). Naturalists in Europe and Asia had been aware, perhaps for centuries, of the interdependence between the spotted (Eurasian) nutcracker (*N. caryocatactes*) and the Swiss (*P. cembra*) and Siberian (*P. sibirica*) stone pines, close relatives of whitebark pine (Turcek and Kelso 1968; Lanner 1982; Chapter 2). For example, Formosof (1933), referring to the Siberian stone pine as "the cedar" wrote, "There is no doubt that the evolution of the nutcracker, belonging to the species *Nucifraga caryocatactes,* went on in the past side by side with that of the cedar, depending directly upon it, and in its turn controlling the actual dissemination of this conifer."

The peoples of Europe and Asia have coexisted far longer with their natural world than have postsettlement Americans. Understanding how whitebark pine seeds are disseminated was clearly a challenge to the infrequent visitor to the western North American subalpine zone. Native Americans may have understood the nutcracker–whitebark pine relationship, but much of their knowledge was not recorded. The respected forester George B. Sudworth, in his classic field guide *Forest Trees of the Pacific Slope,* published in 1908, wrote that whitebark pine cones "dry out and open slowly in high, cold situations where this pine grows." More recently, Krugman and Jenkinson (1974), in their widely cited chapter on the genus *Pinus,* described seed dispersal in whitebark pine in a footnote as "Seeds are dispersed when the detached cone disintegrates." However, the growing evidence in the late 1970s and the 1980s, both ecological and genetic, confirmed that whitebark pine does not "shed" its seeds, and that Clark's nutcrackers are the primary seed dispersers of whitebark pine. Nutcrackers remove ripe seeds from the nearly closed cones of whitebark pine and bury thousands of seeds per bird throughout the forest landscape. Patterns of seed dispersal by nutcrackers account for the elevational and geographical occurrence of whitebark pine, the pioneering status of whitebark pine following disturbance, and the population genetic structure of whitebark pine on local and regional scales.

This chapter describes the adaptations of both the nutcracker and the whitebark pine for their mutualistic relationship, the seed-caching behavior of the nutcracker, and the influence of nutcracker seed dispersal on the ecology and population biology of whitebark pine. It also discusses the historical origin of the association between Clark's nutcracker and whitebark pine. More detailed accounts of the nutcracker-pine interaction are available in Lanner (1980, 1996), Tomback (1983), and Tomback and Linhart (1990). Tomback (1998) is recommended for an overview of the life history of Clark's nutcracker.

Adaptations for Interaction

The Clark's nutcracker and whitebark pine are considered to be coevolved mutualists (e.g., Tomback 1982). That is, in theory they have a mutually beneficial interaction that resulted from reciprocal selection pressures (e.g., Abrahamson 1989). Whitebark pine is actually an obligate or nearly obligate mutualist of Clark's nutcracker (Tomback and Linhart 1990), which is considered unusual in plant-seed disperser systems (Wheelwright and Orians 1982). In other words, whitebark pine depends on nutcrackers almost exclusively for seed dispersal. This interaction works only because nutcrackers are less dependent on whitebark pine than the pine is on them. Nutcracker populations are buffered somewhat against whitebark pine cone crop failures by their behavioral flexibility in seed use (Tomback and Linhart 1990). Nutcrackers also harvest and store the seeds of other pines that are considered mutualists, including limber pine (*Pinus flexilis*) (e.g., Lanner and Vander Wall 1980; Vander Wall 1988), southwestern white pine (*P. strobiformis*) (Benkman et al. 1984; S. Samano and D. F. Tomback, unpublished data), singleleaf piñon (*P. monophylla*) (Tomback 1978; Vander Wall 1988), and Colorado piñon (*P. edulis*) (Vander Wall and Balda 1977). In addition, nutcrackers harvest and store the seeds of several wind-dispersed pines (see Chapter 12), especially ponderosa pine (*P. ponderosa,* Giuntoli and Mewaldt 1978; Torick 1995), and they feed on other foods such as insects, vegetable material, carrion, and small vertebrates (Giuntoli and Mewaldt 1978; Tomback 1978).

Clark's Nutcracker

Clark's nutcracker possesses several morphological and behavioral adaptations for a life history that centers on the year-round use of fresh and stored pine seeds. First, its bill is long, pointed, and sturdy (Figure 5-1). This enables the nutcracker to open both unripe and ripe whitebark pine cones by severing and

Figure 5-1. Clark's nutcracker (*Nucifraga columbiana*) harvesting seeds from whitebark pine (*Pinus albicaulis*) cones. Photo credit: Diana F. Tomback.

ripping off cone scales (Tomback 1978). Unripe cones require considerable effort to open: A bird will initially use powerful jabs of the bill that have the force of the entire body behind them, to separate the scales on intact cones (Tomback 1978).

In the case of limber pine and Colorado piñon pine, and rarely for whitebark pine, nutcrackers sometimes detach the cone first and carry it to an "anvil"—tree crotch, branch, fallen tree, or other support for opening (Vander Wall and Balda 1977; Tomback and Taylor 1987). For pines with cone scales that open when ripe, and even for the slightly open ripe whitebark pine cones, the long bill also enables nutcrackers to reach down behind the scales and remove the seeds (Vander Wall and Balda 1977; Tomback 1978).

Pine squirrels (*Tamiasciurus* spp.) compete with nutcrackers for whitebark pine seeds by cutting down cones for storage as a winter food supply (Tomback 1978; Hutchins and Lanner 1982). In the Rocky Mountains, nutcrackers steal freshly cut whitebark pine cones from red squirrel (*T. hudsonicus*) middens. They fly to a midden, land, quickly select a cone, and fly off with the cone in their bill, despite loud scolding and threats by the squirrel. Nutcrackers then fly to a branch, fallen tree, or rock, where they remove the seeds (Torick 1995; Rockwell 1995; D. F. Tomback, unpublished observations; S. F. Arno, unpublished observations).

Unique to the genus *Nucifraga* is the sublingual pouch or throat pouch, which is formed by a saclike extension of the floor of the mouth, with the opening to the pouch under the tongue (Bock et al. 1973). The pouch is used by nutcrackers to transport seeds from source trees to caching sites and from caches to nestlings or juveniles. When the pouch is filled with seeds, the throat appears to bulge greatly (Vander Wall and Balda 1977; Tomback 1978). Based on the volume capacity of the sublingual pouch, Hutchins and Lanner (1982) estimated that a full pouch consisted of a mean of 93 whitebark pine seeds. Tomback (1978) observed nutcrackers (n = 13) fill their pouches with 35 to 150 whitebark pine seeds, with a mean and median of 77 and 58 seeds, respectively. In a second study, nutcrackers (n = 12) carried 28 to 97 seeds per pouchload, with a mean of 54 seeds (D. F. Tomback, unpublished data). Pouch capacity may vary among individuals, and birds may not typically carry maximum loads.

Clark's nutcrackers possess a remarkable spatial memory, which has been studied in both the field (Tomback 1980) and the laboratory (e.g., Vander Wall 1982; Kamil and Balda 1985). Each nutcracker remembers the precise locations of the thousands of seed caches that it stores in late summer and fall, using environmental features such as rocks, trees, and logs as landmarks or visual cues. In an elegant series of laboratory experiments that controlled for cache-site preferences, nutcrackers displayed high accuracy in retrieving caches and retained memory of cache locations for as long as 285 days, although the accuracy decreased after about 180 days (Kamil and Balda 1985; Balda and Kamil 1992). In the field, nutcrackers have been observed retrieving whitebark pine seed caches from the previous fall more than 9 months after making them (Tomback 1978; Vander Wall and Hutchins 1983). Nutcrackers also do well at spatial memory tasks unrelated to seed caching; in a series of maze experiments, they learned more rapidly than other seed-caching corvids (Balda and Kamil 1989; Kamil et al. 1994).

Nutcrackers nest earlier than any other songbird (Order Passeriformes), beginning courtship as early as December, with first egg dates as early as January but peaking in March (Tomback 1998). They feed their nestlings primarily with seeds recovered from caches made the previous late summer and fall and some insect material (Mewaldt 1956). In addition, both male and female nutcrackers develop an incubation patch, a swollen, highly vascularized bare area in the abdominal region, that enables either parent to incubate eggs and brood young while the other retrieves seeds from its caches (Mewaldt 1952). In most other songbirds only females develop incubation patches. Early nesting probably provides the young birds a long developmental and learning period, which they spend with their parents so that they are prepared to cache seeds independently at the end of their first summer (Vander Wall and Balda 1977; Tomback 1978).

Whitebark Pine

Whitebark pine is classified in the pine subgenus *Strobus,* which represents the white or soft pines (haploxylon pines), and in subsection *Cembrae* (Critchfield and Little 1966; Price et al. 1998; Chapter 2 and 8). *Cembrae* pines are specialized for seed dispersal by nutcrackers: The spotted nutcracker of Eurasia disseminates the seeds of the Swiss stone pine, the Siberian stone pine, the Korean stone pine (*Pinus koraiensis*), and the Japanese stone pine (*P. pumila*). Whitebark pine is the only North American member of this group. All pines of subsection *Cembrae* have both large, wingless seeds and indehiscent cones. Winged seeds are the typical and probably ancestral condition among conifers (Lanner 1980; Critchfield 1986), enabling wind to disperse seeds from open cones, and prolonging descent of seeds so they maximize horizontal travel distance. In *Cembrae* pines the cone scales do not open during ripening (indehiscence), or may pull only slightly away from the core of the cone, so that the seeds are fixed in place and cannot be dislodged by wind. Instead, the wingless seeds must be freed and dispersed from the parent tree by a foraging animal, usually the nutcrackers. Nutcrackers may have played a role in the evolution of wingless seeds, favoring trees that had seeds with reduced seed wings (e.g., Tomback 1983; Tomback and Linhart 1990). When they forage on the seeds of wind-dispersed pines, nutcrackers take the trouble to shake off or sever the seed wings before placing the seeds in their sublingual pouch (Tomback 1978; Torick 1995).

Whitebark pine seeds, which weigh about 175 milligrams each, are large for pines and contain by weight about 52 percent fat, 21 percent carbohydrates, and 21 percent protein (Lanner and Gilbert 1994). They provide nutcrackers a nutritious, energy-rich food source (Tomback and Linhart 1990; Lanner and Gilbert 1994; Chapter 1). The seeds of *Cembrae* pines weigh from 92 milligrams (Japanese stone pine) to 553 milligrams per seed (Korean stone pine); and the seeds of *Cembrae* pines in particular and white pines in general are much larger than the seeds of pines in the subgenus *Pinus,* which tend to be small and winged (Table 3 in Tomback and Linhart 1990).

The *Cembrae* pines (except for the matlike Japanese stone pine), along with limber pine, and the piñon pines assume a growth form that may reflect dependence on corvids for seed dispersal. Tree branching is profuse, producing a flat-topped, shrubby or "lyrate" canopy (Chapter 3). Branches are vertically oriented with horizontally oriented cones at the tips, an arrangement that is believed to make the cones more visible and easily accessible to the birds (Vander Wall and Balda 1977; Tomback 1978; Lanner 1980, 1982). Nutcrackers, in fact, use the same or an adjacent whitebark pine cone as a platform as they harvest seeds (e.g., Tomback 1978).

Nutcrackers begin to break into whitebark pine cones and take pieces of unripe seeds as early as mid-July. Whitebark pine seeds, and probably those of other *Cembrae* pines, develop mature seed coats as early as August 15 to August 25; at this point, nutcrackers begin to harvest and cache seeds in large quantities (Tomback 1978; Hutchins and Lanner 1982). These seed-dispersal dates are early for pines, which do not typically release their seeds until September or later (Krugman and Jenkinson 1974). This may explain why a high percentage of whitebark pine seeds have underdeveloped embryos and delay seed germination for two or more years (Chapter 6). Between the short growing season and the early seed dispersal by nutcrackers, embryos may not have sufficient time to develop.

The Caching Behavior of Clark's Nutcracker

The following descriptions are based specifically on Clark's nutcrackers caching whitebark pine seeds.

Distances between seed source and storage areas. After filling their sublingual pouch, nutcrackers will store seed caches throughout the nearby terrain at any distance from a few meters to several hundred meters from the seed source, fly to one or more seed storage areas at subalpine elevations used communally by the local population of nutcrackers, or fly even greater distances to lower-elevation communal storage areas. For example, Hutchins and Lanner (1982), working in Squaw Basin in the Absaroka Mountains east of Grand Teton National Park, Wyoming, observed nutcrackers cache seeds either within 100 meters of the source trees or transport seeds 3.5 kilometers to the Breccia Cliffs, where they cached seeds. In the Mammoth Mountain area of the eastern Sierra Nevada range, which is in the vicinity of Mammoth Lakes, California, nutcrackers had many communal storage locations (Tomback 1978): During the early morning and early evening, they stored many of their seeds in a communal storage area adjacent to subalpine forest on the west slope of Mammoth Mountain or in terrain adjacent to the source trees. During the day, they routinely traveled 2.5 kilometers down the west slope to a storage area about 200 meters elevation lower, or they headed 12.5 kilometers northeast downslope to a communal storage area at Casa Diablo, about 500 meters elevation lower. In the Tioga Pass area, Yosemite National Park, nutcrackers stored seeds among groves of whitebark pine trees on hillocks above meadows, or transported seeds 8 to 10 kilometers northeast to Lee Vining Canyon, descending about 800 meters in elevation.

Characteristics of communal seed storage areas and cache sites. In general, the communal seed storage areas used by a population of nutcrackers are steep, often windswept, south-facing slopes that accumulate little snow and experience rapid snowmelt. Typically, the substrate is rocky, gravelly, or volcanic. For example, the Breccia Cliffs in the Hutchins and Lanner (1982) study are steep and south-facing. The three communal storage areas around Mammoth Mountain ranged in slope exposure from southeast to southwest, and in steepness from 22 degrees to 30 degrees. In Lee Vining Canyon northeast of Tioga Pass, nutcrackers stored seeds on the steep, south-facing slope of the canyon. At subalpine elevations, the communal slopes often supported whitebark pine trees of different ages. However, the lower-elevation storage areas were well below the elevational limits of whitebark pine, and no whitebark pine seedlings were evident.

Whether caching seeds within whitebark pine communities or on communal storage slopes, nutcrackers selected a similar variety of microsites. In the Mammoth Mountain area, the cache sites most frequently used were around the base of trees or under tree canopies, which on more level terrain experience faster snowmelt in spring than adjacent terrain (Tomback 1978). The next most frequent category of cache placement was in open terrain, followed by cache sites next to rocks, among tree roots, next to fallen trees or large branches; in cracks, holes, and under bark on trees; and among plants (Figure 10 in Tomback 1978). The substrates where nutcrackers cached included mineral soil, gravel, pumice, and forest litter. In meadows at Tioga Pass, nutcrackers stored seeds on the rocky hillocks adjacent to trees. On the west slope of the Wind River Range near Pinedale, Wyoming, nutcrackers stored seeds on very steep northwest- and southeast-facing slopes among rocks and vegetation and in rock fissures; they also stored seeds at the edges of meadows among grasses, sedges, and rocks (D. F. Tomback and J. W. Knowles, unpublished data). On Cathedral Peak in Yosemite National Park, nutcrackers cached seeds at tree line among the krummholz (matlike) patches of whitebark pine (Tomback 1986). Hutchins and Lanner (1982) noted that nutcrackers cached their seeds on all slope exposures, and in meadows, on a streambank, and "even in a puddle of water."

Clark's nutcrackers also cache their seeds in open, disturbed terrain and, particularly, recent burns. Nutcrackers cached seeds on Cathedral Peak in the charred soil of a krummholz whitebark pine community that had burned five years previously, transporting seeds up from the forest below (Tomback 1986). In 1989, one year after the fires in the Greater Yellowstone Area, on Mt. Washburn nutcrackers harvested and cached whitebark pine seeds in burned terrain and transported seeds to other storage areas (Figure 5-2) (D. F. Tomback, unpublished data). W. W. McCaughey (personal communication) observed nutcrackers storing seeds in a recent subalpine forest clear-cut on Palmer Mountain, Gallatin National Forest, north of Yellowstone National Park; and nutcrackers cached seeds in small forest openings created by selec-

Figure 5-2. Stand-replacing burn from the Storm Creek fire, 1988 fires in the Greater Yellowstone Area. Late-successional whitebark pine (*Pinus albicaulis*) community, dominated by subalpine fir (*Abies lasiocarpa*), Henderson Mountain, Gallatin National Forest, northeast of Yellowstone National Park. The first post-fire whitebark pine regeneration appeared throughout this burn in summer 1991 (Tomback 1994). Seed caching by Clark's nutcrackers (*Nucifraga columbiana*) following fire returns successional whitebark pine communities to an early stage where whitebark pine is an important component. Photo credit: Diana F. Tomback.

tive cutting for a whitebark pine restoration project at Smith Creek in the Bitterroot Range (R. E. Keane, personal communication; Chapter 18).

Seed caching. Nutcrackers typically store from 1 to 15 seeds per cache, with averages of 3 to 5 seeds per cache. The seeds are buried from 1 to 3 centimeters deep in substrate (Hutchins and Lanner 1982; Tomback 1982). Data are reasonably consistent from area to area: In Squaw Basin, nutcrackers cached from 1 to 14 seeds per cache, with a mean of 3.2 seeds; single-seed caches represented 35 percent of all observations, followed in frequency by 2-seed caches and 3-seed caches, which represented 18.5 and 18 percent of all observations, respectively (n = 157 caches, Hutchins and Lanner 1982). On Mammoth Mountain, nutcrackers stored 1 to 10 seeds per cache, with a mean and median of 5 seeds per cache. The most frequent cache sizes were 2- and 3-seed caches combined (33.3 percent) and 6- and 7-seed caches combined (37.5 percent), but the sample size was small (n = 24, Tomback 1982). By watching nutcrackers retrieve caches at Tioga Pass, Tomback (1982) determined cache sizes to be from 1 to 15 seeds with a mean and median of 2.6 and 2 seeds per cache, respectively (n = 30). Nutcrackers caching seeds in burned terrain on Mt. Washburn, Yellowstone National Park, stored 1 to 15 seeds per cache, with a mean of 5.2 seeds (n = 23 caches, D. F. Tomback, unpublished data). Nutcrackers also occasionally store caches of more than 30 seeds (Torick 1995). It is possible that nutcrackers vary cache size with terrain type, snow cover, and the likelihood that rodents may dig up seeds.

Caching nutcrackers tend to be fairly quiet and purposeful, often working alone and more rarely in pairs, although many birds may be coming, going, and caching in a communal storage area simultaneously (Tomback 1978; Hutchins and Lanner 1982). The following description comes from Tomback (1978): After traveling to a seed storage area, a nutcracker lands and hops along the ground, turning its head from side to side, apparently examining the terrain until it selects a cache site. With mineral soil, the nutcracker uses side-swiping bill motions to dig a shallow trench. It removes a seed from the sublingual pouch by tossing its head backward slightly. Held in the bill tip, one at a time each seed is placed in the trench in a tight cluster. Then, again with side-swiping bill motions, the trench is covered with soil so as to appear undisturbed, and the bird surveys the cache site. In pumice or more gravelly substrate, a nutcracker selects a cache site and pushes each seed, one at a time, into the same place. After making a cache, a nutcracker usually makes one to several more caches nearby, spacing them from 10 to 300 centimeters apart. Often, the nutcracker flies to another location in the same area and makes another series of caches, and so on until the sublingual pouch is empty. Sometimes nutcrackers store seeds temporarily in caches, and return at a later time to move them elsewhere.

Number of seeds stored per nutcracker. The total number of whitebark pine seeds cached by an individual nutcracker during the late summer and fall has been estimated by both Hutchins and Lanner (1982) and Tomback (1982), making somewhat different assumptions. In Squaw Basin, the nutcracker population cached whitebark pine seeds from August 15 to November 2, a total of

80 days; Hutchins and Lanner (1982) also used the mean full pouch capacity of 93 whitebark pine seeds that had been calculated from pouch volume and assumed that nutcrackers cached intensively for 9 hours a day, pausing 7.5 to 15 minutes each seed storage trip for maintenance behavior and social interactions. Altogether, they estimated that a single nutcracker could cache as many as 98,000 whitebark pine seeds in a season. At a mean of 3.2 seeds per cache, this represents more than 30,600 caches made per bird.

On Mammoth Mountain, the majority of nutcrackers stored whitebark pine seeds for about 38 days in 1974 and for about 47 days in 1975 (Tomback 1982). Because daytime temperatures were usually high, nutcrackers stored seeds only during the cooler hours of the day, with a conservative estimate of 4.5 hours total. In addition, the average pouchload observed, 77 seeds, was used for estimates. It was assumed that each nutcracker made at least one trip per day to store seeds at lower elevations. With these assumptions, each nutcracker stored a total of 35,000 whitebark pine seeds per year in about 9,500 caches at a mean of 3.7 seeds per cache. About 32,000 seeds were stored at subalpine elevations. After storing whitebark pine seeds, however, nutcrackers in both years descended to the Jeffrey pine (*Pinus jeffreyi*) forest where they resumed caching seeds until early December, another 6 weeks.

Tomback (1982) calculated how many whitebark pine seeds a nutcracker needed to sustain itself from April to July, roughly the time that whitebark pine seed caches are retrieved in the Mammoth Mountain area. Based on calculations of nutcracker energetic requirements, the maximum number of seeds required for this period is about 10,000 in roughly 3,000 seed caches—about a third of the seeds that nutcrackers were estimated to store. However, nutcrackers also rely on their caches to feed nestlings and dependent juveniles. Making assumptions about the percentage of breeding birds and frequency of different numbers of young, Tomback (1982) estimated that a local population of nutcrackers might use 55 percent of the seeds cached to feed themselves and their young. Of the remaining caches, many might be lost to rodents or in sites unfavorable to seed germination. However, if a local nutcracker population consisted of 25 birds, and each bird stored 32,000 whitebark pine seeds at subalpine elevations, an excess of 45 percent of the cached seeds represents about 350,000 seeds in 95,000 unused caches. If only 1 percent of these caches eventually produced trees in this area, whitebark pine recruitment after each cone crop would be considerable indeed.

Retrieval of caches. Nutcrackers tend to overwinter and breed in forests below the subalpine zone, where the climate is less harsh and there is easier access to seed caches, although individuals do wander up and down slope (Mewaldt 1956; Tomback 1978). Nutcrackers begin to arrive back at subalpine elevations by the beginning of June, often in family groups consisting of an adult pair and two or three young (Tomback 1978; Vander Wall and Hutchins 1983). In 1975, late storms and unusually deep snowpack delayed the return of nutcrackers to the subalpine zone until early July. In early summer, nutcrackers retrieve seeds from ground newly exposed by melting snow, areas of minimal snow cover such as under tree canopies, or steep, exposed locations such as communal storage slopes (Tomback 1978). Vander Wall and Hutchins

(1983) occasionally observed nutcrackers breaking through snow and ice at the edge of a snowbank to recover seeds.

Nutcrackers either hop along the ground looking for a cache site, or perch in the branches of a tree and then fly down to the site (Tomback 1978; Vander Wall and Hutchins 1983). The nutcracker then thrusts its bill down through the snow, forest litter, or substrate, probing for the seeds. Sometimes the cache position is miscalculated, and the bird changes its position and probes again. If seeds are present, the nutcracker digs with sideswipes of the bill, exposing the cache. Nutcrackers retrieving caches are often accompanied by begging juveniles.

Unretrieved caches begin to germinate as the snow melts, and new seedlings appear throughout the summer (see Chapter 6). Vander Wall and Hutchins (1983) describe how nutcrackers search for germinating seeds, eat the partially depleted seed on the stem, and probe the site for ungerminated seeds. Those germinating seed caches that escape this fate produce single seedlings and seedling clusters that may lead to whitebark pine regeneration.

How Nutcrackers Shape Forest Landscape

There are far more consequences to the ecology and biology of whitebark pine from seed dispersal by nutcrackers than the simple "planting" of trees. These consequences range from tree growth form and population genetic structure at fine and coarse scales, and from the elevational and site-specific distribution of whitebark pine to its successional status in the northwestern United States and southwestern Canada.

Tree Growth Form and Population Genetic Structure

The fine-scale population genetic structure of whitebark pine is affected in several ways by Clark's nutcracker (Chapter 8; Tomback et al. 1990; Tombark and Linhart 1990). First of all, many whitebark pine trees grow as clumps of trunks, often with their bases fused. For example, the tree clump growth form occurred for 90 percent of the tree sites surveyed on Mammoth Mountain, and for 58 percent of the tree sites surveyed on Palmer Mountain (Table 4 in Tomback and Linhart 1990). Genetic analysis reveals that many of these "clumps" are composed of more than one individual (i.e., different genotypes) which is the direct consequence of nutcrackers placing more than one seed in a cache (Linhart and Tomback 1985; Furnier et al. 1987; Rogers et al. 1999). This growth form has been specifically referred to as a *tree cluster* (Tomback et al. 1990). Tree clusters are found among wind-dispersed pines but at a much lower frequency, and may provide evidence for regeneration from seeds cached by animals (Torick et al. 1996).

To complicate matters, it cannot be assumed that a tree clump has arisen from a multiseed cache; genetic analysis is necessary to confirm this. A single genotype can produce a tree clump as well, which appears to be the result of "forking" in the seedling stage or later—branching as a result of weak apical dominance (Chapter 3 and 6). This growth form has been designated a *multi-trunk tree* (Tomback et al. 1990). Furnier et al. (1987) found that about 34 percent of tree clumps were actually of single genotypes, or multitrunk trees. In addition, whitebark pine trees are often found growing as a solitary stem. The tree clumps tend to be more prevalent on severe sites and the single-stem trees on better sites, where whitebark pine and its competitors grow into tall trees (Chapter 3).

The widespread occurrence of the tree cluster growth form has implications for the population biology of whitebark pine. First of all, it represents an extremely clumped population dispersion pattern. Second, further genetic work has shown that the individuals within tree clusters tend to be close relatives—half siblings to full siblings to selfed (the product of self-pollination in a parent tree) individuals (Rogers et al. 1999). This is the consequence of nutcrackers caching together seeds harvested from the same cone or tree (Tomback 1988; Tomback and Linhart 1990; Tomback and Schuster 1994). Third, genetic work shows that individuals within tree clusters are more genetically related to one another than to individuals in neighboring tree clusters (Furnier et al. 1987; Rogers et al. 1999). This results from different nutcrackers caching seeds from different parent trees in random fashion throughout the forest terrain (Tomback and Linhart 1990). As a result of nutcracker caching, there are related individuals within tree clusters but not within a local area, whereas in many wind-dispersed pines there is a family structure to the local population—local aggregations of related, single-trunk individuals (e.g., Linhart et al. 1981). The fine-scale population genetic structure of whitebark pine raises important management questions: Should we attempt to duplicate it if whitebark pine seedlings are planted widely for restoration treatments? What natural selection pressures are relaxed if we do not plant whitebark pine in this fashion?

Nutcracker caching also influences the population genetic structure of whitebark pine on a larger scale, which is discussed by Bruederle et al. (Chapter 8,). In essence, whitebark pine populations are not well differentiated, and most of the genetic variation is the result of individual differences within populations (Jorgensen and Hamrick 1997; Bruederle et al. 1998). One possible explanation for this is that long-distance seed dispersal by nutcrackers genetically homogenizes populations. This population structure also has management implications for seed zone transfer guidelines and restoration efforts for stands impacted by white pine blister rust (*Cronartium ribicola*): It suggests that seeds taken from a defined geographic region may be planted anywhere within that region without altering population structure (Chapters 15, 16, and 17).

Whitebark Pine Distribution and Successional Status

The cache-site preferences of Clark's nutcrackers along with whitebark pine's site requirements determine where whitebark pine grows, and the following discussion is extrapolated from this information. Nutcrackers are particularly attracted to steep slopes, open-canopied forests, open terrain, and recent burns (e.g., Tomback 1978; Baud 1993; Hutto 1995). Nutcrackers cache whitebark pine seeds at and above tree line; and, similarly, they cache seeds below the current elevational distribution of whitebark pine. Consequently, whitebark pine distribution can change in elevation in response to climate warming or cooling.

In seral whitebark pine communities, nutcracker seed dispersal following disturbance, particularly after fire, leads to an early-successional or " pioneering" status for the species (Lanner 1980; Tomback 1982; Tomback and Linhart 1990). Whitebark pine seedlings are unusually robust and tolerant of burned, water-limited soils, and exposed microsites (Arno and Hoff 1990; Chapters 1 and 6). This is particularly important in the northwestern United States and southwestern Canada, where whitebark pine communities are successional throughout part of the upper subalpine zone and dependent on fire for renewal (Arno 1986; Chapter 4). Studies in the Sleeping Child and Saddle Mountain burns in western Montana indicated that whitebark pine becomes established within a few years of fire, and on both north-facing and south-facing exposures (Tomback et al. 1990; Tomback et al. 1993). Similarly, in the Greater Yellowstone Area following the 1988 fires, whitebark pine regeneration appeared widely in 1991 in both moist and dry communities that had been burned by stand-replacing fires (Figure 5-2) (Tomback 1994; D. F. Tomback, unpublished data).

In successional communities, seed dispersal by nutcrackers provides whitebark pine two major advantages in large-scale burns over its wind-dispersed, shade-tolerant competitors subalpine fir (*Abies lasiocarpa*) and Engelmann spruce (*Picea engelmannii*):

1. Nutcrackers disperse whitebark pine seeds much farther than wind-dispersed seeds usually travel. In the Sleeping Child burn (11,350 hectares), Sapphire Range, western Montana, whitebark pine regeneration occurred on plots as far as 8 kilometers from the only whitebark pine seed source.
2. Nutcrackers can disperse whitebark pine seeds against prevailing winds (Tomback et al. 1990). In the Sleeping Child burn, the seed source for whitebark pine, subalpine fir, and Engelmann spruce was on the east side of the burn, opposite from the prevailing winds. Whitebark pine regeneration density in the burn, a quarter century after fire, was an order of magnitude greater than the density of either fir or spruce. In contrast, in the smaller Saddle Mountain burn (1,240 hectares), Bitterroot Mountains, western Montana, the seed source for all three conifers was on the west side of the burn, in line with prevailing winds. The regeneration densities of whitebark pine and subalpine fir were comparable a quarter century after fire (Tomback et al. 1990). Thus, under certain conditions, seed dispersal by nutcrackers gives whitebark pine a particular advantage over its competitors early in succession.

The Origin of the Nutcracker-Pine Interaction

Lanner (1980, 1990) presented evidence that the nutcracker-pine interaction originated in Eurasia, and that the subsection *Cembrae* coevolved with *Nucifraga* in the Old World (see also Chapter 2, this volume). Whereas in North America there is only one *Cembrae* pine, in Eurasia there are four; and similarly, whereas in North America there is only one nutcracker form (Clark's nutcracker is monotypic—has no described subspecies), in Eurasia there are at least ten described subspecies, which correspond to different pine regions. Lanner (1980) suggests that this diversity of stone pine and nutcracker forms in Eurasia corresponds to a long presence on the continent. Furthermore, Lanner (1980) suggests that an ancestral nutcracker form dispersed seeds from Asia across a Beringian land bridge into the New World. Subsequent isolation of these new populations led to speciation.

Based on genetic distances from allozyme data, Krutovskii et al. (1994) estimated that the separation between the Eurasian stone pines and whitebark pine dates back 0.6 million to 1.3 million years. This, they stated, corresponds well with the existence of the Bering Strait land bridge, but the actual migration event may date to the Pliocene, when the land bridge first appeared. Tomback (1983) further suggested that after entering North America, the nutcracker rapidly expanded its range south and began dispersing the seeds of other western North American pines—for example, limber, southwestern white, piñons, which had originally coevolved with seed-storing jays. Thus, Clark's nutcrackers are coevolved with whitebark pine but coadapted with these other pines.

The competing hypotheses are that the stone pine group *Cembrae* is polyphylectic (has multiple origins) and that whitebark pine originated in North America from pine ancestors in subsection *Strobi* through convergent evolution (i.e., indehiscent cones and large wingless seeds arose more than once) (Critchfield 1986). Alternatively, it could be suggested that Clark's nutcracker and whitebark pine are ancestral to the Eurasian stone pines and nutcracker; and, Eurasia may have multiple forms of both nutcrackers and stone pines because of greater geographic area, greater ecological diversity, and more glacial refugia during the Pleistocene. Additional insight into the evolutionary origin of the nutcracker–stone pine interaction may come from gene-sequencing studies that are currently underway for the *Cembrae* and *Strobi* pines (R. A. Price, University of Georgia, personal communication; A. Liston, Oregon State University, personal communication).

Whatever the sequence of origin, Clark's nutcrackers have acted as an important ecological force throughout their range in shaping the western North American montane landscape.

Acknowledgments

I am grateful to Katherine C. Kendall for reviewing this chapter and providing helpful suggestions. I also thank Stephen F. Arno and Robert E. Keane for their excellent editorial comments. Many undergraduate and graduate students and other colleagues have contributed to studies of the nutcracker-pine relationship throughout the years, and I am most appreciative of their interest in and enthusiasm for this fascinating subject.

Literature Cited

Abrahamson, W. G. 1989. Plant-animal interactions: An overview. Pages 1–22 *in* W. G. Abrahamson, editor. Plant-animal Interactions. McGraw-Hill, New York.

Arno, S. F. 1986. Whitebark pine cone crops: A diminishing source of wildlife food? Western Journal of Applied Forestry 1:92–94.

Arno, S. F., and R. J. Hoff. 1990. *Pinus albicaulis* Engelm. Whitebark pine. Pages 268–279 *in* R. M. Burns and B. H. Honkala, technical coordinators. Silvics of North America. USDA Forest Service, Agriculture Handbook 654, Washington, D.C.

Balda, R. P., and A. C. Kamil. 1989. A comparative study of cache recovery by three corvid species. Animal Behavior 38:486–495.

———. 1992. Long-term spatial memory in Clark's nutcracker, *Nucifraga columbiana*. Animal Behavior 44:761–769.

Baud, K. S. 1993. Simulating Clark's nutcracker caching behavior: Germination and predation of seed caches. M.A. thesis. University of Colorado at Denver.

Benkman, C. W., R. P. Balda, and C. C. Smith. 1984. Adaptations for seed dispersal and the compromises due to seed predation in limber pine. Ecology 65:632–642.

Bock, W. J., R. P. Balda, and S. B. Vander Wall. 1973. Morphology of the sublingual pouch and tongue musculature in Clark's nutcracker. Auk 90:491–519.

Bruederle, L. F., D. F. Tomback, K. K. Kelly, and R. C. Hardwick. 1998. Population genetic structure in a bird-disperesed pine, *Pinus albicaulis* (Pinaceae). Canadian Journal of Botany 76:83–90.

Critchfield, W. B. 1986. Hybridization and classification of the white pines (*Pinus* section *Strobus*). Taxon 35:647–656.

Critchfield, W. B., and E. L. Little, Jr. 1966. Geographic distribution of the pines of the world. USDA Forest Service, Miscellaneous Publication 991. Washington, D.C.

Formosof, A. N. 1933. The crop of cedar nuts, invasions into Europe of the Siberian nutcracker (*Nucifaga caryocatactes macrorhynchos*) and fluctuations in numbers of the squirrel (*Sciurus vulgaris*). Journal of Animal Ecology 2:70–81.

Furnier, G. R., P. Knowles, M. A. Clyde, and B. P. Dancik. 1987. Effects of avian seed dispersal on the genetic structure of whitebark pine populations. Evolution 41:607–612.

Giuntoli, M., and L. R. Mewaldt. 1978. Stomach contents of Clark's nutcracker collected in western Montana. Auk 95:595–598.

Hutchins, H. E., and R. M. Lanner. 1982. The central role of Clark's nutcracker in the dispersal and establishment of whitebark pine. Oecologia 55:192–201.

Hutto, R. L. 1995. Composition of bird communities following stand-replacement fires in northern Rocky Mountain (U.S.A.) conifer forests. Conservation Biology 9:1041–1058.

Jorgensen, S. M., and J. L. Hamrick. 1997. Biogeography and population genetics of whitebark pine, *Pinus albicaulis*. Canadian Journal of Forest Research 27: 1574–1585.

Kamil, A. C., and R. P. Balda. 1985. Cache recovery and spatial memory in Clark's nutcracker (*Nucifraga columbiana*). Journal of Experimental Psychology: Animal Behavior Processes 11:95–111.

Kamil, A. C., R. P. Balda, and D. J. Olson. 1994. Performance of four seed-caching corvid species in the radial-arm maze analog. Journal of Comparative Psychology 108:385–393.

Krugman, S. L., and J. L. Jenkinson. 1974. *Pinus* L. Pine. Pages 598–638 *in* C. S. Schopmeyer, technical coordinator. Seeds of woody plants in the United States. USDA Forest Service, Agriculture Handbook No. 450, Washington, D.C.

Krutovskii, K. V., D. V. Politov, and Y. P. Altukov. 1994. Genetic differentiation and phylogeny of stone pine species based on isozyme loci. Pages 19–30 *in* W. C. Schmidt and F. -K. Holtmeier, compilers. Proceedings—International workshop on subalpine stone pines and their environment: The status of our knowledge. USDA Forest Service Intermountain Research Station, General Technical Report INT-GTR-309, Ogden, Utah.

Lanner, R. M. 1980. Avian seed dispersal as a factor in the ecology and evolution of limber and whitebark pines. Pages 15–48 *in* B. P. Dancik and K. O. Higginbotham, editors. Proceedings of Sixth North American Forest Biology Workshop. University of Alberta, Edmonton, Alberta, Canada.

———. 1982. Adaptations of whitebark pine for seed dispersal by Clark's nutcracker. Canadian Journal of Forest Research 12:391–402.

———. 1990. Biology, taxonomy, evolution, and geography of stone pines of the world. Pages 14–24 *in* W. C. Schmidt and K. J. McDonald, compilers. Proceedings—Sympo-

sium on whitebark pine ecosystems: Ecology and management of a high-mountain resource. USDA Forest Service Intermountain Research Station, General Technical Report INT-270, Ogden, Utah.
———. 1996. Made for each other: A symbiosis of birds and pines. Oxford University Press, New York.
Lanner, R. M., and B. K. Gilbert. 1994. Nutritive value of whitebark pine seeds, and the question of their variable dormancy. Pages 206–211 *in* W. C. Schmidt, and F. -K. Holtmeier, compilers. Proceedings—International workshop on subalpine stone pines and their environment: The status of our knowledge. USDA Forest Service Intermountain Research Station, General Technical Report INT-GTR-309, Ogden, Utah.
Lanner, R. M., and S. B. Vander Wall. 1980. Dispersal of limber pine seed by Clark's nutcracker. Journal of Forestry 78:637–639.
Linhart, Y. B., J. B. Mitton, K. B. Sturgeon, and M. L. Davis. 1981. Genetic variation in time and space in a population of ponderosa pine. Heredity 46:407–426.
Linhart, Y. B., and D. F. Tomback. 1985. Seed dispersal by Clark's nutcracker causes multi-trunk growth form in pines. Oecologia 67:107–110.
Mewaldt, L. R. 1952. The incubation patch of the Clark nutcracker. Condor 54:361.
———. 1956. Nesting behavior of the Clark nutcracker. Condor 58:3–23.
Price, R. A., A. Liston, and S. H. Strauss. 1998. Phylogeny and systematics of *Pinus*. Pages 49–68 *in* D. M. Richardson, editor. Ecology and biogeography of *Pinus*. Cambridge University Press, Cambridge, United Kingdom.
Rockwell, D. 1995. Glacier National Park. Houghton-Mifflin Company, New York.
Rogers, D. L., C. I. Millar, and R. D. Westfall. 1999. Fine-scale genetic structure of whitebark pine (*Pinus albicaulis*): Associations with watershed and growth form. Evolution 53:74–90.
Sudworth, G. B. 1908. Forest trees of the Pacific slope. USDA Forest Service Tree Bulletin, Washington, D.C.
Tomback, D. F. 1978. Foraging strategies of Clark's nutcracker. Living Bird 16:123–161.
———. 1980. How nutcrackers find their seed stores. Condor 82:10–19.
———. 1982. Dispersal of whitebark pine seeds by Clark's nutcracker: A mutualism hypothesis. Journal of Animal Ecology 51:451–467.
———. 1983. Nutcrackers and pines: Coevolution or coadaption? Pages 179–223 *in* M. H. Nitecki, editor. Coevolution. University of Chicago Press, Chicago.
———. 1986. Post-fire regeneration of krummolz whitebark pine: A consequence of nutcracker seed caching. Madroño 33:100–110.
———. 1988. Nutcracker-pine mutualisms: Multi-trunk trees and seed size. Pages 518–527 *in* H. Ouellet, editor. Acta XIX Congressus Internationalis Ornithologici, Vol. 1. University of Ottawa Press, Ottawa, Ontario, Canada.
———. 1994. Effects of seed dispersal by Clark's nutcracker on early postfire regeneration of whitebark pine. Pages 193–198 *in* W. C. Schmidt and F. -K. Holtmeier, compilers. Proceedings—International workshop on subalpine stone pines and their environment: The status of our knowledge. USDA Forest Service Intermountain Research Station, General Technical Report INT-GTR-309, Ogden, Utah.
———. 1998. Clark's nutcracker (*Nucifraga columbiana*), No. 331. *In* A. Poole and F. Gill, editors. The birds of North America. The Birds of North America, Inc., Philadelphia.
Tomback, D. F., and Y. B. Linhart. 1990. The evolution of bird-dispersed pines. Evolutionary Ecology 4:185–219.
Tomback, D. F., and W. S. F. Schuster. 1994. Genetic population structure and growth form distribution in bird-dispersed pines. Pages 43–50 *in* W. C. Schmidt and F. -K. Holtmeier, compilers. Proceedings—International workshop on subalpine stone pines and their environment: The status of our knowledge. USDA Forest Service Intermountain Research Station, General Technical Report INT-GTR-309, Ogden, Utah.

Tomback, D. F., and C. L. Taylor. 1987. Tourist impact on Clark's nutcracker foraging activities in Rocky Mountain National Park. Pages 158–172 *in* F. Singer, editor. Towards the year 2000: Conference on science in the National Parks. George Wright Society and U.S. National Park Service, Washington, D.C.

Tomback, D. F., L. A. Hoffmann, and S. K. Sund. 1990. Coevolution of whitebark pine and nutcrackers: Implications for forest regeneration. Pages 118–129 *in* W. C. Schmidt and K. J. McDonald, compilers. Proceedings—Symposium on whitebark pine ecosystems: Ecology and management of a high-mountain resource. USDA Forest Service Intermountain Research Station, General Technical Report INT-270, Ogden, Utah.

Tomback, D. F., S. K. Sund, and L. A. Hoffmann. 1993. Post-fire regeneration of *Pinus albicaulis:* Height-age relationships, age structure, and microsite characteristics. Canadian Journal of Forest Research 23:113–119.

Torick, L. L. 1995. The interaction between Clark's nutcracker and ponderosa pine, a "wind-dispersed" pine: Energy efficiency and multi-genet growth forms. M.A. thesis. University of Colorado at Denver.

Torick, L. L., D. F. Tomback, and R. Espinoza. 1996. Occurrence of multi-genet tree clusters in "wind-dispersed" pines. American Midland Naturalist 136:262–266.

Turcek, F. J., and L. Kelso. 1968. Ecological aspects of food transportation and storage in the Corvidae. Communications in Behavioral Biology, Part A, 1:277–297.

Vander Wall, S. B. 1982. An experimental analysis of cache recovery in Clark's nutcracker. Animal Behavior 30:84–94.

———. 1988. Foraging of Clark's nutcracker on rapidly changing pine seed resources. Condor 90:621–631.

Vander Wall, S. B., and R. P. Balda. 1977. Coadaptations of the Clark's nutcracker and piñon pine for efficient seed harvest and dispersal. Ecological Monographs 47:89–111.

Vander Wall, S. B., and H. E. Hutchins. 1983. Dependence of Clark's nutcracker, *Nucifraga columbiana,* on conifer seeds during postfledging period. Canadian Field-Naturalist 97:208–214.

Wheelwright, N. T., and G. H. Orians. 1982. Seed dispersal by animals: Contrasts with pollen dispersal, problems of terminology, and constraints on coevolution. American Naturalist 119:402–413.

Chapter 6

The Natural Regeneration Process

Ward W. McCaughey and Diana F. Tomback

Understanding the natural regeneration process for whitebark pine (*Pinus albicaulis,* subgenus *Strobus,* subsection *Cembrae*) is essential for managing whitebark pine communities. It provides insight into the successional dynamics of whitebark pine, how succession varies geographically, and particularly how fire may be used as a management tool to return whitebark pine communities to earlier successional stages (Chapters 9 and 18). Knowledge of the regeneration process is integral to all management strategies for whitebark pine, including approaches to managing for white pine blister rust (*Cronartium ribicola*), whether for enhancing natural resistance or for growing rust-resistant seedlings (Chapter 17).

The natural regeneration process for whitebark pine may take up to five years to complete. There are three distinct phases: (1) cone and seed initiation, development, and maturation; (2) seed dissemination; and (3) seed germination. Whitebark pine requires more than two years for the first phase, which is typical of pine species worldwide, including the European and Asian stone pines of the *Pinus* subsection *Cembrae* (Lanner 1990). Male and female cone production begins in whitebark pine trees between the ages of twenty and thirty years (Table 6-1) (Krugman and Jenkinson 1974).

Several developmental stages of the cones and seeds occur during the first phase of regeneration, starting with the initiation of staminate and ovulate cone buds and ending with mature cones and seeds. The second phase involves the dispersal and burial of seeds, primarily by Clark's nutcrackers (*Nucifraga columbiana*), which is briefly summarized here. The final phase of the natural regeneration process includes the physiological development leading up to seed germination, the emergence process, and establishment of young seedlings.

Table 6-1. Height, age of first cone production, interval between large cone crops, and ripe cone color of some seed-bearing trees from the genus *Pinus* (data from Krugman and Jenkinson 1974).

Species	Mature Tree Height (meters)	Age (years)	Interval[1] (years)	Ripe Cone Color
P. albicaulis[2]	33	20–30	3–5	Purple-brown
P. cembra[2]	23	25–30	6–10	Purple-brown
P. koraiensis[2]	46	15–40	3–5	Yellow-brown
P. pumila[2]	3	—	—	Red-brown
P. sibirica[2]	40	25–35	3–8	Violet-brown
P. lambertiana	69	40–80	3–5	Greenish-brown
P. contorta var. *contorta*	12	4–8	1	Yellow-brown
P. monticola	62	7–20	3–7	Yellow-dark
P. aristata	15	20	102	Chocolate

[1]Interval between large cone crops.
[2]Pines of the subsection *Cembrae,* the stone pines.

Cone and Seed Development

The natural regeneration process begins with the differentiation of bud cells after vegetative growth slows in mid- to late summer. Male and female cone development has not been studied in detail for whitebark pine. Mirov (1967) and Krugman and Jenkinson (1974) provide generalized descriptions of the initiation process for the genus *Pinus.* Seed and pollen cone initiation for whitebark pine, like most pines, probably occurs during or just prior to winter bud formation from mid-July through mid-September, with the timing varying with elevation, latitude, and climate (Schmidt and Lotan 1980). Cone buds develop earlier at lower elevations, which are warmer than high elevations. Year-to-year differences in temperature and precipitation affects cone development. Owens (1991) found that buds of trees in the pine subgenus *Strobus,* which includes whitebark pine, initiate pollen cones prior to dormancy and seed cones just after dormancy but prior to pollination. Cone buds at the time of differentiation are only distinguishable with the aid of a microscope and appear as tiny cone-shaped masses of off-white-colored cells (Lanner 1996). Pines are monoecious, with male and female cones, or strobili, on the same tree (Krugman and Jenkinson 1974).

Male Cone Development

Male cone buds overwinter and develop into pollen cones from April through early June, depending on elevation and climate. Male cones are distributed throughout the tree crown, but are borne predominantly on the lower side branches of the current year's growth (Eggers 1986; Arno and Hoff 1990). Mature pollen cones measure 1 centimeter by 1 centimeter (cm) (0.4 inch by 0.4 inch) and are bright red in color, distinguishing them from the yellow pollen cones of limber pine (*Pinus flexilis*), a tree often confused with whitebark pine (Arno and Hoff 1990). The clusters of male strobili are arranged in indistinct spirals along the branch tips and range in length from 1.25 cm to 5 cm (0.5–2 in) (Young and Young 1992). Pollen is shed and wind-dispersed from May through mid-August, varying in date with elevation, latitude, and climatic conditions (Schmidt and Lotan 1980). For example, male cones matured and released pollen during late June and early July 1992, at 2,500 to 2,600 meters (8,200 to 8,500 feet) elevation in southwestern Montana (W. W. McCaughey, unpublished data), but in May and June at 1,550 to 1,650 meters elevation (5,100 to 5,400 feet) in northern Idaho (R. J. Hoff, unpublished data). Although most pollen typically fertilizes nearby trees, it can be dispersed hundreds of miles by wind under certain climatic conditions such as windstorms (Mirov 1967). Long-distance pollen dispersal has been confirmed by several genetic studies (Schuster et al. 1989; Latta and Mitton 1997).

Female Cone Development

Female buds overwinter and begin development in April and May, depending on elevation and climate. The cones, or strobili, are attached directly to the branch (i.e., without a cone peduncle [stem], occur in clusters of two to five, develop at the base of the current year's leader growth, and are found in the upper crown branches (Mirov 1967; McCaughey and Schmidt 1990). When they are pollen receptive, the female cones are 1 centimeter wide by 2 centimeters long (0.4 inch by 0.8 inch). The cone scales open during the pollination period, exposing two white ovules (megasporangia, the future seeds) on the upper face of each scale (Lanner 1996). Pollen grains enter between the scales and lodge on micropilar exudates, droplets of a sugary liquid near the ovules; and the cone scales close after pollination (Mirov 1967; Lanner 1996). The female cones remain on the tree for further growth and development, whereas the male pollen cones shrivel and fall off after pollen release. Female cones grow to a first-year size of about 3 cm long by 2 cm wide (1.2 in by 0.8 in), and the cone scales turn dark brown (Figure 6-1).

Cone and Seed Maturation

Fertilization of egg cells occurs in the spring after pollination, followed by rapid cone enlargement in June and July (Krugman and Jenkinson 1974; Schmidt and Lotan 1980). The ovule goes through a series of mitotic divisions following fertilization, at which time the female cone grows rapidly and changes color from dark brown to deep purple (Table 6-1). In the second summer, cones and seeds mature between mid-August and mid-September (Young and Young 1992), which is signaled by the seed coats turning medium brown in color and the cone scales separating slightly from the core of the cone (Tomback 1978, 1981). Mature cones weigh about 56 grams (2 ounces), are oblong in shape, and measure 5 to 8 cm (2 to 3 in) in length and 2 to 5 cm (0.8 to 2 in) in width; they are covered with beads of oleoresin (pitch) secreted from resin ducts in the swollen apophyses (the exposed portions of the cone scales), which gives the cone a shiny appearance (Lanner 1982; Arno and Hoff 1990).

Figure 6-1. First-year female cones (strobili) of whitebark pine in mid-July following pollination. Dime inserted for size comparison. Photo credit: Ward W. McCaughey.

The pitch sticks to anything that comes in contact with the cones, including bears, birds, small mammals, and humans. After harvesting seeds, nutcrackers routinely bill-wipe to remove the pitch; their throat and breast feathers typically become stained pink from contact with anthocyanins (red pigment) from the broken ends of cone scales (Tomback 1978). In early September, much of the oleoresin hardens, decreasing the stickiness, and cone scales become more brittle (Tomback 1978).

Mature whitebark pine cones are unique among the North American pines, because they have cone scales that remain partly closed (indehiscent) after ripening (Figure 6-2) (Mirov 1967). Indehiscence differs from cone serotiny, which is an adaptation for fireprone ecosystems (Keeley and Zedler 1998). Serotinous cones from trees such as lodgepole pine (*Pinus contorta*) remain sealed, encased by resin, until the heat of fire melts the resin and breaks the seal, resulting in seed release (Lanner 1998). Conversely, the cone scales of most pines flex open widely on the tree shortly after ripening, releasing the seeds, which are dispersed by wind. Scale flexing is caused by differential contraction of two tissue systems: woody strands of short, thick-walled, tracheid-like cells extending from the cone axis to the tip of the scale, and thick-walled sclerenchyma cells underneath the scale (Young and Young 1992). Whitebark pine cones lack the sclerenchyma cells, allowing the overlapping cone scales to open slightly (only about 8 millimeters [0.03 inch]) and expose but not release the seeds (Tomback 1981; Lanner 1982). This partly closed ripening configu-

Figure 6-2. Mature cones of whitebark pine in mid-July during the bumper cone crop year 1999, near Yellowstone National Park. Photo credit: Phillip E. Farnes.

ration holds the seeds in place until they are harvested by nutcrackers. Whitebark pine cone scales have a terminal umbo (i.e., point at the tip of the scale) and a massive apophysis (triangular scale tip that bears the umbo), which tapers to a thin cross-section (fracture zone) beneath the seed-bearing cavities (Harlow et al. 1979). The scales of ripe cones are easily broken off along this thin fracture zone, which facilitates access to seeds by nutcrackers (Lanner 1996). Whitebark pine seeds lack the terminal, membranous wing that enables wind-dispersed seeds to be blown away from parent trees (Krugman and Jenkinson 1974).

Two large, wingless seeds of whitebark pine develop on each cone scale. The mature seed consists of an embryo embedded in the endosperm (female gametophyte), the food storage tissue, surrounded by a thick seed coat. The embryo comprises the cotyledons, plumule, radicle suspensor, and micropyle (Mirov 1967). A whitebark pine cone contains on average about 75 wingless seeds that are light brown to light rust brown in color and measure 7.6 millimeters (0.3 inch) long; there are 1,000 to 2,050 seeds per kilogram (2,200 to 4,500 seeds per pound), with each seed weighing approximately 14 milligrams (0.005 ounce) (Krugman and Jenkinson 1974; McCaughey 1994a).

Cone Production

Whitebark pine trees are first capable of producing seed cones at twenty to thirty years of age, although large cone crops do not typically occur until the trees are at least sixty to eighty years of age (Table 6-1) (Day 1967; Krugman and Jenkinson 1974). This age of first seed bearing is similar to that for most of the other stone pines, such as Swiss stone pine (*Pinus cembra*) (twenty-five to thirty years), Korean stone pine (*P. koraiensis*) (fifteen to forty years), and Siberian stone pine (*P. sibirica*) (twenty-five to thirty-five years). First seed-bearing ages for other North American pines range widely from four to eight years for lodgepole pine (*P. contorta*) to forty to eighty years for sugar pine (*P. lambertiana*) (Young and Young 1992).

The intervals between large cone crops for pines range from one year for lodgepole pine to about a century for bristlecone pine (*P. aristata*) (Table 6-1). The interval of three to five years for whitebark pine is similar to the intervals for other stone pines, except that Swiss stone pine has an interval of six to ten years (Table 6-1). Cone production frequency also varies geographically within species. For example, Tomback (1978) observed moderate to heavy whitebark pine cone crops in four consecutive years on the eastern slope of the Sierra Nevada, whereas in the Greater Yellowstone Area, good cone production occurred every three or four years (Morgan and Bunting 1992). Cone production is influenced by tree canopy size, which in turn is influenced by competition; trees produce larger crowns in more open environments (Kipfer 1992). Large-crowned trees have a large number of fertile shoots and tend to produce the largest cone crops on whitebark pine (Weaver and Forcella 1986; Spector 1999). The duration of cone production peaks near 250 years in whitebark pine but can extend to over 1,000 years in some parts of its range (Arno and Hoff 1990; Perkins and Swetnam 1996).

Managers need to predict whitebark pine cone production for a variety of activities such as timing of postfire regeneration or for determining potential conflicts between grizzly bears and humans. When cone crops are large, bears stay in high-elevation areas where human activity is minimal. In years of small or no cone crop production, bears move into lower elevations in search of food, increasing conflicts with humans and bear mortality (Knight et al. 1987; Chapter 7). Past cone production can be assessed by counting cone scars on branches that were produced over the previous six to twelve years (Weaver and Forcella 1986; Morgan and Bunting 1992).

Studies indicated that year-to-year variation in cone crops within a region appear to be both internally and externally controlled (Weaver and Forcella 1986). Good cone years are usually preceded by poor cone years, suggesting internal control, whereas significant relationships between growth and climatic conditions along with insects and diseases during reproductive bud formation, pollination, growth, and development indicated external control (Weaver and Forcella 1986; McCaughey 1994b). Correlations of cone crops with measured stand and topographic characteristics were minimal. The variables that were the best predictors of cone production were canopy cover, canopy cover with fallen cones, canopy cover with stand size, crown area and volume (Weaver and Forcella 1986; Spector 1999).

Seed Dispersal

Whitebark pine seeds are dispersed primarily by the Clark's nutcracker (Tomback 1978, 1982; Hutchins and Lanner 1982; Chapter 5), whereas whitebark pine's most common forest associates have seeds that are dispersed by wind. Nutcrackers have long, sturdy, pointed bills, enabling them to dig into the tough, fibrous tissues of unripe cones, break into ripe cones, and extract seeds from cone scales (Tomback 1978; Lanner 1982).

Details of the interaction between nutcrackers and whitebark pine are presented in Tomback (1998; Chapter 5). Nutcrackers can carry more than 100 whitebark pine seeds in the throat pouch (Tomback 1978), and have been observed to fly with whitebark pine seeds as far as 8 to 12 kilometers (5 to 7.5 miles) from the seed source to cache sites, but more typically a few meters to a few kilometers (Tomback 1982). Nutcrackers store seeds by scatter-hoarding (making many small caches throughout the forest terrain) in soil or other substrate, placing the seeds at depths of 1 to 3 centimeters (0.4 to 1.2 inches). They bury 1 to 15 or more seeds in each cache, with an average of 3 or 4. One nutcracker stores an estimated 35,000 to 98,000 whitebark pine seeds in good seed crop years (Hutchins and Lanner 1982; Tomback 1982). Clark's nutcrackers cache in a variety of substrates, from mineral soil to rocky rubble to recently burned terrain (Hutchins and Lanner 1982; Tomback 1978, 1982, 1998; Tomback et al. 1990). They frequently cache seeds near the base of trees, roots, logs, rocks, plants, or in cracks and fissures in trees and logs. They also cache outside of tree canopies on open slopes, and particularly south-facing slopes and rocky outcrops that remain reasonably snow-free (Tomback 1978).

Seeds cached by nutcrackers are used as winter, spring, and even summer food. Nutcrackers may return to their cache sites up to one year later to retrieve seeds for themselves and their young (Tomback 1978; Vander Wall and Hutchins 1983). Seeds not recovered by nutcrackers or eaten by other seed foragers are available to germinate (e.g., Tomback 1982). After snowmelt, from June through early September, seedlings arise from nutcracker cache sites as single individuals and in clusters (Tomback 1982; McCaughey 1990). Regeneration commonly occurs on burned litter seedbeds following natural and prescribed fires (Tomback 1986, 1994; Morgan and Bunting 1990).

Several other scatter-hoarding species, such as Steller's jay (*Cyanocitta stelleri*), deer mice (*Peromyscus maniculatus*), and chipmunks (*Tamias* spp.), probably cache whitebark pine seeds and may contribute to a small percentage of regeneration. However, a number of birds, including mountain chickadees (*Parus gambeli*), ravens (*Corvus corax*), and the pine grosbeak (*Pinicola enucleator*), and small mammals—particularly pine squirrels (*Tamiasciurus* spp.), which cut quantities of cones for storage in middens—consume whitebark pine seeds, thereby depleting the number available to the nutcracker for caching (Tomback 1978; Hutchins and Lanner 1982).

Germination and Early Seedling Growth

Seed germination and seedling growth are unusual in whitebark pine. For example, the seeds require compound stratification, which often results in delayed germination. Growing seedlings may spontaneously branch or "fork" at early stages, affecting tree growth form.

Stratification and Germination

Several factors appear to inhibit complete maturation and germination in whitebark pine seeds, including underdeveloped embryos, physiological dormancy, and mechanical barriers in the seed coat (Leadem 1986; Pitel and Wang 1980, 1990). For pines in general, climatic conditions during cone maturity control development of embryos and endosperm. Warm weather accelerates embryo development, whereas cold temperatures inhibit development (Leadem 1986). In the case of whitebark pine, other factors contribute to seed immaturity. Nutcrackers harvest and cache seeds as early as mid-August before embryos are fully developed (Tomback 1978; Hutchins and Lanner 1982; Tomback 1998). Whitebark pine seeds in cones increase in weight through August and September (Hutchins and Lanner 1982), indicating that seeds dispersed in August are not fully developed. An August seed dispersal date is early for the genus *Pinus:* According to the seed dispersal dates listed for sixty pines by Krugman and Jenkinson (1974), fifty-five disperse their seeds in September or later; five pines disperse their seeds from August to October.

Farmer (1997) places whitebark pine seeds with a group of pine species characterized by hard but permeable seed coats and primary dormancy. He defines primary dormancy as the need for seeds to spend time in a "conditioning environment" to develop the ability to germinate. Primary dormancy may result from an impermeable seed coat or from some "combination of seed coat and embryo characteristics which prevent germination." Cold stratification of seeds overwintering in the ground helps to overcome physical and physiological barriers to germination by decreasing inhibitors, increasing growth promoters, and by breaking down the seed coat, enabling the seeds to imbibe water and oxygen (Pitel and Wang 1980; Young and Young 1992). Whitebark pine seeds require a minimum of forty-five to sixty days of cold stratification to overcome the physiological barriers to growth. Often this length of time is insufficient to result in high germination rates (Leadem 1986; Jacobs and Weaver 1990). A multistep treatment and stratification protocol for whitebark pine seeds, that appears to reduce physiological and seed coat inhibition and results in high germination percentages has been developed by the USDA Forest Service, Coeur d'Alene Nursery (Coeur d'Alene, Idaho) (Chapter 16).

The first visual sign of germination is a longitudinal splitting of the seed coat near the narrow end of the seed caused by imbibition of water and later by the growing embryo (Mirov 1967; McCaughey 1994b). Germination begins with swelling and growth of the primary root, elongation of the hypocotyl (stem), and development of the cotyledons (first leaves) that are still inside the seed coat.

Because seeds are buried up to 3 centimeters (1.2 inches) in the ground, the seedlings may take a few days, or possibly weeks to emerge. Seed coats are typically pushed off within the first few days of emergence, although seed coats holding the cotyledons together may remain until early September (W. W. McCaughey, unpublished data). Mirov (1967) wrote that whitebark pine has seven to nine cotyledons. However, observations from more than 200 seedlings from planted seeds north of Yellowstone National Park, Montana, showed that germinants had five to twelve cotyledons. By the end of the first growing season, they had attained heights of 3 to 5 centimeters (1.2 to 2 inches) (McCaughey and Schmidt 1990).

Primary leaves or needles may develop during the first growing season but usually not until the second (Mirov 1967). Mature needles vary in length from 7 to 10 centimeters (2.7 to 3.9 inches) with one to three rows of stomata (openings that allow gas exchange) both on the outer surface of the triangular-shaped needle and on the internal surfaces as well. Stem diameters of germinants ranged from 2 to 4 millimeters (0.08 to 0.16 inch) for whitebark pine compared to half that diameter for germinants of associated species, such as lodgepole pine, Engelmann spruce (*Picea engelmannii*), and subalpine fir (*Abies lasiocarpa*) (W. W. McCaughey, unpublished data).

In laboratory studies, germinant root growth occurred between 10°C and 45°C (50°F and 113°F) with the highest growth rates between 25°C and 35°C (77°F and 95°F). Root growth rates reached 5 to 15 millimeters (0.2 to 0.6 inch) per day (Jacobs 1989; Jacobs and Weaver 1990). In field studies, McCaughey (1990, 1994b) observed whitebark pine seedlings emerging immediately after snowmelt, when soil temperatures were below 10°C (50°F). Whitebark pine seedlings continued emerging from June through early September (McCaughey 1990, 1994b). First-year root growth ranged between 5 centimeters and 18 centimeters (2 inches and 7.1 inches) (W. W. McCaughey, unpublished data).

Recently emerged seedlings are vulnerable to a number of damaging agents. Thick stem diameters do not protect unshaded seedlings from heat damage (solar heat scorching at ground surface), which is a common cause of seedling death in conifers (McCaughey 1990; Kimmins 1997). Scorching of whitebark pine seedlings was higher on nonshaded mineral, litter, and burned seedbeds than on partially shaded (25 and 50 percent) seedbeds in regeneration studies in southwest Montana (McCaughey 1990). New seedlings are susceptible to burying by pocket gophers (*Thomomys talpoides*), uprooting by chipmunks or grazing by birds and small mammals, including Clark's nutcracker (McCaughey 1994b; Vander Wall and Hutchins 1983).

Forking and Tree Clusters

Whitebark pine trees commonly have multiple stems originating at the tree base, which are caused by either (1) loss of apical dominance in the stem and the resultant "forking" into two or more branches, or (2) the emergence of two or more seedlings from one nutcracker cache (Tomback and Linhart 1990; Chapter 5). These two growth forms are referred to as *multitrunk trees* and *tree clusters,* respectively, and require genetic analysis to differentiate them (Tomback and Linhart 1990; Chapter 5). The stems within a multitrunk tree have identical genotypes, whereas the stems of tree clusters have different genotypes (e.g., Linhart and Tomback 1985).

Spontaneous forking of seedlings may result from a lack of strong apical dominance and also from mechanical or grazing damage to apical meristems (e.g., Weaver and Jacobs 1990). Under greenhouse conditions, Weaver and Jacobs observed that 85 percent of whitebark pine seedlings planted in separate pots spontaneously branched, supporting the former explanation. They proposed that this tendency toward branching in whitebark pine may lead to the " lyrate" canopy shape (branching profusely and widely at higher nodes) of mature trees, which is distinctive from the conical canopy shape of most conifers.

The percentage of whitebark pine trees with multiple stems of either growth form appears highest in open, upper subalpine stands and lowest in less exposed, more favorable sites where conifer density is higher (Tomback and Linhart 1990; Weaver and Jacobs 1990). In the northern Rocky Mountains, seral whitebark pine trees grow as tall, single-stemmed trees, whereas trees on exposed sites in climax stands are shrubby and multistemmed (Weaver and Jacobs 1990; S. F. Arno and D. F. Tomback, unpublished observations). Surveying nineteen whitebark pine stands in Montana and Wyoming, Weaver and Jacobs (1990) found that the percentage of trees with multiple stems varied from 8 percent to 79 percent. In the harsh, upper subalpine zone on Mammoth Mountain in the eastern Sierra Nevada, California, 90 percent of the whitebark pine trees surveyed had multiple stems (Tomback and Linhart 1990). Slow growth rates, stressful environments, and reduced competition from other conifers may contribute to the survival of multiple stems at the higher elevations.

Microsite Factors

Seed germination success is influenced by a number of biotic and microsite factors such as shade cover, seedbed type, sowing depth, and predator densities. McCaughey (1990) studied regeneration of whitebark pine on research plots under three shade levels, three seedbed types, two sowing depths, and four predator exclusion levels. Germination increased significantly when shade covered 25 and 50 percent of field research plots.

Germination was significantly higher on mineral seedbed than on litter or burned seedbeds (McCaughey 1990). Surface temperatures were highest on litter seedbeds, up to 79°C (175°F). Heat scorching of seedling stems at ground surface and drought were the major causes of mortality on litter seedbeds (McCaughey 1990).

Overall, germination percentage was significantly lower for whitebark pine seeds that were surface-sown and protected with exclosures to prevent seed predation (11.5 percent), than for seeds buried 2 to 4 centimeters (0.8 to 1.6 inches) deep (33.7 percent). For two successive seasons, all seeds sown directly on the ground surface and not protected with exclosures were eaten by seed foragers (McCaughey 1990). Seed predators were common in the area: Eight rodent species were trapped adjacent to the germination study plots. The two most abundant species trapped were deer mice (*Peromyscus maniculatus*) and Southern red-backed voles (*Clethrionomys gapperi*), both known to forage on conifer seeds. The two species alone comprised 54 and 23 percent, respectively, of the total number of trapped individuals.

Seedling survival may be influenced by microsite factors, such as vegetation and nearby objects, which may modify moisture and temperature relations. In the Sleeping Child burn, Sapphire Range, west-central Montana, whitebark pine seedlings and saplings twenty-six years after fire were disproportionately associated with grouse whortleberry (*Vaccinium scoparium*), smooth wood-rush (*Luzula hitchcockii*), and two sedges (*Carex geyeri, C. rossii*), relative to plant frequency, and rarely associated with the more common beargrass (*Xerophyllum tenax*) (Tomback et al. 1993). Environmental features including fallen trees, branches, wood pieces, rocks, snags, or stumps occurred within 15 centimeters of 85 percent of all whitebark pine regeneration surveyed in the burn. It is possible that these particular features provided some shade and increased soil moisture for seedlings during their early, vulnerable years.

Delayed Seed Germination and Moisture Requirements

Underdeveloped embryos, physiological dormancy, and mechanical barriers to seed germination in whitebark pine that frustrate nursery growers also lead to unusual patterns in naturally regenerating seedlings (Leadem 1986; Pitel and Wang 1980, 1990). The effects of seed dormancy on germination were demonstrated in a subalpine fir clear-cut on the Gallatin National Forest in Montana by McCaughey (1990, 1993, 1994b). Seeds buried 2 to 4 centimeters deep in 1987 experienced 11 percent germination in summer 1988, 45 percent in 1989, and 11 percent in 1990, demonstrating delayed germination. Of surface-sown seeds protected by animal exclosures, only 1 percent germinated in the first summer and 6 percent in the second, with no subsequent germination. Studying natural patterns of whitebark pine germination after the 1988 fires in the Greater Yellowstone Area, Tomback (1994, unpublished data) found synchronous delayed germination in whitebark pine seedlings in two different study areas, one in the north-central region of Yellowstone National Park, and the other outside the northeast park boundary. The first whitebark pine cone crop after the 1988 fires occurred in 1989, but the first whitebark pine seedlings did not appear in the burned study plots until the summer of 1991 in both study areas, following poor cone production in the fall of 1990.

Several pines within subsection *Cembrae* show a similar pattern of delayed seed germination, which appears to be related both to the caching mode of seed dispersal and the short growing season at high elevations. Delayed seed germination is well studied for both Swiss stone pine (Krugman and Jenkinson 1974) and for Japanese stone pine (*Pinus pumila*) (Kajimoto et al. 1998). Delayed germination in whitebark and the other stone pines may be an adaptation that allows seeds to lay dormant and germinate when conditions such as moisture and temperature are favorable.

Sufficient moisture in the uppermost soil layers may facilitate or stimulate whitebark pine seed germination. More whitebark pine seedlings germinated on shaded plots and in years of higher summer precipitation (McCaughey 1993; Tomback et al. 1993). Similarly, Kajimoto et al. (1998) found that episodes of Japanese stone pine recruitment in northern Japan were correlated with higher levels of early summer precipitation.

Whitebark pine is also capable of reproducing vegetatively by rooting from branches that are in continuous contact with the ground surface or have been buried by soil. This process, known as layering, rarely occurs, and the vast majority of reproduction of whitebark pine is from seed germination. Vegetative reproduction from layering of matlike (krummholz) whitebark pine was originally observed at treeline elevations in western Montana (Arno and Hoff 1990).

Factors Limiting Natural Regeneration

The potential for natural regeneration by whitebark pine is reduced by cone and seed insects, fungal diseases, bark beetles, and fire suppression. Whitebark pine cones and seeds are exposed to insects, diseases, and climatic conditions that reduce cone and seed production during their developmental period (Bartos and Gibson 1990). Some of the most common cone and seed insects that damage whitebark pine are cone worms (*Dioryctria* spp. and *Eucosma* spp.), cone beetles (*Conophthorus* spp.) (Bartos and Gibson 1990), and seed chalcids (*Megastigmus* spp.). A seedborne fungus, *Sirococcus strobilinus,* that kills whitebark pine seedlings in nurseries and in natural stands, and *Calocypha fulgens,* referred to as a seed or cold fungus, may cause preemergence seed loss (Hoff and Hagle 1990). Preemergence damping-off disease is caused by fungal decay of seeds and young germinants by *Fusiarium* spp., resulting in extensive mortality, especially in slowly emerging seedlings (Landis et al. 1990).

The introduced fungal disease white pine blister rust attacks seedlings and mature trees, and is currently posing a serious threat to the survival of whitebark pine in parts of its range (Chapters 10 and 11). Whitebark pine appears to be one of the species most susceptible to blister rust, which attacks only five-needled white pines (Hoff et al. 1994). To date, blister rust has caused extensive upper canopy damage and tree mortality in stands in the moist mountain regions of northwestern Montana, northern Idaho, and the Washington Cascades, and has spread through most of the drier regions of whitebark pine's range (Arno 1986; Chapters 10 and 11). Obviously, cone production is reduced wherever trees are infected and killed. Another damaging biotic agent is limber pine dwarf mistletoe (*Arceuthobium cyanocarpum*), which can heavily parasitize stands of whitebark pine and eventually kill trees (Hoff and Hagle 1990; Chapter 11).

Two other major mortality factors are mountain pine beetle (*Dendroctonus ponderosae*) and fire suppression (Arno 1980, 1986; Arno and Hoff 1990; McCaughey and Schmidt 1990). Large numbers of mature whitebark pine are killed when mountain pine beetle populations reach epidemic levels in lower-elevation lodgepole pine stands and move upward into whitebark stands (Bartos and Gibson 1990; Chapter 11).

Because of fire suppression, whitebark pine in some areas of the northern Rocky Mountains of the United States and southern Canada is being replaced by shade-tolerant species, such as subalpine fir and Engelmann spruce. Historical fire intervals in whitebark pine stands ranged from 50 to 300 years or more, depending on site conditions (Arno 1980; Chapters 4 and 9). The exclusion of natural fire from whitebark pine communities for most of the twentieth century has reduced the area burned annually, creating conditions where shade-tolerant species have taken over what were once seral whitebark pine stands and severely limiting the natural regeneration process (Arno 1986).

Acknowledgments

We acknowledge Steve Arno, Robert Keane, and Raymond Shearer (USDA Forest Service Rocky Mountain Researh Station) for their technical reviews of this chapter. We also thank Harold Ziolkowski with the USDA Animal and Plant Health Inspection Service in Bozeman, Montana, for his help in preparing our photographs. We are grateful to Judy O 'Dwyer with the USDA Forest Service, Rocky Mountain Research Station, Bozeman, Montana, for her editorial review throughout all phases of chapter preparation.

Literature Cited

Arno, S. F. 1980. Forest fire history in the northern Rockies. Journal of Forestry 78:460–465.

———. 1986. Whitebark pine cone crops—A diminishing source of wildlife food? Western Journal of Applied Forestry 1:92–94.

Arno, S. F., and R. J. Hoff. 1990. *Pinus albicaulis* Engelm. Whitebark pine. Pages 268–279 *in* R. M. Burns and B. H. Honkala, technical coordinators. Silvics of North America, Vol. 1, Conifers. USDA Forest Service, Agriculture Handbook 654, Washington, D.C.

Bartos, D. L., and K. E. Gibson. 1990. Insects of whitebark pine with emphasis on mountain pine beetle. Pages 171–178 *in* W. C. Schmidt and K. J. McDonald, compilers. Proceedings—Symposium on whitebark pine ecosystems: Ecology and management of a high-mountain resource. USDA Forest Service, Intermountain Research Station, General Technical Report INT-270, Ogden, Utah.

Craighead, J. J., G. B. Scaggs, and J. S. Sumner. 1982. A definitive system for analysis of grizzly bear habitat and other wilderness resources. Wildlands Institute Monograph, No. 1, University of Montana, Missoula.

Day, R. J. 1967. Whitebark pine in the Rocky Mountains of Alberta. Forestry Chronicle 43:278–282.

Eggers, D. E. 1986. Management of whitebark pine as potential grizzly bear habitat. Pages 170–175 *in* G. P. Contreras and K. E. Evans, compilers. Proceedings—Grizzly bear habitat symposium. USDA Forest Service, Intermountain Research Station, General Technical Report INT-207, Ogden, Utah.

Farmer, R. E. Jr., 1997. Seed ecophysiology of temperate and boreal zone forest trees. St. Lucie Press, Delray Beach, Florida.

Harlow, W. M., E. S. Harrar, and F. M. White. 1979. Textbook of dendrology, 6th edition. McGraw-Hill, New York.

Hoff, R. J., and S. Hagle. 1990. Diseases of whitebark pine with special emphasis on white pine blister rust. Pages 179–190 *in* W. C. Schmidt and K. J. McDonald, compilers. Proceedings—Symposium on whitebark pine ecosystems: Ecology and management of a high-mountain resource. USDA Forest Service, Intermountain Research Station, General Technical Report INTO-270, Ogden, Utah.

Hoff, R. J., S. K. Hagle, and R. G. Krebill. 1994. Genetic consequences and research challenges of blister rust in whitebark pine forests. Pages 118–126 *in* W. C. Schmidt and F.-K. Holtmeier, compilers. Proceedings—International workshop on subalpine stone pines and their environment: The status of our knowledge. USDA Forest Service, Intermountain Research Station, General Technical Report INT-309, Ogden, Utah.

Hutchins, H. E., and R. M. Lanner. 1982. The central role of Clark's nutcracker in the dispersal and establishment of whitebark pine. Oecologia 55:192–201.

Jacobs, J. S. 1989. Temperature and light effects on seedlings performance of *Pinus albicaulis*. Masters thesis, Montana State University, Bozeman.

Jacobs, J., and T. Weaver. 1990. Effects of temperature and temperature preconditioning on seedling performance of whitebark pine. Pages 134–139 *in* W. C. Schmidt and K. J. McDonald, compilers. Proceedings—Symposium on whitebark pine ecosystems: Ecology and management of a high mountain resource. USDA Forest Service, Intermountain Research Station, General Technical Report INT-270, Ogden, Utah.

Kajimoto, T., H. Onodera, S. Ikeda, H. Daimaru, and T. Seki. 1998. Seedling establishment of stone pine *Pinus pumila* by nutcracker seed dispersal on Mt. Yumori, northern Japan. Arctic and Alpine Research 30:408–417.

Keane, R. E., S. F. Arno, J. K. Brown, and D. F. Tomback. 1990. Modeling stand dynamics in whitebark pine (*Pinus albicaulis*) forests. Ecological Modeling 51:73–95.

Keeley, J. E., and P. H. Zedler. 1998. Evolution of life histories in *Pinus*. Pages 219–250 *in* D. M. Richardson, editor. Ecology and biogeography of *Pinus*. Cambridge University Press, Cambridge, United Kingdom.

Kendall, K. C. 1983. Use of pine nuts by grizzly and black bears in the Yellowstone area. International Conference on Bear Research and Management 5:166–173.

Kimmins, J. P. 1997. Forest ecology: A foundation for sustainable management. Prentice-Hall, Upper Saddle River, New Jersey.

Kipfer, T. R. 1992. Post-logging stand characteristics and crown development of whitebark pine (*Pinus albicaulis*). Masters thesis, Montana State University, Bozeman.

Knight, R. R., B. M. Blanchard, and D. J. Mattson. 1987. Yellowstone grizzly bear investigations. Report of the Interagency Study Team. USDI National Park Service, Bozeman, Montana.

Krugman, S. L, and J. L. Jenkinson. 1974. *Pinus* L. Pine. Pages 598–638 *in* C. S. Schopmeyer, editor. Seeds of woody plants in the United States. USDA Forest Service, Agriculture Handbook 450, Washington, D. C.

Landis, T. D., R. W. Tinus, S. E. McDonald, and J. P. Barnett. 1990. The biological component: Nursery pests and mycorrhizae. Vol. 5, The container tree nursery manual. USDA Forest Service, Agriculture Handbook 674, Washington, D.C.

Lanner, R. M. 1980. Avian seed dispersal as a factor in the ecology and evolution of limber and whitebark pines. Pages 15–48 *in* B. P. Dancik and K. O. Higginbotham, editors. Sixth North American forest biology workshop proceedings, University of Alberta, Edmonton, Alberta, Canada.

———. 1982. Adaptations of whitebark pine for seed dispersal by Clark's nutcracker. Canadian Journal of Forest Research 12:391–402.

———. 1990. Biology, taxonomy, evolution and geography of stone pines of the World. Pages 14–24 *in* W. C. Schmidt and K. J. McDonald, compilers. Proceedings—Symposium on whitebark pine ecosystems: Ecology and management of a high-mountain resource. USDA Forest Service, Intermountain Research Station, General Technical Report INTO-270, gden, Utah.

———. 1996. Made for each other: A symbiosis of birds and pines. Oxford University Press, New York.

———. 1998. Seed dispersal in *Pinus*. Pages 281–295 *in* D. M. Richardson, editor. Ecology and biogeography of *Pinus*. Cambridge University Press, Cambridge, United Kingdom.

Lanner, R. M., and S. B. Vander Wall. 1980. Dispersal of limber pine seed by Clark's nutcracker. Journal of Forestry 78:637–639.

Latta, R. G., and J. B. Mitton. 1997. A comparison of population differentiation across four classes of gene marker in limber pine (*Pinus flexilis* James). Genetics 146:1153–1163.

Leadem, C. L. 1986. Seed dormancy in three *Pinus* species of the Inland Mountain West. Pages 117–124 *in* R. C. Shearer, compiler. Conifer tree seed in the Inland Mountain West Symposium. USDA Forest Service, Intermountain Research Station, General Technical Report INT-203, Ogden, Utah.

Linhart, Y., and D. F. Tomback. 1985. Seed dispersal by nutcrackers causes multi-trunk growth form in pines. Oecologia 67:107–110.

McCaughey, W. W. 1990. Biotic and microsite factors affecting *Pinus albicaulis* establishment and survival. Doctoral Dissertation, Montana State University, Bozeman.

———. 1993. Delayed germination and seedling emergence of *Pinus albicaulis* in a high-elevation clearcut in Montana, USA. Pages 67–72 *in* D. G. Edwards, compiler. Proceedings—Symposium: Seed dormancy and barriers to germination. Forestry Canada, IUFRO project group P2.04-00, Victoria, British Columbia, Canada.

———. 1994a. Seasonal maturation of whitebark pine seed in the Greater Yellowstone Ecosystem. Pages 221–229 *in* P. Schullery and D. Despain, compilers. Proceedings—Symposium on plants and their environments. Yellowstone National Park, Mammoth Hot Springs, Wyoming.

———. 1994b. The regeneration process of whitebark pine. Pages 179–187 *in* W. C. Schmidt and F.-K. Holtmeier, compilers. Proceedings—International workshop on subalpine stone pines and their environment: The status of our knowledge. USDA Forest Service, Intermountain Research Station, General Technical Report INT-309, Ogden, Utah.

McCaughey, W. W., and W. C. Schmidt 1990. Autecology of whitebark pine. Pages 85–96 *in* W. C. Schmidt and K. J. McDonald, compilers. Proceedings—Symposium on whitebark pine ecosystems: Ecology and management of a highmountain resource. USDA Forest Service, Intermountain Research Station, General Technical Report INT-270, Ogden, Utah.

Mirov, N. T. 1967. The genus *Pinus*. The Ronald Press, New York.

Morgan, P., and S. C. Bunting. 1990. Fire effects in whitebark pine forests. Pages 166–177 *in* W. C. Schmidt and K. J. McDonald, compilers. Proceedings—Symposium on whitebark pine ecosystems: Ecology and management of a high-mountain resource. USDA Forest Service, Intermountain Research Station, General Technical Report INT-270, Ogden, Utah.

———. 1992. Using cone scars to estimate cone crops of whitebark pine. Western Journal of Applied Forestry 7:71–73.

Owens, J. M. 1991. Flowering and seed set. Pages 247–271 *in* A. S. Raghavendra, editor. Physiology of trees. John Wiley, New York.

Perkins, D. L., and T. W. Swetnam. 1996. A dendroecological assessment of whitebark pine in the Sawtooth—Salmon River region, Idaho. Canadian Journal of Forest Research 26:2123–2133.

Pitel, J. A., and B. S. P. Wang. 1980. A preliminary study of dormancy in *Pinus albicaulis* seeds. Bi-Monthly Research Notes 36:4–5.

———. 1990. Physical and chemical treatments to improve germination of whitebark pine seeds. Pages 130–133 *in* W. C. Schmidt and K. J. McDonald, compilers. Proceedings—Symposium on whitebark pine ecosystems: Ecology and management of a high-mountain resource. USDA Forest Service, Intermountain Research Station, General Technical Report INT-270, Ogden, Utah.

Reinhart, D. P., and D. J. Mattson. 1990. Red squirrels in the whitebark zone. Pages 256–263 *in* W. C. Schmidt and K. J. McDonald, compilers. Proceedings—Symposium on whitebark pine ecosystems: Ecology and management of a high-mountain resource. USDA Forest Service, Intermountain Research Station, General Technical Report INT-270, Ogden, Utah.

Schmidt, W. C., and J. E. Lotan. 1980. Phenology of common forest flora of the northern Rockies—1928–1937. USDA Forest Service, Intermountain Forest and Range Experiment Station, Research Paper INT-259, Ogden, Utah.

Schuster, W. S., D. L. Alles, and J. B. Mitton. 1989. Gene flow in limber pine: Evidence from pollination phenology and genetic differentiation along an elevational transect. American Journal of Botany 76:1395–1403.

Spector, D. A. 1999. The influence of forest structure on cone production in whitebark pine throughout the Greater Yellowstone Ecosystem. Masters thesis, Montana State University, Bozeman.

Steele, R., S. V. Cooper, D. M. Ondov, D. W. Roberts, and R. D. Pfister. 1983. Forest habitat types of eastern Idaho–western Wyoming. USDA Forest Service, Intermountain Forest and Range Experiment Station, General Technical Report INT-144, Ogden, Utah.

Tomback, D. F. 1978. Foraging strategies of Clark's nutcracker. Living Bird 16:123–160.

———. 1981. Notes on cones and vertebrate-mediated seed dispersal of *Pinus albicaulis* (Pinaceae). Madroño 28:91–94.

———. 1982. Dispersal of whitebark pine seeds by Clark's nutcracker: A mutualism hypothesis. Journal of Animal Ecology 51:451–467.

———. 1986. Postfire regeneration of krummholz whitebark pine: A consequence of nutcracker seed caching. Madroño 33:100–110.

———. 1994. Effects of seed dispersal by Clark's nutcracker on early postfire regeneration of whitebark pine. Pages 193–198 *in* W. C. Schmidt and F. -K. Holtmeier, compilers. Proceedings—International workshop on subalpine stone pines and their environment: The status of our knowledge. USDA Forest Service, Intermountain Research Station, General Technical Report INT-GTR-309, Ogden, Utah.

———. 1998. Clark's nutcracker (*Nucifraga columbiana*). *In* A. Poole and F. Gill, editors. The birds of North America, No. 331. The Birds of North America, Inc., Philadelphia.

Tomback, D. F., and Y. B. Linhart. 1990. The evolution of bird-dispersed pines. Evolutionary Ecology 4:185–219.

Tomback, D. F., L. A. Hoffmann, and S. K. Sund. 1990. Co-evolution of whitebark pine and nutcrackers: Implications for forest regeneration. Pages 118–129 *in* W. C. Schmidt and K. J. McDonald, compilers. Proceedings—Symposium on whitebark pine ecosystems: Ecology and management of a high-mountain resource. USDA Forest Service, Intermountain Research Station, General Technical Report INT-270, Ogden, Utah.

Tomback, D. F., S. K. Sund, and L. A. Hoffmann. 1993. Post-fire regeneration of *Pinus albicaulis:* Height-age relationships, age structure, and microsite characteristics. Canadian Journal of Forest Research 23:113–119.

Vander Wall, S. B., and R. P. Balda. 1977. Co-adaptations of the Clark's nutcracker and the piñon pine for efficient seed harvest and dispersal. Ecological Monographs 47:89–111.

Vander Wall, S. B., and H. E. Hutchins. 1983. Dependence of Clark's nutcracker, *Nucifraga columbiana,* on conifer seeds during postfledging period. Canadian Field-Naturalist 97:208–214.

Weaver, T., and F. Forcella. 1986. Cone production in *Pinus albicaulis* forests. Pages 68–76 *in* R. C. Shearer, compiler. Proceedings—Conifer tree seed in the Inland Mountain West symposium. USDA Forest Service, Intermountain Research Station, General Technical Report INT-203, Ogden, Utah.

Weaver, T., and J. Jacobs. 1990. Occurrence of multiple stems in whitebark pine. Pages 156–159 *in* W. C. Schmidt and K. J. McDonald, compilers. Proceedings—Symposium on whitebark pine ecosystems: Ecology and management of a high-mountain resource. USDA Forest Service, Intermountain Research Station, General Technical Report INT-270, Ogden, Utah.

Young, J. A., and C. G. Young. 1992. Seeds of woody plants in North America. Dioscoridies Press, Portland, Oregon.

Chapter 7

Whitebark Pine, Grizzly Bears, and Red Squirrels

David J. Mattson, Katherine C. Kendall, and Daniel P. Reinhart

Appropriately enough, much of this book is devoted to discussing management challenges and techniques. However, the impetus for action—the desire to save whitebark pine (*Pinus albicaulis*)—necessarily arises from the extent to which we cherish it for its beauty and its connections with other things that we value. Whitebark pine is at the hub of a fascinating web of relationships. It is the stuff of great stories (cf. Quammen 1994). One of the more interesting of these stories pertains to the dependence of certain grizzly bear (*Ursus arctos horribilis*) populations on its seeds, and the role that red squirrels (*Tamiasciurus hudsonicus*) play as an agent of transfer between tree and bear.

The Geographic Distribution of Pine Seed Use by Bears

Most people probably think of berries and spawning salmon when they think of brown bear (*U. arctos*) foods. However, if one were to consider the entire hemispheric range of brown bears, of which grizzly bears are a North American subspecies, the seeds of stone pines (species of the subsection *Cembrae* of the genus *Pinus*) must be added to the list of archetypical bear foods. Seeds of these species are consumed in large quantities by brown and grizzly bears wherever their geographical ranges include abundant cone-producing stone pines (Mattson and Jonkel 1990). This is especially evident in Siberia (Figure 7-1a), where brown bears eat the seeds of three stone pine species—*Pinus sibirica, P. pumila,* and *P. koraiensis*. The first of these pines is widespread in interior Siberia, the second is abundant along the Pacific coast and inland, while the

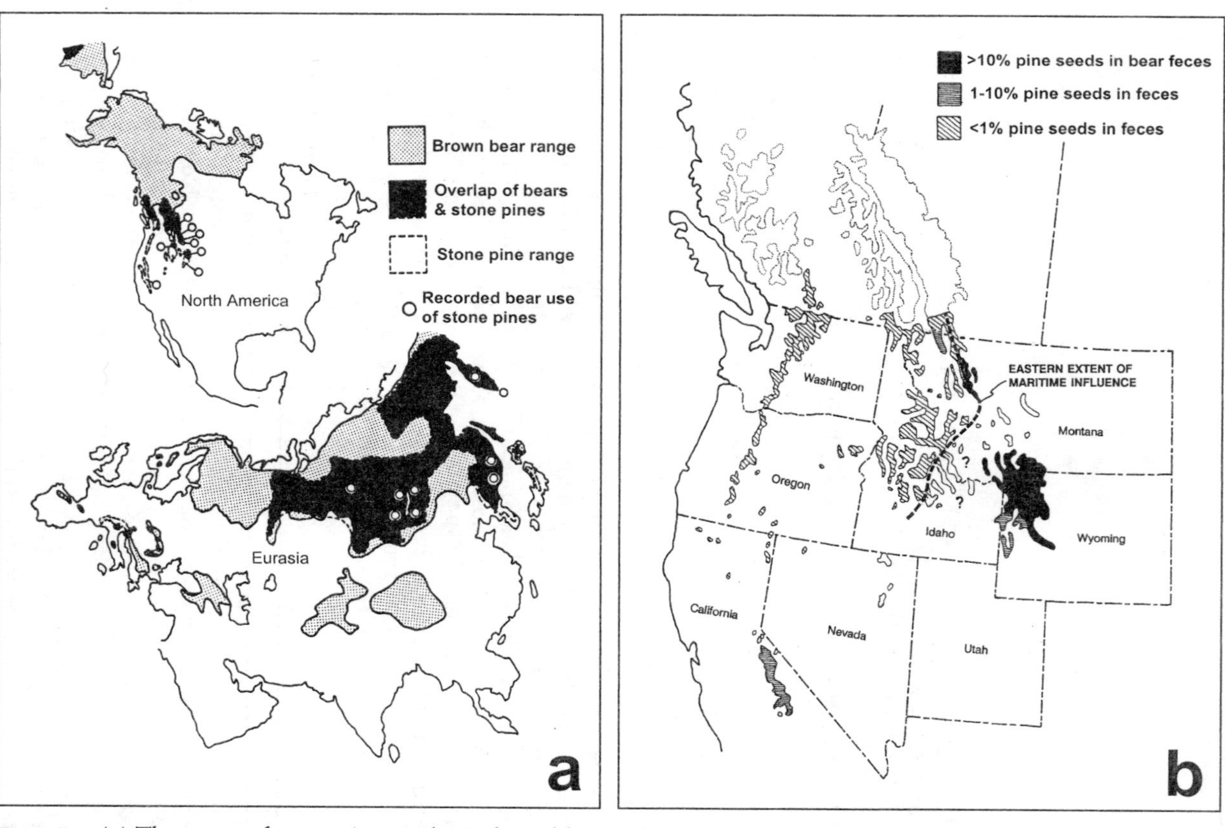

Figure 7-1. (a) The range of stone pines and grizzly and brown bears in Eurasia and North America. Most of stone pine range overlaps with the range of brown bears, as indicated by the areas in black. (b) The range of whitebark pine in North America with levels of use of pine seed by black and grizzly bears shown by different types of shading. The area in Canada delimited by stippling is where whitebark pine is present but infrequent in forest communities.

third has a restricted distribution in temperate maritime regions of far southeastern Siberia and adjacent China and Korea. There is currently a marked association between the southernmost limits of stone pine distribution and the southernmost limit of the main range of brown bears in Asia. One can only speculate about the reasons for this association, given that brown bears have been extirpated by humans from a much more extensive range. It may be that stone pines and brown bears both fare well in regions that are inhospitable to humans.

Grizzly bears consume substantial amounts of pine seeds in only a small part of their North American range (Figure 7-1b) (Mattson and Reinhart 1994). Heaviest use occurs in the Greater Yellowstone Area and the east front of the Montana Rocky Mountains—in regions that experience a continental climate. There are probably several reasons for this geographically limited consumption of pine seeds. For one, whitebark pine is the only stone pine in North America. Limber pine (*P. flexilis,* of the subsection *Strobi*) has seeds comparable in size to those of whitebark pine and occurs in southern parts of grizzly bear range, but is infrequently used by grizzly bears primarily because few squirrels live in most limber pine stands (the importance of squirrels will become obvious later in this chapter). Whitebark pine also is currently abundant and productive primarily in parts of its range that experience continental climates. Whitebark pine was formerly more abundant and productive in inland maritime regions of British Columbia, Idaho, and Montana, but has substantially declined in these areas because of white pine blister rust (*Cronartium ribicola*) and successional replacement by competing trees (Kendall 1995; Chapter 11). In addition, berries are often more abundant in inland maritime regions, which may lead bears in these areas to consume berries as an alternative to pine seeds that are increasingly scarce at higher elevations due to the effects of blister rust.

The Importance of Pine Seeds to Bears

Stone pine seeds have several features that make them a valuable bear food. They are large, especially compared to the seeds of other conifers that share their ranges (Chapter 1). Whitebark pine seeds weigh an average 180 milligrams, compared to 3–13 milligrams for the seeds of species such as subalpine fir (*Abies lasiocarpa = A. bifolia*), Engelmann spruce (*Picea engelmannii*), and lodgepole pine (*Pinus contorta*) (McCaughey et al. 1986). This large size predictably makes the seed easier and more energetically rewarding for grizzly bears to handle and consume. The seeds, and the nutrients contained within, are also less perishable compared to many other bear foods. Seeds that mature in September of one year are sometimes eaten by bears up to twelve months later, especially if the seeds are in cones stored in the cool and moist recesses of a red squirrel's midden (Kendall 1983; Mattson et al. 1994).

Whitebark pine seeds are also a rich source of dietary fat. The average 30 to 50 percent fat content of pine seeds is exceeded in the Yellowstone ecosystem only by that of three other bear foods: adult army cutworm moths (*Euxoa auxiliaris*), certain ant colonies, and the meat of ungulates such as elk (*Cervus elaphus*) and moose (*Alces alces*) late in the growing season after these ungulates have accumulated substantial body fat (Craighead et al. 1982; Lanner and Gilbert 1994; Mattson et al. 1999). Dietary fat is important to grizzly bears not only because it contains concentrated energy but also because it is more efficiently converted to body fat than any other nutrient (McDonald et al. 1988). This conversion of fat from pine seeds to adipose tissue, or body fat, predictably promotes the survival and reproduction of female grizzly bears who rely on adipose reserves not only to hibernate but also to support lactation (Hellgren 1998).

Whitebark pine seeds are a major source of energy for grizzly bears in the Greater Yellowstone Area and along the east slopes of the Montana Rocky Mountains. Recent calculations suggest that grizzlies in the Yellowstone area obtain one-quarter to two-thirds of their net digested energy from pine seeds, depending on the relative abundance of whitebark pine and alternative high-quality foods such as ungulates (*e. g.*, elk, bison [*Bison bison*], and moose), cutthroat trout (*Oncorhychus clarki*), and army cutworm moths (Mattson et al. 1999). Pine seeds are so important that consumption by bears of other high-quality foods after midsummer is apparently largely compensatory to small pine seed crops (Mattson and Reinhart 1994; Mattson 1997).

Pine seeds, however, are not equally important to all bears: on average, females eat roughly twice as many pine seeds as do males (Mattson 2000). This is what we would expect if females were maximizing their fitness. More than for males, it appears the reproductive success of females is contingent on the accumulation of abundant adipose reserves to sustain not only themselves but also their offspring during the prolonged fast of hibernation and posthibernation lethargy. There is no better way to accumulate body fat than to eat fatty foods such as pine seeds.

Foraging Behavior of Bears and Squirrels

Grizzly bears obtain pine seeds in areas like Yellowstone and the east front of the Rockies almost exclusively from cone caches made by squirrels (Mattson and Jonkel 1990; Mattson and Reinhart 1994). In the Yellowstone area between 1977 and 1996, 93 percent of all instances where bears were observed to feed on pine seeds involved the excavation of cones from red squirrel middens. Mature grizzly bears are not good tree climbers; we observed only a handful of instances where grizzlies had apparently done so to procure pine seeds. Thus, because whitebark pine cones rarely spontaneously fall from trees before the seeds within are consumed by birds and chipmunks (*Tamias* spp.), red squirrels perform a critical function for bears by harvesting cones and bringing them to ground level (Hutchins and Lanner 1982). Squirrels also do bears the favor of concentrating cones in large caches or "larder hordes" contained in middens of spongy cone debris accumulated over many years of squirrel activity.

Grizzlies search out squirrel caches and excavate the whitebark pine cones. Initial orientation to squirrel middens is probably by audio cues; that is, pursuit of a chattering squirrel typically leads to that squirrel's midden. Some middens may be visited by bears as many as seven times in a given year (Kendall 1981, 1983). Excavations are often extensive (median = 23 m^2, range 0.15–94 m^2) and are potentially the source of many cones (median = 52, range 0–2069). Once bears procure a pine cone, they are surprisingly fastidious in their consumption of the seeds. Their feces rarely contain cone scales or other cone remnants. The bears apparently achieve this by breaking away the cone scales with their claws, coupled with the facile extraction of seeds with their tongues (Kendall 1981, 1983). The consumption of significant amounts of cone debris seems to occur only when bears are scavenging seeds from cones that are scattered on the forest floor (Mattson and Jonkel 1990).

Seasonal Patterns of Bear Use

Grizzly bears begin consuming newly matured seed crops soon after squirrels start caching cones (Kendall 1981, 1983). Squirrels do not start making large caches until seeds are mature and abundant enough to satiate their immediate appetites (Hutchins and Lanner 1982). Thus, although use of newly matured seeds by grizzly bears has occurred as early as late July, heavy use typically does not begin until the last week of August or the first week of September and persists until the bears den, usually in late October or early November. If a substantial number of cones remain in squirrel middens, bear use resumes typically by early June of the next year and lasts until the seeds are depleted. Whenever and wherever bears feed on pine seeds, they feed on virtually nothing else (Mattson and Reinhart 1994). This is reflected in the fact that bear feces containing pine seeds usually contain little else (Figure 7-2).

As a result of this pattern of exploitation, the average frequency of pine seeds in bear feces is greatest in September through October and is at moderate levels during the remaining months of the year (Figure 7-2) (Craighead et

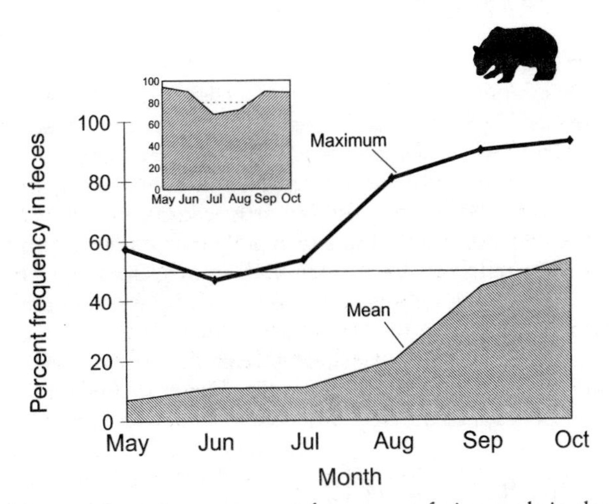

Figure 7-2. Mean and maximum percent frequency of pine seeds in the feces of Yellowstone grizzly bears by month, 1977–1996. The inset diagram presents results only for feces in which pine seeds were present and shows the mean fraction of such feces that consisted of pine seeds.

al. 1982, 1995). Even so, pine seeds can comprise the majority of the grizzly bear diet during any month that bears are active. This was most notably the case in the Yellowstone area during 1979 following the production of a bumper cone crop in 1978. Grizzly bears ate virtually nothing besides pine seeds during most of 1979, transitioning from use of seeds that matured in 1978 to use of a moderately large new crop in August and September. A similar phenomenon occurred in 1990 following a very large seed crop in 1989.

Grizzly bears typically do not exploit the same parts of the whitebark pine zone during their use of successive large seed crops. This is illustrated by patterns of bear foraging on three successive crops in the Yellowstone area, 1985–1990 (Mattson et al. 1994). Exploitation of the 1985 crop, lasting into 1986, occurred primarily on east and west aspects at mid-elevations (2,600–2,750 meters). Exploitation of the 1987 crop, which depleted most middens during that same year, occurred at the lowest elevations of the whitebark pine zone (2,380–2,500 meters), primarily on west and north slopes. Finally, exploitation of the 1989 crop occurred at the very highest elevations (>2,750 meters), primarily on east and south aspects. These patterns suggest two key facts. First, as borne out by cone production on fixed transects (Mattson et al. 1994), successive large crops and related bear use did not overlap with respect to elevation and aspect. This was probably due to the long time it takes individual trees to replenish carbohydrate reserves compared to the interval between successive large seed crops at the scale of the entire whitebark pine zone. Second, over time, grizzly bears make full use of the whitebark pine zone—something that might be missed during the course of a short-term study (Mattson et al. 1994).

Factors Affecting Bear Use of Pine Seeds

Annual Size of the Cone Crop

Grizzly bear consumption of whitebark pine seeds, as indicated by relative frequency of pine seeds in September through October feces, is closely tied to the size of the cone crop. As shown in Figure 7-3, this relationship is S-shaped (*i.e.*, sigmoidal) (Mattson and Reinhart 1994; Mattson et al. 1994), which implies several things. First, grizzly bear use is partly contingent on a threshold of availability that, once surpassed, leads to rapidly escalating consumption. This threshold probably reflects a combination of squirrel foraging behavior and grizzly bear choice among alternative foods. In addition, the rapidly attained cap on frequency of use (i.e., the asymptote) probably reflects a saturation of demand or a saturation of foraging sites by bears at very large crop sizes. This interpretation is supported by the fact that the asymptote for frequency of pine seeds in bear feces is approximately 80 percent—much less than 100 percent.

Density of Active Red Squirrel Middens

Several factors substantially influence the extent to which grizzly bears use pine seeds at any given time and place. Foremost of these is the abundance of red squirrels and their middens (Mattson and Jonkel 1990; Mattson and Reinhart

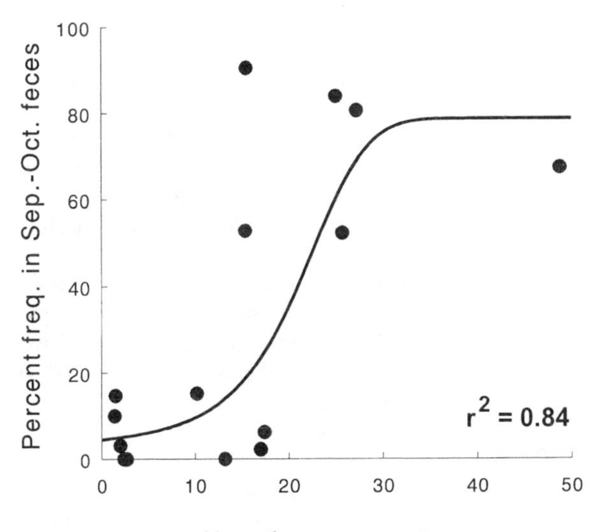

Figure 7-3. Relationship between relative frequency of pine seeds in grizzly bear feces in the Yellowstone ecosystem during September and October and mean number of cones counted per tree on fixed transects in the ecosystem, 1980–1996. Two years were excluded: 1987, because cone counts occurred after many of the cones had already been exploited by seed predators; and 1982, because the sample of bear feces was too small.

1997). As might be expected, the density of midden excavations is strongly dependent on density of middens, which in turn is strongly positively correlated with density of squirrels. This strong association between midden and squirrel densities follows from the fact that red squirrels are highly territorial; they partly define their territories by the presence of usually a single, large, central midden (Smith 1968; Mattson and Reinhart 1996).

Within the whitebark pine zone, midden densities have a strong empirical relationship with lodgepole pine abundance (positive) and steepness of slope (negative) (Mattson and Reinhart 1997). We interpret the positive association with lodgepole pine primarily to be an artifact of strong correlations between lodgepole pine and site features that might favor squirrels (Mattson and Reinhart 1997). Lodgepole pine is more abundant under natural conditions on warmer and less wind-exposed sites that also probably favor higher squirrel densities (Mattson and Reinhart 1990). To a lesser extent, red squirrels may be positively associated with lodgepole pine abundance, because lodgepole pine is a constant provider of food that may become important to squirrels during years when whitebark pine seeds are scarce (Smith 1970). In general, squirrels and their middens are more abundant in stands with high basal areas of mixed conifer species that produce more constant supplies of squirrel food compared to pure whitebark pine stands that produce highly variable seed crops (Reinhart and Mattson 1990; Mattson 2000). Having said this, squirrel densities can be relatively high in high-elevation whitebark pine stands during the infrequent times when whitebark pine seeds are abundant (Kendall 1983). Most of these squirrels are probably juveniles that dispersed from lower elevations rather than progeny of the few resident squirrels.

Given that a squirrel midden is present, several factors determine whether it will be occupied by a squirrel and thus be a candidate for excavation by a bear (Mattson and Reinhart 1997); without squirrels, no cones are cached, and there is no food for a bear to find. First, midden size along with the probability of occupancy declines with increasing elevation. Second, the probability of occupancy also is positively related to abundance of lodgepole pine. Third, a midden is more likely to be occupied if it is large. Last, grizzly bears limit their own future foraging opportunities by reducing the likelihood that a midden will continue to be used by squirrels once it has been excavated; that is, bears not only deprive squirrels of hard-won provender, they also occasionally eat squirrels. The relationships of occupancy to elevation and lodgepole pine abundance again reflect the probable effects of correlated conditions of the physical environment. Squirrels do not fare well on very high and harsh windswept sites (Reinhart and Mattson 1990); their middens also are smaller under these conditions (Mattson and Reinhart 1997). Small middens probably reflect a high turnover in resident squirrels, ephemeral territories, and annually highly variable squirrel densities (Mattson and Reinhart 1996, 1997).

Abundance of Whitebark Pine Seeds in Middens

Given that a midden is present and being used by a squirrel, the likelihood that the midden will be excavated by a grizzly bear depends on the size of the current and previous whitebark pine seed crops, the abundance of mature whitebark pine in the surrounding stand, and the size of the midden (Mattson and Reinhart 1997). As might be expected, size of the current crop has a greater positive effect than does size of the previous crop. Density of mature whitebark pine trees determines how many cones are potentially available to be cached in a midden. All else equal, midden size indicates the number of cones available to be excavated, given that the number of cones taken by bears is strongly and positively correlated with midden size (Mattson and Reinhart 1997). Regardless of the reason, midden size has a strong effect on bear use. Very small middens virtually never get used by bears, while very large middens (>100 m^2) stand about an 80 percent chance of being excavated during any given year, with that chance nearly 100 percent during years with large seed crops. All of these factors affect the size of the food reward for a bear: Grizzly bears are efficient foragers strongly prone to excavate middens when and where they are most likely to obtain a large meal for their efforts.

Pine Seeds and Grizzly Bear Demographics

Because whitebark pine seeds are such an important source of energy and dietary fat in places like Yellowstone, we predict that they affect the reproductive success of grizzly bears, especially females. In fact, reproductive females who ate more pine seeds were more likely to have a surviving three-cub litter (as opposed to one-or two-cub litters) and more likely to reproduce in any given year compared to females who ate fewer pine seeds (Mattson 2000). The effects of diet on reproductive success have been only rarely documented elsewhere for populations of brown or grizzly bears (Mattson 2000). This is to be expected, given that the entire grizzly bear life history strategy, entailing large ranges, large body sizes, and highly flexible behavior, is seemingly "designed" to buffer them against the annual and seasonal vagaries of their foraging environment. The paucity of documented effects of diet on reproduction of grizzly bears further emphasizes the potential importance of pine seeds to maintaining current reproductive rates in the Yellowstone region.

There is a strong relationship between whitebark pine seed crop size and grizzly bear survival, at least in the Greater Yellowstone Area. During years when bears eat few pine seeds, conflicts with humans escalate dramatically (Blanchard 1990; Mattson et al. 1992; Blanchard and Knight 1995). This is illustrated by the strong, negative relationship between number of bears trapped to resolve bear-human conflicts and the frequency of pine seeds in grizzly bear feces during September and October (Figure 7-4). This relationship is curvilinear, with exponentially increasing numbers of bears trapped when few

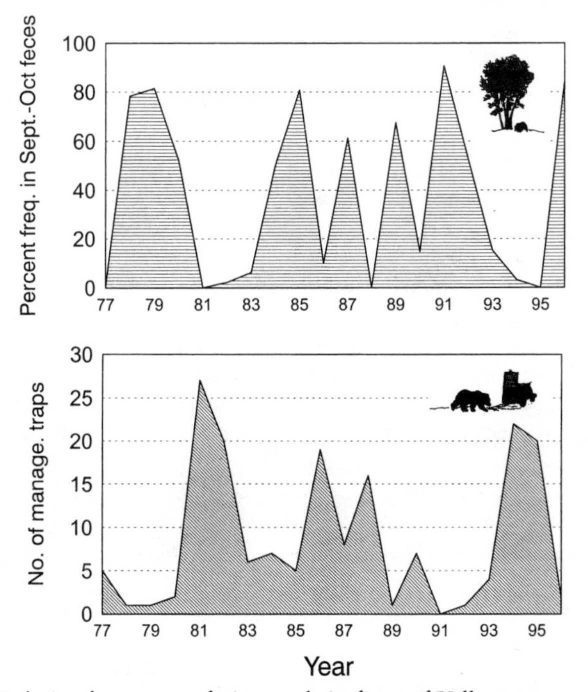

Figure 7-4. Relative frequency of pine seeds in feces of Yellowstone grizzly bears during September and October (top graph) and number of bears trapped to resolve human-bear conflicts during late July through October (bottom graph), 1977–1996. The count of management-trapped bears did not include dependent young such as cubs of the year.

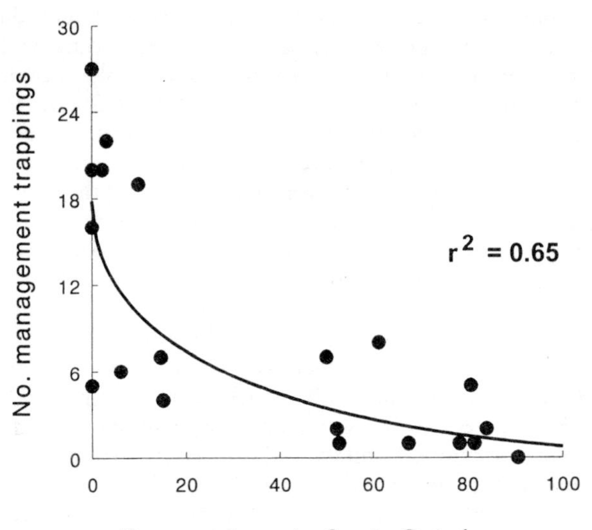

Figure 7-5. Relationship between number of grizzly bears trapped to resolve human-bear conflicts during late July through October and relative frequency of pine seeds in grizzly bear feces during September and October in the Yellowstone ecosystem, 1977–1996.

pine seeds are consumed (Figure 7-5). Of greater consequence is the near doubling of death rate among bears during years when few pine seeds are consumed (Mattson et al. 1992; Mattson 1998; Pease and Mattson 1999). Almost all of this increased mortality is caused by humans (Mattson 1998). Thus, not only do grizzlies get into more trouble with humans during years when pine seeds are scarce, they also are killed by humans at a greater rate. As a consequence, during the years when Yellowstone's grizzly bears are intensively using pine seeds, the population increases, whereas during the years when they are not, the population declines (Pease and Mattson 1999).

Differences in death rates of bears between years of small and large seed crops are ultimately attributable to differences in the distributions of whitebark pine and human facilities such as roads and towns (Mattson et al. 1992). Whitebark pine occurs primarily at high elevations, typically above 2,500 meters. Most resorts or towns and related human activities occur in valleys at lower elevations, typically below 2,400 meters. When pine seed crops are large, grizzly bears are where whitebark pine is abundant—at high elevations far from most humans. Conversely, when pine seed crops are small, grizzly bears tend to concentrate at lower elevations, nearer to humans. Historically, some of the attraction to human facilities was a consequence of those bears conditioned to eat human foods seeking out such foods where they occurred. However, increasingly, the concentration of grizzly bears near people during years when seed crops are small is simply the result of bears seeking alternative natural foods, such as clover (*Trifolium* spp.) or yampa (*Perideridia gairdneri*) roots, that occur at lower elevations near human facilities (Mattson et al. 1992; Gunther 1994). Regardless, when grizzlies are nearer people, numbers of conflicts escalate along with numbers of dead bears.

The availability of abundant whitebark pine seeds thus creates a demographically important refuge for grizzly bears in places like Yellowstone. During years when pine seeds are abundant, grizzlies are attracted to the remote haunts of whitebark pine, where they incur a much lower risk of being killed by humans. A similar effect could be postulated for other high-quality foods that occur in remote areas, such as alpine-dwelling army cutworm moths. However, any refuge attributable to grizzlies using these foods is entirely contingent on a dearth of humans and human facilities in alpine and whitebark pine habitats. If these areas were heavily used by humans, eating pine seeds would actually put bears at risk by attracting them to areas where they would more likely encounter humans and die. If conservation of grizzly bears is the goal, there is thus an imperative to restrict construction of roads or other human facilities and to carefully monitor human activities in the whitebark pine zone (Mattson et al. 1992; Mattson and Reinhart 1994).

Threats to Bears

Grizzly bear populations that rely on whitebark pine seeds for a major part of their energy are vulnerable to anything that might happen to cone-producing trees. This vulnerability is predictably heightened, as in the Yellowstone area, where the bear population is chronically threatened by factors such as human-caused mortality or isolation from other bears. White pine blister rust, global climate warming, and successional replacement by subalpine fir are among the known threats to whitebark pine in interior regions. (Threats to whitebark pine posed by blister rust and subalpine fir are treated extensively in Chapters 1, 9, and 10).

Large Fires

Although fire is identified as being important for the perpetuation of whitebark pine, paradoxically, large fires also may pose a short-term threat to vulnerable bear populations reliant on whitebark pine seeds. This point is highlighted by recent events in the Yellowstone area. Fires burned much of the core of Yellowstone's grizzly bear range during 1988. Many cone-producing whitebark pine trees were killed in the process. Examined by the quadrants of grizzly bear range, losses ranged from nearly 30 percent of cone-producing stands in the north to 12 percent in the east. Based on revisits to a sample of telemetry locations obtained from radio-marked bears prior to the fires, canopy cover of whitebark pine was reduced by 54 percent on such sites (Knight et al. 1990). These figures lead one to expect an impact on overall use of pine seeds by grizzly bears in the wake of the 1988 fires. In fact, given cone crops of the same size pre- and postfire, frequency of use dropped by more than 20 percent throughout the ecosystem during 1989 through 1992 compared to 1977 through 1987 (Mattson 2000). More important, much of this loss occurred in the core of grizzly bear range prior to an upsurge in sightings of bears on the periphery of the ecosystem. In short, during the hiatus between the 1988 fires and the return of cone-producing whitebark pine to previously occupied sites, grizzly bears in the Yellowstone ecosystem likely will be harmed by the removal of so many seed-producing trees (Craighead et al. 1995; Blanchard and Knight 1996).

Recent research has revealed interesting details about how fires might impact red squirrels and grizzly bears (Podruzny et al. 2000). In a 10-km^2 area of which about 5 km^2 burned during the 1988 fires, red squirrel densities (as inferred from active middens) declined relatively little, even though no squirrels were found in burned stands. Apparently, the estimated 300 red squirrels were packing into the remnant unburned forest and occupying smaller territories. At the same time, midden sizes decreased, on average, to about one-half what they were before the fires. As a consequence of the slight drop in squirrel densities and the substantial decrease in midden sizes, the frequency of grizzly bear excavations for pine seeds dropped by 64 percent—a drop that precisely fit the predictions of previous research (Mattson and Reinhart 1997). Thus, at least at the scale of this study, the severity of impacts on bears exceeded the proportional area of a burn as a consequence of synergistic effects on cone-producing trees, squirrel densities, and the sizes of squirrel middens.

Global Climate Warming

The effects of global climate warming potentially complicates any prognosis regarding the effects of fires or white pine blister rust. Recently, scientists have reached near consensus regarding the occurrence of climate warming (e.g., Hansen et al. 1998). Debate now centers more on the rate of warming and its practical consequences. Whitebark pine is vulnerable to climate warming (Romme and Turner 1991; Bartlein et al. 1997; Koteen 1999). The mechanism of loss is probably through competition at stand initiation with early-successional competitors such as lodgepole pine and Douglas-fir that do better on warmer sites (Mattson and Reinhart 1990; Weaver 1994)—modified by changes in fire frequency and size. If such is the case, then whitebark pine may never reestablish as cone-producing trees on some lower subalpine sites burned during fires such as those that occurred in the Yellowstone ecosystem in 1988. Some of the losses due to stand-replacing fire may thus be essentially permanent.

Global climate warming may also lead to accelerated rates of spread by blister rust (Koteen 1999). Conditions conducive to infection are likely to become more common, whether because of increased ambient warming or because of increased variability in the weather conditions (Koteen 1999). The resulting waves of infection would lead to accelerating attrition of whitebark pine, preceded by losses of whitebark pine seed crops. Although predictions such as these are inherently uncertain, they highlight the risks to grizzly bears in Yellowstone, whose status is precarious because of increasing numbers of humans in and near grizzly bear habitat (Pease and Mattson 1999), even without the threat of major perturbations such as fire, climate change, and blister rust.

As in any complex natural system, the effects of climate warming on grizzly bear use of pine seeds will probably not be simple. Given that, currently, squirrels seem to be limited by the harsh climate at highest elevations of the whitebark pine zone (Reinhart and Mattson 1990; Mattson and Reinhart 1996, 1997), squirrel densities may increase with warming. Cone crops on producing whitebark pine trees also may increase, at least in the short term. Preliminary analysis of weather effects on cone crops in the Yellowstone ecosystem suggests, all else equal, that cone production is greater following a series of years with above-average growing season temperatures (Mattson, unpublished results). Such increases in cone crops and squirrels would likely lead to short-term (i.e., several decades long) increases in grizzly bear consumption of pine seeds. However, over longer periods of time, accelerated successional replacement by shade-tolerant competitors such as subalpine fir (Arno and Hoff 1989; Chapter 9) and competitive exclusion early in succession by warmer-site species such as lodgepole pine and Douglas-fir (Mattson and Reinhart 1990; Weaver 1994) would ultimately lead to loss of cone-producing whitebark pine and declines in use of pine seeds by bears.

Some Implications

The relationships among grizzly bears, red squirrels, and whitebark pine, and the effects on the three by factors such as fire, white pine blister rust, and global climate warming, highlight an important general point. Natural systems are complex and open. The enduring conservation of species such as grizzly bears and whitebark pine is contingent on understanding the many important factors affecting them at several spatial and temporal scales. Even so, scientists and managers are likely to be surprised, even with decades of research at their behest. Perturbations from, for example, exotic species and humans are especially troublesome and unpredictable. Global climate change and white pine blister rust were not seen as major problems for grizzly bears in places such as Yellowstone by even the most prescient individuals until recently. Given the somewhat obtuse but important interactions among human distributions, whitebark pine seed crops, and grizzly bear death rates, it is easy to imagine unforeseeable changes in the human or natural environments that would profoundly change this dynamic, with major impacts on grizzly bears living in whitebark pine ecosystems. The complexity of these ecosystems and our limited abilities to predict the consequences of perturbations should encourage humility among the humans trying to manage them.

A tension also arises between the need to act and the advisability of caution. Insofar as we understand them, the means for conserving whitebark pine and grizzly bears may be somewhat incompatible. Grizzly bears flourish only in large areas free from human activity and human features, such as roads, that promote the presence of humans (Craighead et al. 1995). As pointed out earlier, grizzly bears also may suffer from the short-term effects of anything such as fire or timber harvest that removes cone-producing whitebark pine, reduces squirrel densities, or reduces the size of squirrel middens. Such losses are of more concern in areas where bear populations are vulnerable and where fire or timber harvest has been extensive. On the other hand, whitebark pine may require extensive management, including prescription fires, silvicultural treatments, and planting of blister rust–resistant strains for it to remain abundant.

Much of this type of management requires that people be active in the woods and that they have access. We may thus be faced with the paradox of compromising the short-term security of grizzly bear habitat to help promote the long-term abundance of an important food. Whether this is advisable will necessarily depend on the status of individual bear populations, the relative importance of pine seeds as a food, and the development of methods to minimize impacts of restoration efforts on bears.

Restoration strategies for whitebark pine in grizzly bear range logically would conform to standards for long-term management of grizzly bear habitat in places like the Greater Yellowstone Area. These evolving standards are currently articulated in a draft conservation strategy and are organized around grizzly bear Bear Management Units (BMUs). They allow human activities that reduce habitat security in a given BMU to occur only once every ten years (BMUs are permanently delineated areas about the size of a female bear's life range [600–1,000 km^2] that are currently used to monitor grizzly bear populations and schedule management activities). Consequently, under standards that are currently being negotiated, treatments to promote restoration of whitebark pine in grizzly bear range would be limited to one every ten years in any given BMU. Also, they optimally would occur when grizzly bears were least likely to be in the whitebark pine zone: prior to the last week in August following small seed crops the previous year; or later in the year when current seed crops are small. With such an approach, the needs of grizzly bear conservation and the needs of whitebark pine restoration stand a good chance of being simultaneously met.

Acknowledgments

Kerry Gunther of the Yellowstone National Park Bear Management Office and Charles Schwartz of the USGS Interagency Grizzly Bear Study Team offered helpful comments on this chapter. The USGS Biological Resources Division supported the contributions of Katherine C. Kendall and David J. Mattson; the U.S. National Park Service supported the contributions of Daniel P. Reinhart.

Literature Cited

Arno, S. F., and R. J. Hoff. 1989. Silvics of whitebark pine (*Pinus albicaulis*). USDA Forest Service, Intermountain Research Station, General Technical Report INT-253, Ogden, Utah.

Bartlein, P. J., C. Whitlock, and S. L. Shafer. 1997. Future climate in the Yellowstone National Park region and its potential impact on vegetation. Conservation Biology 11:782–792.

Blanchard, B. M. 1990. Relationships between whitebark cone production and fall grizzly bear movements. Pages 223–236 *in* W. C. Schmidt and K. J. McDonald, compilers. Proceedings—Symposium on whitebark pine ecosystems: Ecology and management of a high-mountain resource. USDA Forest Service, Intermountain Research Station, General Technical Report INT-270, Ogden, Utah.

Blanchard, B. M., and, R. R. Knight. 1995. Biological consequences of relocating grizzly bears in the Yellowstone ecosystem. Journal of Wildlife Management 59:560–565.

———. 1996. Effects of wildfire on grizzly bear movements and food habits. International Journal of Wildland Fire 6:117–122.

Craighead, J. J., J. S. Sumner, and G. B. Scaggs. 1982. A definitive system for analysis of grizzly bear habitat and other wilderness resources. Wildlife-Wildlands Institute Monograph No. 1. University of Montana Foundation, Missoula.

Craighead, J. J., J. S. Sumner, and J. A. Mitchell. 1995. The grizzly bears of Yellowstone. Island Press, Washington, D.C.

Gunther, K. 1994. Bear management in Yellowstone National Park, 1960–93. International Conference on Bear Research & Management 9:549–560.

Hansen, J. E., M. Sato, R. Ruedy, A. Lacis, and J. Glascoe. 1998. Global climate data and models: A reconciliation. Science 281:930–932.

Hellgren, E. C. 1998. Physiology of hibernation in bears. Ursus 10:467–477.

Hutchins, H. E. and R. M. Lanner. 1982. The central role of Clark's nutcracker in the dispersal and establishment of whitebark pine. Oecologia 55:192–201.

Kendall, K. C. 1981. Bear use of pine nuts. M.S. Thesis. Montana State University, Bozeman.

———. 1983. Use of pine nuts by grizzly and black bears in the Yellowstone area. International Conference on Bear Research & Management 5:166–173.

———. 1995. Whitebark pine: Ecosystem in peril. Pages 228–230 *in* E. T. LaRoe, G. S. Farris, C. E. Puckett, P. D. Doran, and M. J. Mac, editors. Our living resources. U.S. National Biological Service, Washington, D.C.

Knight, R. R., B. M. Blanchard, and D. J. Mattson. 1990. Yellowstone grizzly bear investigations: Report of the Interagency Study Team 1989. U.S. National Park Service, Interagency Grizzly Bear Study Team, Bozeman, Montana.

Koteen, L. 1999. Climate change, whitebark pine, and grizzly bears in the Greater Yellowstone Ecosystem. M.S. Thesis. School of Forestry and Environmental Studies, Yale University, New Haven, Connecticut.

Lanner, R. M., and B. K. Gilbert. 1994. Nutritive value of pine seeds, and the question of their variable dormancy. Pages 206–211 *in* W. C. Schmidt and F. -K. Holtmeier, compilers. Proceedings—International workshop on subalpine stone pines and their environment: The status of our knowledge. USDA Forest Service, Intermountain Research Station, General Technical Report INT-GTR-309, Ogden, Utah.

Mattson, D. J. 1997. Use of ungulates by Yellowstone grizzly bears *Ursus arctos*. Biological Conservation 81:161–177.

———. 1998. Changes in the mortality of Yellowstone grizzly bears. Ursus 10:129–138.

———. 2000. Causes and consequences of dietary differences among Yellowstone grizzly bears. Ph.D. Dissertation. University of Idaho, Moscow.

Mattson, D. J., and C. Jonkel. 1990. Stone pines and bears. Pages 223–236 *in* W. C. Schmidt and K. J. McDonald, compilers. Proceedings—Symposium on whitebark pine ecosystems: Ecology and management of a high-mountain resource. USDA Forest Service, Intermountain Research Station, General Technical Report INT-270, Ogden, Utah.

Mattson, D. J., and D. P. Reinhart. 1990. Whitebark pine on the Mount Washburn massif, Yellowstone National Park. Pages 106–117 *in* W. C. Schmidt and K. J. McDonald, compilers. Proceedings—Symposium on whitebark pine ecosystems: Ecology and management of a high-mountain resource. USDA Forest Service, Intermountain Research Station, General Technical Report INT-270, Ogden, Utah.

———. 1994. Bear use of whitebark pine seeds in North America. Pages 212–220 *in* W. C. Schmidt and F. -K. Holtmeier, compilers. Proceedings—International workshop on subalpine stone pines and their environment: The status of our knowledge. USDA Forest Service, Intermountain Research Station, General Technical Report INT-GTR-309, Ogden, Utah.

———. 1996. Indicators of red squirrel (*Tamiasciurus hudsonicus*) abundance in the whitebark pine zone. Great Basin Naturalist 56:272–275.

———. 1997. Excavation of red squirrel middens by Yellowstone grizzly bears. Journal of Applied Ecology 34:926–940.

Mattson, D. J., B. M. Blanchard, and R. R. Knight. 1992. Yellowstone grizzly bear mortality, human habituation, and whitebark pine seed crops. Journal of Wildlife Management 56:432–442.

Mattson, D. J., D. P. Reinhart, and B. M. Blanchard. 1994. Variation in production and bear use of whitebark pine seeds in the Yellowstone area. Pages 205–220 *in* D. G. Despain, editor. Plants and their environments: Proceedings of the first biennial scientific conference on the Greater Yellowstone Ecosystem. U.S. National Park Service Technical Report NPS/NRYELL/NRTR, Denver, Colorado.

Mattson, D. J., K. Barber, R. Maw, and R. Renkin. 1999. Coefficients of habitat productivity for Yellowstone's grizzly bear habitat. Technical Report. USGS Forest & Rangeland Ecosystem Science Center, Corvallis, Oregon.

McCaughey, W. W., W. C. Schmidt, and R. C. Shearer. 1986. Seed-dispersal characteristics of conifers in the inland mountain west. Pages 50–62 *in* R. C. Shearer, compiler. Proceedings—Conifer tree seed in the inland mountain west symposium. USDA Forest Service, Intermountain Research Station, General Technical Report INT-203, Ogden, Utah.

McDonald, P., R. A. Edwards, and J. F. D. Greenhalgh. 1988. Animal nutrition, 4th edition. Longman Scientific & Technical, New York.

Pease, C. M., and D. J. Mattson. 1999. Demography of the Yellowstone grizzly bears. Ecology 80:957–975.

Podruzny, S. R., D. P. Reinhart, and D. J. Mattson. 2000. Fire, red squirrels, whitebark pine and Yellowstone grizzly bears. Ursus, in press.

Quammen, D. 1994. Unburied seeds: Henry Thoreau and the Yellowstone grizzly. Outside 19:23–26.

Reinhart, D. P. and D. J. Mattson. 1990. Red squirrels in the whitebark zone. Pages 256–263 *in* W. C. Schmidt and K. J. McDonald, compilers. Proceedings—Symposium on whitebark pine ecosystems: Ecology and management of a high-mountain resource. USDA Forest Service, Intermountain Research Station, General Technical Report INT-270, Ogden, Utah.

Romme, W., H. and M. G. Turner. 1991. Implications of global climate change for biogeographic patterns in the greater Yellowstone ecosystem. Conservation Biology 5:373–386.

Smith, C. C. 1968. The adaptive nature of social organization in the genus of tree squirrels *Tamiasciurus*. Ecological Monographs 38:31–63.

———. 1970. The coevolution of pine squirrels (*Tamiasciurus*) and conifers. Ecological Monographs 40:349–371.

Weaver, T. 1994. Vegetation distribution and production in Rocky Mountain climates—with emphasis on whitebark pine. Pages 142–152 *in* W. C. Schmidt and F. -K. Holtmeier, compilers. Proceedings—International workshop on subalpine stone pines and their environment: The status of our knowledge. USDA, Forest Service, Intermountain Research Station, General Technical Report INT-GTR-309, Ogden, Utah.

Chapter 8

Population Genetics and Evolutionary Implications

Leo P. Bruederle, Deborah L. Rogers, Konstantin V. Krutovskii, and Dmitri V. Politov

Natural selection acts on the genetic variation maintained by populations, adapting them to prevailing environmental conditions. Thus, the evolutionary potential of a species is directly related to its genetic diversity, such as allelic variation at a genetic locus. While plant species differ in the amount of genetic variation that they maintain, pine species (Family Pinaceae) typically have high levels of genetic diversity and are among the most variable species (Ledig 1998). This genetic diversity has been attributed to aspects of their evolutionary history, life history, and biogeography (Hamrick and Godt 1990). Contributing to that genetic diversity is the spatial pattern of genetic variation within and among populations, that is, genetic structure. The latter is expected to vary with those factors affecting the movement of genes (gene flow) within and among populations (Gibson and Hamrick 1991).

In this chapter, we examine genetic diversity and population genetic structure in whitebark pine (*Pinus albicaulis*), a species characterized by a life history trait that is unusual among pines—that is, its seeds are dispersed by birds. Specifically, we ask how much genetic diversity is maintained by natural populations of whitebark pine. We also ask how dispersal of pine seeds by Clark's nutcracker (*Nucifraga columbiana*) influences gene flow and, consequently, the distribution of genetic diversity within and among whitebark pine populations.

A knowledge of the natural levels and distribution of genetic diversity, as well as an understanding of the processes that maintain them, are particularly important to a discussion of whitebark pine management. Management today must consider, in particular, mitigation of the problems associated with white pine blister rust (*Cronartium ribicola*) and successional replacement caused by fire exclusion (Chapters 9, 15, and 17). We begin by briefly addressing the evolutionary origin and affinities of whitebark pine, which provides a frame of reference for our subsequent discussion of genetic diversity and structure in this species.

Evolution of Whitebark and Other Bird-Dispersed Pines

Whitebark pine is one of five bird-dispersed pines currently recognized as comprising *Pinus* subgenus *Strobus* section *Strobus* subsection *Cembrae,* the stone pines (Critchfield and Little 1966; Price et al. 1998; Chapter 2). However, bird-dispersal of pine seeds is not restricted to the *Cembrae* pines (Table 8-1). The independent evolution (convergence) of elements of this complex trait within the genus has been well documented. The majority of bird-dispersed pines are soft pines (*Pinus* subgenus *Strobus*), representing two sections and seven subsections of the subgenus (Price et al. 1998). All have wingless or effectively wingless seeds, a morphology that has been correlated with animal-dispersal of seeds. Although several hard pines (*Pinus* subgenus *Pinus*) have reduced or rudimentary seed wings (Lanner 1998), the sole wingless hard pine is the Italian or Mediterranean stone pine (*P. pinea*).

Table 8-1. Taxonomy of pines (*Pinus*) in subgenus *Strobus* (modified from Price et al. 1998).

Section *Strobus*		Section *Parrya*	
SUBSECTION *CEMBRAE*	SUBSECTION *STROBI*	SUBSECTION *CEMBROIDES*	SUBSECTION *GERARDIANAE*
P. albicaulis[1,2]	*P. armandii*[1,2]	*P. cembroides*[1]	*P. bungeana*[1,2]
P. cembra[1,2]	*P. ayacahuite*	*P. culminicola*[1]	*P. gerardiana*[1,2]
P. koraiensis[1,2]	*P. bhutanica*	*P. discolor*[1]	*P. squamata*
P. pumila[1,2]	*P. chiapensis*	*P. edulis*[1]	
P. sibirica[1,2]	*P. dabeshanensis*[1]	*P. johannis*[1]	Subsection *Rzedowskianae*
	P. dalatensis	*P. juarezensis*[1]	*P. rzedowskii*
	P. fenzeliana[1]	*P. maximartinezii*[1]	
	P. flexilis[1,2]	*P. monophylla*[1]	Subsection *Balfourianae*
	P. lambertiana	*P. nelsonii*[1]	*P. aristata*
	P. monticola	*P. pinceana*[1]	*P. balfouriana*
	P. morrisonicola	*P. remota*[1]	*P. longaeva*[2]
	P. parviflora[1,2]		
	P. peuce[1,2]	Subsection *Krempfianae*	
	P. strobiformis[1,2]	*P. krempfii*	
	P. strobus		
	P. wallichiana		
	P. wangii[1]		

[1]Pine species with wingless or functionally wingless seeds that have been documented from the literature.
[2]Pine species for which there is some evidence of bird-dispersal of seeds (Lanner 1998).

The basis for relegating the five stone pines to subsection *Cembrae* within the genus is the combination of wingless seeds and indehiscent cones that, by definition, do not open when ripe to release the seeds; these two diagnostic traits have presumably evolved as a result of seed dispersal by nutcrackers (*Nucifraga* spp., Family Corvidae). The closely related subsection *Strobi* comprises primarily wind-dispersed species with winged seeds that are released from dehiscent cones that open upon ripening. However, the seeds of some *Strobi* pines, such as limber pine (*P. flexilis*) and southwestern white pine (*P. strobiformis*), are adapted to dispersal by corvids, as well (Lanner 1990, 1996). The origin of a complex set of traits adapted to bird-mediated seed dispersal is one of the central taxonomic and evolutionary problems in pines. Specifically for the white pines (section *Strobus*), at issue is whether the five traditionally recognized stone pine species form a natural or monophyletic group, that is, one that includes all of the descendants of a single, most recent common ancestor.

Until recently, the origin of stone pines and their relationships with other white pines were studied primarily using morphological characters subject to convergent evolution. Therefore, putative relationships based exclusively on morphological traits have not been able to solve this problem unambiguously. Anatomical and karyological (chromosome) studies have also provided ambiguous results (Critchfield 1986; Lanner 1990). However, allozymes, which are the multiple molecular forms of enzymatic proteins resulting from variation at a single gene, and DNA-based genetic markers are being used to assess genetic relatedness and phylogenetic relationships, thereby providing a more precise picture of white pine evolution, including the stone pines.

Phylogenetic studies based upon variation in chloroplast DNA (Strauss and Doerksen 1990; Wang and Szmidt 1993; Krutovskii et al. 1995; Robert A. Price, unpublished data), allozymes (Belokon et al. 1998), and ribosomal DNA (Liston et al. 1999) have shown a high degree of genetic similarity between subsections *Cembrae* and *Strobi*. However, this phylogenetic research does not support the recognition of distinct subsections within section *Strobus*, which is apparently a natural group. Instead, these data indicate that the *Cembrae* pines comprise a portion of a natural group (clade), which also includes representatives of subsection *Strobi*. Although phylogenetic relationships remain unresolved with respect to many pines within section *Strobus*, variation in chloroplast DNA (Wang et al. 1999; Robert A. Price, unpublished data) has revealed Japanese white pine (*P. parviflora*), and especially limber pine, western white pine (*P. monticola*), and eastern white pine (*P. strobus*) to be more distantly related members of the subgenus (Wang et al. 1999; Robert A. Price, unpublished data). Collectively, these data agree with the widely used classification of Critchfield and Little (1966), which placed the two subsections together in section *Strobus*. However, the *Cembrae* pines are clearly derived from within the white pines, and are essentially morphologically and ecologically specialized white pines with a particular syndrome of seed dispersal (Robert A. Price, personal communication).

With regard to the stone pines, specifically, Krutovskii et al. (1990) confirmed a close genetic relationship between Siberian and Swiss stone pines (*P. sibirica* and *P. cembra,* respectively), to which whitebark pine has strong genetic affinities (Krutovskii et al. 1994, 1995; Belokon et al. 1998). Although the Korean stone pine (*P. koraiensis*) and the Japanese stone pine (*P. pumila*) were found to form a second species pair, recent evidence from natural hybridization of Siberian stone pine and Japanese stone pine further attests to the close genetic relationships among all the stone pines (Politov et al. 1999).

Two forces have apparently influenced the complex evolution of the stone pines and their close relatives: adaptation and introgressive hybridization. Diversifying selection for extreme phenotypes may accentuate morphological differences between closely related species, such as the Korean and Japanese stone pines. Alternatively, the similar selection pressures expected from comparable environments may result in convergence between more distantly related species; such species may look more similar and, therefore, more closely related. Thus, selection may obscure the evolutionary history of a group. In addition to convergence, species that co-occur geographically (sympatric species) can also exchange genes through introgressive hybridization. In this process, hybridization is followed by backcrossing to one of the parentals, resulting in the introduction of novel genes into that species. Gene exchange draws hybridizing populations together genetically, homogenizing them, so that they may appear more closely related than they are by origin. For example, a variety of Japanese white pine, specifically *P. parviflora* var. *pentaphylla*, has large, wingless seeds that could have been introduced via introgressive hybridization with the sympatric Japanese stone pine (Watano et al. 1995, 1996). Alternatively, this form may share a common origin with Japanese stone pine (Belokon et al. 1998; Wang et al. 1999).

Population Genetic Diversity and Structure

The genetic diversity and structure of natural pine populations are shaped over time by four processes: natural selection, gene flow, genetic drift, and mutation (Chapter 15). However, the overall influence is a combination of these processes—which may act antagonistically or in concert. *Natural selection* occurs when there are genetically based differences in fertility and viability among individuals in a population. Consequently, the higher survival rate of some individuals relative to others changes the genetic composition of the population. *Gene flow* is the process by which genetic variation is exchanged within and among populations. In pines, this may occur through dispersal of pollen and seeds. However, dispersal events do not result in gene flow unless they contribute to the next generation. Random fluctuations in gene frequencies are known as *genetic drift*. As one generation of plants contains only a sample of the genes from the parental generation, changes can occur and accumulate over time simply due to chance. *Mutations* are heritable alterations in the genetic material and are thus the ultimate source of all genetic variation subject to natural selection.

In general, the processes of gene flow among populations and mutation tend to increase genetic diversity within populations, whereas genetic drift tends to reduce genetic diversity. For most processes, their effects on genetic differentiation and genetic diversity among populations are expected to be opposite. For example, gene flow tends to have a homogenizing effect, reducing genetic structuring among populations and increasing genetic diversity, whereas genetic drift tends to increase genetic differentiation. The effect of natural selection on genetic structure is less certain and can be very different depending on the form of selection (stabilizing, diversifying, etc.) and environmental differences among populations. Of the four processes, only mutation, in theory, is expected to have the same effect on genetic structure as on genetic diversity—increasing both.

Understanding how evolutionary history, life history, and biogeography can affect these evolutionary processes allows us to predict and interpret the levels of genetic diversity and structure in pine species. For example, species with wide geographic ranges that encounter diverse environmental conditions, and hence varying types of natural selection, might be expected to harbor high levels of genetic diversity (Chapter 15). As gene flow reduces genetic structure among populations, species that have features that enhance gene flow—for example, widely transported pollen—might be expected to show less interpopulation differentiation.

The population genetics literature reveals significant relationships between a number of traits and the level and apportionment of allozyme diversity in plant species (Hamrick and Godt 1990). Breeding system, in particular, and geographic range are highly correlated with levels of genetic diversity. For example, mixed-mating, wind-pollinated species maintain considerably higher levels of genetic diversity within populations than do selfing species. A broad geographic range is correlated with high levels of genetic diversity, as well. Genetic structure within species exhibits similar correlations. Breeding system is again the most strongly correlated trait—self-pollinated species exhibit the highest and outcrossing species the lowest level of genetic structure. Life-form, such as annual or long-lived and woody, is the second most significant trait, followed by seed-dispersal mechanism. The highest levels of genetic structure occur in species with seeds that are dispersed by gravity, moderate levels occur in animal-dispersed species, and the lowest levels occur in those species with seeds dispersed by both gravity and animals (via attachment).

Genetic structure is also affected by the interaction of seed dispersal with other ecological and genetic processes. The effect of seed dispersal on genetic structure depends on (1) the extent to which seed dispersal results in colonization of new habitats versus recruitment into existing populations; (2) the density of reproducing adults; and (3) the characteristics of the agent of seed dispersal, for example, the distance that a disperser travels and other behaviors (Hamrick et al. 1993).

The processes influencing the amount and structure of genetic diversity *interact,* thereby making predictions difficult. Whitebark pine has a suite of characteristics that individually have expected, and possibly contradictory, effects on amount and distribution of genetic variation. Collectively, the net result can be ambiguous. For example, the species' long-lived woody perennial life-form, wind-dispersed pollen, and broad current natural distribution would all suggest high levels of genetic diversity. However, within section *Strobus,* whitebark pine is believed to be relatively recently evolved (Lanner 1982), and this, together with small population size, could confer lower levels of genetic diversity. Similar difficulties occur in attempting to deduce the level of genetic structure among populations that, in whitebark pine, is expected to be influenced variously by bird-dispersal of seeds, among other characteristics.

Research on other pine species suggests that theory may not serve us well for predicting genetic diversity and structure in whitebark pine. For example, the pronounced genetic structuring in ponderosa pine (*Pinus ponderosa*)—a species with many characteristics correlated with weak genetic structure—may reflect the diversifying effect of natural selection overpowering the homogenizing effect of widespread pollen dispersal (Mitton et al. 1977). Clearly, the consequences of a complex evolution and life history on genetic diversity and structure can be difficult to predict; as such, speculation abounds.

Genetic Diversity in Whitebark Pine

Allozyme analysis has been used to estimate genetic diversity in whitebark pine both regionally in the Greater Yellowstone Area (Bruederle et al. 1998) and throughout the geographic distribution of the species in North America (Jorgensen and Hamrick 1997). This research has revealed relatively low levels of genetic variation in whitebark pine, at both the population and the species level. While this may not be surprising at the population level—populations are typically disjunct and often isolated—it was not anticipated for the species, which is widespread. These data suggest that most whitebark pine populations maintain much of the genetic variation present in the entire species, at least for those genes encoding allozyme markers (Chapter 15).

The genetic loci that have been examined maintain relatively few genetic variants (Jorgensen and Hamrick 1997); these alleles are typically rare, occurring at very low frequencies in whitebark pine populations. This has contributed to low expected heterozygosities, a measure of genetic diversity that is based on Hardy-Weinberg Law. Hardy-Weinberg Law predicts that genotypes will reach an equilibrium frequency following one generation of random mating and thereafter remain at that frequency. In fact, deviations from expectations assuming Hardy-Weinberg equilibrium revealed heterozygote deficiencies, which were suggestive of inbreeding in most of the whitebark pine populations examined by Jorgensen and Hamrick (1997). This result supports the predictions of Furnier et al. (1987) but contrasts with the heterozygote excesses commonly reported for other pine populations (e.g., Farris and Mitton 1984) that presumably result from selection against selfed progeny. In explanation of the heterozygote deficiencies reported for whitebark pine, Jorgensen and Hamrick (1997) suggested that the tree cluster growth form that often occurs in whitebark pine from the caching of genetically related seed progeny (Chapter 5), may result in inbred (not selfed) progeny on which selection pressures are presumably reduced.

Population means reported for common measures of genetic diversity, such as expected heterozygosity, are lower but similar to those obtained for most of the stone pines except the Japanese stone pine, which has been shown to maintain considerably more diversity (summarized in Jorgensen and Hamrick 1997). Although values for expected heterozygosity can vary markedly among pine species, ranging from 0.000 to 0.590, and even among populations within a species, comparisons with other closely related white pines support the finding of low genetic diversity in whitebark pine. For example, the mean expected heterozygosities of 0.096 (Jorgensen and Hamrick 1997) and 0.152 (Bruederle et al. 1998) reported for whitebark pine fall below the mean of 0.219 (ranging from 0.154 to 0.275) calculated from the data summarized by Ledig (1998) for pines within subgenus *Strobus* that are presumably wind-dispersed (winged seeds).

The low levels of genetic diversity observed in whitebark pine are consistent with the phylogenetic hypothesis that this species is relatively recently evolved. However, hypotheses have been proposed that explain the low levels of genetic diversity with historical biogeographic events—for example, drift due to genetic bottlenecks (sudden decreases in population size) involving Pleistocene refugia—and these events have very likely influenced genetic diversity in whitebark pine.

Genetic Differentiation among Whitebark Pine Populations

In whitebark pine, the majority of the genetic diversity that is present in the species may be attributed to differences among individuals within populations. In other words, whitebark pine populations are poorly differentiated genetically. Published estimates of population differentiation range from lows of 2.5 percent (Bruederle et al. 1998) and 3.4 percent (Jorgensen and Hamrick 1997) to 8.8 percent (Yandell 1992), indicating that a small proportion of all genetic diversity present in this species is due to differences among populations. Well over 90 percent of the genetic diversity maintained by whitebark pine resides within populations, which is typical of pines in general (Ledig 1998).

Whereas genetic differentiation of pine populations has been shown to be low typically (Mitton et al. 1979; Jorgensen and Hamrick 1997), there is considerable variation among wind-dispersed and bird-dispersed pine species. Values reported for genetic differentiation of pine populations range from lows of 1–2 percent (Ledig 1998) to 22 percent for Bishop pine (*P. muricata*) (Millar et al. 1988). Whitebark pine is well within that range. However, when compared to western white pine, a closely related species of subsection *Strobi* and the only other North American white pine for which both regional and rangewide data are available, whitebark pine exhibits less population genetic differentiation. In western white pine, 14.8 percent of all genetic diversity was due to differences among populations, with regional values ranging from 3.17 percent in the mountains of southeastern Oregon and northern California to 12.40 percent in coastal British Columbia and Washington (Steinhoff et al. 1983). This finding is further supported by comparisons with other closely related North American white pines as well, specifically Mexican white pine (*P. ayacahuite*) and limber pine (Ledig 1998); in contrast, eastern white pine has been reported to exhibit levels of genetic differentiation similar to that of whitebark pine (Beaulieu and Simon 1994).

Despite the low levels of genetic differentiation in whitebark pine, there is evidence for some geographic differentiation. Jorgensen and Hamrick (1997) reported a statistically significant, negative relationship between geographic distance and genetic similarity, both across the range of the species and within regions. Furthermore, they reported genetic differences between the western region, comprising the Cascades and Sierra Nevada, and the eastern region, comprising the Rocky Mountains and Great Basin. While statistically significant, only 7.7 percent of the genetic variation present in the species was attributable to genetic differences between the western and eastern regions. Population differentiation was also reported among mountain ranges, with 15.4 percent of the variation in whitebark pine due to differences among ranges within the two regions. Estimates of gene flow ranging from four to nine migrants per generation have been reported for whitebark pine (Jorgensen and Hamrick 1997; Bruederle et al. 1998), enough to overcome the effects of genetic drift.

While the low population genetic differentiation in whitebark pine may seem surprising, given the island nature and small size of many populations, it is possible that the distribution of whitebark pine was more continuous in the relatively recent past. In a comparison of populations representing Pleistocene glaciated and unglaciated areas, Jorgensen and Hamrick (1997) reported differences in the manner in which genetic variation was apportioned. Whereas whitebark pine populations occupying unglaciated and glaciated sites exhibited similar allelic variation, populations occupying unglaciated sites were not as well differentiated as those populations occupying glaciated sites. Jorgensen and Hamrick (1997) attributed this to the nutcrackers' historic migration patterns, with glaciated populations presumably established from cached seeds that derived from different, unglaciated source populations.

Latta and Mitton (1997) previously demonstrated that gene flow in limber pine, which is also nutcracker dispersed, may be attributed largely to pollen movement. However, the data summarized here reveal that populations of whitebark pine are more poorly differentiated genetically than populations of closely related white pines with wind-dispersed seeds, both of which have wind-dispersed pollen. Recruitment into whitebark pine populations is a dynamic phenomenon that may involve different seed sources and long-range movement by different birds over time, as caching sites become available as a result of intermittent disturbance. While these phenomena can be expected to further homogenize populations genetically, other explanations for the limited genetic differentiation, such as stabilizing selection, cannot be excluded.

Population Subdivision in Whitebark Pine

The spatial pattern of genetic variation within populations is known as fine-scale or local genetic structure, or patch structure. Within plant populations, fine-scale genetic structure should be particularly common. Mature, reproductive plants are immobile in adult form, and seed and pollen distribution may be restricted (Dewey and Heywood 1988). However, in tree species—particularly conifers, with their large stature and wind-dispersed pollen—the homogenizing effect of gene flow is often thought to dominate other factors that might lead to fine-scale genetic structure. Given these two opposing factors, other life history traits may determine the extent of fine-scale structure. For example, in bird-dispersed pines, caching of pine seeds can be expected to influence fine-scale genetic structure (Chapter 5).

Fine-scale genetic structure can result from several factors, including isolation in small patches, limited pollen and seed dispersal, and selection acting at the microhabitat level (Berg and Hamrick 1995). Whereas the development of genetic structure among isolated patches of individuals is intuitively understood, Sewell Wright (1943) provided both theoretical and empirical support for the development of genetic structure even in a continuous population simply by means of limited gene flow (local distribution of seeds and/or pollen). In addition to restrictions in gene flow leading to localized genetic drift, fine-scale genetic structure may also be the result of variation in natural selection regimes even over short distances if there are substantial microhabitat differences. Such microhabitat differences might be expected if the species has a broad range and occupies diverse habitats, as is the case with whitebark pine (Chapters 3 and 4).

The measurement of fine-scale genetic structure at a site involves sampling individuals that are increasingly distant from one another; relatedness among individuals is subsequently estimated from the sample using genetic markers. In whitebark pine, the sprawling and intertwining nature of stems within krummholz thickets (matlike tree forms) at tree-line elevations in some parts of the species' range, the possibility of root grafting, and the close associations of trees in some regions into tight clusters demand that care be taken to distinguish the relationship between the sampling unit and the genetic unit when interpreting genetic structure.

The determination and characterization of fine-scale genetic structure is vulnerable to sampling design—more so than for studies at regional or rangewide scales. Without *a priori* insight into a sampling scale that might be biologically meaningful, fine-scale structure might be assessed at a scale that fails to detect the underlying pattern of genetic variation (Dewey and Heywood 1988). In whitebark pine, natural groups exist (e.g., tight clusters of trees, krummholz thickets, elevationally defined groups, etc.) that invite exploration of fine-scale genetic structure.

Two other aspects of fine-scale genetic structure in pines may lead to misinterpretation, especially when comparing species-specific data from two different sources. First, genetic structure can vary among age classes within a population. Second, there may be different genetic structures related to seed-mediated versus pollen-mediated gene flow. For example, when maternally (mitochondrial DNA) and paternally (chloroplast DNA) inherited genetic markers are considered separately, contrasting views of genetic structure within a population may result (Latta et al. 1998).

Fine-scale genetic structure has been traditionally measured with the same statistics as broader-scale studies, such as Wright's F-statistics (Wright 1951) or various measures of genetic distance, such as that of Nei (1973). Frequently, maps are used to show the physical location of individuals and their corresponding genotypes. More recently, local spatial genetic structures have been studied using spatial autocorrelation statistics (Epperson and Li 1997).

Three studies of fine-scale genetic structure exist for whitebark pine. (Table 8-2). Linhart and Tomback (1985) characterized genotypes within six naturally occurring clumps of whitebark pine in Wyoming and found that five of the six clumps comprised more than one genetic individual. This observation confirmed the assumption that many of the clumps were indeed due to the germi-

Table 8-2. Studies of fine-scale genetic structure within whitebark pine populations relative to other bird- and wind-dispersed pines.

Pine Species	Sampling Scale (meters)	Correlates with Genetic Structure	Statistic	Reference
Whitebark	>30 m	Among tree clumps	F_{ST} = 0.451[1]	Rogers et al. 1999
	>30 m	Among krummholz thickets	F_{ST} = 0.244	
	100–500 m	Between growth forms/ elevation differences	F_{ST} = 0.051	
	ca. 100 m	Tree clumps	R = 0.075[2] (within clumps) R = 0.264 (among clumps)	Furnier et al. 1987
		Tree clumps	Multiple genets in most sampled clumps	Linhart and Tomback 1985
OTHER BIRD-DISPERSED PINES				
Limber		Tree clumps	r = 0.43[3](half to full sibs) within clumps r = 0.01 (unrelated) between clumps	Carsey and Tomback 1994
	Adjacent trees and clumps	Tree clumps	r = 0.19 (approx. half-sibs) within clumps	Schuster and Mitton 1991
Swiss stone		Tree clumps	Multiple genets in 70% of sampled clumps	Tomback et al. 1993
OTHER WESTERN U.S. PINE SPECIES, WIND-DISPERSED				
Bristlecone	>100 m	Elevational zones within populations	D_e = 0.011[4]	Hiebert and Hamrick 1983
Lodgepole	ca. 15 m	No fine-scale structure	Autocor.[5]	Epperson and Allard 1989
Ponderosa	1–25 m	All trees within population	Autocor. No cpDNA structure; Patchy mtDNA structure	Latta et al. 1998
	1–50 m	Family structures within a population	F_{ST} = 0.041	Linhart et al. 1981
	>100 m	Slopes of different aspects	Changes in gene frequency	Mitton et al. 1977

[1]F_{ST} = genetic differentiation; represents the correlation between random gametes within a given group relative to gametes within the whole population; increases from zero to a maximum value of one (Wright 1951).
[2]R = Rogers genetic distance; increases from zero with increasing genetic distance (Rogers 1972).
[3]r = relatedness within clusters; increases from zero with increasing relatedness (Quellar and Goodknight 1989).
[4]D_e = Gene diversity among elevational zones within populations; increases from zero with increasing genetic diversity (Nei 1973).
[5]Autocor. = autocorrelation analysis.

nation of multiple seeds—seeds cached in close proximity to one another by the Clark's nutcracker—and not merely the result of a single tree branching at or near the base. The term "*cluster*" was applied to denote the multigenet clumps (Tomback and Linhart 1990). The second study, that of two populations in the Rocky Mountains near Calgary, Alberta, in Canada, examined genetic relationships within and among clumps of whitebark pine (Furnier et al. 1987). This study confirmed the results of Linhart and Tomback (1985), and further indicated that individuals within a cluster are much more closely related than are individuals from different clusters, suggesting a family structure that is probably related to seed-caching behavior (Linhart and Tomback 1985). Furthermore, their analysis revealed no correlation between genetic relatedness and distance; that is, individuals in neighboring clumps did not appear to be any more, or less, closely related than were individuals in distant clumps. This contrasts with the family structure characteristic of many wind-dispersed pines (Furnier et al. 1987), in which genetic relatedness typically decreases with increasing distance from a seed source.

The third and most recently reported study builds on the observations of the earlier studies by confirming that in a third region, the eastern Sierra Nevada range in California, the natural clumps of whitebark pine commonly are composed of individuals with a family structure, often with full-sibling to selfed relationships (Rogers et al. 1999). Genetic differentiation among the clumps is strong (mean $F_{ST} = 0.451$); in other words, 45.1 percent of all *sampled genetic diversity* at a site was attributable to differences among clumps. A more modest degree of genetic structure is related to the differences in elevation and growth form within this region: $F_{ST} = 0.051$ between the upper-elevation prostrate krummholz thickets and the lower-elevation upright tree clumps. These observations are consistent with the seed-caching behavior of Clark's nutcracker (Chapter 5).

In each of the aforementioned three studies, tree clumps were more often than not composed of multiple individuals and, as such, clusters. However, this also implies that some clumps may actually be one individual with multiple branches or stems. An investigation of the genetic composition of the krummholz thicket growth form of whitebark pine revealed that all the sampled thickets were composed of more than one genetic individual (Rogers et al. 1999). The upper-elevation zone, in this part of the species range, was not associated with less genetic diversity than the lower-elevation areas. Relationships among individuals within a thicket were also suggestive of family structure, often half- to full-sibling.

Although fine-scale genetic structure has not been widely studied in other bird-dispersed pines (Table 8-2), findings consistent with those documented for whitebark pine have been reported. In limber pine, another western North American species that has seeds dispersed by Clark's nutcracker, a similar family structure was observed within tree clumps (Schuster and Mitton 1991; Carsey and Tomback 1994); while in Swiss stone pine, which is dispersed by the spotted nutcracker (*Nucifraga caryocatactes*), the majority of tree clumps (~70 percent) were reported to be composed of multiple individuals (Tomback et al. 1993). In the Siberian stone pine, in which the tree cluster growth form is uncommon as one individual per cache typically survives, Krutovskii and Politov (unpublished data) found no significant family structure.

The comparison of fine-scale genetic structure in whitebark pine with that of other pine species in western North America is less clear because of the variation in sampling design and statistical measures employed (Table 8-2). Generalizing substantially, there is a range in fine-scale genetic structure from none to modest, with the presence of structure associated with elevation, slope, or family groups. Although this is a small sample size of all western pines, it suggests that those wind-dispersed species whose ranges overlap with whitebark pine show considerably less fine-scale genetic structure than does whitebark pine.

The presence of striking fine-scale genetic structure in a species that displays only modest broad-scale structure may seem counterintuitive. It is worthy of emphasis that geographic distance is not necessarily correlated with strength of selection influences. Differences in soil moisture, elevation, or aspect, for example, may be associated with strong selection differences over short geographic distances. Furthermore, similar suites of conditions, or their collective genetic impacts, may be repeated over larger geographic distances, thus harmonizing selection pressures and lessening genetic distances.

In conclusion, there is striking fine-scale genetic structure in populations of whitebark pine within those areas of its range where it has been studied. This structure predominantly reflects strong family relationships within groups (where it grows in tight clumps), consistent with bird-dispersed (and cached) seeds, and strong differentiation among clumps. At higher elevations, where the species assumes a prostrate growth form, the krummholz thickets contain multiple genotypes, also with family structure within them, which can be associated with caching behavior by Clark's nutcracker, as well. Interestingly, although Tomback (1986) has shown that seeds are brought up from upright trees just below treeline, Rogers et al. (1999) reported modest genetic differentiation between the higher-elevation krummholz growth form and the lower-elevation upright tree form.

Two caveats to this generalization are offered. A comparison of maternal versus paternal genetic markers would be valuable in elucidating the relative roles of pollen and seeds in gene flow. Given the wide range of the species, and the sometimes disjunct distribution of populations, this strong fine-scale structure may not occur throughout the range of whitebark pine. In some areas, especially where single trees are more prevalent as a consequence of different physical factors and/or where microenvironmental factors may provide a stronger selective force, this structure may not develop (Chapter 3).

Generalizing to Other Bird-Dispersed Pine Species

Effective seed randomization by birds coupled with wind pollination can be expected to result in panmictic (randomly interbreeding) or nearly panmictic population structure in pines, especially where their range is geographically continuous. However, estimating genetic differentiation for a species over a large spatial scale can be controversial. Taking into consideration all ecological information, several very different predictions can be made concerning the impact, in general, of bird-mediated seed dispersal on population structure:

- Populations of bird-dispersed pines are less well differentiated as compared to wind-dispersed pines, due to a more efficient mechanism of seed dispersal leading to population randomization.
- As pioneer species, new or recently established populations of bird-dispersed pines are better differentiated compared to wind-dispersed pines; the "founder effect" of caching seeds may play a significant role in establishing scattered isolated stands.
- Bird-mediated dispersal is only one factor influencing genetic diversity and structure, and plays only a moderate, if any, role in patterns of differentiation on the larger geographic scale, which are determined by other ecological and evolutionary factors.
- There are no unique features in the pattern of genetic variation in bird-dispersed pines relative to wind-dispersed pines. Genetic structure is determined by a combination of factors in each species of both wind-dispersed and bird-dispersed pines, and even over different parts of a species' range.

It is very difficult to test predictions such as these. Traditionally, it has been done by comparing levels of genetic diversity and differentiation between groups of species, for example, bird- and wind-dispersed pines. However, this procedure is associated with many problems. First, the two groups comprise species, each with a different evolutionary history, life history, and biogeography. Second, most population genetic data consist of allozyme markers that may be subject to selection (e.g., Altukhov 1990, 1991; Schuster et al. 1989). Furthermore, the intensity of natural selection can be expected to vary with ecological conditions. In an optimal environment, most allelic variation is likely to be neutral or balanced by slight heterozygote advantage. Thus, balancing selection can also account for genetic homogeneity in the main part of a species' ranges where conditions are optimal. Broadening sampling to include marginal and ecologically diverse populations could be expected to greatly increase the level of population subdivision (Li and Adams 1989). Finally, population genetic data obtained from different genetic loci, as well as differing numbers of loci, may further confound and even invalidate comparisons. Thus, it is important to compare the levels of population differentiation within the same spatial scale, with a similar sampling intensity, over the same subset of genetic loci, and among closely related species.

Keeping the aforementioned reservations in mind, population differentiation in bird-dispersed pines is generally lower when compared to wind-dispersed pines. However, considerable variation exists among the species comprising the two groups of pines, and as such, the statistical variance is large. While some authors have emphasized the role of gene flow in the establishment of both wind- and bird-dispersed conifers, dispersal mechanisms alone are unable to account entirely for the differences in population genetic structure reported for pine species. The potential for other factors causing differentiation should be evaluated for each species separately, taking into account all available data pertaining to the ecology and evolutionary history of the species.

Implications for Management

Ongoing phylogenetic studies are providing remarkable insight into the evolutionary origin and relationships of whitebark pine. Although our traditional taxonomies are being challenged, these data enable us to better understand the evolution and genetic relationships among the pines comprising section *Strobus,* including the *Cembrae* pines. These data potentially have many practical applications, for example, providing a genetic context for future breeding programs, such as those being discussed conferring blister rust resistance to whitebark pine. Phylogenetic data also provide a context for comparative research, such as the aforementioned studies describing genetic diversity and structure in whitebark pine.

Population genetic data support many of the expectations based upon the ecology and evolution of whitebark pine and its mutualistic bird dispersers. Despite the limited amount of genetic data that are currently available for adaptive traits, such as duration of shoot elongation and blister rust resistance (Chapter 15), insights into the genetic diversity and structure maintained by this species are being corroborated by preliminary data collected from the outplanting of whitebark pine seeds (Howard 1999). As such, population genetic data are already providing some tools by which to make informed decisions regarding the conservation and restoration of whitebark pine, as well as other management activities (Chapter 15). For example, in order to secure a significant amount of the allelic diversity in this species, Jorgensen and Hamrick (1997) suggested sampling open-pollinated cones from throughout the range (both ecological and geographical) of whitebark pine.

These data also provide direction for future research. For example, should fine-scale genetic structure be duplicated by planting caches of genetically related seeds or clusters of seedlings? Are small populations of whitebark pine more tolerant of the influences of inbreeding and genetic drift? Clearly, there are many practical applications for the genetic data discussed herein.

Acknowledgments

The authors thank Robert A. Price (University of Georgia) and Aaron Liston (Oregon State University) for providing assistance with the interpretation of phylogenetic data, Steven J. Brunsfeld (University of Idaho) for his comments on this manuscript, and Diana F. Tomback for her expert attention to both scientific and editorial details. We also thank James P. Riser II for his assistance in the preparation of Table 8-1.

Literature Cited

Altukhov, Y. P. 1990. Population genetics, diversity, and stability. Harwood Academic Publishers, London.

———. 1991. The role of balancing selection and overdominance in maintaining allozyme polymorphism. Genetics 85:79–90.

Beaulieu, J. and J. -P. Simon. 1994. Genetic structure and variability in *Pinus strobus* in Quebec. Canadian Journal of Forest Research 24:1726–1733.

Belokon, M. M., D. V. Politov, Y. S. Belokon, K. V. Krutovskii, O. P. Maluchenko, and Y. P. Altukhov. 1998. Genetic differentiation in white pines of the section *Strobus:*

Isozyme analysis data. Doklady Akademii Nauk 358:699–702. [in Russian]. [English translation published in Doklady Biological Sciences 358:81–84.]

Berg, E. E., and J. L. Hamrick. 1995. Fine-scale genetic structure of a turkey oak forest. Evolution 49:110–120.

Bruederle, L. P., D. F. Tomback, K. K. Kelly, and R. C. Hardwick. 1998. Population genetic structure in a bird-dispersed pine, *Pinus albicaulis* (Pinaceae). Canadian Journal of Botany 76:83–90.

Carsey, K. S., and D. F. Tomback. 1994. Growth form distribution and genetic relationships in tree clusters of *Pinus flexilis,* a bird-dispersed pine. Oecologia 98:402–411.

Critchfield, W. B. 1986. Hybridization and classification of the white pines (*Pinus* section *Strobus*). Taxon 35:647–656.

Critchfield, W. B., and E. L. Little. 1966. Geographic distribution of the pines of the world. USDA Forest Service, Miscellaneous Publication No. 991, Washington, D.C.

Dewey, S. E., and J. S. Heywood. 1988. Spatial genetic structure in a population of *Psychotria nervosa.* I. Distribution of genotypes. Evolution 42:834–838.

Epperson, B. K., and R. W. Allard. 1989. Spatial autocorrelation analysis of the distribution of genotypes within populations of lodgepole pine. Genetics 121:369–377.

Epperson, B. K. and T-Q Li. 1997. Gene dispersal and spatial genetic structure. Evolution 51:672–681.

Farris, M. A., and J. B. Mitton. 1984. Population density, outcrossing rate, and heterozygote superiority in ponderosa pine. Evolution 38:1151–1154.

Furnier, G. R., P. Knowles, M. A. Clyde, and B. P. Dancik. 1987. Effects of avian seed dispersal on the genetic structure of whitebark pine populations. Evolution 41:607–612.

Gibson, J. P., and J. L. Hamrick. 1991. Genetic diversity and structure in *Pinus pungens* (Table Mountain pine) populations. Canadian Journal of Forest Research 21:635–642.

Hamrick, J. L., and M. J. W. Godt. 1990. Allozyme diversity in plants. Pages 43–63 *in* A. H. D. Brown, M. T. Clegg, A. L. Kahler, and B. S. Weir, editors. Population genetics, breeding and germplasm resources in crop improvement. Sinauer Press, Sunderland, Massachusetts.

Hamrick, J. L., D. A. Murawski, and J. D. Nason. 1993. The influence of seed dispersal mechanisms on the genetic structure of tropical tree populations. Vegetatio 107–8:281–297.

Hiebert, R. D., and J. L. Hamrick. 1983. Patterns and levels of genetic variation in Great Basin bristlecone pine, *Pinus longaeva.* Evolution 37:302–310.

Howard, J. 1999. Transplanted whitebark pine regeneration: The response of different populations to variation in climate in field experiments. Master's Thesis, University of Montana, Missoula.

Jorgensen, S. M., and J. L. Hamrick. 1997. Biogeography and population genetics of whitebark pine, *Pinus albicaulis.* Canadian Journal of Forest Research. 27:1574–1585.

Krutovskii, K. V., D. V. Politov, and Y. P. Altukhov. 1990. Interspecific genetic differentiation of Eurasian stone pines for isoenzyme loci. Genetika 26:694–707 [in Russian]. [English translation published by Plenum Publishing Corp. in Soviet Genetics, 1990 October:440–450.]

———. 1994. Genetic differentiation and phylogeny of stone pine species based on isozyme loci. Pages 19–30 *in* W. C. Schmidt and F. -K. Holtmeier, compilers. Proceedings—International workshop on subalpine stone pines and their environment: The status of our knowledge. USDA Forest Service, Intermountain Research Station, General Technical Report INT-309, Ogden, Utah.

———. 1995. Isozyme study of population genetic structure, mating system, and phylogenetic relationships of the five stone pine species (subsection *Cembrae,* section

Strobus, subgenus *Strobus*). Pages 279–304 *in* Ph. Baradat, W. T. Adams, and G. Müller-Starck, editors. Population genetics and genetic conservation of forest trees. SPB Academic Publishing, Amsterdam.

Lanner, R. M. 1982. Adaptations of whitebark pine for seed dispersal by Clark's nutcracker. Canadian Journal of Forest Research 12:391–402.

———. 1990. Morphological differences between wind-dispersed and bird-dispersed pines of subgenus *Strobus*. Pages 371–372 *in* W. C. Schmidt and K. J. McDonald, compilers. Proceedings—Symposium on whitebark pine ecosystems: Ecology and management of a high-mountain resource. USDA Forest Service, Intermountain Research Station, General Technical Report INT-270, Ogden, Utah.

———. 1996. Made for each other: A symbiosis of birds and pines. Oxford University Press, New York.

———. 1998. Seed dispersal in *Pinus*. Pages 281–295 *in* D. M. Richardson, editor. Ecology and Biogeography of *Pinus*. Cambridge University Press, Cambridge, UK.

Latta, R. G., and J. B. Mitton. 1997. A comparison of population differentiation across four classes of gene marker in limber pine (*Pinus flexilis* James). Genetics 146:1153–1163.

Latta, R. G., Y. B. Linhart, D. Fleck, and M. Elliot. 1998. Direct and indirect estimates of seed versus pollen movement within a population of ponderosa pine. Evolution 52:61–67.

Ledig, F. T. 1998. Genetic variation in *Pinus*. Pages 251–280 *in* D. M. Richardson, editor. Ecology and Biogeography of *Pinus*. Cambridge University Press, Cambridge, UK.

Li, P., and W. T. Adams. 1989. Range-wide patterns of allozyme variation in Douglas-fir (*Pseudotsuga menziesii*). Canadian Journal of Forest Research 19:149–161.

Linhart, Y. B., and D. F. Tomback. 1985. Seed dispersal by nutcrackers causes multi-trunk growth form in pines. Oecologia 67:107–110.

Linhart, Y. B., J. B. Mitton, K. B. Sturgeon, and M. L. Davis. 1981. Genetic variation in space and time in a population of ponderosa pine. Heredity 46:407–426.

Liston, A., W. A. Robinson, D. Piñero E. R. Alvarez-Buylla. 1999. Phylogenetics of *Pinus* (Pinaceae) based on nuclear ribosomal DNA internal transcribed spacer region sequences. Molecular Phylogenetics and Evolution 11:95–109.

Millar, C. I., S. H. Strauss, M. T. Conkle, and R. D. Westfall. 1988. Allozyme differentiation and biosystematics of the Californian closed-cone pines (*Pinus* subsect. *Oocarpae*). Systematic Botany 13:351–370.

Mitton, J. B., Y. B. Linhart, J. L. Hamrick, and J. S. Beckman. 1977. Observations on the genetic structure and mating system of ponderosa pine in the Colorado Front Range. Theoretical and Applied Genetics 51:5–13.

Mitton, J. B., Y. B. Linhart, K. B. Sturgeon, and J. L. Hamrick. 1979. Allozyme polymorphisms detected in mature needle tissue of ponderosa pine. Journal of Heredity 70:86–89.

Nei, M. 1973. Analysis of gene diversity in subdivided populations. Proceedings of the National Academy of Sciences USA 70:3321–3323.

Quellar, D. C., and K. F. Goodknight. 1989. Estimating relatedness using genetic markers. Evolution 43:258–275.

Politov, D. V., M. M. Belokon, O. P. Maluchenko, Y. S. Belokon, V. N. Molozhnikov, L. E. Mejnartowicz, and K. V. Krutovskii. 1999. Genetic evidence of natural hybridization between Siberian stone pine, *Pinus sibirica* Du Tour, and dwarf Siberian pine, *P. pumila* (Pall.) Regel. Forest Genetics 6(1):41–48.

Price, R. A., A. Liston, and S. H. Strauss. 1998. Phylogeny and systematics of *Pinus*. Pages 49–68 *in* D. M. Richardson, editor. Ecology and Biogeography of *Pinus*. Cambridge University Press, Cambridge, UK.

Rogers, D. L., C. I. Millar, and R. D. Westfall. 1999. Fine-scale genetic architecture of whitebark pine (*Pinus albicaulis*): Associations with watershed and growth form. Evolution 53:74–90.

Rogers, J. S. 1972. Measures of genetic similarity and genetic distance. University of Texas Publications 7213:145–153.

Schuster, W. S. F., and J. B. Mitton. 1991. Relatedness within clusters of a bird-dispersed pine and the potential for kin interactions. Heredity 67:41–48.

Schuster, W. S. F., D. L. Alles, and J. B. Mitton. 1989. Gene flow in limber pine: Evidence from pollination phenology and genetic differentiation along an elevation transect. American Journal of Botany 76:1395–1403.

Steinhoff, R. J., D. G. Joyce, and L. Fins. 1983. Isozyme variation in *Pinus monticola.* Canadian Journal of Forest Research 13:1122–1132.

Strauss, S. H., and A. H. Doerksen. 1990. Restriction fragment analysis of pine phylogeny. Evolution 44:1081–1096.

Tomback, D. F. 1986. Post-fire regeneration of krummholz whitebark pine: A consequence of nutcracker seed caching. Madroño 33:100–110.

Tomback, D. F., F.-K. Holtmeier, H. Mattes, K. S. Carsey, and M. Powell. 1993. Tree clusters and growth form distribution in *Pinus cembra,* a bird-dispersed pine. Arctic and Alpine Research 25:374–381.

Tomback, D. F., and Linhart, Y. B. 1990. The evolution of bird-dispersed pines. Evolutionary Ecology 4:185–219.

Wang, X.-R., and A. E. Szmidt. 1993. Chloroplast DNA-based phylogeny of Asian *Pinus* species. Plant Systematics and Evolution 188:197–211.

Wang, X.-R., Y. Tsumura, H. Yoshimaru, K. Nagasaka, and A. E. Szmidt. 1999. Phylogenetic relationships of Eurasian pines (*Pinus,* Pinaceae) based on chloroplast *rbcL, MATK, RPL20-RPS18* spacer, and *TRNV* intron sequences. American Journal of Botany 88:1742–1753.

Watano, Y., M. Imazu, and T. Shimizu. 1995. Chloroplast DNA typing by PCR-SSCP in the *Pinus pumila—P. parviflora* var. *pentaphylla* complex (Pinaceae). Journal of Plant Research 108:493–499.

———. 1996. Spatial distribution of cpDNA and mtDNA haplotypes in a hybrid zone between *Pinus pumila* and *P. parviflora* var. *pentaphylla* (Pinaceae). Journal of Plant Research 109:403–408.

Wright, S. 1943. Isolation by distance. Genetics 28:114–138.

———. 1951. The genetical structure of populations. Annals of Eugenics 15:323–354.

Yandell, U. G. 1992. An allozyme analysis of whitebark pine (*Pinus albicaulis* Engelm.). Master's Thesis, University of Nevada, Reno.

Part III

Whitebark Pine Communities: Threats and Consequences

The preceding chapters provided important background information about whitebark pine, enabling readers to understand its taxonomic relationships, distributional limits, geographic variation in community successional status, key wildlife interactions, regeneration biology, and population genetic structure. Part III addresses the two major threats to whitebark pine and their ongoing and potential consequences: (1) fire exclusion, which has resulted in widespread successional replacement of whitebark pine in its center of abundance—the United States inland west and adjacent regions in Canada, and (2) the introduced fungal disease, white pine blister rust (*Cronartium ribicola*), which has resulted in major losses of whitebark pine in the northwestern United States and southwestern Canada, with imminent declines throughout most of the range of whitebark pine. White pine blister rust is a potent destructive force on its own, infecting not only whitebark pine but all five-needled white pines; where it acts synergistically with successional replacement, the losses to whitebark pine have been greatly accelerated.

Chapter 9, "Successional Dynamics: Modeling an Anthropogenic Threat," by Robert E. Keane, shows exactly how fire suppression and blister rust are impacting whitebark pine. It first describes in detail the natural successional progression in different whitebark pine community types, both seral and climax, providing flow models of the structural stages. The chapter then presents the results of several simulation models that have been applied to whitebark pine communities to examine the effects of altered fire regimes and blister rust infection. Several findings are of note: As forests become dominated by more shade-tolerant species, such as subalpine fir (*Abies lasiocarpa*), the fire regime is converted from mixed-severity to a severe, stand-replacement regime; whitebark pine becomes locally extinct in the absence of fire; blister rust hastens the successional replacement of whitebark pine; and a broad range of fire return intervals will maintain whitebark pine on the landscape. All simulation models indicate that the most effective management action, even in the presence of blister rust, is to maintain or reintroduce fire to the whitebark pine landscape.

Next, the focus is on the history, effects, and population trends in white pine blister rust. Chapter 10 "Blister Rust: An Introduced Plague," by Geral I. McDonald and Raymond J. Hoff, summarizes the complex life cycle of white pine blister rust, which alternates between shrubs of the genus *Ribes* and white pines. It relates the history of how white pine blister rust was introduced to eastern and western North America from western Europe. Despite widespread government efforts to control blister rust by destroying *Ribes* shrubs, the disease persisted and spread. McDonald and Hoff examine the epidemiology of blister rust within the context of theoretical principles, leading to models and predictions about infection rates. They point out that there are more than fifty *Ribes* species continentwide and more than nine species of five-needled white pines, with the potential for a "continentwide pathosystem." The only means to save white pine communities, and particularly whitebark pine, is to increase the occurrence of genetically based blister rust resistance in populations and to control blister rust infection rates by means of integrated management techniques.

In Chapter 11, "Whitebark Pine Decline: Infection, Mortality, and Population Trends," Katherine C. Kendall and Robert E. Keane, examine the impact of all major disturbance factors, including blister rust, mountain pine beetle (*Dendroctonus ponderosae*), mistletoe (*Arceuthobium* spp.), and fire exclusion, on whitebark pine communities. They discuss in detail the factors associated with the local intensification and spread of blister rust, the relative susceptibility of pines at different elevations, and the vulnerability of blister rust–infected trees to the other disturbance factors. Finally, they examine the decline of whitebark pine and other white pines region by region and predict future trends. They underscore the importance of promoting blister rust resistance in populations by providing opportunities for regeneration.

Diana F. Tomback and Katherine C. Kendall, in Chapter 12, "Biodiversity Losses: The Downward Spiral," discuss the implications for western biodiversity of the decline of whitebark pine communities and the concurrent losses of other white pine communities. They review the importance of whitebark pine as a high-elevation keystone species, the considerable elevational and geographic variation in the structure and composition of whitebark pine communities, and the importance of whitebark pine as wildlife habitat. Based on current information, they speculate about the changes in subalpine forest structure and declines in biodiversity that would result from the loss of whitebark pine, including impacts on grizzly bear populations and on the seed disperser Clark's nutcracker. They show how whitebark pine is linked to other white pine communities through the Clark's nutcracker, and how a long-term decline in seed availability will result in regional declines in nutcrackers and loss of their seed-dispersal services. Finally, Tomback and Kendall describe how several processes work synergistically in small populations to drive them toward extinction in a short time frame; the decline of whitebark pine may be more rapid than predicted from the rate of spread of blister rust.

The last chapter in Part III, "Threatened Landscapes and Fragile Experiences: Conflict in Whitebark Pine Restoration" by Stephen F. McCool and Wayne A. Freimund, takes us in a new but extremely relevant direction: First, it explores the aesthetic and symbolic role of whitebark pine in the recreational experience, and then it examines the philosophical problems and social perceptions associated with potential whitebark pine restoration efforts in designated wilderness areas. A "defining feature of wilderness, as stated in the Wilderness Act of 1964" is that these areas remain "untrammeled by man." The major dilemma is that the introduction of white pine blister rust initially and the exclusion of fire for decades from high-elevation whitebark pine communities were, in fact, trammeling. Scientists argue that restoring whitebark pine is necessary to maintain some semblance of naturalness in these areas. However, actions to restore these communities—that is active manipulation of the landscape—are again trammeling. McCool and Freimund ask, How much compromise in both goals is acceptable?

Chapter 9

Successional Dynamics: Modeling an Anthropogenic Threat

Robert E. Keane

Successional development in whitebark pine (*Pinus albicaulis*) forests provides a temporal ecological context for planning and designing restoration activities. The rate and direction of the ecophysiological, compositional, and structural changes during upper subalpine stand development will dictate the design and implementation criteria of restorative treatments. Moreover, the types and severity of disturbances that occur during the successional cycle will guide the selection of treatments and the details of implementation. The primary purpose of this chapter is to describe succession and disturbance processes in whitebark pine ecosystems as a scientific basis for planning and designing restoration treatments. First, a set of conceptual models of succession and disturbance based on stand composition and structure are presented to illustrate the primary successional trajectories in whitebark pine ecosystems. Then, important changes in ecosystem processes and characteristics that occur during successional development are described from empirical data and research study results. And last, a review of simulation modeling in whitebark pine ecosystems relates simulated results to successional implications in whitebark pine restoration.

Terminology in this chapter is clearly defined to avoid confusion: *Succession* is the sequence of plant, animal, and microbial communities that occupy an area over a period of time in the absence of disturbance (West et al. 1980). *Primary succession* occurs after those rare disturbances, such as glaciation, mining, and landslides, that remove all vegetation, most nutrient pools, and plant propagules from the site. This chapter will deal primarily with *secondary succession,* which is vegetational development after endemic disturbances such

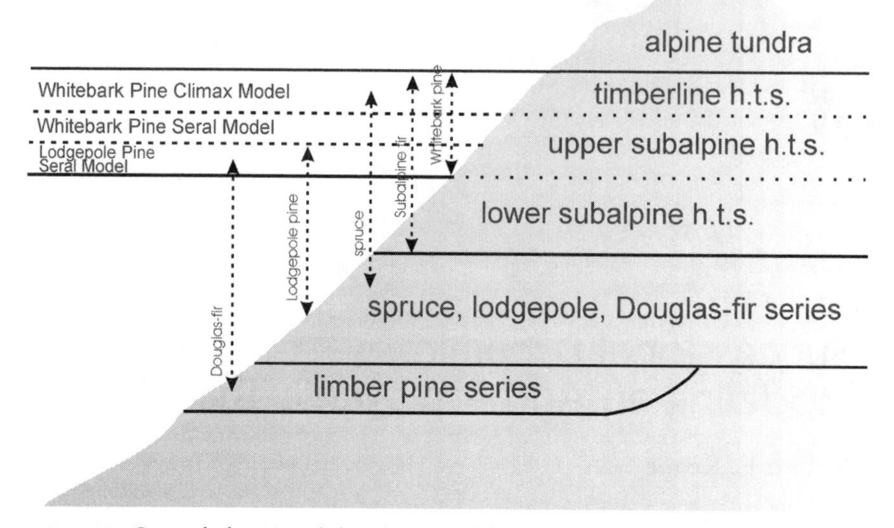

Figure 9-1. General elevational distribution of forest tree species (arrows) and whitebark pine site types described in this chapter. Each site type is assigned the appropriate model (from Arno and Weaver 1990).

as fire, beetles, and disease. *Successional processes* are the causal mechanisms that effect developmental changes in ecosystems such as growth, regeneration, transpiration, photosynthesis, and respiration. *Native fire regimes* occur when fire processes are in harmony with current environmental conditions, while *historical fire regimes* are described by summaries of past fire events (Agee 1993). It is assumed that historical fire regimes were native fire regimes. *Seral species* are typically shade-intolerant plant species usually inhabiting the early and middle phases of the successional cycle, while the shade-tolerant, *late seral* species dominate in the latter stages of succession (Franklin and Mitchell 1967; Noble and Slatyer 1977).

This chapter is limited in scope to upper subalpine sites that support erect, tall (greater than 10 meters) forests where whitebark pine is either the major seral species or the indicated climax species (Figure 9-1). These two sites support the majority of whitebark pine forests that are rapidly disappearing because of blister rust and advancing succession (Chapter 11). Successional dynamics of krummholz communities (shrub or prostrate growth form), timber atolls, and elfin forests that occur at or above timberline are not discussed because of their limited occurrence and complex successional dynamics (Tomback 1986; Arno and Weaver 1990). Discussions in this chapter center primarily on secondary succession in whitebark pine ecosystems at the stand-level with respect to tree species development. Implications of successional advancement across the landscape are also presented.

The Conceptual Model

In general, successional development in whitebark pine ecosystems can be described as the replacement of the moderately shade-intolerant, colonizer tree species whitebark pine with the shade-tolerant subalpine fir (*Abies lasiocarpa*) in the absence of fire, except on those sites that cannot support subalpine fir. This chapter details this simple successional process using three diagrammatic models that represent the temporal changes in stand composition and structure throughout the geographical range of whitebark pine (e.g., Figures 9-2, 9-3, and 9-4). This modeling approach is derived from Kessell and Fischer's (1981) fire effects model, which was based on the successional modeling concepts developed by Noble and Slatyer (1977) and used by Cattelino et al. (1979). Kessell and Fischer's (1981) conceptual model links seral vegetation communities along multiple pathways of successional development that eventually converge to a stable or climax plant community. Many authors have since expanded the Kessell and Fischer (1981) approach to predict fire succession by potential vegetation types for many northern Rocky Mountain forests (e.g.,

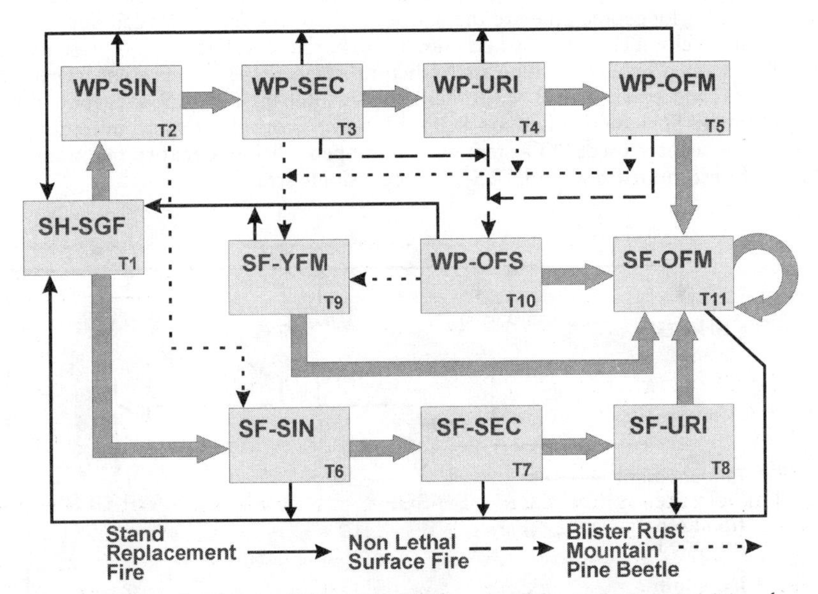

Figure 9-2. The Seral Whitebark Pine Model diagram for succession on sites where whitebark pine is the major seral species. Each box represents a successional class defined by a cover type and structural stage. Thick shaded lines represent successional development, and dashed lines are disturbances. T1–T11 represent transition times or the time it takes for one succession class to transition to the next successional class in the pathway (Table 9-1). Cover types are SH–shrub-herb, WP–whitebark pine, LP–lodgepole pine, and SF–subalpine fir. Structural stages are SGF–shrub-grass-forb; SIN–stand initiation, SEC–stem exclusion closed, SEO–stem exclusion open, URI–understory reinitiation, OFM–old forest multi-strata, and OFS–old forest single strata.

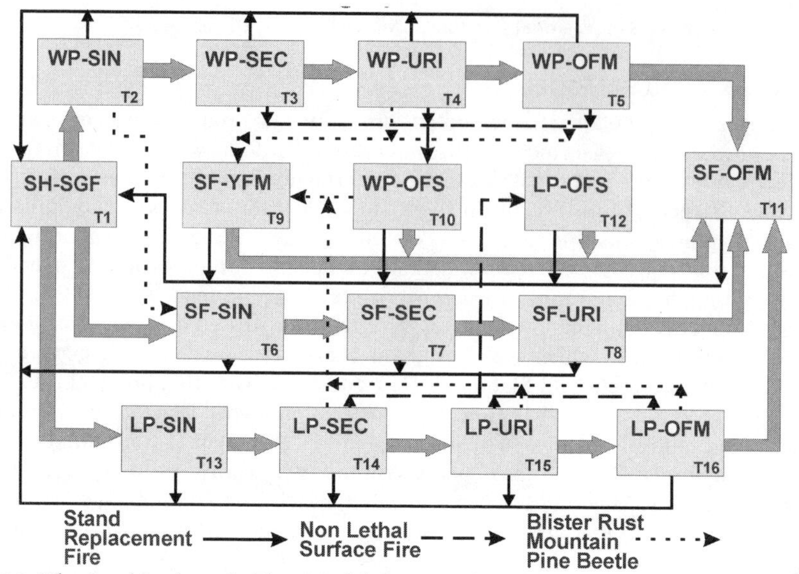

Figure 9-3. The Seral Lodgepole Pine Model diagram for succession on sites where whitebark pine and/or lodgepole pine are the major seral species. Thick shaded lines represent successional development, and dashed lines are disturbances. Each box represents a successional class defined by a cover type and structural stage. T1–T16 represent transition times (Table 9-1). Cover types are SH–shrub-herb, WP–whitebark pine, LP–lodgepole pine, and SF–subalpine fir. Structural stages are SGF–shrub-grass-forb; SIN–stand initiation, SEC–stem exclusion closed, SEO–stem exclusion open, URI–understory reinitiation, OFM–old forest multistrata, and OFS–old forest single strata.

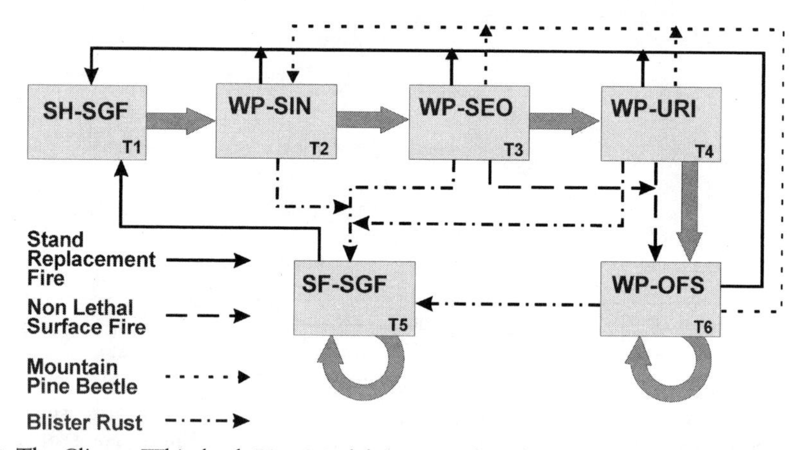

Figure 9-4. The Climax Whitebark Pine Model diagram for succession on severe upper subalpine sites where whitebark pine is the indicated climax species. Thick shaded lines represent successional development. Each box represents a successional class defined by a cover type and structural stage. T1–T6 represent the transition times (Table 9-2). A subalpine fir cover type is shown for those cases where fir coverage is greater than 10 percent after whitebark pine death. Cover types are SH–shrub-herb, WP–whitebark pine, LP–lodgepole pine, and SF–subalpine fir. Structural stages are SGF–shrub-grass-forb, SIN–stand initiation, SEC–stem exclusion closed, SEO–stem exclusion open, URI–understory reinitiation, OFM–old forest multistrata, and OFS–old forest single strata.

Davis et al. 1980; Crane and Fischer 1986; Bradley et al. 1992). Keane et al. (1996b) then extended this concept to include disturbances other than fire and ecosystems other than forests to model coarse-scale succession across the Interior Columbia River Basin. This approach assumes a somewhat constant climate; so, if the climate changes, the cover types, structural stages, transition times, and pathways presented here also may change (Romme and Turner 1991; Bartlein et al. 1997). However, it is probably reasonable to assume that climate change will redistribute biophysical settings across a landscape, and the successional processes within these settings will remain somewhat consistent.

I have refined this approach to simulate fine-scale, stand-level successional dynamics in whitebark pine communities. Only fire, mountain pine beetle (*Dendroctonous ponderosae*), and blister rust (*Cronartium ribicola*) were included as the major disturbances in these models for simplicity. Mountain pine beetles tend to kill most large, cone-bearing whitebark pine in times of epidemics, whereas the exotic blister rust kills nearly all whitebark pine trees regardless of tree size and age. Cutting treatments were not incorporated because they can be designed to convert a successional class to many structural stage and cover-type combinations. Other insect and disease perturbations were not included, because their effects probably would not be manifest at this scale of simulation. For example, root diseases in subalpine fir cover types may only prolong the length of time spent in a successional stage, but they will rarely result in a conversion to a whitebark pine or lodgepole pine cover type unless there is a fire.

Model Basics

In these models, a *succession class* is a distinct successional community defined by a *cover type,* which describes species composition, and a *structural stage,* which describes vertical stand structure. The time spent in a succession class, called *transition time,* depends on the shade tolerance and life spans of the dominant species (Noble and Slatyer 1977), and the productivity of that site (Forcella and Weaver 1977; Arno and Weaver 1990). Basically, transition time is the interval required for the cover type and/or structural stage to change (i.e., life span of a succession class). Transition time is noted within the succession class box in Figures 9-2, 9-3, and 9-4 by the letter "T" and a number. These transitions were estimated from the literature and are presented in Tables 9-1 and 9-2, organized by geographical region. The time spans are only coarse estimates and should be adjusted for local conditions.

Cover type is named for the plant species or life-form having the most basal area (Eyre 1980). It does not describe characteristics of the understory trees or undergrowth plants. There are four general cover types defined in this model. The shrub-herb cover type (SH) describes communities where trees provide less than 5 percent cover, and shrub and herbaceous species dominate plant cover. The whitebark pine (WP) cover type is primarily dominated by whitebark pine but can contain minor components of several other species depending on geographic region, such as Douglas-fir (*Pseudotsuga menziesii*), lodgepole pine (*Pinus contorta*), limber pine (*P. flexilis*), alpine larch (*Larix lyallii*), subalpine fir, Engelmann spruce (*Picea engelmannii*) and white spruce (*Picea glauca*), and

Table 9-1. Successional time span parameters (range of years) for the Seral Whitebark Pine and Seral Lodgepole Pine Models (see Figures 9-2 and 9-3). Cover types are SH–shrub-herb, WP–whitebark pine, LP–lodgepole pine, and SF–subalpine fir. Structural stages are SGF–shrub-grass-forb, SIN–stand initiation, SEC–stem exclusion closed, SEO–stem exclusion open, URI–understory reinitiation, OFM–old forest multistrata, and OFS–old forest single strata.

Successional Class	Northern Rocky Mountains	Central Rocky Mountains	Canadian Rocky Mountains	Cascade Range	Sierra Nevada Range
SH-SGF (T1)	10–30	10–40	5–30	10–40	10–30
WP-SIN (T2)	20–40	20–40	25–40	15–50	10–50
WP-SEC (T3)	50–100	50–100	40–80	30–75	50–120
WP-URI (T4)	50–100	50–125	50–100	50–120	100–150
WP-OFM (T5)	100–200	100–300	75–200	70–150	100–300
WP-OFS (T10)	30–75	50–75	30–80	20–70	40–70
LP-SIN (T13)	5–20	5–50	5–25	5–20	5–25
LP-SEC (T14)	50–75	40–60	50–100	30–75	30–70
LP-URI (T15)	50–75	20–40	30–50	20–50	20–60
LP-OFM (T16)	30–50	30–60	20–50	20–40	20–50
LP-OFS (T12)	20–70	20–70	30–50	20–40	30–70
SF-SIN (T6)	10–30	20–40	10–30	10–30	10–40
SF-SEC (T7)	30–70	30–60	30–60	20–50	20–60
SF-URI (T8)	30–50	30–60	20–50	30–50	30–60
SF-YFM (T9)	20–50	30–50	20–30	20–40	20–50
SF-OFM (T11)	500+	160+	400+	300+	500+
References (see below)	3,4,5,6,7,8,14, 15,19,20,29,30, 35,36,37,38,47, 48,53,59,60,63, 65,66	3,4,6,7,9,10,11, 16,17,27,30,39, 40,42,43,44,45, 47,49,52,53,55, 58,61,62,69	6,7,9,21,34,50, 54	1,2,6,7,13,18, 22,23,24,25,26, 28,33,41,57	6,7,12,51,56,64

1. Agee and Smith (1984)
2. Agee (1994)
3. Alexander (1985)
4. Arno (1980)
5. Arno and Habeck (1972)
6. Arno and Hoff (1990)
7. Arno and Weaver (1990)
8. Arno et al. (1993)
9. Bailey (1975)
10. Bradley et al. (1992)
11. Butler (1986)
12. Clausen (1965)
13. Cole (1982)
14. Cooper et al. (1991)
15. Craighead et al. (1982)
16. Crane and Fischer (1986)
17. Crouch (1987)
18. Dahlgreen (1984)
19. Daubenmire (1981)
20. Davis et al. (1980)
21. Day (1967)
22. del Moral (1979)
23. del Moral and Fleming (1979)
24. Dickman and Cook (1989)
25. Douglas and Ballard (1971)
26. Douglas and Bliss (1977)
27. Eggers (1986)
28. Fahnestock (1977)
29. Fischer and Bradley (1987)
30. Fischer and Clayton (1983)
31. Forcella and Weaver (1977)
32. Forcella (1978)
33. Franklin and Mitchell (1967)
34. Halliday (1937)
35. Hartwell (1997)
36. Keane et al. (1990)
37. Keane et al. (1994)
38. Keane and Morgan (1994)
39. Kipfer (1992)
40. Loope and Gruell (1973)
41. Lueck (1980)
42. Marston and Anderson (1991)
43. Mattson (1997)
44. Mattson and Reinhart (1990)
45. Mattson and Reinhart (1997)
46. Morgan and Bunting (1990)
47. Morgan et al. (1994)
48. Moseley and Bernatas (1992)
49. Murray (1996)
50. Ogilvie (1990)
51. Parker (1988)
52. Perkins and Swetnam (1996)
53. Pfister et al. (1977)
54. Pojar (1985)
55. Reed (1976)
56. Riegel et al. (1990)
57. Rochefort and Peterson (1996)
58. Romme (1982)
59. Smith and Fischer (1997)
60. Steele (1960)
61. Steele et al. (1981)
62. Steele et al. (1983)
63. Thompson and Kuijt (1976)
64. Tomback (1986)
65. Tomback et al. (1993)
66. Tomback et al. (1995)
67. Weaver and Dale (1974)
68. Weaver et al. (1990)
69. Weaver and Forcella (1986)

Table 9-2. Succession time span parameters (years) for Climax Whitebark Pine Model (Figure 9-4). Cover types are SH–shrub-herb, WP–whitebark pine, and SF–subalpine fir. Structural stages are SGF–shrub-grass-forb, SIN–stand initiation, SEO–stem exclusion open, URI–understory reinitiation, and OFS–old forest single strata.

Successional Class	Northern Rocky Mountains	Central Rocky Mountains	Canadian Rocky Mountains	Cascade Range	Sierra Nevada Range
SH-SGF (T1)	10–40	10–50	10–50	20–60	10–40
WP-SIN (T2)	40–100	30–110	30–90	20–100	50–100
WP-SEO (T3)	100–300	100–140	100–250	50–200	100–200
WP-URI (T4)	100–200	100–300	100–200	75–200	100–200
SF-SGF (T5)	100+	100+	100+	100+	100+
WP-OFS (T6)	200+	400+	200+	200+	300+
References (see Table 9-1)	3,4,6,7,15,15,20, 29,30, 32,36,45, 51,57,61,67	3,4,6,7,10,6,17, 27,30,31,32,40, 42,43,44,45,47, 50,51,53,56,59, 60,65,66,67	6,7,21,34,48,52	1,6,7,13,22,25, 26,41,55	6,7,12,49,54,62

mountain hemlock (*Tsuga mertensiana*). The lodgepole pine (LP) cover type occurs when disturbed stands regenerate primarily to lodgepole pine after disturbance, but other species mentioned above can also be present in the stand. The subalpine fir (SF) cover type is dominated by shade-tolerant, upper subalpine species—namely, subalpine fir in most cases—but mountain hemlock or spruce can also dominate depending on the region (Table 9-1).

Structural stage categories are based on stand-development processes as described by Oliver and Larson (1980) and revised by O'Hara et al. (1996). Descriptions and parameters for these categories were modified to represent more realistic development of upper subalpine forests (Table 9-3). The shrub-grass-forb structural stage (SGF) occurs when forest regeneration is impeded by inhospitable site conditions created by a severe disturbance, such as a stand-replacement wildfire; and only nontree species colonize the site. This stage, missing in the O'Hara et al. (1996) classification, was included because a "lag time" of ten to fifty years without conifer regeneration has been observed to occur after intense, high-elevation fires (Agee and Smith 1984; Keane et al. 1990a; Little and Peterson 1991; Agee 1993; Keane et al. 1994; Tomback et al. 1995). The nonforest vegetation that dominates after severe fires (i.e., shrubs, grasses, or forbs) is usually dictated by predisturbance composition, disturbance severity, and site conditions (Arno and Weaver 1990).

The stand initiation stage (SIN) occurs when trees are established as seedlings and saplings, either in clumps or as scattered individuals (Table 9-3). Stem exclusion stages (SEC, SEO) occur when sapling and pole-sized trees start to compete for resources such as light, water, and nutrients. The stem exclusion open canopy (SEO) occurs where severe environmental conditions, such as

Table 9-3. Descriptions of the structural stages used in this chapter.

Name	Abbr.	Description
Shrub-Grass-Forb	SGF	No apparent tree establishment, or establishment widely scattered. Shrubs and/or herbaceous plants dominate plant community.
Stand Initiation	SIN	Growing space is reoccupied following a stand-replacement disturbance. Successful establishment of tree species, ecesis, may be broken or continuous. Mostly seedlings or saplings. Shrubs, forbs, and grass may still dominate.
Stem Exclusion Open Canopy	SEO	Trees start competing for resources (light, water, nutrients), and underground competition limits establishment. Large sapling and pole stand. Open canopy because limited resources prevents crown closure.
Stem Exclusion Closed Canopy	SEC	Trees compete for resources as above but new individuals are excluded by competition for light and underground resources.
Understory Reinitiation	URI	Initiation of new cohort as older cohort occupies less than full growing space. A significant understory layer develops, two stratus stand. Usually shade-tolerant species in understory.
Young Forest Multistrata	YFM	Two or more cohorts present through establishment after periodic disturbances. Occurs only because overstory was killed by insects, disease, or harvesting. Usually old trees are removed.
Old Forest Multistrata	OFM	Two or more cohorts and strata present, including large trees. Multi-aged stand with assortment of tree sizes and canopy strata. Usually shade-tolerant species dominate understory and gain entrance to overstory.
Old Forest Single Stratum	OFS	Single stratum of medium to large, old trees of one or more cohorts. Structure is maintained by nonlethal underburns or timber management. Understory usually absent or scattered.

deep snow, seasonal water deficits, or limited nutrients, restrict forest canopy closure, but there is inter-tree competition (O'Hara et al. 1996). Sites where whitebark pine is the indicated climax species usually have an SEO rather than an SEC structural stage because of adverse site conditions. Understory reinitiation (URI) occurs when trees become established in the understory and provide a seedling and sapling pool that can fill forest gaps. These trees are usually shade-tolerant species able to compete under dense, shaded canopies. Old forest multistrata (OFM) structures usually have two or more well-developed canopy strata of different aged and sized shade-tolerant trees.

Two other structural stages are created entirely by disturbance. The young forest multistrata (YFM) occurs when disturbances such as blister rust, beetle, and timber harvest kills or removes the overstory stratum and leaves a young, uneven-aged layer of shade-tolerant tree species. The old forest single stratum

(OFS) is created by minor disturbances, often low-severity surface fires, that kill the small, understory trees but leave the large, overstory stratum intact.

Disturbances disrupt successional development and either delay or accelerate time spent in a succession class. But, more often, disturbance causes an immediate shift to another succession class. Only three disturbances are discussed in this chapter—wildland fire, blister rust, and mountain pine beetle—represented by the thin, solid, and dashed lines in Figures 9-2 and 9-3. The models and parameters presented here are only generalizations of successional processes in whitebark pine forests. Application of models and parameters to local situations may require adjustments of cover types and transition times, and the inclusion of additional disturbances (see Keane et al. 1996b).

Whitebark Pine Succession Models

Three models are used to describe successional dynamics in whitebark pine forests across the species' entire range, with each model designed for a specific environment. The Seral Whitebark Pine Model (Figure 9-2) presents successional dynamics for those upper subalpine sites where whitebark pine is the major seral species and is successionally replaced by shade-tolerant species. These sites support upright, closed-canopy forests but occur between the lower limits of krummholz stands and the upper-elevational limit of lodgepole pine (Figure 9-1). Nine tree species are often found on these sites, varying in importance across whitebark pine's range. Shade-intolerant species, including whitebark pine, limber pine, lodgepole pine, Douglas-fir, and alpine larch, occupy early- to midsuccession stages. The shade-tolerant subalpine fir, Engelmann spruce, and mountain hemlock eventually replace the shade intolerants (Minore 1979; Arno and Hoff 1990).

The Seral Lodgepole Pine Model (Figure 9-3) is identical to the Seral Whitebark Pine Model, except that there is an additional pathway for those sites where lodgepole pine can become a major component, such as in the Cascades and the greater Yellowstone ecosystem (Pfister et al. 1977; Arno and Weaver 1990). These sites are lower in elevation and can be somewhat warmer than the Seral Whitebark Pine Model sites (Figure 9-1). This model is used for those areas within whitebark pine's range where abundant lodgepole pine seeds are available either in the burned overstory or surrounding communities. The same nine tree species exist in this successional cycle, but favorable site conditions may allow lodgepole pine to dominate a stand. Alpine larch, mountain hemlock, and limber pine rarely occur as minor components of succession classes in this model.

The Climax Whitebark Pine Model (CWPM) (Figure 9-4) represents succession on those sites where whitebark pine is the indicated climax species, because it is the only conifer able to reproduce and mature in these severe environments (Clausen 1965; Pfister et al. 1977). Sites where the CWPM applies often correspond to the highest upper subalpine sites, or relatively cold, high snow environments, where trees may occur in clusters, groves, or tree islands (Arno and Hammerly 1984; Arno and Weaver 1990). Subalpine fir can occur on these sites, but only as scattered individuals with truncated growth forms (Pfister et al. 1977; Arno and Hoff 1990; Arno and Weaver 1990). Whitebark pine also occurs in krummholz form on treeline sites (Tomback 1986; Arno and Hoff 1990), and as a minor seral component in lower subalpine sites (Pfister et al. 1977; Cooper et al. 1991), but these sites are not discussed in this chapter. Habitat types where these models would apply are further discussed by Arno (Chapter 4).

Pathways of these three models are assumed to remain constant across the range of whitebark pine, but the parameters that quantify these pathways will vary by geographic area or climatic zone. Therefore, transition time parameters in Tables 9-1 and 9-2 were stratified by five regions (Figure 9-5). The Northern Rockies region includes those lands in the Rocky Mountains south of Canada, north of Missoula, Montana (47° N latitude), and west of the Continental Divide to eastern Washington (Arno 1979). The Central Rockies are those lands in whitebark pine's range south of Missoula and east of the Continental Divide. Lands that support whitebark pine north of the U.S. border comprise the Canadian Rockies region. Whitebark pine sites along the Cascade Crest and adjacent ranges make up the Cascade region; and whitebark pine

Figure 9-5. Geographical regions corresponding to the stratification of succession transition time parameters presented in Tables 9-1 and 9-2.

forests in the Sierra Nevada and Great Basin ranges are in the Sierra Nevada region.

Successional Dynamics

The shrub-herb–shrub-grass-forb (SH-SGF) succession class occurs after severe disturbances in upper subalpine forests, which modify the site so trees cannot become immediately established. It takes time for nontree species to ameliorate these severe site conditions (Agee and Smith 1984; Kendall and Arno 1990; Keane et al. 1994). The length of time spent in this class mostly depends on the severity of the fire and the harshness of the site, with the longest transition times occurring in the CWPM (Tables 9-1 and 9-2). Transition from the SH-SGF to the next succession class in the seral models depends on three factors—seed proximity, seed abundance, and site harshness. If subalpine fir has frequent and large cone crops, there are abundant seed sources close to the disturbed area (McCaughey et al. 1986), and time in the SH-SGF class is short (less than ten years), there is a high probability that the next class will be subalpine fir–stand initiation (SF-SIN). However, if subalpine fir seed sources are distant (i.e., large burn), or the site was severely burned and lag times are long, then the whitebark pine successional pathway will probably result from nutcracker caching (Figure 9-2). Whitebark pine colonizes severely altered sites better than fir (McCaughey and Schmidt 1990; Arno and Hoff 1990). Moreover, additional whitebark pine cone crops will occur the longer a disturbed site remains in the SH-SGF class, which means there will be a greater chance nutcrackers will cache seeds on the disturbed site. Mixed-severity burns often create small patches on the landscape that can be colonized by a mixture of these two species (Arno and Weaver 1990; Arno et al. 1993). In the Seral Lodgepole Pine Model, presence of lodgepole pine seeds dictates the appropriate pathway after disturbance: If there are lodgepole pines in the overstory, or there are abundant lodgepole pines with semiserotinous cones surrounding the burn, then it is likely that the lodgepole pine pathway will result (Figure 9-3).

The whitebark pine– or lodgepole pine–stand initiation (WP-SIN or LP-SIN) class is made up of seedlings and saplings that will grow to mature pine trees in the seral models (Figures 9-2 and 9-3). As these trees grow, their roots and crowns compete for light, water, and nutrients, triggering the passage into the stem exclusion structural stage (SEC or SEO, depending on seral or climax model). Tree crowns start to close as sunlight is captured by the most competitive individuals in the stand (SEC structural stage) (Warren et al. 1997). Many trees are not able to compete for light or other belowground resources, so they die, creating gaps in the overstory canopy that allow limited light and water to the forest floor, resulting in regeneration of shade-tolerant species (e.g., subalpine fir) to create the whitebark pine– or lodgepole pine–understory reinitiation class (WP-URI or LP-URI) (Figures 9-2 and 9-3). As succession advances, shade-tolerant understory regeneration fills canopy gaps, creating multistoried stands where tall, old whitebark pines become fewer in the overstory and subalpine fir becomes dominant in both the understory and overstory, thereby defining the subalpine fir–old forest multistrata class (SF-OFM). Eventually, all whitebark pine die, creating a pure subalpine fir cover type that replaces itself in perpetuity. Essentially identical stand development processes occur in the lodgepole pine (LP) and subalpine fir (SF) pathways of both seral models, except fir assumes the role of whitebark pine in the SF pathway.

Successional pathway direction after the shrub-herb–shrub-grass-forb class (SH-SGF) is dependent on many landscape and ecosystem characteristics. If the predisturbance stand was mixed fir, spruce, and pine, and the disturbance did not affect a large area, then a mixed stand will probably develop after the disturbance. The tree species that will dominate these mixed communities, and therefore define the cover type, depend on the timing of cone crops, degree of serotiny, postdisturbance seedling survival, and distance to seed sources. For example, a SH-SGF site with several good subalpine fir cone crops in adjacent undisturbed stands and little pine cone production will probably develop along the subalpine fir (SF) pathway. However, large, stand-replacement fires (>1,000 hectares) near abundant whitebark pine seed sources generally follow the whitebark pine (WP) pathway, because the wind-dispersed fir, spruce, and lodgepole pine seeds have limited dispersal distances (McCaughey et al. 1986). Whitebark pine clearly has the colonization advantage in large burns because nutcrackers have been observed dispersing whitebark pine seeds over 10 kilometers (Tomback et al. 1990; Chapter 5). Lodgepole pine usually has the competitive advantage over whitebark pine on warmer upper subalpine sites because of its fast growth, serotiny, and copious seed production. However, if a severe fire kills immature lodgepole pine that have not produced many cones, then the stand will probably progress along the WP pathway (Cattelino et al. 1979; Kessell and Fischer 1981; Bradley et al. 1992).

A slightly different stand-development process occurs in the Climax Whitebark Pine Model (CWPM), since subalpine fir is unable to dominate these sites. Instead, whitebark pine regeneration fills gaps made by the dying overstory. But whitebark pine, which is only moderately shade-tolerant (Minore 1979), has difficulty growing under any forest canopy (Weaver and Dale 1974; Arno and Hoff 1990; Mattson and Reinhart 1990). So, rates of gap replacement are much slower and the size of the gaps are much larger than those observed for the seral models. Seldom do stunted understory whitebark pine that are suppressed for more than fifty years release after gaps occur to become part of the overstory canopy (Keane et al. 1994). As a result, only the open, stem exclusion structural stage (SEO) is represented in the CWPM successional cycle. However, we have added the WP-URI class to represent those stands where a whitebark pine understory is sparse but present. The final stage in the CWPM is the whitebark pine–old forest single strata (WP-OFS) class. This class occurs as a single stratum rather than multiple strata because of the paucity of understory trees.

Disturbance can alter cover types, structural stages, or succession classes in the seral models (Figures 9-2 and 9-3). As mentioned earlier, stand-replacement fires kill most trees and reset successional development to the shrub-herb–shrub-grass-forb (SH-SGF) class. On more productive upper subalpine sites, trees can become established immediately after fire; so, the SH-SGF class is often skipped or shortened (Tomback et al. 1990, 1993). Mixed- and low-severity, nonlethal surface fires kill most small trees and those large trees unable to survive the fire (i.e., subalpine fir) to create single-storied stands of the whitebark pine–old forest single strata (WP-OFS) succession class.

Blister rust and mountain pine beetles often have the opposite effects of fire. These agents tend to kill the seral overstory species, leaving a young, shade-tolerant, multistrata forest composed of tree species that are not the hosts of rust or beetle (i.e., subalpine fir or the SF-YFM class). This accelerates succession to the subalpine fir–old forest multistrata (SF-OFM) class. Blister rust is especially fatal to young whitebark pine, but infection does not seem to occur until these saplings become tall enough to catch rust spores entrained in the prevailing wind stream (Keane et al. 1994; Tomback et al. 1995). If whitebark pine rust resistance is low, local whitebark pine extinctions would probably occur over long time periods, creating landscapes of subalpine fir stands in different stages of succession. Blister rust has not yet been observed to cause extensive damage on sites represented by the climax model (CWPM), but if it does, the subalpine fir–shrub-grass-forb (SF-SGF) successional class would probably be created (Figure 9-4). In theory, this community would be composed of stunted subalpine fir in very small amounts scattered among the whitebark pine snags, which is why it is given a SGF structural stage. Beetle epidemics do not severely impact the smaller whitebark pine trees, so young stands are often spared heavy mortality.

Successional Effects

Effects of advancing succession are extensive and widely varied with many ecosystem properties directly affected by successional development. Presented in this section is a summary of the effects of succession on four important ecosystem properties: ecosystem processes (e.g., disturbance, productivity), floristics (e.g., vegetation composition), ecosystem characteristics (e.g., fuels, vegetation structure), and landscape characteristics (e.g., spatial pattern).

Ecosystem Processes

Some interesting and complex changes in major ecological processes result as subalpine fir replaces whitebark pine in the upper subalpine successional cycle (see summary in Table 9-4). Stand leaf area generally increases with succession because subalpine fir has longer needle retention, higher leaf area to sapwood ratios, and higher leaf mass in the crown (Brown et al. 1985; Peterson et al. 1989; White et al. 1990; Keane et al. 1996c; Callaway et al. 1998). This fundamental species- and stand-level change triggers complex responses in the biogeochemical cycles of the stand (Table 9-4). Transpiration and canopy interception increase with higher leaf areas, which results in periodic seasonal depletion of soil water, increased canopy evaporation, and decreased streamflow (Kaufmann et al. 1987; Troendle and Kaufmann 1987; Hann 1990; Waring and Running 1998). High leaf areas reduce understory light, which can slow decomposition rates, reduce litter and duff drying, limit snow accumulation, and delay soil thaw (Bazzaz 1979; Huston and Smith 1987; Kaufmann et al. 1987; White et al. 1990). Reduced decomposition can change carbon and nutrient cycling, resulting in high woody fuel and duff accumulations that fos-

Table 9-4. Successional changes in important ecophysiological, structural, and ecological characteristics as seral whitebark pine stands are replaced by shade-tolerant subalpine fir.

Attribute	Whitebark Pine Stands	Subalpine Fir Stands
ECOLOGICAL PROCESSES		
Leaf area	low	high
Interception	low	moderate
Transpiration	low	moderate
Autotrophic Respiration	moderate	moderate
Productivity	high	moderate
Decomposition	moderate	low
Net ecosystem productivity	moderate	low
Streamflow	moderate	low
Fire regime	mixed	stand-replacement
STRUCTURAL CHARACTERISTICS		
Duff and litter depth	low	high
Number of logs	low	high
Fuel loading	low	high
Understory cover	high	low
Hiding cover	low	high
Thermal cover	low–moderate	high

ter high-severity wildfires with deep soil heating and high plant mortality. White et al. (1990) found high nitrogen and magnesium levels in whitebark pine litter, whereas subalpine fir and Engelmann spruce litter had high calcium and lower lignin contents. They also found that decay rates were similar under whitebark pine and subalpine fir canopies of equivalent leaf area.

Callaway et al. (1998) investigated the changes in carbon allocation patterns as seral whitebark pine are successionally replaced by subalpine fir. They found that whitebark pine trees stored large amounts of water in a thick sapwood layer. This, combined with low leaf mass (Brown et al. 1985), results in low leaf area to sapwood volume ratios, which enables whitebark pine to survive through seasonal water shortages but decreases its competitiveness because the thicker sapwood layer incurs higher respirative losses. Subalpine fir has higher leaf area to sapwood volume ratios but lower photosynthetic rates (Ryan 1987; Callaway et al. 1998). Callaway et al. (1998) found that upper subalpine sites dominated by whitebark pine reached peak net primary productivities (NPP) at around 150 years. Sources in the ecophysiological literature contend that NPP will decline thereafter until old-growth stands produce just enough carbon to meet respiration requirements (Waring and Running 1998). However, they assume the same tree species are present throughout stand development. Callaway et al. (1998) found that NPP decreased only slightly as subalpine fir trees replaced whitebark pine in the stand, because the fir's thin sapwood layer reduced stemwood respiration, thereby maintaining high productivities into the late stages of succession.

Perhaps the most important disturbance process altered by advancing succession is the fire regime. Fire regimes in whitebark pine forests are typically low to mixed-severity burns occurring at 30- to 500-year intervals (Arno 1986; Brown et al. 1994; Morgan et al. 1994). This is primarily because fuel loads and duff depths are low (Brown and See 1981; Brown and Bevins 1986), and tree crowns are heat porous and high off the ground (Arno and Hoff 1990). However, fire severity and return intervals change drastically as fires are excluded from the landscape by humans, especially in the settings where whitebark pine succeeds to fir. Encroachment by subalpine fir creates multilayered canopies with low crowns and high crown bulk densities (kg biomass m^{-3}) because of higher leaf areas. Slow decomposition rates and higher foliar biomass in subalpine fir forests can result in fuel accumulations that precipitate higher surface fire intensities and flame lengths (Lasko 1990). The higher flame lengths coupled with lower and thicker subalpine fir crowns hasten the transition of a surface fire to a crown fire. And once a crown fire has started, the high crown biomass ensures the propagation of the crown fire throughout the stand and across the landscape. Therefore, the chief long-term consequence of fire exclusion in whitebark pine ecosystems is the conversion of a low- to moderate-severity, mixed fire regime to a stand-replacement, crown fire regime (Steele 1960; Loope and Gruell 1973; Arno et al. 1993; Hartwell 1997). Fires in this new regime will tend to be larger and more intense, and therefore more difficult to control (Keane et al. 1996d).

Murray (1996) notes that whitebark pine landscapes in small, island range wilderness settings are especially vulnerable to fire exclusion. Few fires now reach high-elevation ecosystems in these small, narrow mountain ranges because low-elevation ignitions are usually quickly extinguished before the fire can spread to the upper subalpine forests. As a consequence, high-elevation landscapes in the West Big Hole Range of Montana and Idaho, composed of lodgepole and whitebark pine communities from 1753 to 1913, are becoming dominated by subalpine fir and spruce (Murray 1996).

Insect and disease disturbance processes are also directly affected by the shift in host tree species across the landscape. Beetle and rust epidemics are replaced by root rot and other subalpine fir diseases as the landscape converts from whitebark pine to subalpine fir cover types (Arno and Hoff 1990; Alexander et al. 1990). Landscapes dominated by one cover type have the greatest potential for epidemic infestations of insects and disease because host species patches are spatially connected. Patchy landscapes are barriers to pathogen dispersal.

Floristics

Successional floristics are somewhat simple in whitebark pine ecosystems and are best described by the Initial Floristics Model of Egler (1953). There are few dramatic shifts in vascular plant composition and cover, because the plant species present prior to the disturbance usually populate the site after disturbance (Weaver and Dale 1974; Forcella 1978; del Moral 1979; Keane et al. 1994). In the seral models, species richness remains low throughout the successional cycle, and many of the same species appear in nearly all seral stages at approximately the same cover as the predisturbance community. Keane et al. (1994) sampled different-aged whitebark pine stands on similar sites in the Bob Marshall Wilderness Complex, a 1.2 million-hectare wilderness where over 40 percent of the area has the potential to support whitebark pine, and found similar plant compositions across most successional communities. However, the cover and number of individuals by species usually decrease under dense subalpine fir canopies (Craighead et al. 1982; Arno and Weaver 1990; Mattson and Reinhart 1990; Keane et al. 1994).

Plant diversity in upper subalpine communities is more related to site conditions than to successional gradients (del Moral 1979; Arno and Weaver 1990). Some timberline plant species can be found in stands representative of the Climax Whitebark Pine Model, probably because of lower overstory canopy cover, seasonal wetness, and the severity of the site, which contributes to higher species richness and diversity (Reed 1976; Pfister et al. 1977; Cooper et al. 1991). But the same species are still present before and after a disturbance (Thompson and Kuijt 1976; del Moral 1979). It may be that successional shifts in plant species are a consequence of site environmental changes rather than direct competition and species facilitation (Forcella 1978; Craighead et al. 1982; Murray 1996).

Ecosystem Characteristics

Many ecosystem attributes change during the successional pathways presented in Figures 9-2, 9-3, and 9-4. Live and dead fuel loadings generally increase, probably because of decreased decomposition rates and higher branchfall and litterfall rates (Table 9-4). There is usually an increase in coarse woody debris (downed logs > 7 centimeters diameter), twigwood, and branchwood (Brown and See 1981). Duff depths generally increase proportionally to the density of subalpine fir. The soils are often classified as Cryochrepts or Cryoboralfs during whitebark pine dominance, but as Cryoborolls under subalpine fir canopies because of higher organic matter in soil horizons (Hansen-Bristow et al. 1990). Whereas mountain hemlock and red fir (*Abies magnifica*) stands had many more trees in the youngest age classes, Parker (1988) found that whitebark pine forests had even-aged diameter distributions. Eggers (1986) mentions that whitebark pine cone production declines with increases in subalpine fir in the understory. In lower subalpine environments where whitebark pine was historically abundant (Hartwell 1997), Arno et al. (1993) found that modern fire regimes have resulted in lodgepole pine outcompeting whitebark pine. Both Arno et al. (1993) and Hartwell (1997) measured a decrease in seedling and sapling whitebark pine stands and an increase in older, mature stands from 1900 to 1991 in the absence of disturbance.

Landscape Characteristics

Succession dynamics in whitebark pine forests will alter landscape composition and structure. Keane et al. (1996a) found that blister rust and fire exclusion had reduced the extent of whitebark pine cover types in the Interior Columbia River Basin by about 9,000 km^2 or a decline of approximately 50 percent over the last ninety years. Historically, whitebark pine stands in this region had diverse stand structures; but now, only the old-forest, single-stratum whitebark pine stands remain, because these occur on severe sites (i.e., Climax Whitebark Pine Model) so far inhospitable to blister rust. In the Bob Marshall Wilderness Complex, 56 percent of the whitebark pine stands were experiencing moderate to heavy rust mortality. As a result, over 22 percent of the upper subalpine landscape is composed of subalpine fir cover types when historically they typically comprised about 5 to 10 percent (Keane et al. 1994). Arno et al. (1993) estimated that whitebark pine stands comprised about 14 percent of the lower subalpine landscape circa 1900, but now these stands have been successionally replaced by subalpine fir, lodgepole, and other tree species. Pre-1900 fires on this landscape were relatively frequent and patchy, creating a fine-grained mosaic of young and mixed-aged lodgepole and whitebark pine communities with few late-successional subalpine fir stands. Since 1900, disturbances have been infrequent, and accelerated succession has resulted in a 34 percent increase in late seral communities. In the Bitterroot Mountains of west-central Montana, subalpine fir increased its extent from 10 percent to 42 percent during 1900 to 1995 above 2,100 meters elevation, while whitebark pine decreased from 32 percent to 17 percent (Hartwell 1997). Stands in these upper-elevation landscapes are also becoming multistoried, with greater subalpine fir in the understory.

Subalpine landscapes in the West Big Hole Range of Montana and Idaho have also experienced a shift in dominance (Murray 1996). Decreases in fire frequency and size have resulted in the conversion of lodgepole pine and whitebark pine forests to subalpine fir, albeit at a rate slower than that observed in the Bitterroot Mountains of Montana because of the lower levels of blister rust infection. Murray (1996) also noted the encroachment of pine and fir into subalpine meadows originally created by fire. This encroachment essentially produces more uniform landscapes with little fuel diversity and natural fuelbreaks.

Successional development without disturbance also has an important effect on landscape pattern. Upper subalpine landscapes tend to become more homogeneous as succession converts whitebark pine stands to subalpine fir stands (Hann 1990). As a result, late seral landscapes generally are less fragmented with lower patch density and patch diversity, and a greater connection to larger patches (Hann 1990; Murray 1996; Keane et al. 1999). This creates an interesting situation: Larger patches and higher homogeneity tend to create more continuous crown and surface fuels, so fires burning in these fuels will tend to be larger and more severe. These large fires will then tend to create more homogeneous landscapes and so on. Patchy landscapes have early seral stands that are less flammable than dense, mature subalpine fir stands. Consequently, it appears that humans have modified fire regimes on some whitebark pine landscapes from frequent, mixed-severity, patchy fires to large, severe, and infrequent crown fires.

Succession Modeling

Few simulation models have been applied to, and still fewer were actually built for, whitebark pine ecosystems. Presented here are some of the major modeling efforts for whitebark pine ecosystems. It is noteworthy that these modeling studies generated nearly the same results as observed in the field: Without fire, whitebark pine will eventually be replaced by subalpine fir, and this replacement is accelerated by mortality from blister rust and mountain pine beetle. The primary differences between the modeling efforts are the approaches used to simulate and describe this successional change. In some cases, we have used these models to simulate new results germane to restoration efforts in whitebark pine.

FIRESUM

Perhaps the most comprehensive simulation of whitebark pine systems uses the stand-level, individual tree, ecosystem process gap model FIRESUM (Keane et al. 1989, 1990a, 1990b). This model indirectly simulates successional processes from annual estimates of available light and water, tree density, and temperature. Published results indicate that the rate of replacement of whitebark pine by subalpine fir in the absence of fire differed by environment (Keane et al. 1990b). It took more than 200 years for subalpine fir to gain overstory dominance in the absence of fire for a high-elevation, moist, seral whitebark pine site in the Bitterroot Mountains of Montana. Other lower-elevation Bitterroot sites required only 100 to 150 years for this successional process to occur. It required about 400 years without fire to achieve subalpine fir dominance on a high-elevation site in the Bob Marshall Wilderness Complex of Montana near the upper limits of the Seral Whitebark Pine Model sites (Keane and Morgan 1994). Although simulated crown fires tended to perpetuate whitebark pine, crown fires coupled with blister rust infections accelerated the conversion of diverse landscapes to primarily subalpine fir cover types (Keane et al. 1994). In addition, whitebark pine vigor and whitebark pine cone production (cones ha^{-1} yr^{-1}) increased in stands with intact fire regimes.

Simulated stands where whitebark pine is the climax species (i.e., Climax Whitebark Pine Model, Figure 9-1) exhibited very different successional behavior (Keane et al. 1994). Whitebark pine remained dominant throughout 500 years of simulation with and without fire, but whitebark pine basal area nearly tripled in stands where fire was excluded. Murray (1996) found a 12 percent increase in basal area for similar stands in the West Big Hole Range. Simulated blister rust infections caused two different trajectories, depending on the severity of the site. After the majority of whitebark pine died from rust, subalpine fir was able to remain in the stand and gain a limited degree of dominance, but at very low levels (<10 m^2 ha^{-1}) on some severe upper subalpine sites. However, high, cold, dry sites were converted from whitebark pine forests to semipermanent shrub-herb cover types once the rust killed all pine trees, because the fir could not become established.

Consequences of fire regime modifications and mountain pine beetle epidemics were also investigated by using FIRESUM for several western Montana subalpine stands (Keane et al. 1990a; Keane and Morgan 1994). Fires that occurred within 50 years of the average fire return interval maintained whitebark pine on the landscape. Fires happening at shorter intervals were typically of low intensity because there was inadequate fuel to carry fire through the stand. Less-frequent fires tended to be of high severity, where nearly all whitebark pine and subalpine fir trees were killed. Another major simulation finding was the importance of a stochastic fire interval, or a fire interval that mimics natural variability. Whitebark pine tended to decline on landscapes that were burned on fixed schedules (i.e., every 150 years). This is because most whitebark pine saplings were unable to grow above lethal scorch heights during the static period between fires (Keane and Morgan 1994). Lodgepole pine was the dominant species with 50-year fixed fire intervals on warmer sites because it was able to produce seeds at younger ages than were most other conifers (Keane et al. 1990a). Stochastically simulated fire intervals (i.e., historical fire intervals) created nearly pure stands of whitebark pine on both seral and climax sites, but the basal area of all species was low (<20 m^2 ha^{-1}). Mountain pine beetle epidemics without fire hastened the successional replacement of whitebark pine by subalpine fir and spruce, but when beetle epidemics occurred on landscapes with fire, whitebark pine was able to regenerate and maintain dominance over time. Blister rust infections accelerated succession by an order of magnitude on seral whitebark pine sites, and even the reintroduction of fire could not offset the damage caused by the rust (Keane and Morgan 1994).

The FIRESUM model was used to simulate three new scenarios using the same site and stand information for the Shale Mountain site in western Montana as reported by Keane and Morgan (1994). First, the stochastic fire return interval was varied around the historical mean of 250 years to investigate the sensitivity of whitebark pine basal area to fire frequency without blister rust (Figure 9-6a). Frequent fires (100-year stochastic intervals) tended to decrease whitebark pine basal area and eventually produce landscapes in perpetual seedling-sapling structural stages. The 200-year fire regime maintained whitebark pine at stable, but low, basal areas. The 300-year fire regime resulted in the highest basal areas, but produced the greatest fluctuations caused by high-intensity fires. There was a general decline in basal area for the 400-year fire interval because of subalpine fir invasion, and fluctuations were much lower in magnitude. Whitebark pine became locally extinct by simulation year 800 for the No-Fire run (Figure 9-6a). The major finding of this simulation exercise was the apparent wide range of fire return intervals to maintain whitebark pine dominance. Fire frequency can be within 10 to 30 percent (±50 to 100 years) of the natural fire return interval and still maintain whitebark pine in the stand.

FIRESUM was then employed to determine what level of rust resistance was needed to conserve whitebark pine. Simulation results indicated that at least 20 to 25 percent of whitebark pine trees need to be resistant to prevent the local extinction of whitebark pine on the landscape (Figure 9-6b). This assumes no gain in rust resistance over future generations. Krebill and Hoff (1995) found that seedlings grown from seeds taken from whitebark pine stands with high rust mortality (greater than 90 percent) have over 40 percent rust resistance, well above the 20 percent threshold indicated by modeling. Thus, there is a good chance that whitebark pine will be present on future landscapes, given regeneration opportunities (Chapter 17).

Last, the effect of fire size on whitebark pine seed dispersal and postfire stand dynamics was investigated with FIRESUM by simulating stand-replacement fires at years 50 and 600 and varying seed-dispersal distances (i.e., distance to nearest seed source) from 100 meters to 100 kilometers by orders of ten (Figure 9-6c). Results indicate that the further the seed source was from the burn, the longer it took for whitebark pine to reach the maximum basal area

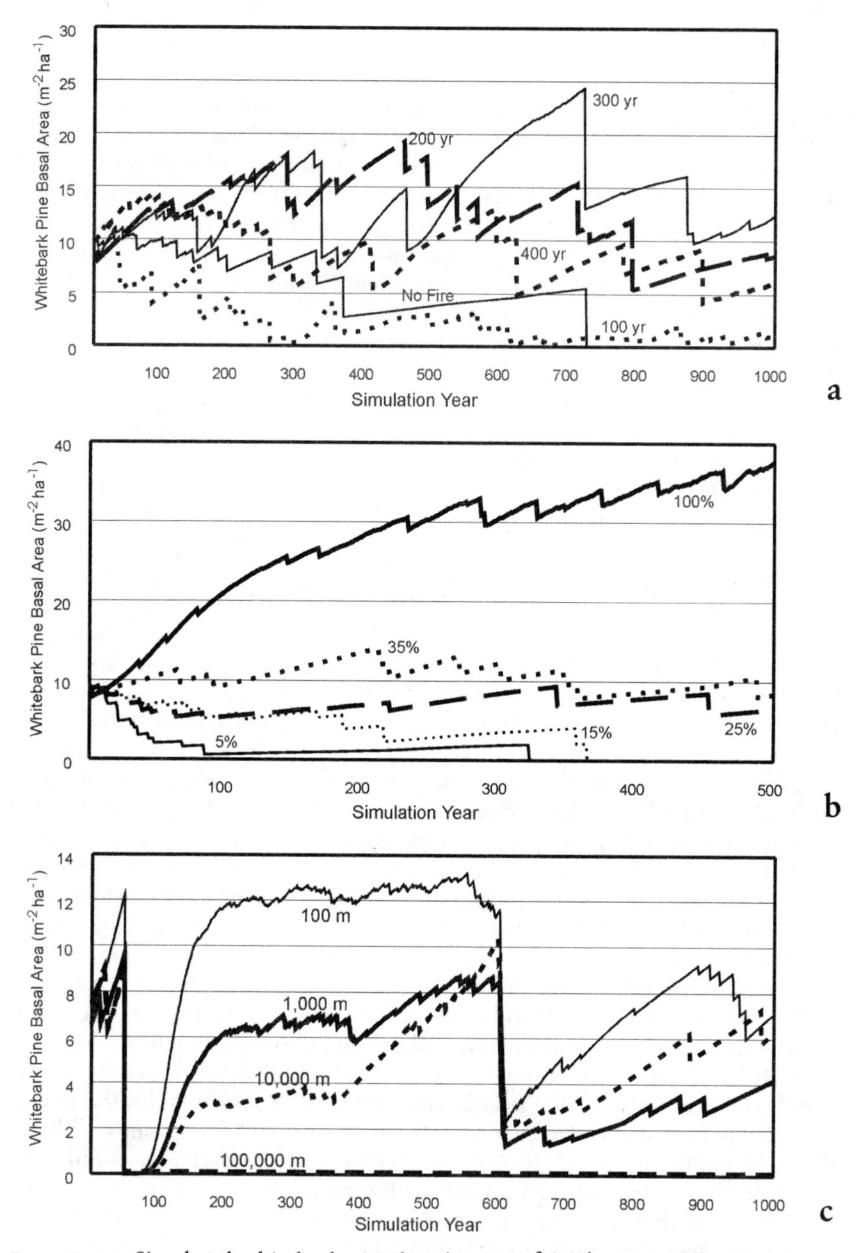

Figure 9-6. **a.** Simulated whitebark pine basal area (m^2 ha^{-1}) using FIRESUM for the Shale Mountain site in Keane and Morgan (1994), where four stochastic fire frequencies (100-, 200-, 300-, and 400-year intervals) were investigated over 1,000 years. The historical fire return interval was approximately 250 years between fires. **b.** Simulated whitebark pine basal area (m^2 ha^{-1}) from FIRESUM for the Shale Mountain site in Keane and Morgan (1994) where four levels of rust resistance (5, 15, 25, and 35 percent resistance) are modeled. **c.** Simulated whitebark pine basal area (m^2 ha^{-1}) from FIRESUM for the Shale Mountain site in Keane and Morgan (1994) where four seed-dispersal distances (100, 1,000, 10,000, and 100,000 meters) are modeled for 1,000 years. A stand-replacement fire is simulated at years 50 and 600.

for crown closure (Figure 9-6c). There seems to be a maximum distance threshold between 10 kilometers and 100 kilometers where whitebark pine cannot colonize large burns. This threshold distance is estimated by simulation to be 15 to 25 kilometers, which is consistent with the longest seed-dispersal distance observed for Clark's nutcrackers (Tomback 1998; Chapter 5). The long establishment periods associated with long seed-dispersal distances may indicate that maximum seed-dispersal distances should *not* be used as a criterion for determining those populations at risk. Rather, 10 kilometers may be more realistic when all stand-development processes are taken into account. These FIRESUM simulations also revealed that subalpine fir did not become established when dispersal distances were greater than 1,000 meters, indicating whitebark pine has a distinct colonization advantage when fires are greater than 300 hectares but less than 30,000 hectares.

Fire-BGC

The spatially explicit biogeochemical model Fire-BGC was used to investigate landscape- and stand-level changes in ecosystem processes and characteristics in fire-dominated environments such as whitebark pine ecosystems (Keane et al. 1996c). Fire-BGC explicitly simulates basic ecosystem processes that govern tree regeneration, growth, and mortality, such as photosynthesis, respiration, and evapotranspiration, for each tree on the landscape at a daily time interval. Fire-BGC was tested on a 20,000-hectare, upper subalpine landscape in the Bob Marshall Wilderness Complex of Montana and produced several interesting results. After 200 to 250 years, the coverage of subalpine fir–dominated stands on the landscape increased from 8 percent to over 90 percent. Simulations with fire at historical fire frequencies (around 200 years) indicated that subalpine fir stands comprised from 8 to 21 percent of the landscape over 500 years (Keane et al. 1996c). The simulated rate of successional advancement depended on the biophysical environment, with the highest, driest sites having the slowest successional development and the northerly, moist slopes quickly converting to fir. Duff, fuel loadings, and leaf areas all increased on landscapes without fire. When the exotic disease blister rust was introduced on fire-excluded landscapes, it took only 70 years for subalpine fir to cover 90 percent of the landscape. Simulated historic fire regimes on rust-infected landscapes only slowed the conversion to subalpine fir by taking approximately 130 simulation years for subalpine fir to cover 90 percent of the landscape. Subalpine fir landscapes were more homogeneous and less fragmented than seral whitebark pine landscapes. Extensive field validation of Fire-BGC revealed that simulated tree diameter growth was within 20 percent of sampled tree cores, but duff depths and fuel load predictions varied between 3 percent and 130 percent of those sampled in the Bob Marshall Wilderness.

The simulated successional changes in species composition and structure affected most other simulated ecosystem processes (Keane et al. 1996d, 1997a). Ecosystem autotrophic and heterotrophic respiration increased as subalpine fir tree density and duff depths increased. Nitrogen tended to accumulate in the duff and litter on the forest floor because of slower decomposition (see Table 9-4). Transpiration and interception increased with rising subalpine fir leaf area, canopy cover, and landscape cover (Keane et al. 1996d; Callaway et al.

1998). Soil water deficits were more common in dense subalpine fir stands; and, as a result, runoff and streamflow tended to decrease with the increasing fir dominance, but this was highly dependent on yearly snowpack. Decomposition rates were much higher in stands dominated by whitebark pine, which contributed to more shallow duff and litter layers.

Climate change and fire regime scenarios were simulated on a large landscape in Glacier National Park using Fire-BGC (Keane et al. 1996d, 1997a, 1999). The 250-year simulations revealed that future fires will be more expansive, intense, and severe under a climate-warming scenario because of higher amounts of fuel in the stand and higher fuel connectivity across the landscape (Keane et al. 1997a). Whitebark pine increased its coverage across the landscape in this new environment in the absence of rust, because it had the superior ability to colonize these larger burns. Apparently, upper subalpine forests and alpine environments will play a large role in future fire dynamics: As the climate warms, these forests will become more productive and move upward in elevation, thereby creating a landscape with more continuous fuels that allow fire to traverse across watersheds (Keane et al. 1996d, 1997a). Simulated Glacier Park landscapes, where all fires are suppressed under warmer climates, will tend to release more carbon dioxide into the atmosphere than will landscapes with intact fire regimes, because of high respiration rates in older stands (Keane et al. 1997a). Landscapes with climate warming and no fire will tend to be more connected, less diverse, and less patchy (Keane et al. 1999).

CRBSUM

A general decline of whitebark pine cover types was also predicted by the coarse-scale spatially explicit model CRBSUM, as applied to the 880,000 km^2 Interior Columbia River Basin (Keane et al. 1996b). CRBSUM simulates succession using the pathway approach shown in Figures 9-2, 9-3, and 9-4, and simulates disturbance using probabilities of occurrence. This model predicted a 90 percent decline in whitebark pine cover types across the region after 100 years due to blister rust epidemics and fire exclusion (Keane et al. 1996a). However, simulated whitebark pine land cover increased by 20 percent from present conditions, when restoration activities, such as the reintroduction of native fire regimes and rust-resistant tree plantings, were implemented across the basin. Keane et al. (1996a) then applied CRBSUM to the Bob Marshall Wilderness Complex in Montana and found that the simulated rate of whitebark pine decline was much faster, with a 90 percent loss of whitebark pine cover types predicted to occur in only 50 years. A similar restoration scenario increased whitebark pine cover, but at a slower rate than that simulated for the Columbia River Basin.

Several programs similar to CRBSUM, but developed for mid- to fine-scale land planning applications, have also been applied to whitebark pine forests. For example, Keane et al. (1997b) created LANDSUM from CRBSUM to investigate the effects of land management practices on landscape composition and structure for four landscapes of different sizes. LANDSUM operates at a polygon level, so it can be used at fine to coarse scales. Keane and Long (Keane et al. 1997b) applied LANDSUM to the Bob Marshall Wilderness Complex at a 30-meter resolution and obtained approximately the same result as the CRB-SUM model. Chew (1997) has developed SIMPPLLE, a more detailed pathway model that simulates landscape changes in composition and structure using a multiple-scale, expert systems approach. The VDDT model, discussed below, was used to quantify input parameters for both LANDSUM and CRBSUM.

VDDT

The Vegetation Dynamics Development Tool (VDDT) is a nonspatial, succession model identical to CRBSUM, but it does not have a landscape implementation (Beukema and Kurz 1995). VDDT simulates only one biophysical setting (i.e., potential vegetation type or habitat type) at a time, but it is an extremely useful computer program for creating, revising, and simulating disturbance and succession for mid- to fine-scale landscapes.

The Seral Whitebark Pine Model presented in Figure 9-2 was coded into the VDDT program to investigate the successional consequences of modified disturbance regimes (results in Figure 9-7). Parameters in Table 9-1 for the Northern Rockies region were used to quantify successional dynamics, and it was assumed the fire rotation on the simulated landscape was approximately 200 to 300 years, depending on cover type and structural stage. It was also assumed that 10 percent of all shrub-herb–shrub-grass-forb (SH-SGF) succession class progressions would branch to the subalpine fir (SF) pathway, while the remainder went to whitebark pine (WP) pathway (Figure 9-3). Six scenarios were simulated for 500 years; three levels of fire occurrence with and without insect and disease disturbances. Fires were modeled as they occurred historically prior to 1900 (200- to 300-year intervals), under a fire exclusion scenario where 50 percent of the fires are extinguished, and under a full-suppression scenario where 90 percent of the fires are suppressed. Mountain pine beetles and blister rust were assumed to be the major insects and disease perturbations in the whitebark pine cover types, whereas root rot was the major disturbance in subalpine fir cover types, and their occurrence probabilities were taken from Keane et al. (1996b). Initial conditions were created by evenly dividing the landscape into every succession class (Figure 9-3), and then running VDDT for 1,000 years under the historical fire regime scenario without insects and disease, and then using the last year of the 1,000-year run as the first year in the simulation.

VDDT simulation results showed a marked decline in whitebark pine as fire was removed from the landscape (Figure 9-7a,b,c). Under the historical fire regime with no insects and diseases, the simulated landscape was somewhat stable, comprised of 60 to 75 percent whitebark pine and around 15 to 25 percent subalpine fir (Figure 9-7a). As fire was suppressed on the landscape, whitebark pine cover decreased as subalpine fir cover increased (Figure 9-7b,c). Under full fire exclusion, subalpine fir dominated the landscape by simulation year 200 as both whitebark pine and shrub-herb cover types rapidly declined at about 1 percent per decade to a low of 10 percent. This seems consistent with the findings of Keane et al. (1994) for the Bob Marshall Wilderness Complex, where they calculated about a 1.5 percent per decade decline rate.

A dramatic decline of whitebark pine is simulated when insects and disease

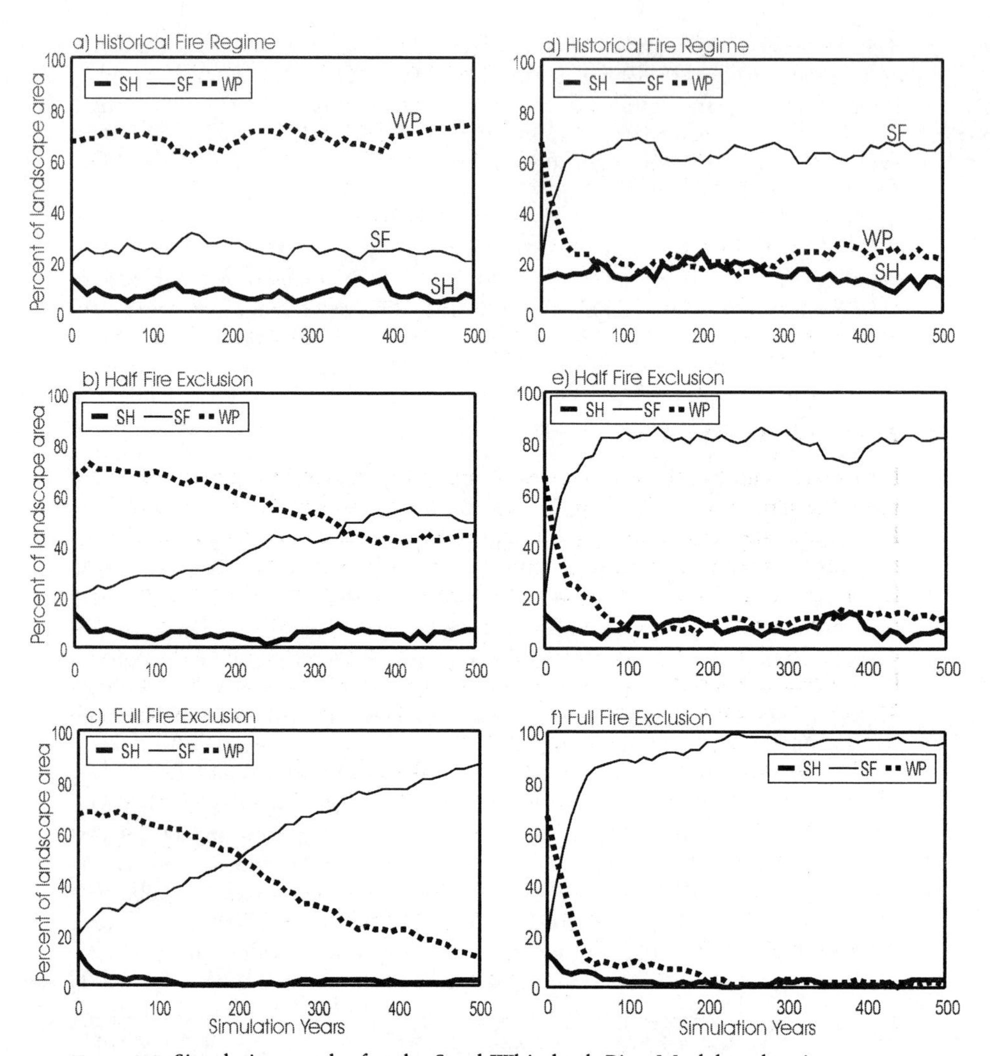

Figure 9-7. Simulation results for the Seral Whitebark Pine Model under six scenarios using the VDDT model parameterized for the Northern Rockies geographic region (Table 9-1). The six scenarios are (a) historical fire occurrence, no insects and disease; (b) fire occurrence under half fire exclusion and no insects and disease; (c) full fire exclusion, no insects and disease; (d) historical fire occurrence with beetles, blister rust, and root rot; (e) half fire exclusion with beetles, rust, and rot; and (f) full fire suppression with beetles, rust, and rot. SH is shrub–herb, WP is whitebark pine, and SF is subalpine fir.

perturbations are included in the VDDT simulation (Figure 9-7d). This decline is accentuated by the exclusion of fire (Figure 9-7e,f). In fact, whitebark pine cover is less than 10 percent of the landscape by simulation year 100 under both fire exclusion scenarios, and less than 1 percent at year 500 under full suppression. Subalpine fir dominates over 60 percent of the landscape by year

50, and its long-term average appears to be dictated by the level of fire exclusion. For example, full fire suppression with insects and diseases creates landscapes that are more than 93 percent subalpine fir (Figure 9-7f).

The extent of shrub-herb cover types is governed mostly by the level of fire exclusion, regardless of insects and disease (Figures 9-7a to 9-7f). Historical fire regimes created more shrub-herb communities and therefore produced more heterogeneous landscapes. Over 40 percent of the landscape tends to collect in the old-forest, multistrata (OFM) structural stage under fire exclusion, and there is often about 20 percent in the understory reinitiation (URI) stage, regardless of the level of insects and disease (see Table 9-3 for descriptions). Historic fire regimes tend to have higher portions of the landscape in the stand initiation (SIN) and stem exclusion (SEC) stages.

Model Overview

It is interesting that the four classes of models discussed here generated essentially the same result. The two ecological process models, FIRESUM and Fire-BGC, simulate ecosystem dynamics from fundamental ecophysiological relationships to the biota. FIRESUM simulates growth, regeneration, and mortality for trees in a stand using abstract representations of primary ecosystem processes, whereas Fire-BGC directly simulates ecophysiological processes and energy flow for all trees in all stands that comprise a landscape. Both models have been extensively tested and verified (Keane et al. 1989, Keane et al. 1996d), represent the most extensive simulations of whitebark pine ecosystems, and are best used for research applications.

The landscape models CRBSUM, LANDSUM, SIMPPLLE, and VDDT simulate succession using deterministic pathways and disturbance from probabilities. All but VDDT have been implemented in a spatial domain, but only SIMPPLLE has integrated spatial relationships in disturbance dynamics. None of these models have been extensively tested because it is difficult to collect suitable validation data (Keane et al. 1996c). These models are best suited for management applications.

Management Implications

This successional modeling exercise and literature review indicate that the most important proactive management action for conserving and restoring whitebark pine on the landscape is to maintain or reintroduce fire in these important high-elevation ecosystems. Mountain pine beetle outbreaks cause a short-term loss of cone-bearing whitebark pine; but, with an intact fire regime, whitebark pine is able to remain on the landscape, because nutcrackers are able to disperse seeds great distances into the burned, beetle-killed areas.

Blister rust epidemics create a similar but more long-lasting effect. Whitebark pine is unable to become dominant in burned stands, because blister rust eventually kills all whitebark pine trees that are not genetically resistant (about 95–99 percent). As a result, whitebark pine will remain on the landscape, but at levels that reflect the degree of rust resistance in local populations. At least two or three generations of whitebark pine must pass before there are sufficient numbers of rust-resistant trees to produce enough seed to colonize burned areas with rust-resistant seedlings. Since whitebark pine does not start producing female cones until 60 to 100 years of age (Weaver and Forcella 1986; Arno and Hoff 1990; McCaughey and Schmidt 1990), it may take centuries for rust-resistant seedlings to grow into the rust-resistant trees that will produce sufficient seeds to maintain whitebark pine dominance in the upper subalpine landscape *providing there is a healthy fire regime*. This assumes that during the rust-resistance-buildup period, whitebark pine populations are able to produce enough seeds so that the nutcracker will not eat all the seeds it harvests (Chapter 12). Preliminary data from Krebill and Hoff (1995) indicate that frequencies of genetic gains in rust resistance are represented in whitebark pine progeny after only one generation, which might indicate that the rust-resistance-buildup period is shorter than the literature or simulation results would indicate. Artificial regeneration using rust-resistant seedlings bred in greenhouses might accelerate this process, but it may be costly and inefficient. Since nutcrackers are extremely efficient at planting whitebark pine seeds, it seems obvious that the most important activity land management agencies can do to prevent the loss of whitebark pine is to increase caching habitat for the bird, and this is most effectively done by maintaining native fire regimes.

Acknowledgments

Special thanks go to Diana Tomback, University of Colorado at Denver; Bob Pfister, University of Montana, Missoula; Dana Perkins, USDA Forest Service; Dave Mattson, USGS Biological Resources Division; Michael Murray, Oregon Natural Heritage Program; Don Despain, USGS Biological Resources Division; Pat Green, USDA Forest Service; John Joy, USDA Forest Service; Regina Rochefort, USGS Biological Resources Division; Steve Arno, USDA Forest Service; and Charlie Johnson, USDA Forest Service, for insightful comments and assistance.

Literature Cited

Agee, J. K. 1993. Fire ecology of Pacific Northwest forests. Island Press, Washington, D.C.

———. 1994. Fire and weather disturbances in terrestrial ecosystems of the eastern Cascades. USDA Forest Service, Pacific Northwest Research Station, General Technical Report PNW-GTR-320, Portland, Oregon.

Agee, J. K., and L. Smith. 1984. Subalpine tree reestablishment after fire in the Olympic Mountains, Washington. Ecology 65:810–819.

Alexander, R. R. 1985. Major habitat types, community types, and plant communties in the Rocky Mountains. USDA Forest Service, Rocky Mountain Research Station, General Technical Report RM-123, Fort Collins, Colorado.

Alexander, R. R., R. C. Shearer, and W. D. Shepperd. 1990. *Abies lasiocarpa* (Hook.) Nutt. Subalpine fir. Pages 60–72 *in* R. M. Burns and B. H. Honkala, technical coordinators. Silvics of North America. Vol. I. Conifers. USDA Forest Service, Agriculture Handbook 654, Washington, D.C.

Arno, S. F. 1979. Forest regions of Montana. USDA Forest Service, Intermountain Research Station, Research Paper INT-218, Ogden, Utah.

———. 1980. Forest fire history in the northern Rockies. Journal of Forestry 78:460–465.

———. 1986. Whitebark pine cone crops—A diminishing source of wildlife food? Western Journal of Applied Forestry 3:92–94.
Arno, S. F., and J. R. Habeck. 1972. Ecology of alpine larch (*Larix lyallii* Parl.) in the Pacific Northwest. Ecological Monographs 42:417–450.
Arno, S. F., and R. P. Hammerly. 1984. Timberline: Mountain and arctic forest frontiers. The Mountaineers, Seattle, Washington.
Arno, S. F., and R. J. Hoff. 1990. *Pinus albicaulis* Engelm. Whitebark Pine. Pages 268–279 *in* R. M. Burns and B. H. Honkala, technical coordinators. Silvics of North America. Vol. I. Conifers. USDA Forest Service, Agriculture Handbook 654, Washington, D.C.
Arno, S. F., and T. Weaver. 1990. Whitebark pine community types and their patterns on the landscape. Pages 118–130 *in* W. C. Schmidt and K. J. McDonald, compilers. Proceedings—Symposium on whitebark pine ecosystems: Ecology and management of a high-mountain resource. USDA Forest Service, Intermountain Research Station, General Technical Report INT-270, Ogden, Utah.
Arno, S. F., E. Reinhardt, and J. Scott. 1993. Forest structure and landscape patterns in the subalpine lodgepole pine type: A procedure for quantifying past and present conditions. USDA Forest Service, Intermountain Research Station, General Technical Report INT-294, Ogden, Utah.
Bailey, D. K. 1975. *Pinus albicaulis.* Curtis's Botanical Magazine. 180:140–147.
Bartlein, P. J., C. Whitlock, and S. L. Shafer. 1997. Future climate in the Yellowstone National Park region and its potential impact on vegetation. Conservation Biology 11:782–792.
Bazzaz, F. A. 1979. The physiological ecology of plant succession. Annual Review of Ecology and Systematics 10:351–371.
Beukema, S. J., and W. A. Kurtz. 1995. Vegetation dynamics development tool user's guide. ESSA Technologies Ltd., Vancouver, British Columbia, Canada.
Bradley, A. F., W. C. Fischer, and N. V. Noste. 1992. Fire ecology of forest habitat types of eastern Idaho and western Wyoming. USDA Forest Service, Intermountain Research Station, General Technical Report INT-290, Ogden, Utah.
Brown, J. K., and C. D. Bevins. 1986. Surface fuel loadings and predicted fire behavior for vegetation types in the northern Rocky Mountains. USDA Forest Service, Intermountain Research Station, Research Note INT-358, Ogden, Utah.
Brown, J. K., and T. E. See. 1981. Downed dead woody fuel and biomass in the northern Rocky Mountians. USDA Forest Service, Intermountain Research Station, General Technical Report INT-117, Ogden, Utah.
Brown, J. K., M. A. Marsden, K. C. Ryan, and E. D. Reinhardt. 1985. Predicting duff and woody fuel consumed by prescribed fire in the northern Rocky Mountains. USDA Forest Service, Intermountain Research Station, Research Paper INT-337, Ogden, Utah.
Brown, J. K., S. F. Arno, S. W. Barrett, and J. P. Menakis. 1994. Comparing the prescribed natural fire program with presettlement fires in the Selway-Bitterroot wilderness. International Journal of Wildland Fire 4:157–168.
Butler, D. R. 1986. Conifer invasion of subalpine meadows, central Lemhi Mountains, Idaho. Northwest Science 60:166–173.
Callaway, R., A. Sala, and R. E. Keane. 1998. Replacement of whitebark pine by subalpine fir: Consequences for stand carbon, water, and nitrogen cycles. Final Report RJVA-INT-95086. On file at the USDA Forest Service, Fire Sciences Laboratory, P.O. Box 8089, Missoula, Montana 59807.
Cattelino, P. J., I. R. Noble, R. O. Slatyer, and S. R. Kessell. 1979. Predicting the multiple pathways of plant succession. Environmental Management 3:41–50.
Chew, J. D. 1997. Simulating vegetative patterns and processes at landscape scales.

Pages 300–309 in Conference Proceedings—GIS 97, 11th Annual Symposium on Geographic Information Systems: Integrating Spatial Information Technologies for Tomorrow. February 17–20, 1997, Vancouver, British Columbia, Canada. GIS World, Inc.

Clausen, J. 1965. Population studies of alpine and subalpine races of conifers and willows in the California High Sierra Nevada. Evolution 19:56–68.

Cole, D. N. 1982. Vegetation of two drainages in Eagle Cap wilderness, Wallowa Mountains, Oregon. USDA Forest Service, Intermountain Research Station, Research Paper INT-288, Ogden, Utah.

Cooper, S. V., K. E. Neiman, and D. W. Roberts. 1991. Forest habitat types of northern Idaho: A second approximation. USDA Forest Service, Intermountain Research Station, General Technical Report INT-236, Ogden, Utah.

Craighead, J. J., J. S. Summer, and G. B. Scaggs. 1982. A definitive system for analysis of grizzly bear habitat and other wilderness resources utilizing LANDSAT multispectral imagery and computer technology. University of Montana Foundation, Wildlife-Wildlands Institute Monograph No. 1, University of Montana, Missoula.

Crane, M. F., and W. C. Fischer. 1986. Fire ecology of the forest habitat types of central Idaho. USDA Forest Service, Intermountain Research Station, General Technical Report INT-218, Ogden, Utah.

Crouch, G. L. 1987. Big game habitat research in subalpine forests in the central Rocky Mountains. Pages 106–111 *in* C. A. Troendle, M. R. Kaufmann, R. H. Hamre, and R. P. Winokur, technical coordinators. Proceedings of a technical conference—Management of subalpine forests: Building on 50 years of research. USDA Forest Service, Rocky Mountain Research Station, General Technical Report RM-149, Fort Collins, Colorado.

Dahlgreen, M. C. 1984. Observations on the ecology of *Vaccinium membranaceum* Dougl. on the southeast slope of the Washington Cascades. Master's thesis, University of Washington, Seattle.

Daubenmire, R. 1981. Subalpine parks associated with snow transfer in the mountains of northern Idaho and eastern Washington. Northwest Science 55:124–135.

Davis, K. M., B. D. Clayton, and W. C. Fischer. 1980. Fire ecology of Lolo National Forest habitat types. USDA Forest Service, Intermountain Research Station, General Technical Report INT-79, Ogden, Utah.

Day, R. J. 1967. Whitebark pine in the Rocky Mountains of Alberta. Forestry Chronicle 43:278–282.

del Moral, R. 1979. High elevation vegetation of the Enchantment Lakes Basin, Washington. Canadian Journal of Botany 57:1111–1130.

del Moral, R., and R. S. Fleming. 1979. Structure of coniferous forest communities in western Washington: Diversity and ecotype properties. Vegetatio 41:143–154.

Dickman, A., and S. Cook. 1989. Fire and fungus in a mountain hemlock forest. Canadian Journal of Botany. 67:2005–2016.

Douglas, G. W., and T. M. Ballard. 1971. Effects of fire on alpine plant communities in the North Cascades, Washington. Ecology 52:1058–1064.

Douglas, G. W., and L. C. Bliss. 1977. Alpine and high subalpine plant communities of the North Cascades Range, Washington and British Columbia. Ecological Monographs 47:113–150.

Eggers, D. E. 1986. Management of whitebark pine as potential grizzly bear habitat. Pages 170–175 *in* G. P. Contreras and K. E. Evans, editors. Proceedings of a Grizzly Bear Habitat Symposium. USDA Forest Service, Intermountain Research Station, General Technical Report INT-207, Ogden, Utah.

Egler, F. E. 1953. Vegetation science concepts—I. Initial floristics composition, a factor in old field vegetation development. Vegetatio 14:412–471.

Eyre, F. H. (Editor). 1980. Forest cover types of the United States and Canada. Society of American Foresters. Washington, D.C.

Fahnestock, G. R. 1977. Interactions of forest fire, flora, and fuels in two Cascade Range wilderness areas. Master's thesis, University of Washington, Seattle.

Fischer, W. C., and A. F. Bradley. 1987. Fire ecology of western Montana forest habitat types. USDA Forest Service, Intermountain Research Station, General Technical Report, INT-223, Ogden, Utah.

Fischer, W. C., and B. D. Clayton. 1983. Fire ecology of Montana forest habitat types east of the Continental Divide. USDA Forest Service, Intermountain Research Station, General Technical Report INT-141, Ogden, Utah.

Forcella, F. 1978. Flora and chorology of the *Pinus albicaulis–Vaccinium scoparium* association. Madroño 25:139–150.

Forcella, F., and T. Weaver. 1977. Biomass and productivity of the subalpine *Pinus albicaulis–Vaccinium scoparium* association in Montana, USA. Vegetatio 35:95–105.

Franklin, J. F., and R. G. Mitchell. 1967. Successional status of subalpine fir in the Cascade Range. USDA Forest Service, Pacific Northwest Research Station, Research Paper PNW-46, Portland, Oregon.

Halliday, W. E. D. 1937. A forest classification for Canada. Canadian Department of Forestry, Resource Development, Forest Service Bulletin No. 89, Ottawa, Ontario, Canada.

Hann, W. J. 1990. Landscape and ecosystem-level management in whitebark pine ecosystems. Pages 335–340 *in* W. C. Schmidt and K. J. McDonald, compilers. Proceedings—Symposium on whitebark pine ecosystems: Ecology and management of a high-mountain resource. USDA Forest Service, Intermountain Research Station, General Technical Report INT-270, Ogden, Utah.

Hansen-Bristow, K., C. Montagne, and G. Schmidt. 1990. Geology, geomorphology, and soils within whitebark pine ecosystems. Pages 62–71 *in* W. C. Schmidt and K. J. McDonald, compilers. Proceedings—Symposium on whitebark pine ecosystems: Ecology and management of a high-mountain resource. USDA Forest Service, Intermountain Research Station, General Technical Report INT-270, Ogden, Utah.

Hartwell, M. 1997. Comparing historic and present conifer species compositions and structures on forested landscape of the Bitterroot Front. Contract completion report RJVA-94928, Rocky Mountain Research Station. On file at the Intermountain Fire Sciences Laboratory, P.O. Box 8089, Missoula, Montana 59807.

Hogg, E. H. 1993. An arctic-alpine flora at low elevation in Marble Canyon, Kootenay National Park, British Columbia. Canadian Field-Naturalist 107:282–292.

Huston, M., and T. Smith. 1987. Plant sucession: Life history and competition. The American Naturalist 130:168–198.

Hutchins, H. E. 1994. Role of various animals in dispersal and establishment of whitebark pine in the Rocky Mountains, U.S.A. Pages 163–171 *in* W. C. Schmidt and F. K. Holtmeier, compilers. Proceedings—International workshop on subalpine stone pines and their environment: The status of our knowledge. USDA Forest Service, Intermountain Research Station, General Technical Report INT-GTR-309, Ogden, Utah.

Kaufmann, M. R., C. A. Troendle, M. G. Ryan, and H. T. Mowrer. 1987. Trees—The link between siliviculture and hydrology. Pages 54–67 *in* C. A. Troendle, M. R. Kaufmann, R. H. Hamre, and R. P. Winokur, technical coordinators. Proceedings of a technical conference—Management of subalpine forests: Building on 50 years of research. USDA Forest Service, Rocky Mountain Research Station, General Technical Report RM-149, Fort Collins, Colorado.

Keane, R. E., and P. Morgan. 1994. Decline of whitebark pine in the Bob Marshall Wilderness Complex of Montana, USA. Pages 245–253 *in* W. C. Schmidt and F.-K. Holtmeier, compilers. Proceedings—International workshop on subalpine stone pines and their environment: The status of our knowledge. USDA Forest Service, Intermountain Research Station, General Technical Report INT-GTR-309, Ogden, Utah.

Keane, R. E., S. F. Arno, and J. K. Brown. 1989. FIRESUM—An ecological process model for fire succession in western conifer forests. USDA Forest Service, Intermountain Research Station, General Technical Report INT-266, Ogden, Utah.

Keane, R. E., S. F. Arno, J. K. Brown, and D. F. Tomback. 1990a. Modeling stand dynamics in whitebark pine (*Pinus albicaulis*) forests. Ecological Modelling 51:73–95.

———. 1990b. Modeling disturbances and conifer succession in whitebark pine forests. Pages 274–288 *in* W. C. Schmidt and K. J. McDonald, compilers. Proceedings—Symposium on whitebark pine ecosystems: Ecology and management of a high-mountain resource. USDA Forest Service, Intermountain Research Station, General Technical Report INT-270, Ogden, Utah.

Keane, R. E., P. Morgan, and J. P. Menakis. 1994. Landscape assessment of the decline of whitebark pine (*Pinus albicaulis*) in the Bob Marshall Wilderness Complex, Montana, USA. Northwest Science 68:213–229.

Keane, R. E., J. P. Menakis, and W. J. Hann. 1996a. Coarse scale restoration planning and design in Interior Columbia River Basin Ecosystems—An example using whitebark pine (*Pinus albicaulis*) forests. Pages 14–20 *in* C. C. Hardy and S. F. Arno, editors. The use of fire in forest restoration. USDA Forest Service, Intermountain Research Station, General Technical Report INT-GTR-341, Ogden, Utah.

Keane, R. E., J. P. Menakis, D. Long, W. J. Hann, and C. Bevins. 1996b. Simulating coarse scale vegetation dynamics using the Columbia River Basin Succession Model—CRBSUM. USDA Forest Service, Intermountain Research Station, General Technical Report INT-GTR-340, Ogden, Utah.

Keane, R. E., P. Morgan, and S. W. Running. 1996c. Fire-BGC—A mechanistic ecological process model for simulating fire succession on coniferous forest landscapes of the northern Rocky Mountains. USDA Forest Service, Intermountain Research Station, Research Paper INT-484, Ogden, Utah.

Keane, R. E., K. Ryan, and S. W. Running. 1996d. Simulating the effect of fire on northern Rocky Mountain landscapes using the ecological process model Fire-BGC. Tree Physiology 16:319–331.

Keane, R. E., C. Hardy, K. Ryan, and M. Finney. 1997a. Simulating effects of fire management on gaseous emissions from future landscapes of Glacier National Park, Montana, USA. World Resource Review 9:177–205.

Keane, R. E., D. G. Long, D. Basford, and B. A. Levesque. 1997b. Simulating vegetation dynamics across multiple scales to assess alternative management strategies. Pages 310–315 *in* Conference Proceedings—GIS 97, 11th annual symposium on Geographic Information Systems—Integrating spatial information technologies for tomorrow. GIS World, Inc., Vancouver, British Columbia, Canada.

Keane, R. E., P. Morgan, and J. D. White. 1999. Temporal pattern of ecosystem processes on simulated landscapes of Glacier National Park, USA. Landscape Ecology 14:311–329.

Kendall, K. C., and S. F. Arno. 1990. Whitebark pine—An important but endangered wildlife resource. Pages 264–274 *in* W. C. Schmidt and K. J. McDonald, compilers. Proceedings—Symposium on whitebark pine ecosystems: Ecology and management of a high-mountain resource. USDA Forest Service, Intermountain Research Station, General Technical Report INT-270, Ogden, Utah.

Kessell, S. R., and W. C. Fischer. 1981. Predicting postfire plant succession for fire management planning. USDA Forest Service, Intermountain Research Station, General Technical Report INT-94, Ogden, Utah.

Kipfer, T. R. 1992. Post-logging stand characteristics and crown development of whitebark pine (*Pinus albicaulis*). Master's thesis, Montana State University, Bozeman.

Krebill, R. G., and R. J. Hoff. 1995. Update on *Cronartium ribicola* in *Pinus albicaulis* in Rocky Mountains, USA. Proceedings 4th IUFRO Rusts of Pines Working Party Conference, Tsukuba:119–126.

Lasko, R. J. 1990. Fire behavior characteristics and management implications in whitebark pine ecosystems. Pages 319–324 *in* W. C. Schmidt and K. J. McDonald, compilers. Proceedings—Symposium on whitebark pine ecosystems: Ecology and management of a high-mountain resource. USDA Forest Service, Intermountain Research Station, General Technical Report INT-270, Ogden, Utah.

Little, R. L., and D. L. Peterson. 1991. Effects of climate on regeneration of subalpine forests after wildfire. Northwest Environmental Journal 7:355–357.

Loope, L. L., and G. E. Gruell. 1973. The ecological role of fire in the Jackson Hole area, northwestern Wyoming. Quaternary Research 3:425–443.

Lueck, D. 1980. Ecology of *Pinus albicaulis* on Bachelor Butte, Oregon. Master's thesis, Oregon State University, Corvallis.

Marston, R. A., and J. E. Anderson. 1991. Watersheds and vegetation of the Greater Yellowstone Ecosystem. Conservation Biology 5:338–346.

Mattson, D. J. 1997. Use of lodgepole pine cover types by Yellowstone grizzly bears. Journal of Wildlife Management 61:480–496.

Mattson, D. J., and D. P. Reinhart. 1990. Whitebark pine on the Mount Washburn Massif, Yellowstone National Park. Pages 106–117 *in* W. C. Schmidt and K. J. McDonald, compilers. Proceedings—Symposium on whitebark pine ecosystems: Ecology and management of a high-mountain resource. USDA Forest Service, Intermountain Research Station, General Technical Report INT-270, Ogden, Utah.

———. 1997. Excavation of red squirrel middens by grizzly bears in the whitebark pine zone. Journal of Applied Ecology 34:926–940.

McCaughey, W. W., and W. C. Schmidt. 1990. Autecology of whitebark pine. Pages 85–96 *in* W. C. Schmidt and K. J. McDonald, compilers. Proceedings—Symposium on whitebark pine ecosystems: Ecology and management of a high-mountain resource. USDA Forest Service, Intermountain Research Station, General Technical Report INT-270, Ogden, Utah.

McCaughey, W. W., and T. Weaver. 1990. Biotic and microsite factors affecting whitebark pine establishment. Pages 140–151 *in* W. C. Schmidt and K. J. McDonald, compilers. Proceedings—Symposium on whitebark pine ecosystems: Ecology and management of a high-mountain resource. USDA Forest Service, Intermountain Research Station, General Technical Report INT-203, Ogden, Utah.

McCaughey, W. W., W. C. Schmidt, and R. C. Shearer. 1986. Seed dispersal characteristics of conifers of the Inland Mountain West. Pages 50–61 *in* R. C. Shearer, compiler. Proceedings—Symposium on Conifer Tree Seed in the Inland Mountain West. USDA Forest Service, Intermountain Research Station, General Technical Report INT-203, Ogden, Utah.

Minore, D. 1979. Comparative autecological characteristics of northwestern tree species: A literature review. USDA Forest Service, Pacific Northwest Research Station, General Technical Report PNW-87, Seattle, Washington.

Morgan, P., and S. C. Bunting. 1990. Fire effects in whitebark pine forests. Pages 166–170 *in* W. C. Schmidt and K. J. McDonald, compilers. Proceedings—Symposium on whitebark pine ecosystems: Ecology and management of a high-mountain resource. USDA Forest Service, Intermountain Research Station, General Technical Report INT-270, Ogden, Utah.

Morgan, P., S. C. Bunting, R. E. Keane, and S. F. Arno. 1994. Fire ecology of whitebark pine (*Pinus albicaulis*) forests in the Rocky Mountains, USA. Pages 136–142 *in* W. C. Schmidt and F. -K. Holtmeier, compilers. Proceedings—International workshop on subalpine stone pines and their environment: The status of our knowledge. USDA

Forest Service, Intermountain Research Station, General Technical Report INT-GTR-309, Ogden, Utah.

Moseley, R. K., and S. Bernatas. 1992. Vascular flora of Kane Lake Cirque, Pioneer Mountains, Idaho. Great Basin Naturalist 52:335–343.

Murray, M. P. 1996. Landscape dynamics of an island range: Interrelationships of fire and whitebark pine (*Pinus albicaulis*). Ph.D. dissertation, College of Forestry, Wildlife and Range Science, University of Idaho, Moscow.

Murray, M. P., S. C. Bunting, and P. Morgan. 1995. Whitebark pine and fire suppression in small wilderness areas. Pages 237–240 *in* J. K. Brown, R. W. Mutch, C. W. Spoon, and R. H. Wakimoto, technical coordinators. Proceedings: Symposium on fire in wilderness and park management. USDA Forest Service, Intermountain Research Station, General Technical Report INT-GTR-320, Ogden, Utah.

Noble, I. R., and R. O. Slatyer. 1977. Postfire succession of plants in Mediterranean ecosystems. Pages 27–36 *in* H. A. Mooney and C. E. Conrad, editors. Proceedings of the symposium environmental consequences of fire and fuel management in Mediterranean climate ecosystems. USDA Forest Service General Technical Report WO-3, Washington, D.C.

O'Hara, K., P. A. Latham, P. Hessburg, and B. G. Smith. 1996. A structural classification for Inland Northwest forest vegetation. Western Journal of Applied Forestry 11:97–102.

Ogilvie, R. T. 1990. Distribution and ecology of whitebark pine in western Canada. Pages 54–60 *in* W. C. Schmidt and K. J. McDonald, compilers. Proceedings—Symposium on whitebark pine ecosystems: Ecology and management of a high-mountain resource. USDA Forest Service, Intermountain Research Station, General Technical Report INT-270, Ogden, Utah.

Oliver, C. D., and B. C. Larson. 1990. Forest stand dynamics. McGraw-Hill, New York.

Parker, A. J. 1988. Stand structure in subalpine forests of Yosemite National Park, California. Forest Science 34:1047–1058.

Perkins, D. L., and T. W. Swetnam. 1996. A dendroecological assessment of whitebark pine in the Sawtooth–Salmon River region, Idaho. Canadian Journal Forest Research 26:2123–2133.

Peterson, D. L., M. J. Arbaugh, and M. A. Lardner. 1989. Leaf area of lodgepole pine and whitebark pine in a subalpine Sierra Nevada forest. Canadian Journal Forest Research 19:401–403.

Pfister, R. D., B. L. Kovalchik, S. F. Arno, and R. C. Presby. 1977. Forest habitat types of Montana. USDA Forest Service, Intermountain Research Station, General Technical Report INT-34 Ogden, Utah.

Pojar, J. 1985. Ecological classification of lodgepole pine in Canada. Pages 77–88 *in* D. Baumgartner, R. G. Krebill, J. T. Arnott, and G. F. Weetman, compilers and editors. Lodgepole pine: The species and its management—Symposium proceedings. Washington State University, Cooperative Extension, Pullman.

Reed, R. M. 1976. Coniferous forest habitat types of the Wind River Mountains, Wyoming. American Midland Naturalist 95:159–173.

Riegel, G. M., D. A. Thornburgh, and J. O. Sawyer. 1990. Forest habitat types of the South Warner Mountains, Modoc County, California. Madroño 37:88–112.

Rochefort, R. M., and D. L. Peterson. 1996. Temporal and spatial distribution of trees in subalpine meadows of Mount Rainier National Park, Washington, U.S.A. Arctic and Alpine Research 28:52–59.

Romme, W. H. 1982. Fire and landscape diversity in subalpine forests of Yellowstone National Park. Ecological Monographs 52:199–221.

Romme, W. H., and M. G. Turner. 1991. Implications of global climate change for biogeographic patterns in the greater Yellowstone ecosystem. Conservation Biology 5:373–386.

Ryan, M. G. 1987. Growth efficiency, leaf area, sapwood volume in subalpine conifers. Pages 180–189 *in* C. A. Troendle, M. R. Kaufmann, R. H. Hamre, and R. P. Winokur, technical coordinators. Proceedings of a technical conference—Management of subalpine forests: Building on 50 years of research. USDA Forest Service, Rocky Mountain Research Station, General Technical Report RM-149, Fort Collins, Colorado.

Smith, J. K., and W. C. Fischer. 1997. Fire ecology of the forest habitat types of northern Idaho. USDA Forest Service, Intermountain Research Station, General Technical Report INT-GTR-363, Ogden, Utah.

Steele, R. W. 1960. The role of forest fire in the Bob Marshall Wilderness Area. Montana Forest and Conservation Experiment Station, School of Forestry, University of Montana Report #1, Missoula.

Steele, R., R. D. Pfister, R. A. Ryker, J. A. Kittams. 1981. Forest habitat types of central Idaho. USDA Forest Service, Intermountain Research Station, General Technical Report INT-114, Ogden, Utah.

Steele, R., S. V. Cooper, D. M. Ondov, D. W. Roberts, and R. D. Pfister. 1983. Forest habitat types of eastern Idaho–western Wyoming. USDA Forest Service, Intermountain Research Station, General Technical Report INT-144, Ogden, Utah.

Thompson, L. S., and J. Kuijt. 1976. Montane and subalpine plants of the Sweetgrass Hills, Montana, and their relation to early postglacial environments of the Northern Great Plains. Canadian Field-Naturalist 90:432–448.

Tomback, D. F. 1986. Post-fire regeneration of krummholz whitebark pine: A consequence of nutcracker seed caching. Madroño 33:100–110.

———. 1998. Clark's nutcracker (*Nucifraga columbiana*), No. 331. *In* A. Poole and F. Gill, editors. The birds of North America. The Birds of North America, Inc., Philadelphia.

Tomback, D. F., L. A. Hoffmann, and S. K. Sund. 1990. Coevolution of whitebark pine and nutcrackers: Implications for forest regeneration. Pages 118–130 *in* W. C. Schmidt and K. J. McDonald, compilers. Proceedings—Symposium on whitebark pine ecosystems: Ecology and management of a high-mountain resource. USDA Forest Service, Intermountain Research Station, General Technical Report INT-270, Ogden, Utah.

Tomback, D. F., S. K. Sund, and L. A. Hoffmann. 1993. Post-fire regeneration of *Pinus albicaulis:* Height-age relationships, age structure, and microsite characteristics. Canadian Journal Forest Research 23:113–119.

Tomback, D. F., J. K. Clary, J. Koehler, R. J. Hoff, and S. F. Arno. 1995. The effects of blister rust on postfire regeneration of whitebark pine: The Sundance burn of northern Idaho (USA). Conservation Biology 9:654–664.

Troendle, C. A., and M. R. Kaufmann. 1987. Influence of forests on the hydrology of the subalpine forest. Pages 68–78 *in* C. A. Troendle, M. R. Kaufmann, R. H. Hamre, and R. P. Winokur, technical coordinators. Proceedings of a technical conference—Management of subalpine forests: Building on 50 years of research. USDA Forest Service, Rocky Mountain Research Station, General Technical Report RM-149, Fort Collins, Colorado.

Waring, R. H., and Running, S. W. 1998. Forest ecosystems: Analysis at multiple scales, 2d edition. Academic Press, San Diego, California.

Warren, S. P., S. Scharosch, and J. S. Steere. 1997. A comparison of forest vegetation structural stage classifications. Pages 216–222 *in* R. Teck, M. Moeur, and J. Adams, compilers. Proceedings: Forest Vegetation Simulator Conference. USDA Forest Service, Intermountain Research Station, General Technical Report INT-GTR-373, Ogden, Utah.

Weaver, T., and D. Dale. 1974. *Pinus albicaulis* in central Montana: Environment, vegetation, and production. American Midland Naturalist 92:222–230.

Weaver, T., and F. Forcella. 1986. Cone production in *Pinus albicaulis* forests. Pages

68–76 *in* R. C. Shearer, compiler. Proceedings—Conifer tree seed in the Inland Mountain West. USDA Forest Service, Intermountain Research Station, General Technical Report INT-203, Ogden, Utah.

Weaver, T., F. Forcella, and D. Dale. 1990. Stand development in whitebark pine woodlands. Pages 40–48 *in* W. C. Schmidt and K. J. McDonald, compilers. Proceedings—Symposium on whitebark pine ecosystems: Ecology and management of a high-mountain resource. USDA Forest Service, Intermountain Research Station, General Technical Report INT-270, Ogden, Utah.

West, D. C., H. H. Shugart, and D. B. Botkin. 1980. Forest succession: Concepts and application. Springer-Verlag, New York.

White, G. J., G. A. Baker, M. E. Harmon, G. B. Wiersma, and D. A. Bruns. 1990. Use of forest ecosystem process measurements in an integrated environmental monitoring program in the Wind River Range, Wyoming. Pages 214–222 *in* W. C. Schmidt and K. J. McDonald, compilers. Proceedings—Symposium on whitebark pine ecosystems: Ecology and management of a high-mountain resource. USDA Forest Service, Intermountain Research Station, General Technical Report INT-270, Ogden, Utah.

Chapter 10

Blister Rust: An Introduced Plague

Geral I. McDonald and Raymond J. Hoff

White pine blister rust is a disease of five-needled pines (commonly called "white pines") caused by the fungus, *Cronartium ribicola*. This pathogen is well known for a complicated life cycle composed of five kinds of spores on two host species. To complete each generation, the fungus produces two types of spores on white pine hosts and three types on woody shrubs of the genus *Ribes,* the alternate host. All native North American white pines are highly susceptible to this rust fungus. Since its introduction to the continent in about 1900, blister rust has made dramatic impacts on most forested ecosystems where white pines are important components. Inland ecosystems at higher elevations are at especially high risk because one or two species of noncommercial white pines usually dominate (Table 10-1). These noncommercial species are whitebark pine (*Pinus albicaulis*), limber pine (*P. flexilis*), southwestern white pine (*P. strobiformis*), bristlecone pine (*P. aristata*), Great Basin bristlecone pine (*P. longaeva*), and foxtail pine (*P. balfouriana*). All North American white pines react similarly toward the rust (Table 10-1). However, most knowledge about the behavior of the blister rust pathosystem in North America is derived from studies of the low- to middle-elevation commercial species of white pine, western white pine (*P. monticola*), eastern white pine (*P. strobus*), and sugar pine (*P. lambertiana*). In this chapter, we first discuss how blister rust came to North America, and then we apply the knowledge gained from the study of commercial species to assess risk to higher-elevation ecosystems where noncommercial species, especially whitebark pine, are important ecosystem components.

Table 10-1. Responses[1] of wild populations of North American species of white pine when artificially inoculated with the blister rust fungus (*Cronartium ribicola*).

Pinus Species	Ecosystem Elevation[2] (meters)	Number Seedlings	Percent Infected	Spots/m of Needle[3]	Percent Killed[3]
P. aristata	2,290–3,500	92	88	36.7	79
P. albicaulis[4]	1,370–3,660	207	67	6.7	92
P. balfouriana	1,830–3,500	92	100	84.4	100
P. flexilis	1,524–3,660	384	89	9.5	89
P. lambertiana	335–3,200	259	91	4.4	91
P. monticola[4]	0–3,000	1,463	76	5.0	68
P. strobiformis	2,000–3,050	323	90	5.3	58
P. strobus	0–1,220	374	99	20.3	87

[1] Data obtained from Hoff et al. (1980).
[2] Elevations from Little (1980).
[3] Rust mortality at three years, based on infected seedlings.
[4] Samples obtained from phenotypically resistant parents.

Biology of Blister Rust

The most prominent external sign of blister rust, the aeciospore stage, occurs in the living bark tissue of all infected white pines (Figure 10-1). These spores resemble 12-by-30-micrometer (1 micrometer = 0. 000039 inches) orange footballs. Millions are produced in hundreds of saclike structures about the size of thimbles (aecia) on the boles and/or branches of each infected pine tree. Aecia resemble blisters; hence, the common name blister rust. The long-lived (weeks) aeciospores are responsible for long-distance spread of blister rust. Viable aeciospores can be carried more than 500 kilometers by the wind (Mielke 1943). Growth processes in pines and *Ribes* synchronize the appearance of spores with emergence and growth of the *Ribes* leaves. The windborne aeciospores land on the undersides of the *Ribes* leaves, where they germinate and establish new colonies. The first visible signs of new colonies are uredinia, orange horseshoe-shaped structures about 0.5 millimeters across, where orange urediniospores develop. These spores, about the same size and shape as aeciospores, multiply the rust on *Ribes* throughout the summer during cool, wet periods (100 percent relative humidity) by spreading the rust on the same or nearby bushes.

The onset of cool nights in August typically triggers production of teliospores, the third kind of spore, within the uredinia. Teliospores are barrel-shaped spores 10 micrometers wide by 40 micrometers long. They are packed in multiple layers, like a roll of paper towels, into tan-colored, hornlike structures called telial columns, which are about 100 micrometers thick and up to 1,500 micrometers long. Telial columns develop during mid- to late summer and await the occurrence of their germination trigger—near 100 percent relative humidity for a period of six to eight hours. Each teliospore packed in these

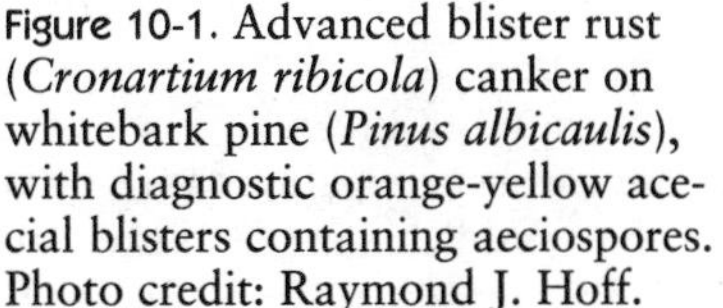

Figure 10-1. Advanced blister rust (*Cronartium ribicola*) canker on whitebark pine (*Pinus albicaulis*), with diagnostic orange-yellow acecial blisters containing aeciospores. Photo credit: Raymond J. Hoff.

columns germinates and produces four basidiospores (a sphere about 5 to 10 micrometers in diameter). Typically, each column produces about 6,000 basidiospores in twenty to forty-eight hours in late summer or early fall during cool, moist periods.

Basidiospores are thin-walled and fragile. Constant high humidity is needed for them to retain viability. They are windborne short distances (usually less than 300 meters) to the needles of pine hosts, where they germinate and infect through stomata. A red or yellow spot (about 1 millimeter square) appears on the surface of the needle at the point of infection in about four to six months. The spot is caused by the growth of a hyphal mass (pseudosclerotium) in the mesophyll tissue of the needle. About the time that a pseudosclerotium becomes visible to the human eye, a few hyphae penetrate the vascular bundle of the needle and grow down the vascular tissue into the stem. Rust mycelia continue to grow in phloem tissues of a branch or stem to produce a canker. After entering the phloem, infections develop into cankers that produce fusiform swelling of stems or branches and pycniospores that breed with neighboring infections to create a new generation of aeciospores. The aecium containing the new aeciospores typically erupts from apparently healthy bark on the surface of the canker after two years and will continue annual eruptions on successive rings of new bark until the stem or branch is girdled (Figure 10-1). This action of the canker causes death of branches, top kill, or death of the entire tree, depending on its location. This disease cycle, coupled with the predictable growth habits of white pine (a single well-defined annual growth cohort on branches and bole), facilitated the accurate aging of cankers (Lachmund 1933). Many findings related to blister rust epidemiology depend on this relationship.

Early Development in Northern Europe

H. A. Dietrich first observed white pine blister rust in 1854 in the Baltic provinces of Russia on European black currant (*Ribes nigrum*), red currant (*R. rubrum*), palm gooseberry (*R. palmatum*), and introduced eastern white pine(*Pinus strobus*) (Spaulding 1911). Dietrich did not recognize that the fungus was the same on both *Ribes* and pines. The realization of a single fungus came in 1888 after Klebahn performed cross-inoculation tests, which showed that the rusts on *Ribes* and pine were the same species (Spaulding 1911).

The rust next appeared on pines in Finland in 1861, in East Germany in 1865 and 1869, and in Denmark in 1883. By 1900, blister rust had spread throughout northern Europe. The rust was especially severe where extensive forests of eastern white pine had been established after introduction of this exotic pine into Europe in 1705 (Bean 1914). Susceptible pine stands were intermingled with patches of cultivated European black currant, a favorite garden currant of Europe. This situation provided the environment for rapid spread by assuring abundance of both hosts in close proximity.

The establishment of blister rust on European species apparently resulted from an introduction of infected Siberian stone pine (*Pinus cembra*) from Siberia (Spaulding 1911, 1922). A note about stone pine taxonomy is in order. A comparison of distribution maps by Spaulding (1922) and Lanner (1990) clearly shows that Spaulding did not distinguish Siberian stone pine (*P. sibirica*) from Swiss stone pine (*P. cembra*) and that he considered Japanese stone pine (*P. pumila*) to be *P. cembra* var. *pumila*. Botanical gardens of Europe advertised standing requests for plants of all kinds from newly explored regions throughout the world and received various species of white pines from around the world. Following the appearance of the rust in northern Europe, extensive examination of existing Eurasian white pine stands soon established that blister rust was present on these native hosts. According to Spaulding (1922), some early investigators concluded that the fungus was endemic to Eurasia; others contended that the rust was endemic to the Swiss Alps, where a canker was found on a single 200-year-old Swiss stone pine in 1903. But pathologists who studied this and other infections in the Alps concluded that the infections were recent and came from outside the Alps (Spaulding 1922). However, Siberian stone pine and Japanese stone pine growing in north-central and eastern Asia are the most probable original hosts (Spaulding 1922).

The introduction into northern Europe allowed an exotic pest to eliminate an exotic host, eastern white pine. The destructive abilities of blister rust were quickly determined. Spaulding (1911) listed some of the disaster statements that several foresters made during this time in northern Europe, showing just how quickly the rust could destroy its pine host. Dietrich, in 1854, observed that blister rust killed most eastern white pine in a few years in the Baltic provinces of Russia. In Finland in 1876, Hisenger found that new infections in thirty-year-old eastern white pine caused rapid mortality. Rostrup, in 1883, reported that blister rust caused more damage than any other bark-inhabiting rust in Denmark. By 1901, infection was so severe on eastern white pines in Belgium that foresters considered it imprudent to continue managing the species. Somerville (1909) stated that the rate of increase of blister rust was so high in England that it seemed hopeless to manage eastern white pine or any other five-needled pine from North America.

There were many forest tree nurseries in Europe at this time because most forest districts grew their own seedlings. They sold their excess seedlings to those who needed them, and when the North American market developed, they began to grow additional millions of eastern white pine seedlings for export (Spaulding 1922). Pathologists in Europe had been warning foresters since the 1890s to prevent the transportation of seedlings around Europe, especially to North America, without thorough inspection for the disease. Inspections were probably attempted because infections of the stem were easily seen and identified. However, there was a major unknown problem. At the turn of the century, since the site of infection on pines was mistakenly thought to be the stem, the erroneous assumption was made that any new infection could be quickly spotted.

The true infection court—secondary needles, primary leaves, and cotyledons—was discovered much later (Clinton and McCormick 1919). As outlined earlier, basidiospores land on a needle, germinate, and penetrate through stomata into the living tissue of the leaf. In six to twelve months, the fungus grows down the needle and into the stem. Specific timing depends on distance from infection site to the base of the needle and on growing conditions. Since stem infection was not easily visible until a year or more after infection, there was at least a one-year delay before infections could easily be seen. No one realized that yellow needle spots observed on needles were the first signs of blister rust. So, many seedlings infected with blister rust were labeled "uninfected" and shipped to North America.

Introduction into Eastern North America

Forest nurseries in France and Germany sold eastern white pine to North American foresters from about 1890 to 1914. Several million seedlings were sold in 1914 alone (Spaulding 1922). Foresters, nurserymen, and pathologists were all well aware of blister rust and were on the lookout for infections on the imported seedlings (Spalding and Fernow 1899). The first indication of blister rust in North America came from infected leaves of golden currant (*Ribes aureum*) collected in 1892 in Kansas (Spaulding 1911). This putative introduction was later shown (Spaulding 1922) to be an occurrence of *Cronartium occidentale,* a stem rust closely related to white pine blister rust that alternates between *Ribes* and piñon pines (*Pinus edulis* and *P. monophylla*). In 1906, true white pine blister rust infections were found on *Ribes* at the State Experiment Station in Geneva, New York (Spaulding 1911). These *Ribes* plants were pulled and burned. Then on June 8, 1909, Perley Spaulding, a pathologist with the Bureau of Plant Industry in Washington, D.C., verified blister rust on some German-grown seedlings of eastern white pine (Spaulding 1909). Spaulding soon determined that several million three-year-old seedlings from this nursery had been imported and distributed to 226 sites in several northeastern states.

Blister rust also was found on seedlings in several other nurseries in Germany and France (Spaulding 1911). Several of these nurseries also had shipped their stock to eastern North America. The infection level on these seedlings ranged from 1 percent to 3 percent. Several other species of white pine also were imported from Europe. These included western white pine, limber pine, sugar pine, Mexican white pine (*P. ayacahuite*), Japanese white pine (*P. parviflora*), Swiss stone pine, and Siberian stone pine (Spaulding 1911). Since most species were imported for use as ornamentals, they probably caused some of the widespread distribution of the fungus.

Spaulding's findings alerted the foresters of eastern North America. Subsequently, all white pines imported from Europe were inspected. Imported diseased trees were destroyed, patches of *Ribes* located near white pine plantations were inspected and, if infected, were destroyed (Spaulding 1922). Pathologists' expectations that these actions would stop the rust were not realized, because too many infected trees had been imported and *Ribes* shrubs were too common. An effort to eradicate *Ribes* was organized, and foresters started systematically destroying all *Ribes* plants, including European black currant that was common in garden plots. This last action met considerable resistance because gardeners were reluctant to give up this highly desirable currant (Maloy 1997). As infection levels increased in the northeastern United States, pathologists noted that some native eastern white pine trees were not infected (Snell 1931; Schreiner 1938; Riker et al. 1953). Farrar (1947) and Heimburger (1956) made the same observation in eastern Canada. Studies showed that some of these trees were resistant to blister rust (Riker et al. 1953; Heimburger 1956). Eventually *Ribes* eradication was deemed an uneconomical way of controlling blister rust throughout eastern forests. However, a successful eradication program was continuously applied in the state of Maine (Ostrofsky et al. 1988).

Introduction into Western North America

In the fall of 1921, blister rust was found at Vancouver, British Columbia, on European black currant (Mielke 1943). Further searching revealed its presence at three locations on Vancouver Island and on European black currant over a considerable area of the lower Fraser Valley, as well as a few infected exotic pines (Mielke 1943). Across the border, in the state of Washington, the rust was found on European black currant near Sumas, Mt. Vernon, Beverly, Parkland, Port Townsend, and on two planted white pines in a Mt. Vernon nursery. Inspections in the spring of 1922 revealed that the rust was widely distributed on western white pine and *Ribes* in western British Columbia, but the infection in Washington was found only on *Ribes*. The source of the blister rust was still unknown.

In 1922, Canadian officials inspected all plantings of white pine imported between 1904 and 1914 (Mielke 1943) and found one lot of infected white pines growing at Point Grey, near Vancouver, Canada. A landowner had imported 1,000 eastern white pine seedlings from Ussy, France, in 1910. At the time of inspection in 1922, 180 trees remained and 68 were infected by blister rust. The oldest canker was located on stem tissue produced in 1910. No other introductions have been documented in western North America.

Understanding early events in the western introduction entailed application of knowledge about the role of environmental factors in the spread of blister rust and the development of infections on *Ribes* and pines. Early scouting for blister rust and the dating of cankers started as soon as the rust was discovered in the fall of 1921. The observation made by foresters and pathologists in the eastern United States and Europe that spread (aeciospore dissemination and infection), intensification (urediospore facilitated multiplication), and subsequent pine infection occurred in certain "wave" years was soon confirmed (Mielke 1943). Many years later, a critical analysis of climate and detailed requirements for infection of both *Ribes* and pines demonstrated the mechanics of wave years (McDonald et al. 1981). Wave years result when aeciospores released into a cool moist spring are followed by frequent cool, moist infection periods during the summer to facilitate many cycles of multiplication by urediospores. Similarly, the infection potential afforded by teliospores is realized by cool, moist weather that facilitates teliospore germination and pine infection by basidiospores.

The eastern white pine imported in 1910 probably did not produce many aeciospores until 1913. Typically, aeciospore production in pine does not occur until the third spring after infection. By chance, suitable weather and aeciospore production coincided in 1913 to produce the first postintroduction wave year. Moist and cool spring and summer weather suitable for blister rust spread and intensification also occurred in 1917, 1921, 1923, 1927, and 1937 (Mielke 1943). Blister rust first appeared on whitebark pine in North America in 1926 in the coast range of British Columbia and spread to northern Idaho stands of whitebark pine by 1938 (Childs et al. 1938).

As in eastern North America, infected trees were destroyed and a *Ribes* eradication program was initiated. However, the fungus had moved rapidly. Back-dating of canker age (Lachmund 1933) on western white pine showed the fungus had spread southward almost to the Columbia River by 1920; it was in southern Oregon by 1929, and halfway down the Sierra Nevada range of California by 1941. It had spread eastward on western white pine into Idaho by 1923 and on into northwestern Montana by 1927. The rust reached whitebark pine growing on the continental divide in Glacier National Park in 1939 (Mielke 1943). It was reported to have reached southern Idaho limber pine by 1945 (Krebill 1964). By 1961, the rust had reached sugar pine in the southern end of the Sierra Nevada (Kinloch and Dulitz 1990). Subsequently, the rust was found on southwestern white pine in southern New Mexico (Hawksworth 1990) and limber pine in the Black Hills of South Dakota (Lundquist et al.1992). The New Mexico introduction was backdated to about 1970 (Geils et al. 1999). There is a single reported occurrence on limber pine planted in the center of North Dakota (Draper and Walla 1993). Meanwhile, the rust has spread through the area between the southern Sierra Nevada and Placerville, California (Kliejunas 1996) and moved slightly southward in central Oregon (Schmitt and Scott 1998). It also made a southerly advance in central Idaho (Smith and Hoffman 1998), and crossed the border on limber pine from

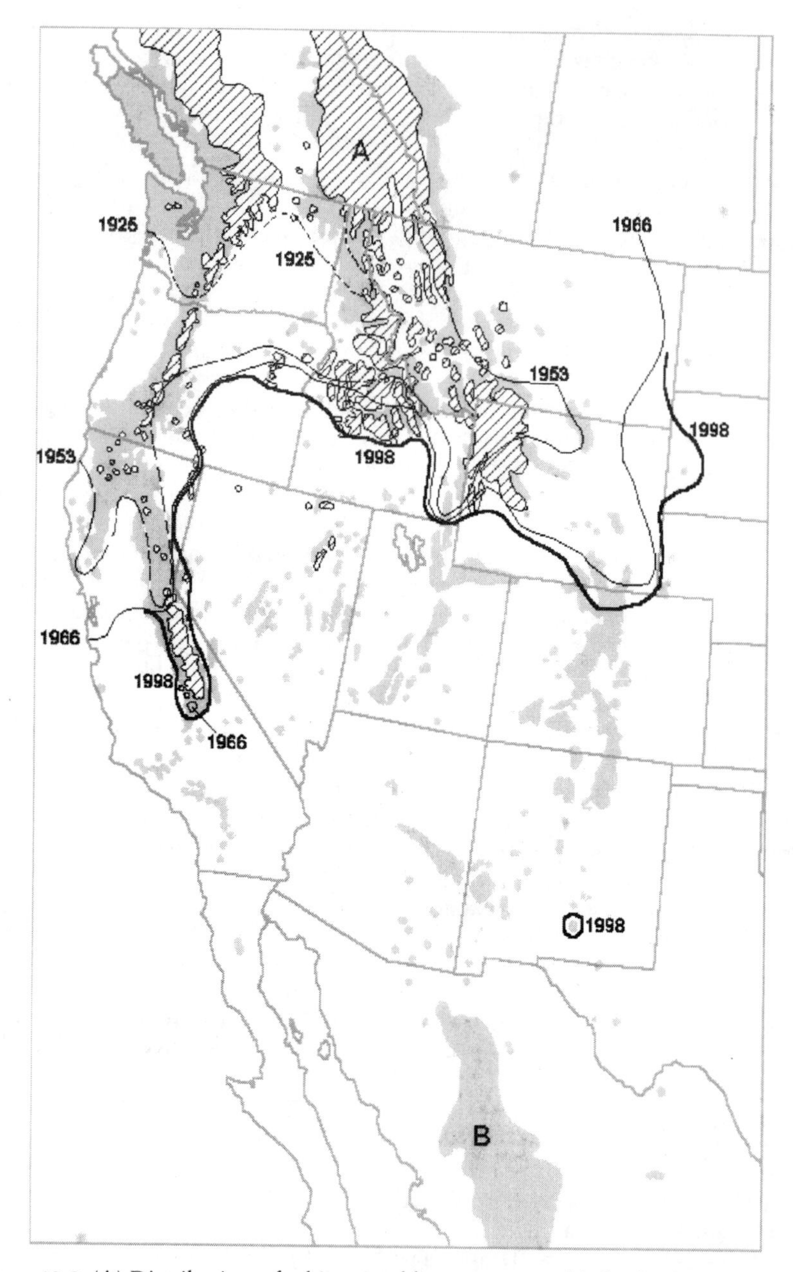

Figure 10-2. (A) Distribution of white pine blister rust on whitebark pine (*Pinus albicaulis*) (hatched range), and (B) on other five-needle white pines in western North America, 15 (1925), 43 (1953), 56 (1966), and 88 (1998) years after introduction at Point Grey near Vancouver, British Columbia, in 1910.

Wyoming to northern Colorado (D. W. Johnson, USDA Forest Service, personal communication). The rust was found on foxtail pine in 1967 in northern California, where it first appeared in 1942 (Miller 1968). The current extent of rust includes nearly the full range of whitebark pine, and the entire range of all associated North American white pines is either occupied or threatened (Figure 10-2).

Blister Rust Control Work in the West

After the discovery of blister rust in the West in 1921, forest workers first began to remove infected pines and soon began *Ribes* eradication. Although the cultivated European black currant was not as extensively planted in the West as it was in the East, many native *Ribes* species were very susceptible. Millions of *Ribes* shrubs were pulled out of creek bottoms, forested uplands, and mountain slopes located in state forests, national forests, national parks, and private lands. Few results were published in journals. However, significant information and insight was reflected in the Blister Rust Control Reports (U.S. Bureau of Plant Quarantine 1921–1953). These reports continued until 1966, after the control program was transferred to the U.S. Forest Service. This eradication effort provided many jobs, enabled countless students to graduate from college, and certainly helped the economy during the Great Depression. Contrary to expectations, the eradication effort did not achieve the desired amount of control of blister rust because abundance of *Ribes* shrubs, longevity of *Ribes* seed in the soil, ability of many *Ribes* species to sprout from rhizomes, and potential for long-distance spread of blister rust by aeciospores necessitated that each area receive repeated treatment. This, in turn, caused a backlog in untreated areas. Meanwhile, the mortality of western white pine was increasing. By 1966, pathologists realized they could not manage the fungus under western conditions by only eradicating *Ribes* (Toko et al. 1967; Neuenschwander et al. 1999). In 1968, the eradication program was officially terminated (Ketcham et al. 1968).

During the *Ribes* eradication effort, others were seeking fungicidal controls. Two promising chemicals were actidione and phytoactin. More than 40,500 hectares were treated with aerial application of actidione in diesel oil. Although actidione was known to be effective against blister rust, it was not effective enough. Actidione was not translocated at the rates anticipated, so each canker had to be treated individually. Pines also translocated another antibiotic, phytoactin, which showed early promise. In the end, however, it too was found to be ineffective. The chemical control program was terminated in 1966 (Ketcham et al. 1968), and the entire program was the subject of a recent review (Maloy 1997). A new chemical, tridimefon, has shown potential as a systemic fungicide protective against *Cronartium ribicola* (Chapter 17).

The immense challenge of controlling blister rust is illustrated by Mielke's data (1943). He pieced together the infection events for *Ribes* and western white pine following introduction of the rust at Point Grey in 1910. Based on dating thousands of cankers, Mielke concluded that, in 1913 (the "good" year for aeciospore production after introduction), the fungus moved from the introduced pines to native *Ribes* 380 kilometers from Point Grey. This is not surprising; in light of the fact that blister rust was even found on *Ribes* 500 kilometers north of the range of western white pine (Mielke 1943). In 1917, the next good season for high spore production, the fungus moved even further. Therefore, by the time pathologists were aware of a problem, it was already too late. The rust had covered much of the range of western white pine and most of the northern portion of the range of whitebark pine and limber pine. Harvesting western white pine accelerated, planting ceased, and pest problems in the remaining western white pines intensified (Monnig and Byler 1992). In the absence of significant restoration efforts, western white pine and its associated communities appear doomed (Neuenschwander et al. 1999)—with the same outlook for whitebark pine communities (Chapter 1).

High-Elevation Epidemiology

Successful restoration of whitebark pine will require the large-scale application of new technologies within the framework of an integrated rust management plan (Chapter 17). Application requires a thorough understanding of the epidemiological behavior of blister rust on high-elevation sites. Some thought has already been applied (Hoff and Hagle 1990; Krebill and Hoff 1995; Van Arsdel and Krebil 1995; Van Arsdel et al. 1998; Geils et al. 1999), but none of these discussions utilizes modern epidemiological concepts, and therefore, many conclusions may be flawed.

Dissemination of Basidiospores

The first issue is the influence of light on the production and germination of sporidia. Some fungi release their spores only at night, whereas others release them only during the day (Panzer et al. 1957). Even though Hirt (1942) had demonstrated that teliospore germination and eastern white pine infection could occur during daylight hours, Van Arsdel (1967) asserted that since blister rust in the Lake States produces basidiospores (teliospore germination) only at night, the dynamics of night breezes explain patterns of spore transport and pine infection, including long-distance spread. Potential for nighttime long-distance spread of sporidia was shown by a single example of silver-iodide particle transport by night breezes in a western mountainous situation (Lloyd et al. 1959). Meanwhile, experiments had shown that western teliospores germinated equally well under continuous darkness or light (Bega 1959), and that sporidia germination was not influenced by light (Bega 1960). Differences in rust ecotypes could cause such disagreements. As for the importance of night breezes in spreading rust in western mountains, Van Arsdel and Krebill (1995) concluded that they have little influence in the distribution of piñon rust (caused by *Cronartium occidentale*) in Nevada.

Serious questions remain unanswered about the relative importance of local and long-distance mechanisms of spread in high-elevation forests. The influence of night breezes or other potential long-distance spread mechanisms such as "cloud" phenomena may provide partial answers. The roles of *Ribes* distribution and microclimate, and the different genotypes of rust, *Ribes,* and hosts are important, too. One way to foster understanding is the application of modern epidemiological concepts. Disease can be visualized as a pyramid (Browning 1978). The base is composed of the host, the pathogen, and the environment, whereas time is represented by elevation. Comparative epidemiological analysis, a powerful tool developed for analysis of disease epidemics in crop plants (Kranz 1978), focuses attention on all five aspects of the disease pyramid. We say five, because time is important in two processes: development of individual epidemics and adaptation of mediating traits in host and pathogen. Some of these techniques have been applied to the analysis of blister rust epidemics (McDonald and Hoff 1982; McDonald et al. 1991, 1994; McDonald and Dekker-Robertson 1998).

Principles of Comparative Epidemiology

Accumulation of diseased plants in a population is described by one of two equations. If the inciting organisms move from plant to plant, the "compound interest" equation is appropriate (van der Plank 1963). On the other hand, if the casual organisms do not travel directly from plant to plant, the "simple interest" equation is appropriate (van der Plank 1963). Since blister rust is a simple interest epidemic (Kinloch and Byler 1981; McDonald and Hoff 1982), the nonlinear monomolecular model of Madden and Campbell (1990) is appropriate,

$$y = K(1 - e^{-rt}) \tag{1}$$

where y = incidence at time t, K = maximum incidence, e = base of natural logarithm, r = absolute infection rate in incidence per year, and t = age of the epidemic in years. Solving Equation 1 for r when the initial amount of disease is zero gives the following equation,

$$r = \ln(K/(K - y))/t \tag{2}$$

where r, K, y, and t are as in Equation 1 and ln = natural logarithm. Parameter r is useful for comparing epidemics and can be calculated when y and t are known. The parameter K is necessary because y does not reach unity in many disease epidemics. Nonrandom distribution of inoculum, infection microclimate, or resistance genes can cause the departure (see McDonald et al. 1991). In the case of blister rust epidemics in wild-type western white pine in northern Idaho, the average value of K is about 0.98 (McDonald et al 1991, Figure 17). A parameter d, known as the "deviation factor," attempts to quantify the deviation of K from 1 and is computed as

$$d = [c - (-\ln(1 - y))]/[c(-(\ln(1 - y)))] \tag{3}$$

where d = quantitative indicator of amount of deviation from random, c = average number of cankers per tree, ln = natural logarithm, and y = rust incidence at c (McDonald et al. 1991). When K = 1, rust incidence in a stand is related to the average number of infections per tree (see McDonald et al. 1991). If one assumes homogeneous susceptibility and microclimate, then d provides information about spore spread. K can be set to 1 or, if d is known, calculated as

$$K = 1 - e^{-[1/(0.002 + d)]} \quad (4)$$

where K = maximum rust incidence and 0.002 is a constant related to c = 500 cankers per tree. If d is close to zero, then spore distribution is random, and K should approach unity. If d is less than 0, then spore distribution is uniform and K should be close to 1. This logic should distinguish between local and long-distance spread. Infections caused by spores coming from nearby shrubs within a stand should be clumped, whereas distribution of infection resulting from spores transported over a long distance should be random or even uniform (d close to or less than 0). These ideas have not been explicitly tested, but we did find suitable data in several reports. First, Van Arsdel et al. (1961) found an isolated line of *Ribes* plants located next to a young eastern white pine stand, and sampled it in 150-meter bands moving away horizontally from the inoculum source. We computed d and r from the published data (Table 10-2). Theoretically, infections should reach a random to uniform distribution, that is K = 1 at some distance from the infection source. In Wisconsin (Van Arsdel et al. 1961), d equaled 3.03 in the first 150 meters and approached 0 after 150 meters (Table 10-2). After ten years of exposure, and assuming K equals 1, r equaled 0.034 up to 150 meters and 0.01 at 600 meters. From this result, we conclude that clumped distribution of sporidia occurs up to a radius of 150 meters from a source of spores, and beyond 150 meters, the distribution of sporidia becomes random to uniform. In theory, a properly designed experiment could separate local spread from low-density sources of inoculum and long-distance spread from a concentration of inoculum sources.

The parameter d has additional potential to increase our understanding of epidemics. For example, experiments have been conducted in western white pine (Buchanan and Kimmey 1938) and sugar pine (Kimmey and Wagener 1961) where infection that resulted from a single shrub in a single year was mapped in concentric circles with the infected *Ribes* plant at the center. The d parameter under these conditions can approach 15 (G. I. McDonald, unpublished data), indicating that a highly clumped distribution of infections is normal for a single cycle of infection. However, if several cycles of infection (wave years) occur on a site with many infection centers, d should tend to 0 and K should approach unity. Stands with low densities of *Ribes* plants and infrequent wave years should yield d values near 0.5 or 1 and K between 0. 86 and 0.63 (from Equation 4 where c = 500). The point is that a K of 1 must be verified, because rust-free pines may be simple escapes, even at high levels of y and c. This means that rust-free trees observed in an area of high disease incidence may have avoided the disease by chance rather than by being phenotypically resistant.

In Idaho, Stillinger (1943) noted one canker per infected tree (uniform distribution) at about 90 to 100 meters from a single *Ribes* plant. Infections occurred up to 150 meters from an individual source of sporidia. Early studies in western white pine indicated a maximum sporidial spread of 275 to 375 meters (Mielke 1943). Under ideal weather conditions and large concentrations

Table 10-2. Epidemiological parameters of blister rust measured on populations of various species of North American white pine.

State	*Pinus* Species	Source	Experiment	Trees	t[1]	p[2]	c[3]	d[4]	r[5]	*Ribes*[6]
Wisconsin	*P. strobus*	Van Arsdel et al. 1961[7]	Spread in 1st 150m band	118	10	.24	1.47	3.03	0.229	NA
			Spread in 2nd 150m band	70		.14	0.14	–0.51	0.019	NA
			Spread in 3rd 150m band	177		.07	0.07	0.63	0.011	NA
			Spread in 4th 150m band	230		.08	0.08	–0.51	0.011	NA
Idaho	*P. monticola*	McDonald & Dekker-Robertson (1998)[8]	Natural forest—high hazard	194	6	.98	2.75	–.012	0.70	3,700
		McDonald (unpublished)[8]	Natural forest—low hazard	216	23	.91	6.86	0.27	0.117	0.25
		Stillinger (1943)[9]	Spread experiment	NA	8	.44	NA	NA	0.072	2.5
				NA		.45	NA	NA	0.074	25
				NA		.55	NA	NA	0.100	250
				NA		.89	NA	NA	0.276	1,235
				NA		.98	NA	NA	0.489	2,500
North Carolina	*P. strobus*	Powers and Stegall (1971)	Natural forest	151	10	.62	NA	NA	0.096	NA
Canada	*P. flexilis*	Gautreau (1963)	Natural forest	NA	10	.88	14.5	0.41	0.322	NA
Colorado	*P. flexilis*	Johnson (unpublished)	Natural forest	NA	5	.50	NA	NA	0.134	NA
New Mexico	*P. strobiformis*	Geils et al. (1999)[10]	Natural forest—low hazard	146	6	.06	0.06	–0.51	0.011	NA
			Natural forest—mid hazard	181	10	.44	2.39	1.30	0.169	NA
			Natural forest—high hazard	228	10	.46	4.18	1.40	0.226	NA

[1] t = period of exposure to blister rust in years.
[2] p = proportion of population infected.
[3] c = average number of cankers per tree.
[4] $d = [c - (-\ln(1 - p))]/[c(-\ln(1 - p))]$, where c = actual cankers/tree at time t and p = actual incidence at time t (see McDonald et al. 1991)
[5] r = absolute infection rate (see Equation 2 in text).
[6] Number of *Ribes* plants per hectare.
[7] Natural stand located in northern Wisconsin where infection was measured in linear plots from a line of *Ribes*.
[8] Seedlings from susceptible seed collections planted in experimental plantations (see McDonald and Dekker-Robertson 1998).
[9] Plots installed in natural reproduction.
[10] Natural stands located in central New Mexico, where hazard was determined by a landscape parameter model (see also Van Arsdel et al. 1998).

of basidiospores, spread distance might achieve 1,600 meters (Mielke 1943). If we assume a limit of 300 meters, a modest r of 0.01, and K = 1, then a single *Ribes* plant could produce 40 percent infection in twenty-five years (from Equation 1) on 28 hectares (area of a circle with radius of 300 meters).

Are these assumptions reasonable? Stillinger (1943) presented data for y, t, and *Ribes* density (species not specified) that we used to calculate r where K = 1 (Table 10-2). After eight years of exposure, r-values varied from 0.072 for 2.5 shrubs/hectare to 0.489 for 2,500 shrubs/hectare (Table 10-2). We have seen K = 1, r = 0.504, and d = –0.12 for susceptible western white pine after six years of exposure in a plantation that supported about 3,700 *Ribes* plants per hectare (McDonald and Dekker-Robertson 1998). This compares well to Stillinger's r = 0.489. A susceptible plantation of western white pine achieved K = 0.98, r = 0.117, and d = 0.27 after twenty-three years of exposure to 0.03 *Ribes* plants/hectare (G. I. McDonald, unpublished data). Natural limber pine stands in Alberta, Canada (r = 0.322) were similar to the northern Idaho value of r = 0.276 for 1,250 *Ribes* plants per hectare (Table 10-2). A native stand of eastern white pine in North Carolina produced an r equal to 0.096 (Powers and Stegall 1971). Now we can apply the above epidemiological analysis to the whitebark pine/blister rust pathosystem found on subalpine sites.

Some published data from Washington, Oregon, and Idaho enable a critical examination of epidemiological behavior of blister rust on high-elevation sites (Table 10-3). One paper (Bedwell and Childs 1943) provides an early epidemiological comparison of western white pine and whitebark pine. Stands containing mixed populations of the two pines were inspected in 1938 (Table 10-3). Sites (averaging 1.9 hectares) were sampled for *Ribes* plants, and 9 to 37 pairs of trees 0.6 to 3.6 meters tall were selected for examination. These authors provided counts of cankers, estimates of exposure time, and proportions of populations infected. The oldest canker on each plot was recorded at all sites to establish the essential time line; these were about ten years old for all plots. They estimated local *Ribes* densities as absent, rare, occasional, and common. To compare sites, we assumed absent = 0, rare = 0.25/hectare, occasional = 2.5/hectare, and common = 25/hectare, and then calculated r and d (Table 10-3).

Some interesting patterns appeared. For the east side of the Cascade Mountains in Washington and the Bitterroot Divide in Idaho, infection rates showed that whitebark pine was 72.5 times more susceptible than associated western white pine (Table 10-3). The same ratio for the combined Olympic Mountains and Mount Hood plots was 4.8 times. Are these early data revealing some kind of founder effect that could differentiate populations of the rust, do host populations differ geographically, or is the difference explained by sampling error? We tend to discount sampling error because sample sizes were reasonable. This result highlights an important knowledge gap, founder effects and rust genetics, which will be discussed later in this chapter.

Data are available from one additional high-elevation site in Washington State. The *Ribes* eradication program conducted in the western United States (Ketcham et al. 1968) provides some good data for one site on Mount Rainier (Riley 1941). The White River blister rust control unit in Mount Rainier National Park was a 1,050-hectare stand of whitebark pine from

Table 10-3. Values calculated for blister rust epidemic parameters in white pines growing at high elevations in western North America. See Table 10-2 for definitions of variables.

State	*Pinus* Species	Sources	Location	Trees	t	p	c	d	r	*Ribes*[1]
Washington	*P. albicaulis*	Bedwell and Childs (1943)	Olympic Mountains	26	10	.92	7.08	0.250	0.253	0.25
	P. monticola			26		.69	2.69	0.48	0.117	
	P. albicaulis		Central Cascades	9	10	.89	5.44	0.27	0.221	25
	P. monticola			9		0.0	—	—	0.0	
	P. albicaulis	Gynn and Chapman (1951)	Mt. Rainier NP	602	10	.55	NA[2]	NA	0.08	30
Idaho	*P. albicaulis*	Bedwell and Childs (1943)	Bitterroot Divide 1	37	10	.59	1.14	0.24	0.089	0.25
	P. monticola			37		.0	—	—	0.0	
	P. albicaulis		Bitterroot Divide 2	17	10	.71	2.12	0.34	0.124	0.0
	P. monticola			17		.06	0.06	−0.51	0.006	
	P. albicaulus	Hoff (unpublished data)	Sawtell Peak	49	40	.51	NA	NA	0.018	NA
Oregon	*P. albicaulis*	Bedwell and Childs (1943)	Mount Hood 1	27	10	.88	4.96	0.27	0.212	0.0
	P. monticola			27		.23	0.35	0.94	0.026	
	P. albicaulis		Mount Hood 2	17	10	.94	22.65	0.31	0.281	0.0
	P. monticola			17		.35	0.71	0.90	0.043	
	P. albicaulis		Wasco County 1	17	10	.71	3.82	0.55	0.124	0.25
	P. monticola			17		.18	0.24	0.90	0.019	
	P. albicaulis		Wasco County 2	10	10	1.0	22.40	0.10	0.691	2.5
	P. monticola			10		.70	1.40	0.12	0.12	
Montana	*P. albicaulis*	Gynn and Chapman (1948)	Glacier NP	470	5	.06	NA	NA	0.013	445
	P. flexilis			303	5	.07	NA	NA	0.014	484
	P. monticola			115	6	.18	NA	NA	0.034	116
Wyoming	*P. albicaulis*	Carlson (1978)	Yellowstone NP	NA	6	.04	NA	NA	0.007	NA
	P. albicaulis	Toko and Dooling (1968)		NA	15	.24	NA	NA	0.018	225

[1]Number of *Ribes* plants per hectare.
[2]No data available.

which 575 *Ribes* plants per hectare were removed on first eradication. Each of four third-pass eradications applied from 1941 to 1951 removed an average of 30 plants/hectare (Riley 1941, 1942, 1944; Gynn and Chapman 1947). From these data, we conclude that the White River unit supported about 30 *Ribes* shrubs/hectare (mostly *R. lacustre*) for the period 1941 to 1951. In 1951, 602 seedlings of whitebark pine less than 1.5 meters tall were examined for rust infection (Gynn and Chapman 1951). If we assume that trees 1.5 meters tall were ten years old, then the calculated r of 0.08 (Table 10-3) is in line with our northern Idaho results for western white pine stands supporting 25 shrubs/hectare (Table 10-2). This result and newer results from northern Idaho (Tomback et al. 1995) indicate that whitebark and western white pine are sometimes equally susceptible, whereas results discussed above show up to a 72.5 times difference. Founder effects are a possible explanation of this paradox.

Pine and *Ribes* Geography

Blister rust expression requires two hosts in close proximity, as well as a suitable environment. Pine and *Ribes* distributions and densities define the potential for disease expression. A survey of white pines and *Ribes* was conducted to determine the potential for damage by blister rust in the central Rocky Mountains (Joy 1934). Strips 20 meters wide were installed at varying densities in most stands supporting white pines in Colorado and Wyoming. In Wyoming, 1.52 million hectares supported 390 whitebark and/or limber pine trees per hectare. The *Ribes* eradication program also provided good data about these species distributions for sites in the Rocky Mountains at Glacier, Yellowstone, and Rocky Mountain National Parks (Gynn and Chapman 1948, 1949, 1950). An important aspect of *Ribes* distribution was demonstrated by results from three stands in Yellowstone National Park. At Mt. Washburn, 1,740 hectares supported 534 shrubs/hectare (*R. lacustre, R. viscosissimum, R. petiolare, R. inerme,* and *R. montigeum*) distributed over the entire unit, whereas the Mammoth unit, 639 hectares in size, supported 163 shrubs/hectare (*R. lacustre, R. viscosissimum, R. petiolare, R. setosum, R. inerme,* and *R. cereum*) (Gynn and Chapman 1949). The Craig Pass unit, however, clearly demonstrated the existence of "*Ribes*-free" zones. Fourteen hundred hectares of this 1,600-hectare unit were free of *Ribes* (Riley 1938). The 200 hectares that required eradication supported 55 shrubs/hectare of *R. lacustre, R. viscosissimum, R. inerme,* and *R. montigeum* (Gynn and Chapman 1949). In Colorado, 222,582 hectares of limber and bristlecone pine were reported, and the Pike National Forest supported 145,690 hectares of the two pines that averaged 329 stems/hectare (Joy 1934). An eradication unit on the Pike National Forest yielded 69 shrubs/hectare (*R. montigeum, R. cereum,* and *R. inerme*) (Joy 1938). In Rocky Mountain National Park, a 6,070-hectare unit supported 193 limber pine trees and 175 *Ribes* shrubs/hectare (Gynn and Chapman 1950). Thus, there is considerable geographic variation in both pine and *Ribes* densities.

A Continentwide Epidemic: Genetic Variation and "Founder Effects" in Blister Rust

So far, we have painted a complex picture of blister rust that contains many players. Members of the two host genera are widely distributed across most of North America, so that a continentwide pathosystem with more than fifty *Ribes* species and more than nine species of five-needled white pines might become a reality. Aeciospores may transport rust genes over thousands of kilometers. This section looks at our rudimentary knowledge about how the rust has adjusted to its new home.

The blister rust reports have demonstrated that most high-elevation ecosystems support one or more species of *Ribes* at sufficient densities to cause major damage, if the alternate host is susceptible and the environment is suitable for the rust. All of the *Ribes* species found in these areas are currently known to be hosts. The only one in doubt is *R. cereum*. Hahn (1928) lists this species as moderately susceptible; however, in the Sacramento Mountains of New Mexico, *R. cereum* was rarely infected (Van Arsdel et al. 1998). Kimmey and Mielke (1944) transplanted *R. cereum* from Sequoia National Forest in the southern Sierra Nevada to northern California, where they were inoculated with local blister rust. Local *R. cereum* plants were highly resistant, but the transplants were judged moderately susceptible. In addition, *R. cereum* was the only infected alternate host found in proximity to blister rust–infected limber pine in the Black Hills of South Dakota (Lundquist et al. 1992).

Other species apparently have exhibited similar variations in susceptibility. One geographic combination of the rust and a specific clone of *R. viscosissimum* produced infections, while another combination did not (McDonald 2000). Kimmey and Mielke (1944) report that *R. montigeum* collected in northern California was moderately susceptible. An additional species found in the Rocky Mountains, *R. pinetorum*, is known to be highly susceptible (Geils et al. 1999). Many questions remain unanswered about relationships among *C. ribicola* and its *Ribes* hosts. Founder effects resulting from the long-distance rain of aeciospores or urediospores might explain some of the patterns of variations.

Recent molecular studies of blister rust show considerable local variation that cannot be tied to a specific geographical area. No differentiation was found across a 1,000-kilometer east-to-west transect in eastern Canada (Hamelin et al. 1995; Et-touil et al. 1999). Similarly, Kinloch et al. (1998) could not distinguish distance-related differences in a north-to-south transect from British Columbia to southern California. In each case, most variation was local. However, eastern and western populations of blister rust were delineated (Hamelin 1998). The consensus at this point is that two large epidemiological units are maintained by gene flow within each and none between. These flows are probably mediated by long-distance spread via aeciospores (Kinloch et al. 1998) or possibly urediospores (Hamelin et al. 1995) and founder effects. It is too early to sort out all the ramifications this model will have on our thinking about the epidemiological behavior of blister rust.

One condition is clear, however. The range of alternate hosts for *Cronar-*

tium ribicola within its center of diversity is much wider than its range in North America (Millar and Kinloch 1991). In Japan, four rust taxa are recognized on Japanese stone pine, of which three are pine-to-pine forms (Imazu and Kakishima 1995). In the Russian Far East, two taxa of blister rust, *C. ribicola* and *C. kamtschaticum,* infect Japanese stone pine (Azbukina 1995). They cycle to *Ribes* or *Pedicularis* and *Castilleja* species, respectively. These results indicate that there is genetic variation within *Cronartium* species for infecting new hosts that could be inadvertently intoduced into either the eastern or western North American blister rust populations. Alternatively, hybridization between the eastern and western populations in North America could result in new assortments of genes that could produce unexpected behavior.

Other factors may be at play in this complex pathosystem. Genetic analysis of populations of single aeciospores showed a strong genetic component for variation in some epidemiological fitness traits for blister rust development on *Ribes* (McDonald 1982). In addition, certain aeciospore infections can be influenced by temperature to switch from production of urediospores to production of teliospores (McDonald and Andrews 1980), and ecotypes exhibiting modified aeciospore germination requirements may develop. For example, an aeciospore "adjusted" for low elevations may have a reduced level of germination at high elevations. The adjustment could be due to either adaptation or phenotypic plasticity (McDonald 1996). Low infection rates would occur in some areas until germination requirements were appropriately modified. Because this "environmental" conditioning of spore germination has apparently occurred in other rusts, we believe it could happen with white pine blister rust (McDonald 1996). This mechanism was hypothesized to explain the movement of blister rust to the southern Sierra Nevada, a relatively dry region, in 1961 (McDonald 1996).

Finally, the new outbreak of blister rust on limber pine in Colorado (D. W. Johnson, personal communication) produced a relatively high r of 0.134 (Table 10-2) versus the baseline r of 0.016. This recent advance in the geographic range of blister rust (Chapter 11) could be the result of new ecotypes that can function more efficiently in colder and drier environments compared to the original founder. If the rust can make or has already made such ecological or genetic adjustments, all North American white pine species should be considered at risk.

Regardless, blister rust has and will continue to cause much damage in the central Rocky Mountains and elsewhere in North America. Unfortunately, few of these historic data included canker counts to enable the calculation of d. Therefore, we cannot analyze spread, except to reinforce the point that central Rocky Mountain forests support relatively large and evenly distributed populations of several susceptible *Ribes* species. Comparative epidemiological analysis can still be applied to r-values. For example, comparing whitebark pine in Yellowstone National Park after fifteen years of exposure to blister rust (Table 10-3) (Toko and Dooling 1968) with a mature whitebark stand on Sawtell Peak, located 76 kilometers to the west, after forty years' exposure (R. J. Hoff, unpublished data), we find r = 0.018 for both populations. The rates for limber pine and whitebark pine at Oldman Lake in Glacier National Park are almost identical at 0.014 (Table 10-3). The average of these four values,

0.016, can serve as a benchmark for high-elevation white pine forests of the central Rocky Mountains. Data from an infection survey at Oldman Lake in 1967 (Toko and Dooling 1968) produced a y of 0.45. If we assume a twenty-seven-year exposure and use Equation 2 to compute r, then 0.022 compares favorably with 0.013 measured in 1948 at the same site (Table 10-3). The baseline rate (0.016) is about 10 times lower than that reported for whitebark pine in the eastern Cascade Mountains and Bitterroot Divide in Idaho and nearly twenty times less than for similar stands in the Olympic Mountains and Oregon Cascades or limber pine in the Canadian Rockies (Tables 10-2 and 10-3). This rate is about five times less than that reported for a *Ribes* density of 250 shrubs/hectare in a northern Idaho stand of western white pine (Tables 10-2 and 10-3). On a more somber note, the baseline rate is also 8.5 times less than that observed in the Colorado limber pine pathosystem located in the cold and dry environment of the central Rockies (Table 10-2). Is the Colorado rust epidemic the result of adaptation, founder selection, or a chance occurrence of a weather-related local spread? Blister rust was found in 1992 on limber pine at Pumpkin Buttes, Wyoming, about 32 kilometers north of the Colorado outbreak (B. W. Geils, personal communication).

The Last Hope: Resistant Pines

While the eradication program was underway, a third front to combat blister rust was active in western North America. As in the East, pathologists observed individual rust-free western white pines, even though trees around them supported hundreds or thousands of cankers (Lachmund 1934; Mielke 1943; Buckland 1946). In 1946, Bingham (1983) confirmed his first sighting of a canker-free tree in Idaho. The canker-free tree was nearly 30 meters tall, 60 years old, and the only rust-free tree in a 380-tree sample obtained from the stand. During the next three years, fourteen more phenotypically resistant trees were located (Bingham 1983). Some of these rust-free trees were located in stands where individual susceptible trees supported more than 2,000 cankers.

In 1950, the U.S. Department of Agriculture (Division of Plant Disease Control of the Bureau of Entomology and Plant Quarantine, U.S. Forest Service Northern Rocky Mountain Forest and Range Experiment Station, and U.S. Forest Service Region 1) funded a five-man team to search for more rust-free trees. These trees were to be used to determine if the phenotypic resistance expressed in such uninfected trees was inherited (Bingham 1983). A. E. Squillace, of the Northern Rocky Mountain Station, and R. T. Bingham directed the program. They visited the Institute of Forest Genetics in Placerville, California, to learn how to perform controlled pollination of western white pine. When Bingham retired in 1974, more than 3,000 resistant candidates had been located, five seed orchards established, and three long-term field tests installed. In addition, Bingham's collaboration with J. W. Hanover, R. J. Hoff, and G. I. McDonald produced numerous publications on the nature of blister rust resistance (Bingham 1983). In northern Idaho and western Montana, more than five million rust-resistant seedlings have been planted each year since 1990.

Family differences for several resistance factors became obvious after the initial artificial inoculations. Some seedlings had needle spots but no stem symptoms, or stem symptoms appeared but soon changed to atypical symptoms. All were potential expressions of different kinds of resistance mechanisms. Symptoms characteristic of the resistance found thus far are listed, along with detailed descriptions and discussion of the usefulness of each, in Hoff and McDonald (1980) and Chapter 17.

However, the story is not complete without a discussion of variation in virulence and/or aggressiveness. Two races of *C. ribicola* have been hypothesized for the Inland Northwest geographic region. One race expressed itself as yellow needle-spots and the second appeared as red spots. The races revealed themselves by differential rates of infection. Seedlings supporting both yellow and red spots had two times more needle infections than seedlings infected by a single spot type (McDonald and Hoff 1975). A dominant gene for resistance found in sugar pine led to discovery of another type of racial variation. A breeding program for resistance to blister rust in California sugar pine was based on this gene (Kinloch et. al. 1970). A few years later, a race of blister rust capable of neutralizing that gene appeared (Kinloch and Comstock 1981). The new race was called "Happy Camp" for the area in northern California where it was found. Another new blister rust race, labeled "Champion Mine," was discovered in the central Oregon Cascade Mountains (McDonald et al. 1984). This race caused an equal increase in rust severity on both resistant and susceptible western white pine families. A major gene for resistance in western white pine was recently reported (Kinloch et al. 1999). These authors hypothesize that the gene is associated with the Champion Mine race. Another indication of the existence of races of blister rust occurs in stem tissue. Often, both typical cankers and cankers with bark reactions—that is, a resistance reaction—will be present on the same seedling. The obvious explanation is that different races caused the different symptoms (Hoff 1986).

Recently, an unexpectedly high rate of infection was observed on a rust-resistant population of western white pine growing on the Merry Creek plantation near Clarkia, Idaho. This experimental plantation was installed in 1970 using four kinds of stock: field-run susceptible (control), first-generation resistant (one generation of selection for resistance = F_1), back-crossed resistant (F_1 pollen crossed to original parents = B_1), and second-generation resistant (selected F_1 crossed to selected F_1 = F_2). Twelve years after planting (1982), rust mortality in the control seedlings was 100 percent. Twenty-six years after planting (1996), the F_1 and B_1 stocks were 100 percent infected, and 93 percent of the F_2 stock were infected (McDonald and Dekker-Robertson 1998).

The obvious classical explanation of the Merry Creek result was that a new race of rust had appeared. However, based on repeated measurement of rust intensification over twenty-six years, McDonald and Dekker-Robertson (1998) hypothesize that the high level of infection in resistant western white pine stock happened not because of a new race of rust, but because resistance in this western white pine population was horizontal rather than vertical (see Chapter 17). In the case of horizontal resistance, differences in rust incidence among generations of selection is characterized by a linear reduction of r (Equation 2), and given enough disease pressure and time the most highly resistant population will become 100 percent infected (van der Plank 1963). In the case of vertical resistance, incidence of disease can vary from low to high, depending on the specific combination of host and pathogen (van der Plank 1963). Racial variation of the pathogen overcomes these kinds of resistance. In addition, r conditioned by horizontal resistance may be influenced by soil conditions (Shaner and Finney 1977). For Merry Creek, r for the hot-burn regions of the plantation was 1.5 times that of cool-burn regions across all stock types (McDonald and Dekker-Robertson 1998). Infection rate (r) for the first cycle of selection (F_I) was reduced three times from the wild population, and r for the second cycle was reduced six times. At Merry Creek, proportional differences in r were related to stand history and exhibited a linear reduction with cycles of selection for blister rust resistance. Both these findings argue that resistance in northern Rockies western white pine is "durable." According to theory, durable resistance cannot be negated by racial variation for virulence in blister rust (see Simmonds 1991 for a discussion of durable resistance in cereal crops). These experiences with western white pine are important, because they may foretell how blister rust–resistant whitebark pine will behave, since its resistance architecture may be similar to that of western white pine (Chapter 17).

Integrated Management of White Pine Blister Rust

There is little doubt that many genes for resistance are present in western white pine and potentially in whitebark and other white pines. In addition, the appearance of races of blister rust that will adversely affect expression of some of these genes is almost certain, and environmental factors can surely influence the genes for resistance and for virulence. Breeding for resistance can play a role in management of white pine (Hoff and McDonald 1972), as can integrated management (Hagle et al. 1989). Forces of natural selection could be harnessed if five-needled white pines are given a chance to regenerate (Hoff et al. 1976; Neuenschwander et al. 1999; Chapters 1 and 9). After all, the Eurasian white pines exhibit resistance to blister rust even to the extent that some, like Swiss stone pine, are considered immune (Hoff et al. 1980). However, natural selection for resistance for any white pine species will not happen unless regeneration sites exposed to full sun are provided and an adequate source of seeds is ensured. In the case of disturbance-dependent whitebark pine, this process can be thwarted by suppression of the major disturbance (fire).

Three to four generations of breeding may increase resistance sufficiently in a white pine species to ensure its sustainability under high rust hazard, that is, $r > 0.20$. In the absence of new technology, the time required for such a breeding program, even in the case of a short-generation-interval species like western white pine, will be forty to fifty years. Meanwhile, can five-needled white pines be managed to maintain their ecological functions and preserve their genes for a generation or two? Here is where comparative epidemiology can help. Delineation of low-rust-hazard sites should be afforded high priority. The distribution of *Ribes* on the Craig Pass control unit at Yellowstone National Park (discussed earlier) illustrates that hazard determination has potential value. Two attempts so far have not been entirely successful (Ostrofsky et al. 1988; Geils et al. 1999) (see Chapter 17 and Table 10–2), but much available new technology remains to be applied. Management of *Ribes* density can be combined with judicious pruning of cankers and adjustable levels of resistance or genetically determined r to achieve a targeted r (McDonald et al. 1991). The aim of *Ribes* management should be to remove enough shrubs to achieve an r appropriate for the host species in its specific rust situation. Pruning visible cankers on the lower 3 meters of saplings will further increase longevity of treated stands (Hunt 1998). Computer models under development will help managers decide if pruning or any other integrated management option is economically feasible or biologically sound for a particular situation (G. I. McDonald, unpublished data). Obviously, the fungus and native five-needle pines would likely reach an accommodation in natural systems, if wildfire and prescription burns provided ample regeneration opportunities. In the absence of regeneration opportunities, prognosis for the rust and pines to reach this accommodation is dim. Active management and creation of regeneration sites by prescribed fire or other means is essential (see Chapters 17 and 18). None of the breeding, integrated management, or naturalization programs directed at North American white pine species has advanced far enough to assure success, but potential solutions do exist.

Acknowledgments

We thank Brian W. Geils and Alan E. Harvey, both of the USDA Forest Service Rocky Mountain Research Station, and Detlev R. Vogler, USDA Forest Service, Pacific Southwest Research Station, for reviewing earlier drafts of this chapter.

Literature Cited

Azbukina, Z. M. 1995. *Cronartium* species on *Pinus* species in the Russian Far East. Pages 65–69 *in* Proceedings of the 4th IUFRO Rust of Pine Working Party Conference, Tsukuba, Japan.

Bean, W. J. 1914. Trees and shrubs hardy in the British Isles. Vol. 2. London.

Bedwell, J. L., and T. W. Childs. 1943. Susceptibility of whitebark pine to blister rust in the Pacific Northwest. Journal of Forestry 41:904–912.

Bega, R. V. 1959. The capacity and period of maximum production of sporidia in *Cronartium ribicola*. Phytopathology 49:54–57.

———. 1960. The effect of environment on germination of sporidia in *Cronartium ribicola*. Phytopathology 50:61–69.

Bingham, R. T. 1983. Blister rust–resistant western white pine for the Inland Empire: The story of the first 25 years of the research and development program. USDA Forest Service, Intermountain Research Station General Technical Report INT-146, Ogden, Utah.

Browning, A. J. 1978. Managing host genes: Epidemiologic and genetic concepts. Pages 191–212 *in* J. G. Horsfall and E. B. Cowling, editors. Plant disease: An advanced treatise, Vol. I: How disease is managed. Academic Press, New York.

Buchanan, T. S., and J. W. Kimmey. 1938. Initial tests of the distance of spread to and intensity of infection on *Pinus monticola* by *Cronartium ribicola* from *Ribes lacustre* and *R. visosissimum*. Journal of Agricultural Research 56:9–30.

Buckland, D. C., 1946. Interim report on the effect of blister rust damage to the management of western white pine in the Upper Arrow Forest. Canadian Dominion, Department Agriculture, Forest Pathology Laboratory Mimeo Report, Victoria, British Columbia, Canada.

Carlson, C. E. 1978. Noneffectiveness of *Ribes* eradication as a control of white pine blister rust in Yellowstone National Park. USDA Forest Service, Northern Region, State and Private Forestry, Report 78-18, Missoula, Montana.

Childs, T. W., J. L. Bedwell, and G. H. Englerth. 1938. Blister rust infection on *Pinus albicaulis* in the northwest. Plant Disease Reporter 22:139–140.

Clinton, G. P., and F. A. McCormick. 1919. Infection experiments of *Pinus strobus* with *Cronartium ribicola*. Connecticut Argicultural Experiment Bulletin 214:428–459.

Draper, M. A., and J. A. Walla. 1993. First report of *Cronartium ribicola* in North Dakota. Plant Disease 77:952.

Et-touil, K., L. Bernier, J. Beaulieu, J. A. Berube, A. Hopkin, and R. C. Hamelin. 1999. Genetic structure of *Cronartium ribicola* populations in eastern Canada. Phytopathology 89:915–919.

Farrar, J. L. 1947. Forest tree breeding in Canada. Dominion Forest Service, reprint of paper prepared for the 5th British Empire Forest Conference.

Gautreau, E. 1963. Effects of white pine blister rust in limber pine stands of Alberta. Canada Department of Forestry, Forest Entomology and Pathology Branch, Bimonthly Progress Report 19:3.

Geils, B. W., D. A. Conklin, and E. P. Van Arsdel. 1999. A preliminary hazard model of white pine blister rust for the Sacramento Ranger District, Lincoln National Forest. USDA Forest Service, Rocky Mountain Research Station, Research Note RMRS-RN-6, Fort Collins, Colorado.

Gynn, J. C., and C. M. Chapman. 1947. Blister rust control, Mount Rainier National Park, 1947. Pages 62–65 *in* White pine blister rust control in the northwestern region, January 1 to December 31, 1947. USDA Bureau of Entomology and Plant Quarantine, Spokane, Washington.

———. 1948. Blister rust control, Glacier National Park, 1948. Pages 80–85 *in* White pine blister rust control in the northwestern region, January 1 to December 31, 1948. USDA Bureau of Entomology and Plant Quarantine, Spokane, Washington.

———. 1949. Blister rust control, Yellowstone National Park, 1949. Pages 83–85 *in* White pine blister rust control in the northwestern region, January 1 to December 31, 1949. USDA Bureau of Entomology and Plant Quarantine, Spokane, Washington.

———. 1950. Blister rust control, Yellowstone National Park, 1950. Pages 92–94 in White pine blister rust control in the northwest region, January 1 to December 31, 1950. USDA Bureau of Entomology and Plant Quarantine, Spokane, Washington.

———. 1951. Blister rust control, Mount Rainier National Park, 1951. Pages 64–66 *in* White pine blister rust control in the northwestern region, January 1 to December 31, 1951. USDA Bureau of Entomology and Plant Quarantine, Spokane, Washington.

Hagle, S. K., G. I. McDonald, and E. A. Norby. 1989. White pine blister rust in northern Idaho and western Montana: Alternatives for integrated management. USDA Forest Service, Intermountain Research Station General Technical Report. INT-261, Ogden, Utah.

Hahn, G. G. 1928. The inoculation of Pacific Northwestern *Ribes* with *Cronartium ribicola* and *C. occidentale*. Journal of Agricultural Research 37: 663–683.

Hamelin, R. C. 1998. Molecular epidemiology of white pine blister rust. Pages 255–259 *in* Proceedings First IUFRO Rust of Forest Trees Working Party Conference, 2–7 Aug. 1998, Saariselka, Finland. Finnish Forest Research Institute Research Papers 712, Helsinki.

Hamelin, R. C., J. Beaulieu, and A. Plourde. 1995. Genetic diversity in populations of *Cronartium ribicola* in plantations and natural stands of *Pinus strobus*. Theoretical and Applied Genetics 91:1214–1221.

Hawksworth, F. G. 1990. White pine blister rust in southern New Mexico. Plant Disease 74:938.

Heimburger, C. 1956. Blister rust resistance in white pine. Northeastern Forest Tree Improvement. Conference Proceedings 3:6–11.

Hirt, R. R. 1942. The relation of certain meteorological factors to the infection of eastern white pine by the blister rust fungus. Bulletin of the New York State College of Forestry, Syracuse University, Technical Publication 59, Syracuse, New York.

Hoff, R. J. 1986. Inheritance of bark reaction resistance mechanism in *Pinus monticola* infected with *Cronartium ribicola*. USDA Forest Service, Intermountain Research Station, Research Note INT-361, Ogden, Utah.

Hoff, R., and S. Hagle. 1990. Diseases of whitebark pine with special emphasis on white pine blister rust. Pages 179–190 *in* W. C. Schmidt and K. J. McDonald, compilers. Proceedings—Symposium on whitebark pine ecosystems: Ecology and management of a high-mountain resource. USDA Forest Service, Intermountain Research Station, General Technical Report INT-270, Ogden, Utah.

Hoff, R. J., and G. I. McDonald. 1972. Stem rusts of conifers and the balance of nature. Pages 525–535 *in* R. T. Bingham, R. J. Hoff, and G. I. McDonald, editors. Biology of rust resistance in forest trees. USDA Forest Service, Miscellaneous Publications 1221, Washington, D.C.

———. 1980. Improving rust-resistant strains of inland western white pine. USDA Forest Service, Intermountain Research Station, Research Paper INT-245, Ogden, Utah.

Hoff, R. J., G. I. McDonald, and R. T. Bingham. 1976. Mass selection for blister rust resistance: A method for natural regeneration of western white pine. USDA Forest Service, Intermountain Research Station, Research Note INT-202, Ogden, Utah.

Hoff, R. J., R. T. Bingham, and G. I. McDonald. 1980. Relative blister rust resistance of white pines. European Journal of Forest Pathology 10:307–316.

Hunt, R. S. 1998. Pruning western white pine in British Columbia to reduce white pine blister rust losses: 10-year results. Western Journal of Applied Forestry 13:60–63.

Imazu, M., and M. Kakishima. 1995. Blister rusts on *Pinus pumila* in Japan. Pages 27–36 *in* Proceedings of the 4th IUFRO Rust of Pine Working Party Conference, Tsukuba, Japan.

Joy, E. L. 1934. White pine–*Ribes* survey in Wyoming and Colorado, 1934. Pages 194–213 *in* Blister rust control in the far west, January 1 to December 31, 1934, USDA Bureau of Entomology and Plant Quarantine, Spokane, Washington.

———. 1938. Blister rust control work in Colorado and Wyoming, 1938. Pages 109–127 *in* Blister rust control in the northwestern region, January 1 to December 31, 1938. USDA Bureau of Plant Industry, Spokane, Washington.

Ketcham, D. E., C. A. Wellner, and S. S. Evans Jr. 1968. Western white pine management programs realigned on northern Rocky Mountain national forests. Journal of Forestry 66:329–332.

Kimmey, J. W., and J. L. Mielke. 1944. Susceptibility to white pine blister rust of *Ribes cereum* and some other *Ribes* associated with sugar pine in California. Journal of Forestry 42:752–756.

Kimmey, J. W., and W. W. Wagener. 1961. Spread of white pine blister rust from *Ribes* to sugar pine in California and Oregon. USDA Technical Bulletin 1251, Washington, D.C.

Kinloch, B. B., and J. W. Byler. 1981. Relative effectiveness and stability of different resistance mechanisms to white pine blister rust in sugar pine. Phytopathology 71:386–391.

Kinloch, B. B. Jr., and M. Comstock. 1981. Race of *Cronartium ribicola* virulent to major gene resistance in sugar pine. Plant Disease 65:604–605.

Kinloch, B. B. Jr., and D. Dulitz. 1990. White pine blister rust at Mountain Home Demonstration State Forest: A case study of the epidemic and prospects for genetic control. USDA Forest Service, Pacific Southwest Research Station, Research Paper PSW-204, Berkeley, California.

Kinloch, B. B. Jr., G. O. Parks, and C. W. Fowler. 1970. White pine blister rust: Simply inherited resistance in sugar pine. Science 167:193–195.

Kinloch, B. B. Jr., R. D. Westfall, E. E. White, M. A. Gitzendanner, G. E. Dupper, B. M. Foord, and P. D. Hodgskiss. 1998. Genetics of *Cronartium ribicola.* IV. Population structure in western North America. Canadian Journal of Botany 76:91–98.

Kinloch, B. B. Jr., R. A. Sniezko, G. D. Barnes, and T. E. Greathouse. 1999. A major gene for resistance to white pine blister rust in western white pine from the western Cascade Range. Phytopathology 89:861–867.

Kliejunas, J. 1996. An update of blister rust incidence on the Sequoia National Forest. USDA Forest Service, Pacific Southwest Region, State and Private Forestry Forest Pest Management Report R96-01, San Francisco, California.

Kranz, J. 1978. Comparative anatomy of epidemics. Pages 33–62 *in* Plant disease: An advanced treatise. Vol. II: How disease develops in populations. Academic Press, New York.

Krebill, R. G. 1964. Blister rust found on limber pine in northern Wasatch Mountains. Plant Disease Reporter 50:532.

Krebill, R. G., and R. J. Hoff. 1995. Update on *Cronartium ribicola* in *Pinus albicaulis* in the Rocky Mountains, USA. Pages 119–126 *in* Proceedings of the 4th IUFRO Rust of Pine Working Party Conference, Tsukuba, Japan.

Lachmund, H. G. 1933. Method of determining age blister rust infection on western white pine. Journal of Agriculture Research 46:675–693.

———. 1934. Damage to *Pinus monticola* by *Cronartium ribicola* at Garibaldi, British Columbia. Journal of Agriculture Research 49:239–249.

Lanner, R. M. 1990. Biology, taxonomy, evolution, and geography of stone pines of the world. Pages 14–24 *in* W. C. Schmidt and K. J. McDonald, compilers. Proceedings—Symposium on whitebark pine ecosystems: Ecology and management of a high-mountain resource. USDA Forest Service, Intermountain Research Station, General Technical Report INT-270, Ogden, Utah.

Little, E. L. 1980. National Audubon Society field guide to North American trees: Western region. Alfred A. Knopf, New York.

Lloyd, M. G., C. A. O'dell, and H. J. Wells. 1959. A study of spore dispersion by use of silver-iodide particles. Bulletin American Meteorological Society 40:305–309.

Lundquist, J. E., B. W. Geils, and D. W. Johnson. 1992. White pine blister rust on limber pine in South Dakota. Plant Disease 76:538.

Madden, L. V., and C. L. Campbell. 1990. Nonlinear disease progress curves. Pages 181–229 *in* J. Kranz, editor. Epidemics of plant disease: Mathematical analysis and modeling. Springer-Verlag, Berlin.

Maloy, O. C. 1997. White pine blister rust control in North America: A case history. Annual Review Phytopathology 35:87–109.

McDonald, G. I. 1982. Genetic variation of epidemiological fitness traits among single-aeciospore cultures of *Cronartium ribicola.* Phytopathology 72:1391–1396.

———. 1996. Ecotypes of blister rust and management of sugar pine in California. Pages 137–147 *in* B. B. Kinloch Jr., M. Marosy, and M. E. Huddleston, editors. Sugar pine: Status, values, and roles in ecosystems. University of California, Division of Agriculture and Natural Resources, Publication 3362, Oakland.

———. 2000. Geographic variation of white pine blister rust aeciospore infection efficiency and incubation period. HortTechnology 10:533–536.

McDonald, G. I., and D. S. Andrews. 1980. Influence of temperature and spore stage on production of teliospores by single aeciospore lines of *Cronartium ribicola.* USDA Forest Service, Intermountain Forest and Range Experiment Station, Research Paper INT-256, Ogden, Utah.

McDonald, G. I., and D. L. Dekker-Robertson. 1998. Long-term differential expression of blister rust resistance in western white pine. Pages 285–295 *in* Proceedings First IUFRO Rust of Forest Trees Working Party Conference, 2–7 Aug. 1998, Saariselka, Finland. Finnish Forest Research Institute, Research Paper 712, Helsinki.

McDonald, G. I., and R. J. Hoff. 1975. Resistance to *Cronartium ribicola* in *Pinus monticola:* An analysis of needle-spot types and frequencies. Canadian Journal of Botany 53:2497–2505.

———. 1982. Engineering blister rust–resistant western white pine. Pages 404–411 *in* Resistance to diseases and pests in forest trees. Pudoc Wageningen, Germany.

McDonald, G. I., R. J. Hoff, and W. R. Wykoff. 1981. Computer simulation of white pine blister rust epidemics: 1 Model formulation. USDA Forest Service, Intermountain Forest and Range Experiment Station, Research Paper INT-258, Ogden, Utah.

McDonald, G. I., E. M. Hansen, C. A. Osterhaus, and S. Sammam. 1984. Initial characterization of a new strain of *Cronartium ribicola* from the Cascade Mountains of Oregon. Plant Disease 68:800–804.

McDonald, G. I., R. J. Hoff, and S. Sammam. 1991. Epidemiologic function of blister rust resistance: A system for integrated management. Pages 235–255 *in* Y. Hiratsuka, J. K Samoil, and P. V Blenis, editors. Rusts of pines, Proceedings of the IUFRO Rusts of pine Working Party Conference. Forestry Canada, Northwest Region, Northern Forestry Centre Information Report NOR-X-317, Edmonton, Alberta, Canada.

McDonald, G. I., R. J. Hoff, T. M. Rice, and R. Mathiasen. 1994. Measuring early performance of second generation resistance to blister rust in western white pine. Pages 133–150 *in* D. M. Baumgartner, D. M. Lotan, J. R. Tonn, editors. Symposium proceedings—Interior cedar-hemlock-white pine forests: Ecology and management. Department of Natural Resources, Washington State University, Pullman.

Mielke, J. L. 1943. White pine blister rust in western North America. Yale University School of Forestry Bulletin 52, New Haven, Connecticut.

Millar, C. I., and B. B. Kinloch. 1991. Epidemiologic function of blister rust resistance: A system for integrated management. Pages 1–38 *in* Y. Hiratsuka, J. K Samoil, and P. V. Blenis, editors. Rusts of pines, Proceedings of the IUFRO Rusts of Pine Working Party Conference. Forestry Canada, Northwest Region, Northern Forestry Centre Information, Report NOR-X-317, Edmonton, Alberta, Canada.

Miller, D. R. 1968. White pine blister rust found on foxtail pine in northwestern California. Plant Disease Reporter 52:391–392.

Monnig, E., and J. Byler. 1992. Forest health and ecological integrity in the northern Rockies. USDA Forest Service, Northern Region, State and Private Forestry, FPM Report 92-7, Missoula, Montana.

Neuenschwander, L. F., J. W. Byler, A. E. Harvey, G. I. McDonald, D. S. Ortiz, H. L. Osborne, G. C. Snyder, and A. Zack. 1999. White pine and the American West: A vanishing species—Can we save it? USDA Forest Service, Rocky Mountain Research Station, General Technical Repert-RMRS-98-197, Fort Collins, Colorado.

Ostrofsky, W. D., T. Rumpf, D. Struble, and R. Bradbury. 1988. Incidence of white pine blister rust in Maine after 70 years of a *Ribes* eradication program. Plant Disease 72:967–970.

Panzer, J. D., E. C. Tullis, and E. P. Van Arsdel. 1957. A simple 24-hour slide spore collector. Phytopathology 47:512–514.

Powers, H. R., and W. A. Stegall Jr. 1971. Blister rust on unprotected white pines. Journal of Forestry 69:165–167.

Riker, A. J., T. F. Kouba, W. H. Brener, and R. F. Patton. 1953. White-pine trees selected for resistance to white-pine blister rust. Pages 322–323 *in* Proceedings, 7th International Botanical Congress, Stockholm.

Riley, M. C. 1938. Preeradication survey, Yellowstone National Park, 1941. Pages 91–92 *in* White pine blister rust control in the northwestern region, January 1 to December 31, 1938. USDA Bureau of Entomology and Plant Quarantine, Spokane, Washington.

———. 1941. Blister rust control work, Mount Rainier National Park, 1941. Pages 92–95 *in* White pine blister rust control in the northwestern region, January 1 to

December 31, 1941. USDA Bureau of Entomology and Plant Quarantine, Spokane, Washington.

———. 1942. Blister rust control work, Mount Rainier National Park, 1942. Pages 91–95 *in* White pine blister rust control in the northwestern region, January 1 to December 31, 1942 USDA Bureau of Entomology and Plant Quarantine, Spokane, Washington.

———. 1944. Blister rust control, Mount Rainier National Park, 1944. Pages 80–86 *in* White pine blister rust control in the northwestern region, January 1 to December 31, 1944. USDA Bureau of Entomology and Plant Quarantine, Spokane, Washington.

Riley, M. C., and C. M. Chapman. 1946. Blister rust control, Yellowstone National Park, 1946. Pages 77–80 *In* White pine blister rust control in the northwestern region, January 1 to December 31, 1946. USDA Bureau of Entomology and Plant Quarantine, Spokane, Washington.

Schmitt, C. L., and D. W. Scott. 1998. Whitebark pine health in northeastern Oregon and western Idaho. USDA Forest Service, Pacific Northwest Region, Blue Mountains Pest Management Zone Report BMZ-99-03, LaGrande, Oregon.

Schreiner, E. J. 1938. Research in forest genetics at the Northeastern Forest Experiment Station. Northern Nut Growers Association Proceedings 29:12–14.

Shaner, G., and R. E. Finney. 1977. The effect of nitrogen fertilization on the expression of slow-mildewing resistance in Knox wheat. Phytopathology 67:1051–1056.

Simmonds, N. W. 1991. Genetics of horizontal resistance to disease of crops. Biological Review 66:1051–1056.

Smith, J., and J. Hoffman. 1998. Status of white pine blister rust in Intermountain Region white pines. USDA Forest Service, Intermountain Region, State and Private Forestry Forest Health Protection, Report R4-98-02, Boise, Idaho.

Snell, W. H. 1931. The Kelm Mountain blister-rust infestation. Phytopathology 21:919–921.

Somerville, W. 1909. *Peridermium strobi,* the blister of Weymouth pine. Quarterly Journal of Forestry 3:232–236.

Spalding, V. M., and B. E. Fernow. 1899. The white pine. USDA Forest Division Bulletin 22, Washington, D.C.

Spaulding, P. C. 1909. European black currant rust on the white pines of America. USDA Bureau of Plant Industry Circular Number 36, Washington, D.C.

———. 1911. The blister rust of white pine. USDA Bulletin Number 206, Washington, D.C.

———. 1922. Investigations of the white pine blister rust. USDA Bulletin Number 957, Washington, D.C.

Stillinger, C. R. 1943. Results of investigations on the white pine blister rust. Pages 127–140 *in* White pine blister rust control in the northwestern region, January 1 to December 31, 1943. USDA Bureau of Entomology and Plant Quarantine, Spokane, Washington.

Toko, H. V., and O. J. Dooling. 1968. An evaluation of the blister rust control program of the National Park Service in the Rocky Mountain area. Unpublished report on file at USDA Forest Service, Northern Region, State and Private Forestry, Missoula, Montana.

Toko, H. V., D. A. Graham, C. E. Carlson, and D. E. Ketchum. 1967. Effects of past *Ribes* eradication on controlling white pine blister rust in northern Idaho. Phytopathology 57:1010.

Tomback, D. F., J. K. Clary, J. Koehler, R. J. Hoff, and S. F. Arno. 1995. The effects of blister rust on post-fire regeneration of whitebark pine: The Sundance burn of northern Idaho (U.S.A.). Conservation Biology 9:654–664.

Van Arsdel, E. P. 1967. The nocturnal diffusion and transport of spores. Phytopathology 57:1221–1229.

Van Arsdel, E. P., and R. G. Krebill. 1995. Climatic distribution of blister rust on pinyon and white pines in the USA. Pages 127–133 *in* Proceedings of the 4th IUFRO Rust of Pine Working Party Conference, Tsukuba, Japan.

Van Arsdel, E. P., A. J. Riker, T. F. Kouba, V. E. Suomi, and R. A. Bryson. 1961. The climatic distribution of blister rust on white pine in Wisconsin. USDA Forest Service, Lake States Forest Experiment Station, Station Paper No. 87, St. Paul, Minnesota.

Van Arsdel, E. P., D. A. Conklin, J. B. Popp, and B. W. Geils. 1998. The distribution of blister rust in the Sacramento Mountains of New Mexico. Pages 275–283 *in* Proceedings First IUFRO Rust of Forest Trees Working Party Conference, 2–7 Aug. 1998, Saariselka, Finland. Finnish Forest Research Institute, Research Papers 712, Helsinki.

van der Plank, J. E. 1963. Plant diseases: Epidemics and control. Academic Press, New York.

Chapter 11

Whitebark Pine Decline: Infection, Mortality, and Population Trends

Katherine C. Kendall and Robert E. Keane

Throughout major parts of their range, whitebark pine (*Pinus albicaulis*) communities have declined dramatically over the past fifty years from the combined effects of disease, insects, and successional replacement (Chapters 1 and 9). Humans are indirectly responsible for much of this deterioration. The most rapid and precipitous reductions are caused by an introduced fungal disease, white pine blister rust (*Cronartium ribicola*). Blister rust has been most devastating in the more mesic parts of whitebark pine range. Although blister rust has taken a smaller toll in drier or colder communities, ultimately all whitebark pine stands are at risk.

In the northern Rocky Mountains of the United States, the long-term policy of excluding natural fires has caused large-scale replacement of whitebark pine, a relatively shade-intolerant species, by other more shade-tolerant trees (Chapters 4 and 9). Whitebark pine is most productive in habitats where it is a seral species, and yet whitebark pine will continue to decline in these areas if fire is not allowed periodically to set back the successional clock. Without fire to create openings in the forest, whitebark pine regeneration eventually is outcompeted by tree species with lower light requirements, such as subalpine fir (*Abies lasiocarpa*) and Engelmann spruce (*Picea engelmannii*). Young age classes are virtually absent in many areas. Climate change and evolution of more lethal strains of blister rust threaten to make the future even more perilous for whitebark pine. In this chapter, we review the historic and current status of whitebark pine, and project the future trends of whitebark pine communities throughout western North America. We include discussions of the mechanisms and temporal scale of decline.

Mechanisms of Decline

White Pine Blister Rust

The interaction of climate and the spatial distribution of its hosts, *Ribes* and five-needled (white) pines, determine the course of blister rust infection in whitebark pine stands. The transmission of blister rust to pine trees requires the fungus to cycle through four of its five spore stages within one year, while passing from pine to *Ribes* and back to pine (Chapter 10). Prevailing winds, local winds, and episodic winds all play an important role in disseminating blister rust spores. In late spring or early summer, the aeciospores produced in cankers on pine trees are released and infect *Ribes.* Aeciospores are quite durable and remain viable across a wide range of weather conditions during wind dispersal. Because of their hardiness, they are capable of disseminating across vast distances (up to 480 kilometers,) (Mielke 1943).

Once a *Ribes* plant is infected by successful germination of aeciospores, a number of life-cycle phases are completed within its leaves. Spread and intensification of rust infection requires that certain moisture and temperature conditions be met during this interval. Three periods are particularly limiting: June and July, when urediniospores are formed on *Ribes* leaves, released, and infect neighboring *Ribes;* August and September, when telial columns form on *Ribes* leaves and produce basidia; and, most important, September and October, when basidiospores travel to pine needles (Mielke and Kimmey 1935). Moderate daytime temperatures (1°C to 28°C) are required for urediniospore and basidiospore formation (Lachmund 1934, Van Arsdel et al. 1956; Van Arsdel 1965). All stages need moisture in the form of rain, dew, or high humidity. Intensification of blister rust infection occurs when frequent, heavy dews allow the formation of several generations of urediniospores during a single summer (Mielke 1943). Basidiospores are produced only if there is sufficient summer precipitation to retain leaves on *Ribes* plants through telia and basidia formation. Finally, twenty-four to forty-eight hours of temperatures below 20°C and high humidity are required for the formation, transport to pine needles, and germination of the fragile basidiospores (Mielke 1943). The frequency of years with which all these conditions are met is a primary factor in determining the rate of spread of blister rust infection.

The density and rust susceptibility of *Ribes* and its physical proximity to white pine trees also influence rust transmission. Blister rust, a mild annual disease on *Ribes,* dies each year when the shrubs drop their leaves. Thus, rust infection of *Ribes* requires a source of inoculum (aeciospores) each spring. However, blister rust is a perennial disease in pine trees. The fungus usually continues to grow within whitebark pine until the tree dies or, rarely, is able to overcome the infection. Large trees may have dozens of separate infections (cankers) from a year of intense infection or from lower levels of exposure to rust over multiple years. Each year an infected tree lives, it has the potential to produce aeciospores that infect *Ribes* and thereby perpetuate the disease.

Because aeciospores rely on wind transport, moderate temperatures, and moist conditions to maintain viability, successful spore dispersal usually occurs only over short distances. Typically, urediniospores infect nearby shrubs, and basidiospore infection range is less than several hundred meters (Benedict 1981), but the presence of blister rust in whitebark pine communities that have no *Ribes* nearby is evidence that these spores can travel much farther (K. C. Kendall, unpublished data). In sites with susceptible *Ribes,* the pace of blister rust intensification increases as *Ribes* density increases and as the distance between *Ribes* patches and pine stands declines. All *Ribes* species are capable of harboring blister rust and spreading infection, but individual species differ in their susceptibility and the amount of teliospore production. In general, of the twenty-seven species tested, *Ribes hudsonianum* var. *petiolare* (formerly *R. petiolare*), *R. sanguineum, R. inerme,* and *R. viscosissimum* support the highest levels of rust among those species commonly found within the range of whitebark pine (Mielke et al. 1937; Kimmey 1938; Mielke 1943; Hitchcock and Cronquist 1976).

Where *Ribes* shrubs are abundant and when summer weather allows multiple urediniospore generations, high levels of spores can accumulate. If conditions permit a large proportion of these spores to mature and then produce basidiospores that reach white pine needles, the result is a wave of new infections in pines. During wave years, rust infection spreads into new areas and intensifies in previously infected stands. In humid, mountain climates, wave years are virtually annual events, because the cool, moist conditions that are favorable to blister rust often support an abundance of susceptible *Ribes.*

The frequency of infection wave years in whitebark pine ecosystems is highly variable, depending on elevation, geographical region, topographical setting, genetic variation in the rust, and wind patterns. The colder temperatures of higher elevations tend to inhibit spore development, but precipitation and humidity are generally higher than at low elevations, favoring the spread of rust. The continental climates of geographical regions, such as Yellowstone National Park, central Idaho, and the Great Basin country (Nevada and vicinity), tend to be drier in late summer and fall, thereby limiting optimal spore germination conditions (Keane and Morgan 1994). However, isolated mountain ranges may experience high humidities during the period of basidiospore dispersal because of orographic weather patterns. Some mountain ranges are drier than neighboring mountains because they lie in the rain shadow of other peaks. Mutation and genetic variation in the rust can also cause regional differences in infection severity. During the past thirty years, numerous virulent races of blister rust have evolved that are able to overcome one or more defense mechanisms of pine trees (Kinloch and Comstock 1981; Hoff and McDonald 1993; chapter 10).

It is not always clear why rust infection has intensified more quickly in some areas than in others. Reasons for differing rust severity are complex. They may include, but are probably not limited to, various combinations of levels of host susceptibility, climate, topographic position, and the spatial distribution of hosts. In general, weather conditions requisite for rust intensification occur more frequently in climates with coastal influences than in dry, continental climates; thus, infection levels build faster. Valley topographic positions also tend to be more susceptible to rust than are higher-elevation sites (Van Arsdel 1965). *Ribes* species most susceptible to blister rust tend to grow along stream courses (Kimmey 1938) and are most abundant in valley bottoms, where they are disjunct from whitebark pine stands (J. P. Smith, unpublished data, School of Forestry, Northern Arizona University, Flagstaff). Upland *Ribes* species tend to be less susceptible than streamside species (Kimmey 1938). As the distance separating concentrations of susceptible *Ribes* from whitebark pine increases, the probability of basidiospore dissemination to trees is apt to decline (Buchanan and Kimmey 1938; Kimmey and Wagener 1961). In some areas, blister rust is unable to complete its life cycle in the subalpine zone due to cold temperatures. The low levels of rust infection in the subalpine zone of arid, continental climates and the restriction of cankers to the upper crowns suggest an off-site infusion of basidiospores from stream valleys (J. P. Smith, personal communication). Host continuity also plays an important role at the landscape level. For example, the Rocky Mountains have a more continuous distribution of white pine and *Ribes,* and higher rust infection levels than the forest "islands" of the Great Basin (J. P. Smith, unpublished data).

After slow initial spread in continental climates, rust infection then may intensify quickly in the subalpine zone. Predictions of the frequency of weather events conducive to rust infection and spread suggest that favorable conditions may occur less frequently at low elevations where most *Ribes* grow than at the higher elevations where whitebark pine is found (Koteen 1999). Observations in the parts of the Greater Yellowstone Ecosystem (GYE) with more humid climates are consistent with this prediction. Paradoxically, Smith (1997) found more low-elevation sites infected with rust, but rust intensity of infected sites increased with elevation in the GYE. He hypothesized that blister rust infection levels in whitebark pine mount slowly as infrequent favorable weather events allow spores to reach the high country. However, once rust becomes established in the subalpine zone, the stage could be set for more rapid intensification of the disease.

The lag time between the arrival of rust in an area and the expansion of the disease in whitebark pine communities in continental climates depends on the ability of blister rust to become established within the subalpine zone. In both dry and humid continental climates, basidiospores probably disseminate from riparian *Ribes*. Drier areas may be unable to support the full rust cycle due to the presence of less susceptible *Ribes,* unfavorable upland temperature and moisture conditions, or larger distances between hosts than those found in more humid regions (J. P. Smith, personal communication). Low levels of blister rust infection at high elevations are unlikely to be reliable sources of aeciospores for riparian *Ribes,* because the spores are produced too late in the season to complete development. J. P. Smith (personal communication) reasons that the delay should decrease with the occurrence of susceptible *Ribes* and other pines, like western white pine (*Pinus monticola*) in Washington and Oregon, at intermediate elevations. Because they grow closer to riparian *Ribes,* mid-elevation species can serve as conduits for blister rust spores to subalpine stands. The decline of sugar pine (*Pinus lambertiana*) in the Sierra Nevada soon after the arrival of rust, and the subsequent infection of its higher-elevation neighbors, western white pine and whitebark pine (Mielke 1943), is consistent with this hypothesis. The five-needled pines that grow closest to valley-bottom

Ribes (sugar pine) became infected with blister rust earlier than those at higher elevations (western white and whitebark pine). Blister rust spores had shorter distances to cover to infect whitebark pine once blister rust was present in sugar and western white pine.

Mountain Pine Beetle and Dwarf Mistletoe

Periodically, whitebark pine forests suffer heavy mortality when mountain pine beetles (*Dendroctonus ponderosae*) spill out from heavy infestations in adjacent lodgepole pine stands, or when whitebark pine stand characteristics are favorable to beetles. Because pine beetles preferentially attack the large, older trees that produce most of the whitebark pine cones, heavy beetle infestations can significantly diminish seed production. Mountain pine beetle epidemics erupted across the west from 1910 through the 1930s and again from the 1970s through the 1980s, creating many whitebark pine "ghost" forests (Arno and Hoff 1990; Perkins and Swetnam 1996). For example, in the West Big Hole Range that straddles the Idaho-Montana border, whitebark pine basal area measurably decreased from 1928 to 1932 as a result of a mountain pine beetle epidemic (Murray 1996). Many large whitebark pine trees in northeast Oregon and northwest Montana were killed during beetle epidemics in the late 1970s. Trees infected with blister rust are stressed and appear to be more attractive to mountain pine beetles (Keane et al. 1994) or more vulnerable to successful attack as the infection decreases the vitality of trees, making them less able to combat invading beetles.

Dwarf mistletoe (*Arceuthobium* spp.) infections cause severe damage to whitebark pine in some localities (Arno and Hoff 1990). Heavily infected trees experience reduced growth rates and seed production, and increased mortality rates and vulnerability to other diseases and insects (Hawksworth and Wiens 1972). Because dwarf mistletoe is controlled by stand-replacing fire, it is now more prevalent than before the advent of wildlands fire-suppression policies (Alexander and Hawksworth 1976). Whitebark pine is a principal host of limber pine dwarf mistletoe (*Arceuthobium cyanocarpum*) (Mathiasen and Hawksworth 1988). On the northwest slopes of Mount Shasta, more than 95 percent of the whitebark pine trees are infected, and half have been killed by dwarf mistletoe (Mathiasen and Hawksworth 1988). Heavy *A. cyanocarpum* infections of whitebark pine also have been reported in the Copper Mountains, near Elko, Nevada, and South Pass in Fremont County, Wyoming (Mathiasen and Hawksworth 1988). Whitebark pine is an occasional host of several other mistletoe species: lodgepole or leafless (*A. americanum*) (Hagle et al. 1987), hemlock (*A. tsugense*) (Hawksworth and Wiens 1972), and larch (*A. laricis*) (Mathiasen 1998). Heavy parasitism by leafless dwarf mistletoe has been suggested as the cause of widespread mortality of whitebark pine on Wizard Island in Crater Lake National Park, Oregon (Jackson and Faller 1973).

Fire Exclusion

In the northern Rocky Mountains of the United States and in adjacent regions in Canada, whitebark pine depends on fire to maintain its dominance or presence on sites where it is a successional species (Chapters 4, 9, and 14). It benefits from low-severity fires, because it often is able to survive them. Its shade-tolerant competitors have thinner bark, shallow roots, and dense low branches that carry flames to their crowns (Arno and Hoff 1990). During moderate fire weather years, surface fires consume scattered ground fuels and kill most fir, spruce, and young whitebark pine, but few larger whitebark pine trees.

The fire exclusion policies of the past eighty years in the northern U.S. Rocky Mountains, have harmed whitebark pine communities in several ways. Fire suppression has increased competition from shade-tolerant species and advanced the age of whitebark pine stands, thereby heightening whitebark pine's vulnerability to blister rust infection, mountain pine beetle attack, and dwarf mistletoe infestation. These stressors have allowed subalpine fir to dominate much of the upper subalpine forest, including the understories of most seral whitebark pine communities, in the Rocky Mountains of Montana and Idaho (Keane and Morgan 1994; Chapters 4 and 9). The same mechanisms are responsible for mountain hemlock (*Tsuga mertensiana*) dominance in high-mountain forests of the Cascade Range in Oregon (Arno and Hoff 1990). Heavy mortality from blister rust has accelerated this conversion. As the frequency of fire has declined, so has the number of openings suitable for nutcracker caching and whitebark pine seedling growth (Keane and Morgan 1994; Chapter 9). With the transition of whitebark pine communities to fir, hemlock, and spruce, it is likely that mounting fuel loads will increase the frequency of large fires.

High-severity fires kill trees of all species and sizes, but whitebark pine often becomes more abundant in postfire communities, because it has a competitive advantage over wind-dispersed species when colonizing extensive burns (Keane et al. 1990; Tomback et al. 1990; Chapter 5). Clark's nutcrackers (*Nucifraga columbiana*), the primary seed dispersers for whitebark pine, use openings created by stand-replacing fires as seed-caching sites (Tomback 1998). Typically, they cache seeds 1 to 3 kilometers or less from harvest sites and sometimes cache only a few meters from seed sources. However, they are capable of carrying whitebark pine seeds up to 22 kilometers. This is 100 times farther than most wind-dispersed seeds of spruce and fir travel (Tomback et al. 1990). Although fire may accelerate the loss of whitebark pine at a local level, it is necessary to perpetuate whitebark pine communities at a landscape scale.

Temporal Scale of Decline

Whitebark pine trees of all sizes are vulnerable to rust infection. Smaller trees, however, are killed more quickly because there is a shorter distance for the fungal hyphae to travel from the infected needles to the main stem, and there is a smaller cambium circumference to girdle in the main stem and branches. Most infected seedlings die within three years (Hoff and Hagle 1990). Red needles usually signal that blister rust has infected and killed the ends of branches (Hoff 1992). Large trees often have many such "flagged" branches. Typically, however, these trees remain alive, albeit not vigorous, for decades after becoming infected with rust, because they die slowly from the top down. Because cones form in the upper third of the crown, cone production ceases long before the tree actually dies (Arno and Hoff 1990; Keane and Arno 1993). Rapid declines can occur, as demonstrated by the deterioration of whitebark pine in western Montana (Keane and Arno 1993). Between 1971 and 1991, 42 percent of the trees died, and all but 10 percent were killed by rust.

The process of trees dying from blister rust infection accelerates when the sugar-laden edges of cankers are fed upon by rodents, speeding girdling of the tree (Hoff 1992; Hunt and Meagher 1992). However, rodent gnawing may also help reduce inoculum potential by decreasing aecial production by up to 15 percent (Mielke 1935). At high-elevation sites, cankers form more slowly, and infected trees take longer to die. The cold, dry environments of whitebark pine communities allow snags and fallen trees to persist for a century or longer in the absence of fire (Schmitt and Scott 1998), and without historical information for a stand, make it difficult to determine when a die-off occurred.

Blister rust infection intensified quickly in the Pacific Northwest after its initial introduction. Several measures to control blister rust were attempted (Chapter 10). For example, *Ribes* eradication began in Mount Rainier National Park in the White River–Sunrise Park area in 1931, and cankers were excised from trees beginning in 1940 (Riley 1932, 1941). By 1952, nearly all the mature trees and 52 percent of the reproduction in this area were infected with rust despite massive control efforts. Meanwhile, outside the rust control area boundaries, western white pine and whitebark pine had been nearly eliminated by the disease (Gynn and Chapman 1953).

The white pine blister rust epidemic initially proceeded in a predictable pattern. The disease intensified earliest where the climate was most favorable to rust and progressed more slowly where the climate was drier and colder (Chapter 10). Heavy mortality in whitebark pine occurred in northern Idaho, for example, by the late 1960s (Kendall and Arno 1990). However, in the past thirty years, the incidence and intensity of white pine blister rust in northern Idaho have increased dramatically (Smith and Hoffman 2000). The southward spread of the disease, though, has slowed during this time (Smith and Hoffman, in press). A similar pattern is found in the Sierra Nevada range. Blister rust reached sugar pine in the southern Sierra Nevada by the 1960s (Kliegunas 1996); however, the southward spread of blister rust in whitebark pine has been slower, presumably because of drought effects during key life-cycle stages in rust. Prospects for blister rust reaching isolated Great Basin forests in Nevada appear unlikely at this time, but it cannot be ruled out in the future (Smith and Hoffman 2000). Blister rust has spread to white pine in other isolated locations, such as limber pine in the Bighorn Mountains in north-central Wyoming (Lundquist 1993) and southwestern white pine (*Pinus strobiformis*) in New Mexico (Geils et al. 1999; Chapter 10).

When the amount of land area currently occupied by whitebark pine cover types is compared to historic coverage, the magnitude of loss from fire exclusion and blister rust becomes alarming. The area of whitebark pine cover types is estimated to have declined 45 percent during the past 100 years in the interior Columbia River Basin and the Bob Marshall Wilderness Complex in Montana (Keane et al. 1996). Moreover, the whitebark pine communities that occur on less harsh sites have been hit hardest. In these more productive communities where whitebark pine is a major seral species, the land area occupied by whitebark pine has plummeted 98 percent (Keane et al. 1996). For example, in the headwaters of Lick Creek on the Bitterroot National Forest in southwestern Montana, seral whitebark pine communities covered 14 percent of the land area circa 1900 (Arno et al. 1993). Now, because of succession and blister rust mortality, there are no stands dominated by whitebark pine, and the extent of stands with cone-bearing trees has declined by half. Most of the remaining whitebark pine stands in the U.S. Intermountain West are climax communities at high-elevation sites with a single-stratum structure, limited capacity to produce cones, and little whitebark pine regeneration. The combination of destroyed seed sources from mountain pine beetle kill, blister rust, and succession has resulted in severely reduced regeneration in some areas (Chapter 1). The low density of whitebark pine regeneration within the large Sundance burn in northern Idaho was attributed to the severe mountain pine beetle and blister rust damage to the seed source on the perimeter of the burn (Tomback et al. 1995). Fewer than 10 percent of whitebark pine stands sampled in the Bob Marshall Wilderness Complex, for example, have any whitebark pine regeneration younger than fifty years old (Keane et al. 1994).

Regional Population Trends

Information about the current status of whitebark pine is incomplete for many areas, but growing awareness of whitebark pine's value has spurred research and produced a more comprehensive picture in recent years. Whitebark pine ecosystems are most compromised in southwestern Canada (Campbell 1998; Stuart-Smith 1998) and the northwestern United States (Arno 1986; Keane and Arno 1993; Keane et al. 1994; Kendall 1999) (Figure 11-1). There, most stands have high levels of mortality, blister rust incidence, and rust severity (as measured by the number of cankers per tree). Whereas death and disease rates vary widely throughout the range of whitebark pine, damage generally declines in areas to the north in Canada (Campbell 1998; Stuart-Smith 1998), and to the east and south in the United States (Kendall et al. 1996a,b; Kendall 1998; Smith and Hoffman 2000). In many areas where blister rust damage is generally light or moderate, elevated levels of rust occur sporadically.

Rocky Mountains

The most complete information on the status of whitebark pine comes from the Rocky Mountain region (Table 11-1). Here, we describe the state of whitebark pine first in the Canadian Rocky Mountains, Montana, and Wyoming, and then in Idaho and eastern Oregon and Washington.

In general, the highest mortality from blister rust in the Rocky Mountains occurs in northwestern Montana, northern Idaho, and the southern Canadian Rockies, where a quarter to half of all whitebark pine trees are dead (Keane et al. 1994; Kendall et al. 1996a; Stuart-Smith 1998). Blister rust infection is responsible for most of this mortality. Of the remaining live trees in this region, 80 to 100 percent currently are infected with blister rust and eventually will die (Kendall et al. 1996a; Campbell 1998; Stuart-Smith 1998). Blister rust infection rates generally decrease with increasing latitude, reaching an average of 17

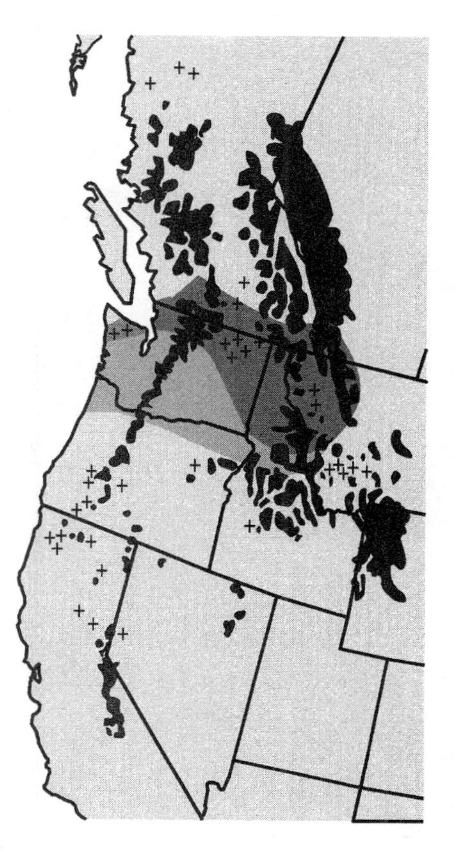

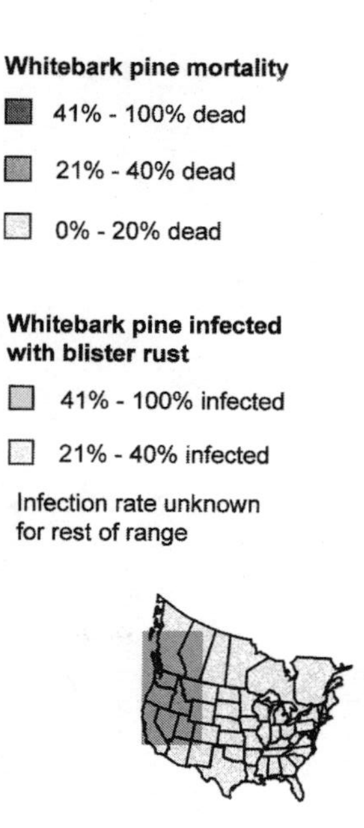

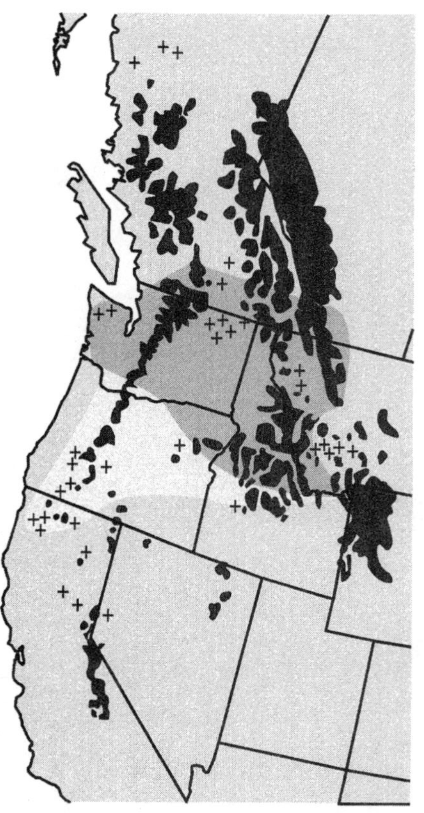

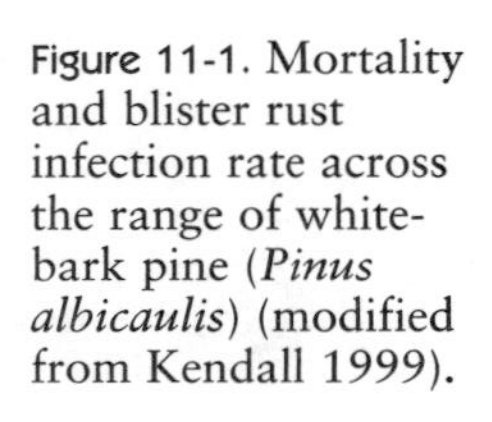

Figure 11-1. Mortality and blister rust infection rate across the range of whitebark pine (*Pinus albicaulis*) (modified from Kendall 1999).

Table 11-1. Mortality, blister rust infection, and crown kill rates in sampled locations throughout the range of whitebark pine.

Location (Source)	Sample Size: No. Plots (Trees)	% Dead (Range)	% Infected with Rust (Range)	% Crown Kill (Range)
Jasper NP AB, Mt. Robson BC (Stuart-Smith 1998)	6 (150)	NA	17 (0–40)	NA
Banff, Kananaskis NPs AB, Invermere, Kootenay, Yoho NPs BC (Stuart-Smith 1998)	14 (350)	NA	17 (0–56)	NA
SW Alberta, Waterton Lakes NP AB, Cranbrook BC (Stuart-Smith 1998)	9 (225)	NA	59 (27–76)	NA
Waterton Lakes NP AB (Kendall et al. 1996a)	14 (1,675)	26 (14–48)	47 (14–62)	13 (3–18)
Glacier NP MT (Kendall et al. 1996a)	204 (6,090)	44 (0–100)	78 (0–100)	26 (0–70)
Bob Marshall Wilderness Complex west-central MT (Keane 1994)	111 (>2,220)	NA	83 (67–93)	33 (15–48)
Western MT (Keane and Arno 1993)	17 (>340)	42	89 (8–100)	46 (8–68)
Blackfeet Reservation northwest MT (Kendall et al. 1996a)	5 (199)	58 (45–85)	85 (69–100)	33 (18–52)
Sweetgrass Hills north-central MT (Kendall 1998)	4 (270)	12 (0–20)	70 (55–89)	23 (12–42)
Gallatin NF southern MT (Kendall et al. 1996b)	75 (6,839)	10 (0–43)	2 (0–41)	NA
Yellowstone NP northwest WY (Kendall et al. 1996a)	84 (7,872)	7 (0–64)	5 (0–50)	1 (0–21)
Shoshone NF northwest WY (Harris 1999)	14	NA	12 (0–60)	NA
Grand Teton NP northwest WY (Kendall et al. 1996a)	32 (2,613)	7 (0–50)	13 (0–54)	1 (0–9)
Bridger–Teton NF northwest WY (Smith and Hoffman 2000)	11 (550)	NA	10	NA
Coeur D'Alene, Kaniksu NFs northern ID, Kootenai, Flathead NFs northwest MT (Kendall 1994)	12	46 (0–96)	48 (4–100)	NA
Nez Perce, Clearwater, St. Joe NFs north-central ID, Lolo NF western MT (Kendall 1994)	15	32 (0–96)	57 (2–100)	NA
Salmon-Selway Ecosystem ID & MT (J. T. Hogg, unpublished data, Craighead Wildlife-Wildlands Institute, Missoula, MT)	32 (1,992)	5 (0–35)	3 (0–16)	7 (0–70)

continues

Location (Source)	Sample Size: No. Plots (Trees)	% Dead (Range)	% Infected with Rust (Range)	% Crown Kill (Range)
Boise, Payette, Targhee NFs south-central ID (Smith and Hoffman 2000)	32 (1,600)	NA	50 (42–59)	NA
Challis, Salmon, Sawtooth NFs southern ID (Smith and Hoffman 2000)	35 (1,750)	NA	6 (0–13)	NA
British Columbia (Campbell 1998)	53 (3,594)	21 (0–64)	36 (0–100)	NA
Mount Rainier NP northern WA (R. Rochefort, unpublished data, North Cascades National Park, WA)	61 (1,700)	41 (0–99)	82 (10–100)	28 (0–99)
Cascade Range eastern WA (Hadfield et al. 1996)	6 (462)	8 (2–14)	20 (13–27)	NA
Kings Canyon, Sequoia NPs east-central CA (Duriscoe 1995)	12	NA	0	0

percent in the rain-shadowed slopes of Jasper National Park (Stuart-Smith 1998). However, rates in some areas do not conform to this trend. For example, just west of Jasper, 40 to 60 percent of whitebark pine trees sampled are infected with rust (Campbell 1998).

Whitebark pine mortality and rust infection rates also decline to the south in the United States, as the summer climate becomes drier and therefore less favorable to the spread of blister rust and the buildup of infection levels (Chapter 10). The proportion of trees infected with rust is lower in the southern portion of the Bob Marshall Wilderness Complex (67 percent) than in the northwestern parts of this area (92 percent) (Keane et al. 1994). Two hundred miles farther south in the Greater Yellowstone Ecosystem, the cause of most whitebark pine mortality is not identified as blister rust, and the mortality rate is less than 10 percent (Kendall et al. 1996a). Only 2 to 5 percent of whitebark pine trees surveyed in the Gallatin National Forest in southwestern Montana and Yellowstone National Park are infected with blister rust; however, the average rust infection rate in Grand Teton National Park is approximately twice as high as in Yellowstone (Kendall et al. 1996a, b). In general, blister rust is intensifying in the subalpine zone of the GYE, and there is little reason to think this trend will not continue (Smith 1997)

Whitebark pine mortality and rust infection levels are elevated in the moist, northern half of Idaho and tend to decrease in the drier south-central and southeast mountains. In the Salmon and Selway Rivers region, 50 percent of trees larger than 10 centimeters in diameter are dead, and blister rust is present in 34 percent of stands sampled (J. T. Hogg, unpublished data, Craighead Wildlife-Wildlands Institute, Missoula, Montana). The West Big Hole Range that lies on the border between central Idaho and Montana has low to moderate levels (<25 percent infection rate) of blister rust and little mortality to date

(Murray et al. 1998). However, farther southeast in the Targhee National Forest along the western boundaries of Yellowstone and Grand Teton National Parks, blister rust is present in all but one of twenty-one stands sampled. On average, 49 percent of trees sampled were infected with blister rust, but on many sites between Monida Pass and Targhee Pass and west of Yellowstone and Grand Teton National Parks, respectively, 70 to 100 percent of the trees are infected (Kendall 1994; Smith and Hoffman 2000). Three hundred miles to the northwest, the cause of extensive whitebark pine mortality in the Payette National Forest in central Idaho has not been conclusively determined but may have been accelerated by interactions between mountain pine beetle, root pathogens, and rust (Smith and Hoffman 2000). In the forests in the southern half of Idaho and northwestern Wyoming, high rust levels are starting to kill many trees (Smith and Hoffman 2000).

The health of stands in northeast Oregon, some of the westernmost ranges of the Rocky Mountains (Figure 11-1), varies widely with location, elevation, and the proximity of *Ribes* (Schmitt and Scott 1998). In some areas, mountain pine beetles have killed most of the large pines, thereby eliminating most cone production. This, coupled with blister rust infection and other factors, leaves many whitebark pine communities in steep decline with limited regeneration capacity (Schmitt and Scott 1998). Stands on the west edge of Hells Canyon Wilderness are decimated by blister rust, and due to large tree diameters and a high density of whitebark pine trees, heavy mortality from mountain pine beetle is expected here in the future (Schmitt and Scott 1998). To the southwest on Elkhorn Ridge, blister rust damage is severe in whitebark pine trees in some areas but much lower in high-elevation stands. Schmitt and Scott (1998) surmise that local conditions, such as the absence of *Ribes* in the high-elevation stands and unfavorable wind patterns, account for these low rates. In nearby Eagle Cap Wilderness and Seven Devils Lake areas, mortality and blister rust damage are moderate to light. Further south in Oregon, in the Strawberry Mountain Wilderness, most losses were caused by a beetle epidemic in the late 1980s. Blister rust is present in western white pine (*P. monticola*) on north slopes but is not evident on east-facing aspects and at higher elevations where whitebark pine grows (Schmitt and Scott 1998).

Whitebark pine communities have been devastated by blister rust in northeastern Washington in the past few decades. On the Colville National Forest, whitebark pine health is generally poor, and vigor is declining because blister rust infection is severe, and mountain pine beetles attack the infected trees (Sniezko et al. 1994). Also, because most whitebark pine seedlings succumb to rust, whitebark pine is being replaced by other tree species in every stand where it occurs. It is now unusual to find subalpine communities in this area still dominated by whitebark pine (Sniezko et al. 1994).

Cascade Range

In the Cascade Range, as in the Rocky Mountains, there is a latitudinal gradient in whitebark pine health. Whitebark pine stands are beleaguered by rust in the north end of the range, but appear to be healthy in the south. Blister rust is present on 20 to 60 percent of the whitebark pine trees sampled in southern

British Columbia (Campbell 1998). In the Wenatchee National Forest in central Washington, whitebark pine communities are declining due to high mortality from blister rust and mountain pine beetle attack. Blister rust also has caused moderate to heavy damage to whitebark pine on the south side of Mt. Hood in northwestern Oregon (S. F. Arno, personal communication). However, the incidence of rust on whitebark pine is low on Mt. Adams in southeastern Washington, despite an abundance of *Ribes*. Stands in central, southern, and southwest Oregon are reported to be healthy, although there is scant information on the status of whitebark pine in the Willamette National Forest of west-central Oregon (Sniezko et al. 1994). Little is known about the vitality of whitebark pine in the Cascades of northern California, where it is a minor forest element. Throughout the Cascades, in areas of low to moderate damage, there are typically some "hot spots," or areas of much higher mortality or rust infection rates.

Coast Range and Olympic Mountains

With the heavy rainfall and high humidity of these coastal ranges and their geographic proximity to the epicenter of the rust epidemic, it is not surprising that whitebark pine stands there suffered some of the earliest and most dramatic declines. Blister rust infection rates average 80 to 100 percent in the Bulkley Range midway up the Coast Mountains in British Columbia (Campbell 1998). In the Olympic Mountains in extreme northwestern Washington, whitebark pine is found in the northeastern rain-shadow zone (Arno and Hoff 1990). Although whitebark pine has never been a major forest component in this region, a lengthy history with blister rust has resulted in heavy mortality and made it increasingly difficult to find live whitebark pine trees in the Olympics.

Sierra Nevada

Whitebark pine, a sparsely scattered stand component in the Sierras of northern California, becomes more abundant in the southern half of the range. Although blister rust occurs throughout the Sierra Nevada, infection thus far has centered on sugar pine and western white pine, which occur at middle elevations. However, levels of blister rust infection in whitebark pine trees range widely. Rust infection rates of whitebark pine near Tioga Pass range from zero (Kendall 1994) to 15 percent (Smith and Hoffman 2000). More than 50 percent of the whitebark pine trees sampled in the Bald Mountains north of Reno, Nevada, and near the summit of Mt. Rose west of Reno are infected with rust (Smith and Hoffman 2000). Smith and Hoffman found no blister rust at two sites they sampled south of Tioga Pass, and Duriscoe (1995) encountered rust in fewer than 1 percent of whitebark pine trees sampled in Sequoia and Kings Canyon National Parks.

Great Basin Ranges, Nevada

Blister rust has not been found on whitebark pine in the forest "islands" that are associated with the Basin and Range region of Nevada, but it occurs at low to moderate levels in several areas in the Carson Range along the western boundary of the state (Smith and Hoffman 2000). The Carson Range, unlike the island ranges, is linked to infected stands in the Sierra Nevada by continuous forest cover.

Future Trends in Whitebark Pine Communities

From our initial understanding of the mechanisms of blister rust spread, it appeared that the cold temperatures of high elevations and droughty summer climates would offer whitebark pine some protection from heavy damage by rust. Studies of the relationship of climate to the spread of blister rust were conducted in Yellowstone National Park, Wyoming, in the 1960s and 1970s (Hendrickson 1970; Carlson 1978). The results supported a "climatic escape" hypothesis: Climates, like Yellowstone's, could be too cold and dry to allow blister rust to become intense. Cold temperatures limit rust basidiospore formation, and low humidity inhibits the spread of rust from *Ribes* to pine; but the persistence of rust in the Greater Yellowstone Ecosystem was evidence that local climate enables rust to become established in pines (Hendrickson 1970). However, Carlson (1978) felt that whitebark pine in Yellowstone would escape heavy damage by blister rust, because ecological conditions would limit the severity of infection. Research on the spread of blister rust in whitebark pine under suboptimal climatic conditions made plant pathologist Richard Krebill (1969) somewhat less optimistic. He concluded that whitebark pine would survive an epidemic where rust was hindered by low humidity and temperatures, but that many stands would be lost in the higher elevations of Yellowstone National Park and southward (Krebill 1969).

Based on what we have learned about the epidemiology of blister rust during the last twenty years, it appears unlikely that any whitebark pine stand is safe from potentially heavy damage by blister rust, regardless of prevailing weather. Whereas blister rust infection disperses and intensifies more rapidly in humid regions with longer growing seasons, environmental conditions generally inhospitable to rust do not preclude its spread. Furthermore, climate is not always a reliable predictor of rust severity. For example, no obvious connection between the intensity of rust infection and climate was found in British Columbia (Campbell 1998) or in the Bob Marshall Wilderness Complex in northwestern Montana (Keane et al. 1994).

In areas where rust infection is presently mild, climate does not rule out increased infection in the future. For example, after being present at low levels for decades, blister rust is beginning to cause widespread mortality along the boundary between southwestern Montana and Idaho (Smith and Hoffman 2000). Even in the high central plateau of Yellowstone National Park, rust infection rates are gradually increasing. Where no rust was detected in 1971 (Carlson 1978), 11 percent of the mature trees sampled in Yellowstone in 1996 (K. C. Kendall, unpublished data) were infected with rust. This estimate is conservative, because evidence of infection in taller, older trees can be hard to see. Infected large trees often take decades to die, and rust can be difficult to identify due to the growth of lichen and weathering of damaged crowns (K. C.

Kendall, unpublished observations). The fate of seedling and sapling whitebark pine trees that were individually tagged in Yellowstone may be more telling. All were in good health and rust-free when originally examined in 1969 (Anonymous, undated). Twenty-five years later, 24 percent of the trees have died, and 29 percent of those still alive are infected with rust (K. C. Kendall, unpublished data). Some of the mortality may be attributed to forest succession. However, the higher infection rates of small trees than of larger ones (5 percent) suggest that blister rust is responsible for at least a portion of the decline, and that the first visible disruption to stands may come in the form of deficiencies in smaller size classes, because young trees succumb to rust more quickly than do larger trees.

If weather always conformed to the climatological average, it is likely that some parts of whitebark pine's range would remain free of rust. But weather is variable, and eventually conditions favorable to the regeneration of rust are likely to occur even in the most hostile climates. This has been documented in the central Sierra Nevada range, where, despite a climate unfavorable to rust, unusual periods of favorable weather created explosive epidemics over large areas in western white and sugar pine (Kinloch and Dulitz 1990). These waves of infection spread the disease into many new areas and intensified infection in older centers. Even in the relatively dry, warm southern Sierra Nevada, weather conditions conducive to the spread of blister rust in sugar pine can occur, albeit rarely (Kinloch and Dulitz 1990).

For disjunct whitebark pine populations, large distances from a source of infection may delay the arrival of blister rust and onset of mortality but does not guarantee protection. The Sweet Grass Hills, an island range surrounded by prairie and wheat fields, represent the easternmost population of whitebark pine in northern Montana. Even though the closest white pine communities are in Glacier National Park almost 160 kilometers (100 miles) to the west, blister rust incidence and crown kill rates in these two areas are comparable (Table 11-1) (Kendall et al. 1996a; Kendall 1998). The discovery of blister rust in southwestern white pine (*Pinus strobiformis*) in southern New Mexico (Hawksworth 1990) illustrates the vulnerability of even the most distant and isolated stands. The nearest known rust occurrence is 1,000 kilometers north in Wyoming and 1,400 kilometers west in California, without continuous forest cover links to the newly infected stands. We do not yet know whether this outbreak is the result of a separate introduction from an unknown local source or long-distance spread from existing infections in California. Climate modeling suggests that episodic long-distance dispersal from California is possible (B. W. Geils, USDA Forest Service, Flagstaff, Arizona, personal communication). The spread of blister rust also may be impeded but not eliminated by low densities of *Ribes* plants or *Ribes* species less susceptible to rust infection.

Widespread heavy mortality of whitebark pine has important implications for long-term whitebark pine conservation. Only a tiny fraction of trees scattered within stands survive intense rust epidemics, because resistance in the original population is extremely low (1 to 5 percent) (Chapter 10). Extreme population bottlenecks (large reductions in population size) typically cause loss of genetic diversity through inbreeding and genetic drift, and result in diminished resilience to environmental change (Nei 1987). However, whitebark pine populations often begin with high levels of inbreeding, because they were founded by a small number of individuals in areas with no other whitebark pine, and because many trees grow in clusters with close relatives (Tomback and Linhart 1990, Hoff et al. 1994; Chapter 5). Over time, lethal and other detrimental genes may have been reduced in frequency, leaving a species that is not adversely affected by inbreeding (Hoff et al. 1994). Stuart-Smith (1998) found higher levels of heterozygosity than expected (from the Hardy-Weinberg model) in whitebark pine at sites with elevated rates of blister rust infection. Presumably, this is because heterozygous individuals are more likely to have resistant genes. From an evolutionary perspective these findings imply that natural selection will increase rust resistance within populations, but this depends on good regeneration opportunities. It is probable that some diminished populations will be lost through large disturbance events, such as mountain pine beetle outbreaks and stand-replacing fires, that can kill the remaining live individuals within a stand. Because there is little genetic differentiation among whitebark pine populations (Jorgensen and Hamrick 1997; Bruederle et al. 1998; Chapter 8), these losses may have little effect on overall genetic diversity, but this illustrates the perils of reducing a population to extremely low levels (Chapter 12). It is important to note that fire exclusion has a far greater negative than positive consequence for whitebark pine. In the absence of fire, atypical amounts of fuel accumulate that foster more fires that are lethal to mature whitebark pine trees.

Continued fire exclusion and heavy mortality from rust will allow subalpine fir and other shade-tolerant trees to increase. Without the restoration of historical fire regimes, the future for whitebark pine is bleak, particularly in moister climates (Keane et al. 1996). Most upper subalpine stands will be converted to dense fir and spruce or mountain hemlock with a minor whitebark pine component in as little as 50 to 100 years (Chapter 9). With widespread blister rust mortality, tree-line and semiarid sites currently dominated by whitebark pine will transition to herbaceous and shrub-dominated communities. The reintroduction of stand-replacing fire would foster whitebark pine regeneration by creating open sites suitable for nutcracker caching and tree establishment (Chapters 9 and 18). It also would help limit losses due to mountain pine beetle epidemics by creating mosaics of multiage stands that are less conducive to large beetle outbreaks (Hessberg et al. 1994).

Future climate change also may not bode well for whitebark pine. Most models predict a decline in whitebark pine cover in the northern Rocky Mountains, as conditions at high elevation that support this species change (Chapter 7). In fact, with a simulated doubling of CO_2 in the Yellowstone region, Bartlein et al. (1997) found whitebark pine to be the most severely affected of all the conifers studied in the region. Furthermore, climate change in the Greater Yellowstone Ecosystem is predicted to increase the frequency of weather conditions that spread rust infection in whitebark pine (Koteen 1999). In contrast, one model predicted that the increased frequency of fire resulting from climate change will cause an expansion of whitebark pine communities in Glacier National Park (Keane et al. 1998). Therefore, the impact of climate change on whitebark pine is inconclusive.

Because whitebark pine trees grow slowly and usually do not produce female cones until they are about sixty-five years old (S. F. Arno, personal communication), they have a long generation time and, consequently, adapt more slowly to environmental conditions compared to many other conifers. Despite a fifty-year history of blister rust in most parts of whitebark pine's range, the epidemic has not yet reached equilibrium, and the rust infection rate and death toll continue to rise. Regeneration will be greatly delayed in areas where infection is high, because seed sources are diminished by crown kill and because most young trees are quickly killed by rust (Hoff and Hagle 1990; Tomback et al. 1995). Even more important is the lack of fire to create suitable openings for whitebark pine regeneration. Large reductions in cone production will also decrease carrying capacity for a number of wildlife species that feed on whitebark pine seeds, including the threatened grizzly bear (*Ursus arctos*) (Chapters 7 and 12). This has serious consequences for future whitebark pine seed production and regeneration. Without intervention, only the small proportion of whitebark pine with some genetic resistance to rust will be able to survive to maturity in areas heavily assaulted by rust (Hoff 1994). They, however, could easily be killed in the new stand-replacing fire regimes that result from advanced succession in seral whitebark pine communities (Chapter 9). The potential for heightened threats to whitebark pine from rust is illustrated by the emergence of a race of blister rust virulent to sugar pine trees carrying a dominant gene that previously imparted rust resistance (Kinloch and Comstock 1981).

Whitebark pine communities will continue to decline as long as fire exclusion limits the availability of sites suitable for nutcracker caches and blister rust kills trees faster than they can regenerate. Currently, there are no apparent limitations to the spread of rust northward in the Canadian Rocky Mountains, and past experience leads us to believe that rust is a threat to whitebark pine across its entire range. Although the geographic extent of blister rust appears to have changed little in the past thirty years, the incidence and intensity of existing rust infections have increased sharply. This higher intensity of infection will translate into an increase in whitebark pine crown kill and mortality in coming decades. Without management intervention, the future for whitebark pine during the next several hundred years is one of continuing decline, functional extinction, and local extirpation (Chapters 1 and 12). The numbers of whitebark pine able to persist eventually will become so low that whitebark pine will no longer be influential at the population, community, or landscape scales.

There are, however, some promising developments that indicate the potential of whitebark pine to persist on the landscape. First, more than 40 percent of the progeny from seeds collected from healthy survivors in heavily infected stands showed some blister rust resistance (Chapters 10 and 17). This suggests that a rapid buildup of rust resistance in whitebark pine is possible, even though large seed crops may be rare. Second, wildland fire policies of natural resource management agencies have been revised in recent years and now call for greater levels of prescribed fire across large expanses of forest with whitebark pine communities (U.S. General Accounting Office 1999). Natural selection will continue to build levels of rust resistance in populations through time, but only if fire and restoration techniques provide suitable areas for regeneration.

Of critical importance is the preservation of the ecological processes that will ensure the presence of whitebark pine on the high-mountain landscape. Natural rust resistance can be fostered by the restoration of fire regimes so that suitable caching conditions are available for Clark's nutcrackers and optimal growing environments are created for whitebark pine regeneration. Selection for rust resistance in whitebark pine could also be enhanced by the establishment of natural "seed orchards" in areas where blister rust damage is severe (Chapter 17). With management intervention on several fronts, natural selection should once again create vital subalpine communities that include whitebark pine as a principal component.

Acknowledgments

J. P. Smith and Brian Geils, USDA Forest Service, Rocky Mountain Research Station, Flagstaff, Arizona, and Steve Arno provided many thoughtful comments and technical information that helped improve this chapter. Kate Kendall's contributions were supported by the U.S. Geological Survey, and Bob Keane's contributions were supported by the U.S. Forest Service.

Literature Cited

Alexander, M. E., and F. G. Hawksworth. 1976. Fire and dwarf mistletoes in North American coniferous forests. Journal of Forestry 74:446–449.

Anonymous. Undated. Unpublished data on blister rust infection rates of whitebark pine in "cluster" plots established in Yellowstone National Park in 1969. Yellowstone National Park Archives, Wyoming.

Arno, S. F. 1986. Whitebark pine cone crops: A diminishing source of wildlife food? Western Journal of Applied Forestry 1:92–94.

Arno, S. F., and R. Hoff. 1990. *Pinus albicaulis* Engelm. Whitebark pine. Pages 268–279 *in* R. M. Burns and B. H. Honkala, technical coordinators. Silvics of North America, Vol. 1. Conifers. USDA Forest Service, Agriculture Handbook 654, Washington, D.C.

Arno, S. F., E. D. Reinhardt, and J. H. Scott. 1993. Forest structure and landscape patterns in the subalpine lodgepole pine type: A procedure for quantifying past and present conditions. USDA Forest Service, Intermountain Research Station, General Technical Report INT-294, Ogden, Utah.

Bartlein, P. J., C. Whitlock, and S. L. Shafer. 1997. Future climate in the Yellowstone National Park region and its potential impact on vegetation. Conservation Biology 11:782–792.

Benedict, W. V. 1981. History of white pine blister rust control—A personal account. USDA Forest Service, FS-355, Washington, D.C.

Bruederle, L. P., D. F. Tomback, K. K. Kelly, and R. C. Hardwick. 1998. Population genetic structure in a bird-dispersed pine, *Pinus albicaulis* (Pinaceae). Canadian Journal of Botany 76:83–90.

Buchanan, T. S., and J. W. Kimmey. 1938. Initial tests of the distance of spread to and intensity of infection on *Pinus monticola* by *Cronartium ribicola* from *Ribes lacustre* and *R. viscosissimum*. Journal of Agricultural Research 56:9–30.

Campbell, E. M. 1998. Whitebark pine forests in British Columbia: Composition, dynamics, and the effects of blister rust. M.S. thesis, University of Victoria, Victoria, British Columbia, Canada.

Carlson, C. E. 1978. Noneffectiveness of *Ribes* eradication as a control of white pine blister rust in Yellowstone National Park. USDA Forest Service, Northern Region, State and Private Forestry Report No. 78–18, Missoula, Montana.

Duriscoe, D. 1995. White pine blister rust in Kings Canyon and Sequoia National Parks: Preliminary results of an extensive survey. On file at USDI, National Park Service, Sequoia and Kings Canyon National Parks, Three Rivers, California.

Geils, B. W., D. A. Conklin, and E. P. Van Arsdel. 1999. A preliminary hazard model of white pine blister rust for the Sacramento Ranger District, Lincoln National Forest. USDA Forest Service, Rocky Mountain Research Station, Research Note RMRS-RN-6, Flagstaff, Arizona.

Gynn, J. C., and C. M. Chapman. 1953. Blister rust control: Mount Rainier National Park, 1952. Pages 54–56 *in* White pine blister rust control: Northwestern Project, January 1 to December 31, 1952. USDA Agricultural Research Administration, Bureau of Entomology and Plant Quarantine, Western Region IV, Blister Rust Control, Spokane, Washington.

Hadfield, J., P. Flanagan, and A. Camp. 1996. White pine mortality survey in the eastern Washington Cascade Range. Nutcracker Notes 7:8. http://www.mesc.usgs.gov/glacier/nutnotes.htm

Hagle, S. K., S. Tunnock, K. E. Gibson, and C. J. Gilligan. 1987. Field guide to diseases and insect pests of Idaho and Montana forests. USDA Forest Service, State and Private Forestry, Northern Region Publication Number R1-89-54, Missoula, Montana.

Harris, J. L. 1999. Evaluation of white pine blister rust disease on the Shoshone National Forest. USDA Forest Service, Rocky Mountain Region, Forest Health Management, Renewable Resources, Rapid City Service Center, Biological Evaluation R2-99-05. Golden, Colorado.

Hawksworth, F. G. 1990. White pine blister rust in southern New Mexico. Plant Disease 74:938.

Hawksworth, F. G., and D. Wiens. 1972. Biology and classification of dwarf mistletoes (*Arceuthobium*). USDA Forest Service, Agricultural Handbook 401, Washington, D.C.

Hendrickson, W. H. 1970. Assessing the potential of white pine blister rust to limber and whitebark pine in Yellowstone National Park. USDI National Park Service, Office of Natural Science Studies, Final Report RSP YELL-N-40, Washington, D.C.

Hessberg, P. S., R. G. Mitchell, and G. M. Filip. 1994. Historic and current roles of insects and pathogens in eastern Oregon and Washington forested lands. USDA Forest Service, Pacific Northwest Research Station, General Technical Report PNW-GTR-327, Seattle, Washington.

Hitchcock, C. L., and A. Cronquist. 1976. Flora of the Pacific Northwest: An illustrated manual. University of Washington Press, Seattle.

Hoff, R. J. 1992. How to recognize blister rust infection on whitebark pine. USDA Forest Service, Intermountain Research Station, Research Note INT-406, Ogden, Utah.

Hoff, R. 1994. Artificial rust inoculation of whitebark pine seedlings—rust resistance across several populations. Nutcracker Notes 4:7–9. http://www.mesc.usgs.gov/glacier/nutnotes.htm

Hoff, R., and S. Hagle. 1990. Diseases of whitebark pine with special emphasis on white pine blister rust. Pages 179–190 *in* W. C. Schmidt and K. J. McDonald, compilers. Proceedings—Symposium on whitebark pine ecosystems: Ecology and management of a high-mountain resource. USDA Forest Service, Intermountain Research Station, General Technical Report INT-270, Ogden, Utah.

Hoff, R. J., and G. I. McDonald. 1993. Variation of virulence of white pine blister rust. European Journal of Forest Pathology 23:103–109.

Hoff, R. J., S. K. Hagle, and R. G. Krebill. 1994. Genetic consequences and research challenges of blister rust in whitebark pine forests. Pages 118–126 *in* W. C. Schmidt and F. -K. Holtmeier, compilers. Proceedings of the International workshop on subalpine stone pines and their environment: The status of our knowledge. USDA Forest Service, Intermountain Research Station, General Technical Report INT-GTR-309, Ogden, Utah.

Hunt, R. S., and M. D. Meagher. 1992. How to recognize white pine blister rust cankers. Fact Sheet, Canada—British Columbia Partnership Agreement on Forest Resource Development II, Forestry Canada, Victoria, British Columbia, Canada.

Jackson, M. T., and A. Faller. 1973. Structural analysis and dynamics of the plant communities of Wizard Island, Crater Lake National Park. Ecological Monographs 43:441–446.

Jorgensen, S. M., and J. L. Hamrick. 1997. Biogeography and population genetics of whitebark pine, *Pinus albicaulis*. Canadian Journal of Forest Research 27:1574–1585.

Keane, R. E., and S. F. Arno. 1993. Rapid decline of whitebark pine in Western Montana: Evidence from 20-year remeasurements. Western Journal of Applied Forestry 8:44–47.

Keane, R. E., and P. Morgan. 1994. Landscape processes affecting the decline of whitebark pine (*Pinus albicaulis*) in the Bob Marshall Wilderness, Montana, USA. Pages 159–209 *in* Proceedings of the 12th Conference of Fire and Forest Meteorology, Society of American Foresters, Bethesda, Maryland.

Keane, R. E., S. F. Arno, J. K. Brown, and D. F. Tomback. 1990. Modeling stand dynamics in whitebark pine (*Pinus albicaulis*) forests. Ecological Modeling 51:73–95.

Keane, R. E., P. Morgan, and J. P. Menakis. 1994. Landscape assessment of the decline of whitebark pine (*Pinus albicaulis*) in the Bob Marshall Wilderness Complex, Montana, USA. Northwest Science 68:213–229.

Keane, R. E., J. P. Menakis, and W. J. Hann. 1996. Coarse-scale restoration planning and design in Interior Columbia River Basin Ecosystems: An example for restoring declining whitebark pine forests. Pages 14–19 *in* C. C. Hardy and S. F. Arno, editors. The use of fire in forest restoration. USDA Forest Service, Intermountain Research Station, General Technical Report INT-GTR-341, Ogden, Utah.

Keane, R. E., K. Ryan, and M. Finney. 1998. Simulating the consequences of altered fire regimes on a complex landscape in Glacier National Park, USA. Pages 310–324 *in* T. L. Pruden and L. A. Brennan, editors. Tall Timbers Fire Ecology Conference Proceedings 20:310–324.

Kendall, K. C. 1994. Whitebark pine monitoring network. Pages 109–118 *in* K. C. Kendall and B. Coen, editors. Workshop proceedings: Research & management in whitebark pine ecosystems. USDI National Biological Service, Glacier National Park, West Glacier, Montana.

———. 1998. Whitebark pine and limber pine status in the Sweetgrass Hills, Montana. Nutcracker Notes 9:10–11. http://www.mesc.usgs.gov/glacier/nutnotes.htm

———. 1999. Whitebark pine. Pages 483–485 *in* M. J. Mac, P. A. Opler, C. E. Puckett-Haecker, and P. D. Doran, editors. Status and trends of the nation's biological resources. U.S. Department of the Interior, Washington, D.C.

Kendall, K. C., and S. F. Arno. 1990. Whitebark pine—An important but endangered wildlife resource. Pages 264–273 *in* W. C. Schmidt and K. J. McDonald, compilers. Proceedings—Symposium on whitebark pine ecosystems: Ecology and management of a high-mountain resource. USDA Forest Service, Intermountain Research Station, General Technical Report INT-270, Ogden, Utah.

Kendall, K., D. Schirokauer, E. Shanahan, R. Watt, D. Reinhart, R. Renkin, S. Cain, and G. Green. 1996a. Whitebark pine health in northern Rockies national park ecosystems: A preliminary report. Nutcracker Notes 7:16. http://www.mesc.usgs.gov/glacier/nutnotes.htm

Kendall, K., D. Tyers, and D. Schirokauer. 1996b. Preliminary status report on whitebark pine in Gallatin National Forest, Montana. Nutcracker Notes 7:19. http://www.mesc.usgs.gov/glacier/nutnotes.htm

Kimmey, J. W. 1938. Susceptibility of *Ribes* to *Cronartium ribicola* in the West. Journal of Forestry 36:312–320.

Kimmey, J. W., and W. W. Wagener. 1961. Spread of white pine blister rust from *Ribes* to sugar pine in California and Oregon. USDA Forest Service, Pacific South West Research Station, PSW Technical Bulletin 1251, Berkeley, California.

Kinloch, B. B. Jr., and M. Comstock. 1981. Race of *Cronartium ribicola* virulent to major gene resistance in sugar pine. Plant Disease 65:604–605.

Kinloch, B. B. Jr., and D. Dulitz. 1990. White pine blister rust at Mountain Home Demonstration State Forest: A case study of the epidemic and prospects for genetic control. USDA Forest Service, Pacific Southwest Research Station, Research Paper PSW-204, Berkeley, California.

Kliegunas, J. 1996. An update of blister rust incidence on the Sequoia National Forest. USDA Forest Service, Pacific Southwest Region, State and Private Forestry, Forest Pest Management Report R96-01, Berkeley, California.

Koteen, L. 1999. Climate change, whitebark pine, and grizzly bears in the greater Yellowstone ecosystem. M.S. thesis, Yale University, New Haven, Connecticut.

Krebill, R. G. 1969. Letter to W. M. Hendrickson, September 10. Yellowstone National Park archives, Yellowstone National Park, Wyoming. (Editor's note: R. G. Krebill was a research scientist with the USDA Forest Service.)

Lachmund, H. G. 1934. Seasonal development of *Ribes* in relation to spread of *Cronartium ribicola* in the Pacific Northwest. Journal of Agricultural Research 49:93–114.

Lundquist, J. E. 1993. Large scale spatial patterns of conifer diseases in the Bighorn Mountains, Wyoming. USDA Forest Service, Rocky Mountain Forest and Range Experiment Station, Research Note RM-523, Fort Collins, Colorado.

Mathiasen, R. L. 1998. Comparative susceptibility of conifers to larch dwarf mistletoe in the Pacific Northwest. Forest Science 44:559–568.

Mathiasen, R. L., and F. G. Hawksworth. 1988. Dwarf mistletoes on western white pine and whitebark pine in northern California and southern Oregon. Forest Science 34:429–440.

Mielke, J. L. 1935. Rodents as a factor in reducing aecial sporulation of *Cronartium ribicola*. Journal of Forestry 33:994–1003.

———. 1943. White pine blister rust in western North America. Yale University, School of Forestry, Bulletin No. 52, New Haven, Connecticut.

Mielke, J. L., and J. W. Kimmey. 1935. Dates of production of the different spore stages of *Cronartium ribicola* in the Pacific Northwest. Phytopathology 25:1104–1108.

Mielke, J. L., T. W. Childs, and H. G. Lachmund. 1937. Susceptibility to *Cronartium ribicola* of the four principal *Ribes* species found within the commercial range on *Pinus monticola*. Journal of Agricultural Research 55:317–346.

Murray, M. P. 1996. Landscape dynamics of an island range: Interrelationships of fire and whitebark pine (*Pinus albicaulis*). Ph.D. dissertation, University of Idaho, Moscow.

Murray, M. P., S. C. Bunting, and P. Morgan. 1998. Fire history of an isolated subalpine mountain range of the Intermountain Region, United States. Journal of Biogeography 25:1071–1080.

Nei, M. 1987. Molecular evolutionary genetics. Columbia University Press, New York.

Perkins, D. L., and T. W. Swetnam. 1996. A dendroecological assessment of whitebark pine in the Sawtooth–Salmon River region, Idaho. Canadian Journal of Forest Research 26:2123–2133.

Riley, M. C. 1932. Blister rust activities, Mount Rainier National Park. Pages 188–198 *in* Blister rust work in the Far West, January 1 to December 31, 1931. USDA Division of Blister Rust Control, Spokane Branch, Spokane, Washington.

———. 1941. White pine blister rust control: Mount Rainier National Park, 1940. Pages 83–86 *in* Blister rust work in the Far West, January 1 to December 31, 1931. USDA Bureau of Entomology and Plant Quarantine, Division of Plant Disease Control, Blister Rust Control, Spokane, Washington.

Schmitt, C. L., and D. W. Scott. 1998. Whitebark pine health in northeastern Oregon and western Idaho. USDA Forest Service, Pacific Northwest Region, Report No. BMZ-99-03, Portland, Oregon.

Smith, J. P. 1997. Insect and disease problems in Intermountain West whitebark pines with emphasis on white pine blister rust disease. M.S. thesis, University of Idaho, Moscow.

Smith, J. P., and J. T. Hoffman. 2000. Status of white pine blister rust in the Intermountain West. Western North American Naturalist. 60:165–179.

Sniezko, R. A., J. Linn, J. Beatty, and S. Martinson. 1994. Whitebark pine information survey results, USDA Forest Service, Region 6. Pages 71–77 *in* K. C. Kendall and B. Coen, editors. Workshop proceedings: Research & management in whitebark pine ecosystems. USDI National Biological Service, Glacier National Park, West Glacier, Montana.

Stuart-Smith, G. J. 1998. Conservation of whitebark pine in the Canadian Rockies: Blister rust and population genetics. M.S. thesis, University of Alberta, Edmonton, Canada.

Tomback, D. F. 1998. Clark's nutcracker (*Nucifraga columbiana*), No. 331. *In* A. Poole and F. Gill, editors. The birds of North America. The Birds of North America, Inc., Philadelphia.

Tomback, D. F., and Y. B. Linhart. 1990. The evolution of bird-dispersed pines. Evolutionary Ecology 4:185–219.

Tomback, D. F., L. A. Hoffmann, and S. K. Sund. 1990. Coevolution of whitebark pine and nutcrackers: Implications for forest regeneration. Pages 118–129 *in* W. C. Schmidt and K. J. McDonald, compilers. Proceedings—Symposium on whitebark pine ecosystems: Ecology and management of a high-mountain resource. USDA Forest Service, Intermountain Research Station, General Technical Report INT-270, Ogden, Utah.

Tomback, D. F., J. K. Clary, J. Koehler, R. J. Hoff, and S. F. Arno. 1995. The effects of blister rust on post-fire regeneration of whitebark pine: The Sundance burn of northern Idaho (USA). Conservation Biology 9:654–664.

U.S. General Accounting Office. 1999. A cohesive strategy is needed to address catastrophic wildfire threats. U.S. General Accounting Office Report to the Committee on Forests and Forest Health. GAO/RCED-99-65. U.S. Government Printing Office, Washington, D.C.

Van Arsdel, E. P. 1965. Micrometeorology and plant disease epidemiology. Phytopathology 55:945–950.

Van Arsdel, E. P., A. J. Riker, and R. F. Patton. 1956. The effects of temperature and moisture on the spread of white pine blister rust. Phytopathology 46:307–318.

Chapter 12

Biodiversity Losses: The Downward Spiral

Diana F. Tomback and Katherine C. Kendall

The dramatic decline of whitebark pine (*Pinus albicaulis*) populations in the northwestern United States and southwestern Canada from the combined effects of fire exclusion, mountain pine beetles (*Dendroctonus ponderosae*), and white pine blister rust (*Cronartium ribicola*), and the projected decline of whitebark pine populations rangewide (Chapters 10 and 11) do not simply add up to local extirpations of a single tree species. Instead, the loss of whitebark pine has broad ecosystem-level consequences, eroding local plant and animal biodiversity, changing the time frame of succession, and altering the distribution of subalpine vegetation (Chapter 1). One potential casualty of this decline may be the midcontinental populations of the grizzly bear (*Ursus arctos horribilis*), which use whitebark pine seeds as a major food source (Chapter 7). Furthermore, whitebark pine is linked to other white pine ecosystems in the West through its seed-disperser, Clark's nutcracker (*Nucifraga columbiana*) (Chapter 5). Major declines in nutcracker populations ultimately seal the fate of several white pine ecosystems, and raise the question of whether restoration is possible once a certain threshold of decline is reached.

Biological diversity is now viewed from a perspective more complex than simply counting numbers of individual species. A complete definition of diversity includes not only the numbers of species but also the genetic diversity represented by the collective populations of a species, and finally, the diversity represented by different ecological communities—their ecosystem structures, processes, and interactions with the physical environment (Primack 1998). From this combined perspective in particular, whitebark pine communities are richly diverse.

Our intentions for this chapter are to explore in general terms (1) community types and biodiversity represented across the range of whitebark pine, (2) the projected consequences to biodiversity, and Clark's nutcracker populations in particular, from loss of whitebark pine and the ramifications for other white pine communities connected to whitebark pine by Clark's nutcracker, and (3) the outlook for restoration once the declines reach a critical threshold. Although some of the latter discussion is speculative, the information accumulated on white pine ecosystems in recent decades lends support to our view.

Biodiversity of Whitebark Pine Communities

Whitebark pine is a keystone species of the upper subalpine zone (see Chapter 1). The ways in which whitebark pine increases biodiversity relate to its hardiness and ability to modify the harsh, upper subalpine zone environment; its early successional status; and the large, nutritious seeds and shelter it provides.

Keystone Functions

Whitebark pine is a hardy species that tolerates exposed sites and poor soils up to tree-line elevations, where it forms a prostrate, or krummholz, growth form (Arno and Hoff 1990). Tolerance of harsh, windswept sites enables it to grow where other conifers cannot. Consequently, whitebark pine plays an important role in regulating stream flow in upper-elevation watersheds, where soils have little water-holding capacity. The shade provided by whitebark pine canopies protracts snowmelt, thus regulating the rate of runoff and reducing soil erosion (Farnes 1990). In addition, its roots stabilize rocky, poorly developed soils, leading to understory vegetation establishment, and further reducing soil erosion.

Once whitebark pine is established on inhospitable sites, it fosters the establishment of other species by mitigating an otherwise extreme environment. At upper subalpine elevations in the northern Rocky Mountains of the United States, whitebark pine facilitates the survival and growth of its competitors, subalpine fir (*Abies lasiocarpa*) and Engelmann spruce (*Picea engelmannii*). Young fir or spruce often cluster around a single whitebark pine "nurse" tree, which shelters them from wind-blasted ice and snow, and moderates the harshness of the local environment (Callaway 1998). In the upper subalpine throughout its range, whitebark pine is typically the lone pioneer tree species to colonize steep slopes, exposed moraines, and rocky cliffs. Once small islands of whitebark pine are established, spruce and fir invade, and a forest understory layer develops (Lanner 1980; Lanner 1996 and references therein).

The same hardiness that leads to pioneering in harsh, upper-elevation environments enables whitebark pine to become established after fire (e.g., Tomback 1986; Tomback et al. 1990, 1993). This tolerance for fire-damaged soils and harsh exposures is particularly important in the northern and central Rocky Mountains of the United States, inland Northwest, and adjacent regions in Canada, where whitebark pine is successional on more than half of the area that it occupies (Arno 1986; Chapter 4). Whitebark pine creates favorable microsites in these harsh environments, which lead to the establishment of other forest and understory species and ultimately even to subalpine fir and whitebark pine's successional replacement.

Whitebark pine seeds are unusually large and nutritious, an adaptation that not only produces robust seedlings (e.g., Chapter 6) tolerant of harsh, droughty environments but also attracts and sustains Clark's nutcracker, the primary seed disperser for the species (Lanner and Gilbert 1994; Chapters 1 and 5). These seeds also comprise an important seasonal food source for many other species of small birds and mammals (discussed later in this chapter), and for black bears (*Ursus americanus*) and grizzly bears, which obtain whitebark pine seeds from cones stored in pine squirrel (*Tamiasciurus* spp.) middens (Mattson et al. 1991; Chapter 7).

Whitebark pine communities also provide valuable wildlife habitat at high elevations, especially where other conifers cannot grow (Chapter 1). These forests offer shelter, nesting sites, burrows, territories, and home ranges. The large, spreading crowns of long-lived trees provide protection from the wind, and long-standing snags provide nest-holes for cavity-nesting species. Krummholz thickets at tree line and whitebark pine stands and forests are home to a number of migrant and resident breeding birds, small mammals, carnivores, mule deer (*Odocoileus hemionus*), elk (*Cervus elaphus*), and blue grouse (*Dendragapus obscurus*) at different seasons (e.g., Lonner and Pac 1990; K. C. Kendall, unpublished data; D. F. Tomback, unpublished data), and a variety of arthropod species that live in trees, understory, and forest litter (e.g., Christiansen and Lavigne 1996).

With the current decline and elimination of whitebark pine in many areas in the northwestern United States and southwestern Canada (e.g., Keane and Arno 1993; Keane et al. 1994; Campbell 1998; Stuart-Smith 1998; see Chapter 11), the interactions outlined above have been diminished, and losses in biodiversity sustained.

Geographic and Structural Diversity

Whitebark pine occurs throughout the higher mountains of western North America, ranging from about 37° N to about 55° N latitude (Arno and Hoff 1990; Chapter 3). Because mountain zones decrease in elevation with latitude, whitebark pine is found as high as 3,050 to 3,660 meters in the southernmost part of its range and as low as 900 meters in the northernmost part of its range (Arno and Hoff 1990; Chapter 2). This broad geographic range of whitebark pine, divided into eastern (Rocky Mountain) and western portions, is subject to different climate regimes, maritime and continental, and orographic effects such as rain shadows. Consequently, whitebark pine communities show great geographic variation in their forest and understory composition, and ecological dynamics, which is reviewed in Chapter 4.

Two contrasting examples illustrate this: In moist, upland sites of the northwestern inland-maritime region, whitebark pine is successional, forming mixed associations with subalpine fir, Engelmann spruce, and sometimes mountain hemlock (*Tsuga mertensiana*) and alpine larch (*Larix lyallii*), with an understory of Labrador tea (*Ledum glandulosum*), red mountain heather (*Phyl-*

lodoce empetriformis), fool's huckleberry (*Menziesia ferruginea*), and white rhododendon (*Rhododendron albiflorum*) (Lackschewitz 1991; Chapter 4). In the dry, volcanic Warner Mountains at the edge of the Great Basin, whitebark pine and, to a lesser extent, lodgepole pole (*Pinus contorta*) form a climax community, with big sagebrush (*Artemisia tridentata*), grasses (*Poa nervosa, Stipa californica*), sedge (*Carex pensylvanica*), and herbs (*Arenaria aculeata,* and *Penstemon* spp.) comprising the sparse understory (Arno and Hammerly 1984, Chapter 4).

Within a region, whitebark pine may occur on most aspects and, within limits, both moist and dry sites (Chapter 4). For example, in the Rocky Mountains of Canada whitebark pine shares dominance of upper subalpine and treeline communities with Engelmann spruce, subalpine fir, and, in some areas, alpine larch. Here, the understory varies from heather (*Phyllodoce empetriformis* where the climate has some maritime influence, to *P. glanduliflora* where the climate is continental) on moist sites to grouse wortleberry (*Vaccinium scoparium*) on well-drained sites, and common juniper (*Juniperus communis*) on the driest sites (Arno and Hammerly 1984; Chapter 4). Thus, in addition to variation in forest and understory composition with geographic range, there is variation with local site characteristics (e.g., dry versus moist) and topography (slope and aspect).

Disturbance and elevation are also factors that increase community diversity. Where whitebark pine forms seral communities, there is variation in both understory and forest community composition with successional stage—for example, with time after—but also with disturbance patch size and adjacent community types (D. F. Tomback, unpublished data), as described for other western forest communities (e.g., Lyon 1984; Stickney 1986; Turner et al. 1997). In the inland northwestern United States and southwestern Canada, whitebark pine forms successional communities in the subalpine zone on favorable sites, self-perpetuating climax communities on exposed sites, timber atolls in the high snow areas, and enduring matlike, or krummholz, communities at tree line (Chapter 4). Even in the northernmost and southernmost regions where whitebark pine occurs, it forms both high-elevation climax communities and krummholz communities at tree line, with different forest associates and different understory plants in each community type. Finally, within a region, whitebark pine communities may form ecotones with other community types, such as meadows, riparian zones, or alpine tundra, futher adding to the structural complexity and plant diversity.

In summary, there is great community diversity, both structural and ecological, across the range of whitebark pine. These communities, in turn, represent a considerable spectrum of understory plant biodiversity (e.g., Pfister et al. 1977; Steele et al. 1983) as well as diversity in mycorrhizal fungi (root symbiotic fungi), free-living fungi, soil and leaf bacterial communities, lichens, mosses, and bryophytes, most of which are not well-known (e.g., Eversman et al. 1990). In Glacier National Park, Montana, for example, more than 190 species of vascular plants occur in whitebark pine communities, and some do not occur in any other community type (K. C. Kendall and K. T. Peterson, unpublished data).

Wildlife Diversity

Whitebark pine communities support a tremendous diversity of vertebrate and invertebrate species, which vary with geographic location, elevation, and successional stage. Bird and mammal species that inhabitat whitebark pine communities may be year-round residents, breeding residents, transients, or seasonal elevational migrants.

The large, nutritious seeds of whitebark pine (Lanner and Gilbert 1994; Chapter 1) attract a number of foraging birds and mammals, which varies somewhat geographically (Table 12-1). Large or moderate cone crops are produced irregularly in the Rocky Mountains, for example, every three to five years (Arno and Hoff 1990; Morgan and Bunting 1992), but may be more regular elsewhere (e.g., Sierra Nevada had moderate or heavy crops in four successive years, Tomback 1978). Both Clark's nutcrackers and pine squirrels (*Tamiasciurus* spp.) begin foraging on unripe seeds as early as July. Nutcrackers dig into the juicy, resinous cones with their long, sharp bills to extract pieces of soft seeds (Tomback 1978; Hutchins and Lanner 1982), and pine squirrels cut down and gnaw open cones to obtain seeds (Kendall 1983). Only when the cones ripen in late August do nutcrackers begin to cache seeds (Chapter, 5) and squirrels cut down quantities of cones to store in middens.

When the seeds ripen, birds and small mammals actively forage in the canopies of whitebark pine trees. Woodpeckers, chipmunks, and golden-mantled ground squirrels are able to break off cone scales themselves and remove seeds, but some of the smaller birds, and usually even the Steller's jays, obtain seeds only from cones that were previously opened by nutcrackers (see Table 12-1 for species list and scientific names) (Tomback 1978; D. F. Tomback, unpublished data; Hutchins and Lanner 1982). Seed-eating pine grosbeaks and red crossbills are transient and unpredictable in occurrence in whitebark pine communities (e.g., Small 1974), but most of the other species are observed feeding daily.

Whitebark pine seeds are a major food source for grizzly bears and black bears in the East Front of the Rocky Mountains, the Greater Bob Marshall Ecosystem, and in the Greater Yellowstone Area (Craighead et al. 1982; Kendall 1983; Mattson et al. 1991; Mattson and Reinhart 1994; Chapters 1 and 7), and historically over much of the inland Northwest in the United States (Kendall and Arno 1990). The bears raid pine squirrel middens, digging up whitebark pine cones, extracting the seeds, and eating the seeds along with the occasional, unfortunate squirrel. After a productive cone crop, the following spring and summer grizzly bears forage almost exclusively on whitebark pine seeds (Kendall 1983; Mattson and Reinhart 1997). Consequently, whitebark pine communities have been identified as critical grizzly bear habitat (e.g., U.S. Fish and Wildlife Service 1997) and essential, for example, to the viability of the Greater Yellowstone Area grizzly bear population (Mattson et al. 1992).

Much of the wildlife diversity in whitebark pine communities, however, is related to resources other than pine seeds. A number of birds are migratory breeding residents, including nighthawks, swifts, hummingbirds, flycatchers, swallows, wrens, warblers, sparrows, and the colorful mountain bluebirds, and western tanagers (see Table 12-2 for list and scientific names); many of these

Table 12-1. Birds and small mammals that feed on whitebark pine (*Pinus albicaulis*) seeds. Locations observed: I = Inyo National Forest, Sierra Nevada (Tomback 1978); Y = Yellowstone National Park, Wyoming (D. F. Tomback, unpublished data); and BT = Bridger-Teton National Forest, Rocky Mountains (Hutchins and Lanner 1982). Observations of bear foraging from Yellowstone National Park and the Greater Yellowstone Area (Kendall 1983; Chapter 7 and references therein). Foraging modes: seeds taken from 1 = unripe cones in trees, 2 = ripe cones in trees, 3 = ripe cones first opened by nutcrackers, 4 = cones cut by pine squirrels (*Tamiasciurus* spp.) under trees, and 5 = fallen seeds under trees.

Bird and Mammal Species	Location	Foraging Mode
FAMILY PICIDAE		
Williamson's sapsucker (*Sphyrapicus thyroideus*)	I	2, 3
Hairy woodpecker (*Picoides villosus*)	I	2, 3
White-headed woodpecker (*Picoides albolarvatus*)	I	2, 3
FAMILY CORVIDAE		
Clark's nutcracker (*Nucifraga columbiana*)	I, Y, BT	1, 2, 3, 4
Steller's jay (*Cyanocitta stelleri*)	I, BT	2, 3, 5
Common raven (*Corvus corax*)	BT	2, 3
FAMILY PARIDAE		
Mountain chickadee (*Poecile gambeli*)	I, BT	3
FAMILY SITTIDAE		
Red-breasted nuthatch (*Sitta canadensis*)[1]	BT	3
White-breasted nuthatch (*Sitta carolinensis*)	I	3
FAMILY FRINGILLIDAE		
Pine grosbeak (*Pinicola enucleator*)	I, Y, BT	3
Cassin's finch (*Carpodacus cassinii*)	I, Y	3
Red crossbill (*Loxia curvirostra*)	I	2, 3
FAMILY SCIURIDAE		
Chipmunks (*Tamias* spp.)	I, BT	2, 3, 5
Golden-mantled ground squirrel (*Spermophilus lateralis*)	I, BT	2, 3, 4
Douglas's squirrel (*Tamiasciurus douglasii*)	I	4
Red squirrel (*Tamiasciurus hudsonicus*)	BT	4
FAMILY MURIDAE		
Deer mouse (*Peromyscus maniculatus*)[2]		5
Southern red-backed vole (*Clethrionomys gapperi*)[2]		5
FAMILY URSIDAE		
Black bear (*Ursus americanus*)	Y	4
Grizzly bear (*Ursus arctos*)	Y, BT	4

[1]Forging success questioned by Hutchins and Lanner (1982).
[2]Known to feed on the seeds of other pines and may also feed on whitebark pine seeds (Smith and Balda 1979).

birds feed on the flush of summer insects and, in the case of hummingbirds, nectar from wildflowers. Birds of prey are common in whitebark pine communities, probably in response to the numbers of birds and mammals that are present. Frequently occurring raptor species include the bird-eating Cooper's hawk and prairie falcon (Table 12-2).

In one study, elk used whitebark pine communities very little in summer for feeding and shelter, but in fall they favored mixed subalpine zone communities with both whitebark pine and subalpine fir and high densities of tree regeneration (Lonner and Pac 1990). Mule deer inhabited whitebark pine communities more in summer than in fall (Lonner and Pac 1990). In winter, blue grouse often roost in the dense crowns of whitebark pine, where they feed on needles and buds while sheltered from wind and concealed from predators. The birds usually move to high elevations, including whitebark pine communities, in late summer and fall and move downslope in spring, although differences in their habits may occur locally (Small 1974; Andrews and Righter 1992). Several bird species, including the olive-sided flycatcher (*Contopus borealis*), three-toed woodpecker (*Picoides tridactylus*), black-backed woodpecker (*P. arcticus*), Clark's nutcracker, and mountain bluebird, are associated with all early post-fire coniferous forest communities, including those that experienced stand-replacing burns (Hutto 1995), as well as established forest.

In general, the more structurally diverse the community, the greater the number of species that it supports (Smith 1990). Whitebark pine communities that form mixed associations with other forest trees, as opposed to pure stands, and have well-developed understory are home to a greater diversity of animal species. Red squirrels, for example, prefer moister whitebark pine habitat types that also support other conifers such as lodgepole pine (Reinhart and Mattson 1990), which may offer cone production in off-years for whitebark pine cone crops. Krummholz or matlike whitebark pine tree-line communities that form an ecotone with tundra may support more wildlife than either community alone. Successional mosaics similarly increase wildlife diversity.

Summed over the range of whitebark pine, there is considerable wildlife biodiversity supported by whitebark pine communities, representing many taxonomic groups and great ecological breadth in lifestyles.

Whitebark Pine Decline: Biodiversity Losses

From the perspective of objective science, there is a vast natural experiment taking place. Without major management intervention, the blister rust epidemic and advanced ecological succession from fire exclusion and the synergism facilitating mountain pine beetle infestations will eliminate much of the whitebark pine in the inland northwestern United States and southwestern Canada (e.g., Arno 1986; Keane and Arno 1993; Keane et al. 1994; Tomback et al. 1995; Campbell 1998; Stuart-Smith 1998). Large-scale losses from this combination of exotic disease and human intervention are already in progress (Chapters 10 and 11). We will be the observers of the decline in the biodiversity of whitebark pine communities from diminishment of the uppermost subalpine forest, increasing homogeneity of subalpine communities (Chapter 9), and loss of a major wildlife food source (Chapter 1).

Table 12-2. (A) Birds observed in whitebark pine (*Pinus albicaulis*) communities June through September at Minaret Summit and on nearby Mammoth Mountain, Inyo National Forest, 2,700–3,000-meter elevation, Sierra Nevada range, California (Tomback 1978; D. F. Tomback, unpublished data). The whitebark pine communities included lodgepole pine (*P. contorta*), mountain hemlock (*Tsuga mertensiana*), red fir (*Abies magnifica*), and western white pine (*P. monticola*), and formed an ecotone with a dry meadow community at Minaret Summit. (B) Mammals observed in the same whitebark pine communities as above (M) (Tomback 1978; D. F. Tomback, unpublished data), or in whitebark pine communities in Yellowstone National Park (Y) (K. C. Kendall, unpublished data).

A. BIRDS	
Cooper's hawk (*Accipiter cooperii*)	Violet-green swallow (*Tachycineta thalassina*)
Northern goshawk (*Accipiter gentilis*)	Mountain chickadee (*Poecile gambeli*)
Red-tailed hawk (*Buteo jamaicensis*)	White-breasted nuthatch (*Sitta carolinensis*)
Golden eagle (*Aquila chrysaetos*)	House wren (*Troglodytes aedon*)
Prairie falcon (*Falco mexicanus*)	Mountain bluebird (*Sialia currucoides*)
Blue grouse (*Dendragapus obscurus*)	American robin (*Turdus migratorius*)
Great horned owl (*Bubo virginianus*)	Yellow-rumped warbler (*Dendroica coronata*)
Common nighthawk (*Chordeiles minor*)	Townsend's warbler (*Dendroica townsendi*)
White-throated swift (*Aeronautes saxatalis*)	Wilson's warbler (*Wilsonia pusilla*)
Allen's hummingbird (*Selasphorus sasin*)	Western tanager (*Piranga ludoviciana*)
Red-breasted sapsucker (*Sphyrapicus ruber*)	Brewer's sparrow (*Spizella breweri*)
Downy woodpecker (*Picoides pubescens*)	Vesper sparrow (*Pooecetes gramineus*)
Hairy woodpecker (*Picoides villosus*)	Fox sparrow (*Passerella iliaca*)
Northern flicker (*Colaptes auratus*)	White-crowned sparrow (*Zonotrichia leucophrys*)
Western wood-pewee (*Contopus sordidulus*)	Dark-eyed junco (*Junco hyemalis*)
Steller's jay (*Cyanocitta stelleri*)	Cassin's finch (*Carpodacus cassinii*)
Clark's nutcracker (*Nucifraga columbiana*)	Red crossbill (*Loxia curvirostra*)
Common raven (*Corvus corax*)	Pine siskin (*Carduelis pinus*)
Horned lark (*Eremophila alpestris*)	Pine grosbeak (*Pinicola enucleator*)
B. MAMMALS	Bushy-tailed woodrat (*Neotoma cinerea*) Y
Snowshoe hare (*Lepus americanus*) Y	Common porcupine (*Erethizon dorsatum*) Y
Chipmunks (*Tamias* spp.) M, Y	Coyote (*Canis latrans*) M, Y
Belding's ground squirrel (*Spermophilus beldingi*) M	Black bear (*Ursus americanus*) Y
Golden-mantled ground squirrel (*Spermophilus lateralis*) M, Y	Grizzly bear (*Ursus arctos*) Y
Douglas's squirrel (*Tamiasciurus douglasii*) M	American marten (*Martes americana*) M, Y
Red squirrel (*Tamiasciurus hudsonicus*) Y	Mule deer (*Odocoileus hemionus*) Y

White pine blister rust is now present throughout the range of whitebark pine (except for a few isolated populations in the Great Basin), western white pine (*Pinus monticola*) and sugar pine (*P. lambertiana*); the disease has spread to limber pine (*P. flexilis*), southwestern white pine (*P. strobiformis*), and foxtail pine (*P. balfouriana*) (Chapters 10 and 11). Other five-needled white pines known to be susceptible to blister rust include the Great Basin bristlecone pine (*P. longaeva*), bristlecone pine (*P. aristata*), and Mexican white pine (*P. ayacahuite*) (Hoff et al. 1980). New understanding of climatic patterns point to long-distance transport of blister rust spores and episodic weather patterns leading to intensification of local infection levels (Chapters 10 and 11). Only pockets of whitebark pine on high, extremely exposed sites may escape infection, and even that is not certain.

The relatively recent discovery of high levels of blister rust infection in southwestern white pine in the isolated Sacramento Mountains of southern New Mexico (Hawksworth 1990) underscores the potential for long-distance dispersal of blister rust spores (Chapter 11, this volume) and raises additional concerns: Southwestern white pine is a potential conduit for the disease to Mexico. The range of southwestern white pine extends from the southwestern United States south through the higher mountain regions of northern Mexico. The gap separating southwestern white pine and Mexican white pine, which comes in at roughly 20° N latitude, could be bridged by long-distance dispersal of blister rust spores. Shrubs of the genus *Ribes,* the obligatory alternate host of white pine blister rust, are present in at least parts of the range of Mexican white pine (G. I. McDonald, personal communication), which continues south into Honduras and El Salvador. Thus, the losses of whitebark pine and other five-needled white pine ecosystems have biodiversity implications at all scales: local, regional, rangewide, and finally, throughout the western montane United States, western Canada, and possibly Mexico.

General Consequences

The consequences of losing whitebark pine as a major subalpine forest species may be extrapolated directly from its keystone functions and ability to tolerate harsh conditions (Chapter 1). We predict the following:

Subalpine forest community structure. The pioneering status of whitebark pine and the hardiness of trees and seedlings are major factors in the time frame and sequence of community development in the subalpine zone. If we lose whitebark pine at the upper-elevational limits of its distribution, where other conifers cannot grow, there will be consequences to both local and regional watershed hydrology. Snowmelt will be more rapid, resulting in early-season flooding, increased soil erosion, and depressed, late-season stream flows (Farnes 1990). Human water uses, such as domestic consumption, irrigation, and livestock watering may be affected.

Clark's nutcrackers cache seeds on steep slopes and rocky ledges and in fissures above the continuous forest (e.g., Tomback and Kramer 1980; Hutchins and Lanner 1982), as well as on bare moraines and hills above moist meadows (Tomback 1978; Lanner 1980), leading to the establishment of new stands. As whitebark pine numbers decline, there will be (1) a decrease in forested area at high elevations, (2) fewer suitable microhabitats available for tree and plant colonization, for example, no whitebark pine "nurse" tree associations with subalpine fir and Engelmann spruce, facilitating their establishment, and (3) lengthened time frame and sequence of community development (e.g., meadow

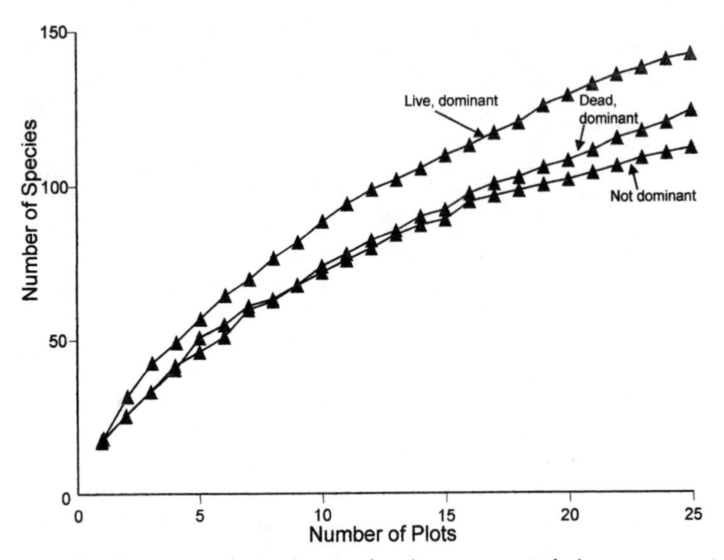

Figure 12-1. Species accumulation curves for three groups of plant communities in Glacier National Park: live whitebark pine trees are the dominant tree species = **Live, dominant;** whitebark pine trees were the dominant tree species and are now dead (primarily from blister rust) = **Dead, dominant;** and whitebark pine trees are a minor component = **Not dominant.** Samples for each category were twenty-five plots drawn at random from larger plot pools (K. C. Kendall and K. T. Peterson, unpublished data).

succession) and altered patterns and lengthened time frame of postfire succession.

Finally, as whitebark pine becomes an increasingly minor element in subalpine forests, these communities will become far more homogeneous in structure and plant and animal taxa. For example, a study in Glacier National Park found plant species richness to be higher in communities where whitebark pine was a major component than in communities where it was a minor component (Figure 12-1). Furthermore, the numbers of unique plant species were lower in whitebark pine communities where the trees had died (K. C. Kendall and K. T. Peterson, unpublished data). Without whitebark pine, the forest composition early in succession will be the same as the composition in climax communities. As some combination of subalpine fir, spruce, and mountain hemlock or other late-successional conifers come to characterize the landscape, fires will tend to become larger, more severe, and generally stand-replacing (Chapter 9).

Wildlife biodiversity. The decline of whitebark pine will result in the loss of whitebark pine seeds, effectively decreasing the carrying capacity of the subalpine zone for pine seed–eating birds and mammals. Population declines may be particularly great for species that harvest and store the seeds in fall for use during winter and spring months—for example, Clark's nutcracker, Steller's jay, pine squirrels, and chipmunks, but perhaps also nuthatches and chickadees. None of these species are threatened with extinction, because all occur in other coniferous forest communities at different elevations throughout the montane West. However, their local and regional population sizes will decline, along with species richness at higher elevations. It is possible that in the absence of whitebark pine, the resilience of some species to poor cone production in other conifers will be weakened.

In contrast, where fruits are less abundant in the Bob Marshall Wilderness Complex and south, especially through the Greater Yellowstone Area, grizzly bear populations may be threatened with extinction or at least experience higher demographic variability from declines in whitebark pine (see Chapter 7). These grizzly bear populations are of questionable viability even now, with encroaching human development and a variety of unnatural disturbances, including recreational use of their range. The inevitable decline in whitebark pine seed production represents loss of a high-quality food source for grizzly bears and decreased reproductive success (Chapter 7). The situation is complicated by red squirrels acting as intermediaries in this foraging relationship: Their decline in the upper subalpine forest may be rapid because of a possible threshold effect as whitebark pine communities disappear. With a reduction in seed availability in the subalpine zone in late summer and fall, prehibernation grizzly bears will wander more widely in search of food, increasing their encounters with people. More management actions will result in higher bear mortality (Mattson et al. 1992).

The decline in whitebark pine communities also implies a reduction in wildlife habitat at the highest elevations—upper subalpine zone and possibly even tree-line communities—and also the loss of structural complexity to subalpine forests in general. The consequences are reduced breeding habitat for migrant and resident species, particularly those that require open-canopied forests, and less shelter and food for seasonally migrating species. This may further depress populations of some neotropical migrants, such as flycatchers, warblers, western tanagers, and finches, which are also losing their wintering grounds (e.g., Pyle et al. 1994). All influences considered, numbers of wildlife species are expected to decline in the subalpine zone, as well as regional population sizes of individual species.

Consequences for Nutcrackers and the Future of White Pine Communities

Nutcracker seed sources. Although whitebark pine is an obligate dependent on Clark's nutcrackers for seed dispersal, nutcrackers use other seed sources in addition to whitebark pine. This plasticity in foraging behavior has enabled nutcrackers to flourish in the montane West despite unpredictable cone production in pines in general (Tomback and Linhart 1990; Chapter 5). Clark's nutcrackers serve as important seed dispersers for the following pines with large, wingless seeds: limber (e.g., Vander Wall 1988), southwestern white (Benkman et al. 1984; S. Samano and D. F. Tomback, unpublished data), and the piñon pines (*P. edulis* and *P. monophylla*) (e.g., Vander Wall and Balda

1977; Vander Wall 1988). These species are white pines, that is, in the pine subgenus *Strobus*. In addition, nutcrackers cache the seeds of several white pines with winged seeds, including Great Basin bristlecone pine (Lanner 1988, Baud 1993), and probably sugar pine. Furthermore, nutcrackers cache the seeds of several other conifers with winged seeds, including the large-seeded yellow pines (subgenus *Pinus*) Jeffrey (*P. jeffreyi*) (Tomback 1978) and ponderosa pine (*P. ponderosa*) (Giuntoli and Mewaldt 1978; Torick 1995), and Douglas-fir (*Pseudotsuga menziesii*) (Giuntoli and Mewaldt 1978). Nutcrackers may also feed on the seeds of foxtail pine and occasionally on the seeds of western white pine, which are very small (Chapter 1). Although the seeds of these species are primarily wind-dispersed, seed dispersal by nutcrackers, jays, and small mammals appears to contribute to their regeneration as well (e.g., Tomback 1978; Lanner 1988; Vander Wall 1992). Nutcrackers also forage on insects, spiders, vegetable matter, and carrion, and prey on small vertebrates (Giuntoli and Mewaldt 1978; Tomback 1998).

Within the range of whitebark pine, several conifers used by nutcrackers may co-occur. For example, in the south-central eastern Sierra Nevada, limber, foxtail, singleleaf piñon (*P. monophylla*), and Jeffrey pine are available as seed sources at different elevations and in different drainages. In the Greater Yellowstone Area, limber pine and Douglas-fir also occur, and in the northern Rocky Mountains limber, ponderosa pine, and Douglas-fir are widely available. However, in all these regions, whitebark pine cones are the first to ripen and the preferred food source of nutcrackers from midsummer through fall (Tomback 1978; Hutchins and Lanner 1982). Each year, nutcrackers arrive in whitebark communities in late spring and early summer and retrieve seed caches, if available, exposed by melting snow; they also begin foraging for unripe seeds in any new cones (Tomback 1978; Vander Wall and Hutchins 1983; Chapter 5). After harvesting and caching whitebark pine seeds in late summer and fall, nutcrackers typically search for other seed sources at lower elevations or in adjacent montane regions (e.g., Tomback 1978). As long as fresh seeds are available, nutcrackers will continue to make seed caches, often as late as December at lower elevations (Tomback 1978; Torick 1995).

Predicted changes in nutcracker populations. Whitebark pine communities in the inland northwestern United States and southwestern Canada are declining rapidly (Chapters 1, 10, and 11). Data from two studies vividly portray the condition of the whitebark pine communities in Montana. A large-scale assessment of whitebark pine communities in the Bob Marshall Wilderness Complex indicated that about 22 percent of the whitebark pine landscape had experienced high mortality from blister rust, an additional 39 percent moderate mortality, and an average of 83 percent of living whitebark pine trees were infected with blister rust (Keane et al. 1994). In a second study, remeasurements of twenty-year-old plots in whitebark pine communities in western Montana clearly illustrated the rapid decline resulting from both succession and blister rust (Keane and Arno 1993): Whitebark pine basal area decreased an average of 42 percent in the seventeen plots remeasured, and blister rust infected an average of 89 percent of the trees in each plot.

Although the rapid decline of whitebark pine in the northwestern United

States and southwestern Canada is partly the consequence of fire exclusion, the blister rust epidemic alone can destroy entire whitebark pine communities in regions and at elevations where whitebark pine is not fire-dependent (Chapters 10 and 11). As blister rust infections intensify in other infected stands throughout the range of whitebark pine, comparable patterns of dead trees and dying trees with dead canopies are expected.

As mortality progresses in whitebark pine stands, we suspect that nutcracker foraging and seed-dispersal patterns may change. Nutcrackers may initially continue to spend late spring and summer at subalpine elevations, foraging both on seed caches made the previous fall and on unripe seeds from new cones. The reserve of seeds cached the previous fall will decline as cone production is steeply curtailed. By mid-July nutcrackers will begin searching whitebark pine communities for new cones, which should be sparse even in "good cone" years, given the damaged canopies and dead trees. When cones are located, nutcrackers will harvest unripe seeds from these for immediate consumption. Throughout the summer, fewer and fewer intact cones will be available as seeds are consumed, and pine squirrels probably cut down what cones remain. Consequently, few seeds will be available for caching, and thus for potential future regeneration. Nutcrackers will probably leave the subalpine zone in August or September in search of cone crops in other pine forests at lower elevations or in adjacent mountain regions. We see this pattern, in fact, in years with failed whitebark pine cone crops (Fisher and Myres 1980; Vander Wall et al. 1981). Depending on how much nutcrackers learn from their parents and other nutcrackers about seed-source selection, it is possible that new generations of nutcrackers may be less inclined to forage in the subalpine zone.

With a decline in whitebark pine trees and seed availability over time, fewer nutcrackers will return to subalpine forest in this region and, instead, will inhabit areas with other viable seed-producing pines. Two major consequences may arise from this scenario: (1) Region-wide nutcracker populations will decrease in number as the carrying capacity, that is, general food availability, declines; and (2) nutcrackers will not cache whitebark pine seeds, even if disturbances such as fire open up suitable terrain. Once cone availability declines sharply, regeneration in whitebark pine communities may nearly cease. These patterns may be virtually impossible to reverse, except with intense, costly restoration efforts.

Linkages to other white pine communities. As described above, Clark's nutcrackers are highly mobile and will search for other cone crops within the same or different regions after harvesting and caching whitebark pine seeds. Nutcrackers from the same region typically move from the higher-elevation seed sources into one or more, if available, lower-elevation seed sources, which they rely on in winter and early spring (Tomback 1978; Chapter 5). All nutcracker-dependent pines—whitebark, limber, southwestern white, and the piñon pines—are linked to some extent by overlap in nutcracker populations, that is, by sharing dispersers (Figure 12-2). When there is a cone-crop failure in one of these pine species, nutcrackers congregate in the other pines present in the same region that produce cones. A decline in long-term seed availability in one pine affects the size of disperser populations available for the other pines. For example, whitebark, limber, and singleleaf piñon pine co-occur at different elevations in the eastern Sierra Nevada, California, and share a regional nutcracker population; and limber pine, southwestern white, and Colorado piñon pine (*P. edulis*) co-occur at different elevations in the southern Rocky Mountains, and similarly share regional populations.

White pine blister rust now occurs in parts of the range of limber pine and southwestern white pine, and further spread and intensification are anticipated (Chapters 10 and 11). The piñon pines are more distantly related and not susceptible to white pine blister rust. We envision the following scenario: As blister rust spreads, white pine cone production declines, and more trees die, regional nutcracker populations will also decline, limiting disperser services for all nutcracker-dependent pines (Figure 12-2). If the decline in blister rust–susceptible nutcracker-dependent pines is severe and cone production is greatly reduced, the consequences may be far-reaching. As in whitebark pine, nutcrackers may consume most of the seed production, with little or none available for caching. With a major decline in rangewide carrying capacity, nutcracker populations may drop precipitously, or vary greatly in size from year to year, and the future of all nutcracker-dependent white pines could become doubtful, without early intervention.

Nutcrackers also harvest and cache the seeds of several white pines with winged seeds. Sugar pine, western white pine, and foxtail pine are infected with blister rust, and the two bristlecone pines are susceptible but not yet infected, to our knowledge. Nutcrackers also harvest and cache the seeds of Jeffrey pine and ponderosa pine, of the yellow pine subgenus. These pines collectively serve as backup, or even "mainstay" seed sources in the case of the widespread ponderosa pine, if preferred seed sources fail in cone production (e.g., Giuntoli and Mewaldt 1978; Torick 1995). Nutcrackers easily move from whitebark and/or limber pine to these wind-dispersed pines, because their cones open later in the fall. These auxilliary pines are linked to nutcrackers in a more diffuse fashion; nutcrackers do not play as critical a role in their regeneration, nor will they dependably harvest and store these seeds if alternatives are available (e.g., Torick 1995).

The auxiliary pines that are not blister rust–susceptible—ponderosa, Jeffrey, and the piñon pines—as well as the susceptible pines not yet infected—for example, bristlecone pines—will become the primary seed resources, and thus the refugia, for Clark's nutcracker in the event of the rangewide decline of whitebark, limber, and southwestern white pine. To date, blister rust has infected limber pine throughout the northern and central Rocky Mountains and is just spreading into the eastern, southern Rocky Mountains, that is, Colorado (Chapter 11). The further spread of blister rust in southwestern white pine will be from limber pine populations in the north, from the Sacramento Mountains in southern New Mexico (Hawksworth 1990), and from long-distance spore dispersal from the west (B. W. Geils, personal communication). Should restoration efforts eventually prove to be successful for these species, nutcrackers may reinvade their former range and again play the role of disperser in these ecosystems.

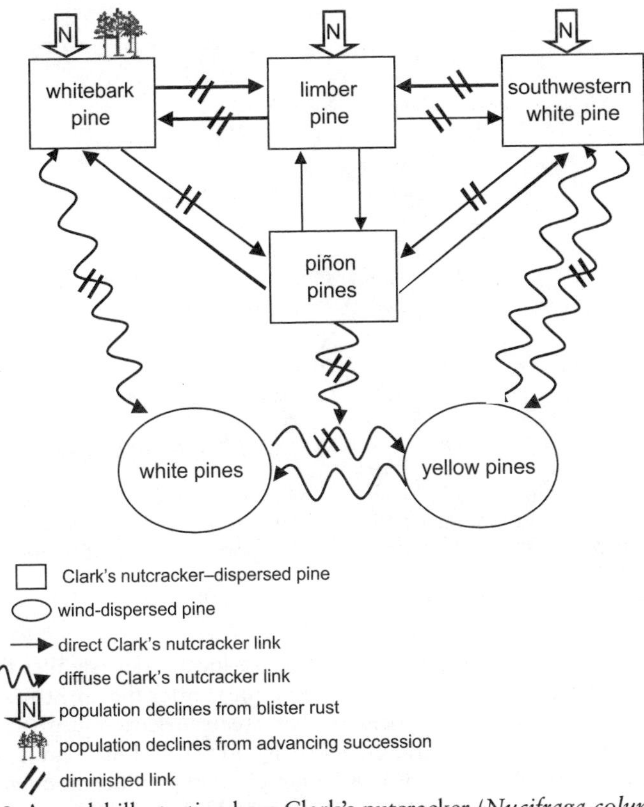

Figure 12-2. A model illustrating how Clark's nutcracker (*Nucifraga columbiana*) populations directly interlink populations of bird-dispersed pines and more diffusely link populations of bird-dispersed pines with wind-dispersed pines. The bird-dispersed pines in the model include whitebark (*Pinus albicaulis*), limber (*P. flexilis*), southwestern white (*P. strobiformis*), and the piñon pines (*P. edulis* and *P. monophylla*). The wind-dispersed pines include sugar (*P. lambertiana*), bristlecone (*P. longaeva* and *P. aristata*), foxtail (*P. balfouriana*), ponderosa (*P. ponderosa*), and Jeffrey pine (*P. jeffreyi*). Whitebark pine populations are declining from a combination of white pine blister rust (*Cronartium ribicola*) and advancing succession; limber and southwestern white pines, as well as sugar pine and foxtail pine, are experiencing the spread of blister rust as well. Loss of pine seed sources lowers the size of nutcracker populations, which further decreases seed-dispersal services and thus regeneration in bird-dispersed pines. See text for details.

Whitebark Pine: The Downward Spiral

We have presented our view of the rangewide consequences to whitebark pine, biodiversity, and several bird-dependent white pine ecosystems from the spread of white pine blister rust. However, there is a cautionary message: The decline of whitebark pine and other white pines regionally may be more rapid than

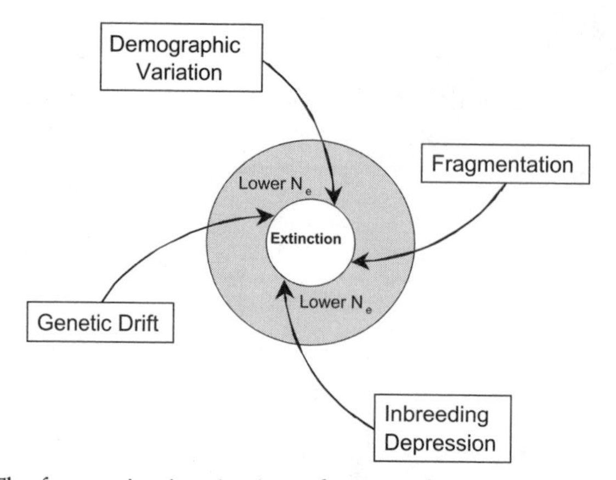

Figure 12-3. The four extinction vortices, *demographic variation, fragmentation, inbreeding depression,* and *genetic drift,* modeled after Gilpin and Soulé (1986). Any single vortex can lead to extinction; however, reduced effective population size N_e leads to more than one or all vortices operating simultaneously. When N_e becomes sufficiently small, the vortex takes the population to extinction.

predicted from the rate of spread of blister rust locally. The smaller the population, the more vulnerable to extinction, and the shorter the time to extinction.

In theory, there are four processes that create negative feedback effects, which hasten a population's decline to extinction—the four "extinction vortices" of Gilpin and Soulé (1986). All four processes reduce population size, particularly the genetic effective population size N_e for a species; and any one vortex can by itself lead to extinction, although all four may work synergistically (Figure 12-3). The first extinction vortex, which we will call *demographic variation,* is the result of chance environmental events that lower population size and increase variance in population growth rate. Whitebark pine has entered the first extinction vortex; the blister rust epidemic has both reduced local population sizes and potential population growth. Regeneration is dependent on Clark's nutcrackers, which adds another dimension of vulnerability not experienced by wind-dispersed pines. Loss of disperser reliability adds to demographic variation or uncertainty.

Decreases in population size and increasing demographic uncertainty lead to patchiness and *fragmentation* within a species, the second extinction vortex. Fragmentation leads to isolation, which increases the probability of local extinction. Abrupt changes in population size through fragmentation reduce N_e. Blister rust is creating population remnants or fragments within a previously more continuous regional distribution of whitebark pine—either pockets of trees that escaped spore transmission, or resistant individuals.

A decrease in population N_e leads to the third vortex, *inbreeding depression* and the loss of heterozygosity. The smaller the N_e, the higher the rate of inbreeding. With few remaining trees within an area or region, pollen availability may be inadequate: Whitebark pine trees may self-pollinate, or related members of tree clusters may cross-pollinate (see Chapter 8). The consequence is an increase in the rate of inbreeding through time, which means increasing homozygosity and the decline of genetic variation in subsequent generations. Further decrease in N_e also leads to the fourth vortex, *genetic drift,* which operates over a longer time frame. Genetic drift, random changes in gene frequencies, results in the loss of genetic variation. Natural selection requires genetic variation to adapt populations to local environments. This vortex implies a long-term loss in adaptiveness in whitebark pine populations, which can lead to extinction, particularly in the face of environmental changes such as climate warming.

Any one of these processes, or all in combination, may rapidly lead to local or broader extinctions of whitebark pine or any of the other nutcracker-dependent white pines infected with blister rust. The genetic and demographic processes represented by the extinction vortices are now operating on whitebark pine populations and argue for immediate intervention, such as restoration treatments. As whitebark pine communities decline, the considerable rangewide biodiversity they support is reduced as well.

Acknowledgments

We thank Stephen F. Arno and Robert E. Keane for past discussions about biodiversity in whitebark pine communities, effects on nutcracker populations, and the future of whitebark pine, and for helpful suggestions for this chapter.

Literature Cited

Andrews, R., and R. Righter. 1992. Colorado birds. Denver Museum of Natural History, Denver, Colorado.

Arno, S. F. 1986. Whitebark pine cone crops—a diminishing source of wildlife food? Western Journal of Applied Forestry 1:92–94.

Arno, S. F., and R. P. Hammerly. 1984. Timberline: Mountain and arctic forest frontiers. The Mountaineers, Seattle, Washington.

Arno, S. F., and R. J. Hoff. 1990. *Pinus albicaulis* Engelm. Whitebark pine. Pages 268–279 *in* R. M. Burns and B. H. Honkala, technical coordinators. Silvics of North America, USDA Forest Service, Agriculture Handbook 654, Washington, D.C.

Baud, K. S. 1993. Simulating Clark's nutcracker caching behavior: Germination and predation of seed caches. M. A. thesis, University of Colorado at Denver.

Benkman, C. W., R. P. Balda, and C. C. Smith. 1984. Adaptations for seed dispersal and the compromise due to seed predation in limber pine. Ecology 65:632–642.

Boyd, R. (ed.). 1999. Indians, fire, and the land in the Pacific Northwest. Oregon State University, Corvallis.

Callaway, R. M. 1998. Competition and facilitation on elevation gradients in subalpine forests of the northern Rocky Mountains, USA. Oikos 82:561–573.

Campbell, E. M. 1998. Whitebark pine forests in British Columbia: Composition, dynamics, and the effects of blister rust. M.S. thesis, University of Victoria, Victoria, British Columbia, Canada.

Christiansen, T. A., and R. J. Lavigne. 1996. Habitat requirements for the reestablishment of litter invertebrates following the 1988 Yellowstone National Park fires. Pages 147–150 *in* J. M. Greenlee, editor. The ecological implications of fire in Greater Yellowstone. International Association of Wildland Fire, Fairfield, Washington.

Craighead, J. J., J. S. Sumner, and G. B. Scaggs. 1982. A definitive system for analysis of grizzly bear habitat and other wilderness resources. Wildlife-Wildlands Institute Monograph 1. University of Montana, Missoula.

Eversman, S., C. Johnson, and D. Gustafson. 1990. Vertical distribution of epiphytic lichens on three tree species in Yellowstone National Park. Pages 367–368 *in* W. C. Schmidt and K. J. McDonald, compilers. Proceedings—Symposium on whitebark pine ecosystems: Ecology and management of a high-mountain resource. USDA Forest Service, Intermountain Research Station, General Technical Report INT-270, Ogden, Utah.

Farnes, P. E. 1990. SNOTEL and snow course data describing the hydrology of whitebark pine ecosystems. Pages 302–304 *in* W. C. Schmidt and K. J. McDonald, compilers. Proceedings—Symposium on whitebark pine ecosystems: Ecology and management of a high-mountain resource. USDA Forest Service, Intermountain Research Station, General Technical Report INT-270, Ogden, Utah.

Fisher, R. M., and M. T. Myres. 1980. A review of factors influencing extralimital occurrences of Clark's nutcracker in Canada. Canadian Field-Naturalist 94:43–51.

Gilpin, M. E., and M. E. Soulé. 1986. Minimum viable populations: Processes of species extinction. Pages 19–34 *in* M. E. Soulé, editor. Conservation biology: The science of scarcity and diversity. Sinauer Associates, Inc., Sunderland, Massachusetts.

Giuntoli, M., and L. R. Mewaldt. 1978. Stomach contents of Clark's nutcracker collected in western Montana. Auk 95:595–598.

Hawksworth, F. G. 1990. White pine blister rust in southern New Mexico. Plant Disease 74:938.

Hoff, R. J., R. T. Bingham, and G. I. McDonald. 1980. Relative blister rust resistance of white pines. European Journal of Forest Pathology 10:307–316.

Hutchins, H. E., and R. M. Lanner. 1982. The central role of Clark's nutcracker in the dispersal and establishment of whitebark pine. Oecologia 55:192–201.

Hutto, R. L. 1995. Composition of bird communities following stand-replacement fires in northern Rocky Mountain (U.S.A.) conifer forests. Conservation Biology 9:1041–1058.

Keane, R. E., and S. F. Arno. 1993. Rapid decline of whitebark pine in western Montana: Evidence from 20-year remeasurements. Western Journal of Applied Forestry 8:44–70.

Keane, R. E., P. Morgan, and J. P. Manakis. 1994. Landscape assessment of the decline of whitebark pine (*Pinus albicaulis*) in the Bob Marshall Wilderness Complex, Montana, USA. Northwest Science 68:213–229.

Kendall, K. C. 1983. Use of pine nuts by grizzly and black bears in the Yellowstone area. International Conference on Bear Research and Management 5:166–173.

Kendall, K. C., and S. F. Arno. 1990. Whitebark pine—An important but endangered wildlife resource. Pages 264–273 *in* W. C. Schmidt and K. J. McDonald, compilers. Proceedings—Symposium on whitebark pine ecosystems: Ecology and management of a high-mountain resource. USDA Forest Service, Intermountain Research Station, General Technical Report INT-270, Ogden, Utah.

Lackschewitz, K. 1991. Vascular plants of west-central Montana—Identification guidebook. USDA Forest Service, Intermountain Research Station, General Technical Report INT-277, Ogden, Utah.

Lanner, R. M. 1980. Avian seed dispersal as a factor in the ecology and evolution of limber and whitebark pines. Pages 14–48 *in* B. P. Dancik and K. O. Higginbotham, editors. Proceedings of Sixth North American Forest Biology Workshop. University of Alberta, Edmonton, Alberta, Canada.

———. 1988. Dependence of Great Basin bristlecone pine on Clark's nutcracker for regeneration at high elevations. Arctic and Alpine Research 20:358–362.

———. 1996. Made for each other: A symbiosis of birds and pines. Oxford University Press, New York.

Lanner, R. M., and B. K. Gilbert. 1994. Nutritive value of whitebark pine seeds and the question of their variable dormancy. Pages 206–211 *in* W. C. Schmidt and F. -K. Holtmeier. Proceedings—International workshop on subalpine stone pines and their environment: The status of our knowledge. USDA Forest Service, Intermountain Research Station, General Technical Report INT-GTR-309, Ogden, Utah.

Lonner, T. N., and D. F. Pac. 1990. Elk and mule deer use of whitebark pine forests in southwest Montana: An ecological perspective. Pages 237–244 *in* W. C. Schmidt and K. J. McDonald, compilers. Proceedings—Symposium on whitebark pine ecosystems: Ecology and management of a high-mountain resource. USDA Forest Service, Intermountain Research Station, General Technical Report INT-270, Ogden, Utah.

Lyon, L. J. 1984. The Sleeping Child Burn—21 years of postfire change. USDA Forest Service, Intermountain Forest and Range Experiment Station, General Technical Report INT-330, Ogden, Utah.

Mattson, D. J., and D. P. Reinhart. 1994. Bear use of whitebark pine seeds in North America. Pages 212–220 *in* W. C. Schmidt and F. -K. Holtmeier. Proceedings—International workshop on subalpine stone pines and their environment: The status of our knowledge. USDA Forest Service, Intermountain Research Station, General Technical Report INT-GTR-309, Ogden, Utah.

———. 1997. Excavation of red squirrel middens by grizzly bears in the whitebark pine zone. Journal of Applied Ecology 34:926–940.

Mattson, D. J., B. M. Blanchard, and R. R. Knight. 1991. Food habits of Yellowstone grizzly bears, 1977–1987. Canadian Journal of Zoology 69:1619–1629.

———. 1992.Yellowstone grizzly bear mortality, human habituation, and whitebark pine seed crops. Journal of Wildlife Management 56:432–442.

Morgan, P., and S. C. Bunting. 1992. Using cone scars to estimate past cone crops of whitebark pine. Western Journal of Applied Forestry 7:71–73.

Pfister, R. D., B. L. Kovalchik, S. F. Arno, and R. C. Presby. 1977. Forest habitat types of Montana. USDA Forest Service, Intermountain Forest and Range Experiment Station, General Technical Report INT-34, Ogden, Utah.

Primack, R. B. 1998. Essentials of conservation biology, 2d edition. Sinauer Associates, Sunderland, Massachusetts.

Pyle, P., N. Nur, and D. F. DeSante. 1994. Trends in nocturnal migrant landbird populations at southeast Farallon Island, California, 1968–1992. Pages 58–74 *in* J. R. Jehl Jr. and N. K. Johnson, editors. A century of avifaunal change in western North America. Cooper Ornithological Society, Studies in Avian Biology No. 15.

Reinhart, D. P., and D. J. Mattson. 1990. Red squirrels in the whitebark pine zone. Pages 256–263 *in* W. C. Schmidt and K. J. McDonald, compilers. Proceedings—Symposium on whitebark pine ecosystems: Ecology and management of a high-mountain resource. USDA Forest Service, Intermountain Research Station, General Technical Report INT-270, Ogden, Utah.

Small, A. 1974. The birds of California. Collier Books, New York.

Smith, C. C., and R. P. Balda. 1979. Competition among insects, birds and mammals for conifer seeds. American Zoologist 19:1065–1083.

Smith, R. L. 1990. Ecology and field biology, 4th edition. Harper and Row, New York.

Steele, R., S. V. Cooper, D. M. Ondov, D. W. Roberts, and R. D. Pfister. 1983. Forest habitat types of eastern Idaho–western Wyoming. USDA Forest Service, Intermountain Forest and Range Experiment Station, General Technical Report INT-144, Ogden, Utah.

Stickney, P. F. 1986. First decade plant succession following the Sundance Fire, northern Idaho. USDA Forest Service, Intermountain Research Station, General Technical Report INT-197, Ogden, Utah.

Stuart-Smith, G. J. 1998. Conservation of whitebark pine in the Canadian Rockies: Blister rust and population genetics. M.S. thesis, University of Alberta, Edmonton, Alberta, Canada.

Tomback, D. F. 1978. Foraging strategies of Clark's nutcracker. Living Bird 16:123–161.

———. 1986. Post-fire regeneration of krummholz whiteback pine: A consequence of nutcracker seed caching. Madroño 33:100–110.

———. 1998. Clark's nutcracker (*Nucifraga columbiana*), No. 331. *In* A. Poole and F. Gill, editors. The birds of North America. The Birds of North America, Inc., Philadelphia, Pennsylvania.

Tomback, D. F., and K. A. Kramer. 1980. Limber pine seed harvest by Clark's nutcracker in the Sierra Nevada: Timing and foraging behavior. Condor 82:467–468.

Tomback, D. F., and Y. B. Linhart. 1990. The evolution of bird-dispersed pines. Evolutionary Ecology 4:185–219.

Tomback, D. F., L. A. Hoffmann, and S. K. Sund. 1990. Coevolution of whitebark pine and nutcrackers: Implications for forest regeneration. Pages 118–129 in W. C. Schmidt and K. J. McDonald, compilers. Proceedings—Symposium on whitebark pine ecosystems: Ecology and management of a high-mountain resource. USDA Forest Service, Intermountain Research Station, General Technical Report INT-270, Ogden, Utah.

Tomback, D. F., S. K. Sund, and L. A. Hoffmann. 1993. Post-fire regeneration of *Pinus albicaulis:* Height-age relationships, age structure, and microsite characteristics. Canadian Journal of Forest Research 23:113–119.

Tomback, D. F., J. K. Clary, J. Koehler, R. J. Hoff, and S. F. Arno. 1995. The effects of blister rust on postfire regeneration of whitebark pine: The Sundance Burn of northern Idaho (U.S.A.). Conservation Biology 9:654–664.

Torick, L. 1995. The interaction between Clark's nutcracker and ponderosa pine, a "wind-dispersed" pine: Energy-efficiency and multi-genet growth forms. M.A. thesis, University of Colorado at Denver.

Turner, M. G., W. H. Romme, R. H. Gardner, and W. W. Hargrove. 1997. Effects of fire size and pattern on early succession in Yellowstone National Park. Ecological Monographs 67:411–433.

U.S. Fish and Wildlife Service. 1997. Grizzly bear recovery in the Bitterroot Ecosystem. Draft Environmental Impact Statement. U.S. Fish and Wildlife Service. Bitterroot Grizzly Bear EIS, P.O. Box 5127, Missoula, MT 59806.

Vander Wall, S. B. 1988. Foraging of Clark's nutcrackers on rapidly changing pine seed resources. Condor 90:621–631.

———. 1992. The role of animals in dispersing a "wind-dispersed" pine. Ecology 73:614–621.

Vander Wall, S. B., and R. P. Balda. 1977. Coadaptations of the Clark's nutcracker and piñon pine for efficient seed harvest and dispersal. Ecological Monographs 47:89–111.

Vander Wall, S. G., and H. E. Hutchins. 1983. Dependence of Clark's nutcracker, *Nucifraga columbiana,* on conifer seeds during postfledging period. Canadian Field-Naturalist 97:208–214.

Vander Wall, S. B., S. W. Hoffman, and W. K. Potts. 1981. Emigration behavior of Clark's nutcracker. Condor 83:162–170.

Chapter 13

Threatened Landscapes and Fragile Experiences: Conflict in Whitebark Pine Restoration

Stephen F. McCool and Wayne A. Freimund

Several years ago, we backpacked into one of those beautiful wilderness subalpine cirques of the northern Rocky Mountains. Our campsite the first night was located near a stand of whitebark pine (*Pinus albicaulis*). The trees were magnificent, their shapes and forms providing intrigue, mystery, and beauty, adding important dimensions to our wilderness experience. The trees could not speak—but if they could, we wondered what stories they would tell. How many storms had ravaged the lake basin in which they struggled? How had they survived decades of long and severe winters? How many recreationists had enjoyed not only their shelter but their splendid contours as well? How many trees had been hit by lightning? How long did they live? We wanted to ask of the trees: Is it those severe and relentless winters that have twisted you so? We concluded that these magnificent, aged whitebark pines must be very strong and unimaginably enduring. The story the tree would tell us is real; it is one that would lend definition and meaning to the landscape. It would provide an important dimension to our wilderness experiences.

We also wondered about their role in the ecosystem. Looking back on that trip, we thought about the mutualistic and potentially synergistic relationship between the Clark's nutcracker (*Nucifraga columbiana*) and whitebark pine. We thought about the trees—such as subalpine fir (*Abies lasiocarpa*)—that might replace the whitebark pine given the declines noted in Chapter 1. Whitebark pine is clearly in an era of decline. The decline has several causes, and many potential biophysical and social consequences. Both white pine blister rust (*Cronartium ribicola*) and fire suppression have been identified as important factors leading to declines in both the numbers and distribution of whitebark pine. Chapter 1 documents the significant, and potentially disastrous, effects of such declines on native flora and fauna. Whitebark pine plays an important role in subalpine ecosystems; its loss carries implications for both birds and animals, particularly the Clark's nutcracker and grizzly bears (*Ursus arctos horribilis*).

In response to such declines, a variety of scientists have proposed restoration actions. These actions themselves involve return of fire in whitebark pine ecosystems, selective cutting, and, potentially, genetic alteration to increase resistance to blister rust. Each of these actions would be considered where restoration is feasible. And some restoration would occur in designated wilderness.

We felt, that while subalpine fir is a beautiful tree in its own right, it wouldn't substitute for those gnarled, twisted-limbed sentries of the mountains. How would the lake basin look without any whitebark pine trees? Would we find their absence acceptable? Would we even notice? How would it look with only subalpine fir? How would it look without any trees?

As scientists who have studied recreationists in wilderness settings, we wondered what those visitors would think about whitebark pine. Would they view the tree as an intrinsic and essential component of the ecosystem? We speculated about their understanding of the species' population dynamics. We considered how much understanding they might have of the intricate relationships among nature's fire, whitebark pine, and the Clark's nutcracker. We speculated how wilderness visitors would view the decline in whitebark pine. We also concluded that the acceptability of such declines and restoration activities in those rugged and frequently remote settings would be problematic. What kinds of trade-offs would visitors and interested publics be willing to make?

These are not insignificant questions. Designated wilderness and the distribution of whitebark pine overlap significantly. Areas outside of designated wilderness with whitebark pine are commonly of similar landscape character—craggy mountaintops, roaring streams, captivating rocks, sharp cliffs, rugged scenery. While social scientists generally do not study recreation and recreationists in terms of biological ecosystem boundaries, we do know that much of the recreation occurring in whitebark pine ecosystems is in wilderness and backcountry. Such recreation experiences are highly dependent on relatively natural and untrammeled landscapes, where ecological processes continue to shape the landscape, with high opportunities for solitude in a remote and largely unmodified setting. What is or is not done about whitebark pine in an ecological sense, we realized, carries implications for the character of recreational and aesthetic experiences in those wonderful subalpine settings.

The whitebark pine issue, like many, has been a long time in development, and is a result of many factors. These issues not only are problematic from a technical restoration perspective but also go to the heart of what wilderness and other similar areas are supposed to be. We explore the dimensions of this controversy to understand the questions involved more fully, the upcoming philosophical debate that will accompany proposals for restoration, and the institutional dimensions influencing restoration. Certainly this controversy foreshadows many more, as we become increasingly aware of the impacts of human activity on natural and wild landscapes.

More specifically, we address three objectives: First, we examine the role of trees, and specifically whitebark pine, in recreational experiences. No doubt trees serve important aesthetic, symbolic, and instrumental purposes in recreational experiences, but our knowledge of these purposes is extremely limited. Yet, understanding these roles is critical to assessing the consequences of both potential losses of whitebark pine and restoration activities to recreational experiences. To understand these effects, we need to review briefly the history of how trees have been depicted in art. Such depictions help us understand human-nature transactions as they are reflected in artistic representations. These representations provide symbolic evidence of the relationships between people and their landscapes and how restoration may affect those relationships.

Second, we turn to the issue of the acceptability of restoration activities. The scientific community seems to have reached a consensus that large portions of the whitebark pine ecosystem are declining, and that active restoration is necessary to maintain healthy whitebark pine communities (Chapters 1, 9, and 11). The process of restoration to address these declines has both technical and political aspects (Light and Higgs 1996; Chapter 19). As such, the question of what constitutes acceptable restoration comes to fore, in both technical and political dimensions. Understanding the parameters that influence the social/political acceptability of restoration activities is a prerequisite to implementing any type of management to counteract whitebark pine declines.

Adding a unique challenge to the restoration of whitebark pine in wilderness is that a defining feature of wilderness, as stated in the Wilderness Act of 1964, is that it exists in areas where "earth and its community of life are untrammeled by man" (PL 88-577). Trammeling refers to the manipulation or alteration of the landscape (Barry 1998) and often serves as an indicator of what makes wilderness unique from other land-use designations. Eighty years of fire exclusion and suppression can be considered trammeling. Likewise, active manipulation to revert the landscape to what might have been can also be considered trammeling. This leads us to the questions of what is lost when such trammeling occurs, whether two trammels (wrongs) make a right, and for how long these acts of trammeling will be necessary to restore conditions to some baseline. Howard Zahniser, the architect of the Wilderness Act, states that our deepest need for wilderness is as an aid in "forsaking human arrogance and courting humility in a respect for the community and with regard for the environment" (Nash 1982; Borrie 1995). Since it took eighty years to recognize the extent of our whitebark pine dilemma, it may be worth taking adequate time now to more fully understand the implications of our response to it. It is important that we have acknowledged as thoroughly as possible the potential impacts of restoration to wilderness (Barry 1998; Worf 1997). We suggest this with a sensitivity to the time element; some scientists argue that a mere decade or two of delay will make restoration even more difficult (e.g., Chapter 1).

Finally, since the types of recreational experiences occurring in whitebark pine systems are largely low-density and nature-dependent ones, many of which are guided by the Wilderness Act of 1964 and other similar protected-landscape legislation, we must review the institutional context for restoration activities suggested to address whitebark pine declines. Such a discussion is important simply because the Wilderness Act represents a compelling philosophical statement about how certain landscapes should be managed. Not only must restoration acknowledge this institutional context, but also the meanings people attach to landscapes are influenced by such symbolic strides as the Wilderness Act. As Galliano and Loeffler (1995) have noted:

> Perhaps one of the most significant characteristics of ecosystem management is its ability to substantiate the emotional and symbolic meanings of natural resources on our country's public lands . . . Peoples' values, as well as their own sense of identity, are shaped by their interaction with the landscape and the emotions they feel about it. Thus, the values and meanings of places serve as significant factors reflecting the human dimensions of ecosystems and serve as linkages between social experiences and geographic areas. (p. 1)

Our discussion of the institutional framework would be incomplete without mentioning policy barriers and scientific limitations to restoration activities in wilderness, although broader observations could be made of public lands in general (see also Chapter 19). Such questions form the larger political context that shapes management, confronts those who propose restoration, and determines whether such proposals are in the broader public interest.

Human Perception of Whitebark Pine: A Transactional Approach to Understanding the Importance of Trees in Wilderness Experiences

Values of forest environments result from transactions between people and a forest's biophysical characteristics within a specific context (Ittelson 1973). Therefore, the aesthetic value of the whitebark pine environment and the aesthetic impact of management actions will vary depending on the physical characteristics of the forest, the characteristics of the people experiencing the environment, and the psychological, social, physical, managerial, and temporal context in which the experience takes place (Pitt and Zube 1987; Ittelson 1973).

While much of human perception involves visual cues, nonvisual sensations play an important role in enriching forest experiences. The aesthetic experience of a whitebark pine stand, for example, is strongly linked to the smell of pine resin, the sound of wind whispering through the boughs, and the feel of pine needles underfoot, as much as it is linked to the visual image of stately, uplifting trunks and open parklike stands. Over time, people learn through their own experiences, social norms, and symbolic associations that a particular combination of smell, touch, sound, and sight is a whitebark pine forest. While they may not know the scientific difference between whitebark pine and other tree species, they are aware of the particular setting that is produced by this combination of patterns and sensations. The transactional definition of the aesthetic value of whitebark pine, therefore, is an experience emanating from a multisensory interaction of humans and whitebark pine that is affected by the meanings that society or particular social groups assign over time to those experiences and settings. These sensations are associated with larger environments and contexts, such that experiencing a whitebark pine forest occurs largely within subalpine settings remote from the contrasting sensations of civilizations and within a policy context that is distant from the active vegetation manipulation that characterizes a substantial portion of the U.S. landscape.

Such transactions often become associated with scenic aesthetic evaluations of landscapes. Thus, a whitebark pine forest not only communicates feelings concerning a particular pattern of sensations and recollections, but may come to be associated with visually attractive scenes.

The Contextual Nature of Human Perception

Aesthetic values perceived by people in the forest will vary depending on their background characteristics, life cycles or lifestyles, and physical, cognitive, or emotional abilities (Anderson et al. 1992). That which is perceived as attractive by a twenty-year-old who grew up in the mountains of Montana may invoke fear and distaste to an eighty-year-old lifelong resident of Bangalore, India. Similarly, an individual's perceived aesthetic values of a specific forested site may vary through repeated visits depending on alterations to the psychological, social, physical, managerial, or temporal context of the visit. When seeking a private experience of solitude, a person's aesthetic sensitivities and expectations of the forest are likely to be quite different from seeking a setting for an extended family picnic (Driver et al. 1987). The gaining of familiarity with a given forest or the development of expertise in forest recreation activities will likely affect the aesthetic values derived by a person over time.

The Contextual Role of Landscape

The experiences people have with forest environments take place over time and space (Litton 1974). Thus, the visual response to any location within an environment is constantly undergoing comparison with other locations. In addition, the visual responses of different parts of an environment are continually being integrated toward an overall impression of the place. The aesthetic values of Yellowstone National Park, for example, have been dramatically altered by the scope of fires that occurred in 1988. It is impossible to visit the park now without the results of fire becoming a dominant part of the experience. While the experience of a recently burned landscape is highly valued by some, the unattractive quality of fire may lead to a diminished appreciation for the geysers, mountains, plants, and wildlife that compose the extent of the park's attributes for others. The sequential and integrative nature of a landscape experience has two implications: (1) Responses at given locations must be framed in the context of responses that could be derived in nearby environments, and (2) responses to vastly different environments (e.g., Montana and Arizona) are not directly comparable (Anderson et al. 1992). Yet, aesthetics has a process or ecological side: Some may see as much beauty in the processes leading to the landscape as in the resulting landscape.

The transactions that occur between people and whitebark pine settings, therefore, are defined by both the ecological and social context of the experience. The merger of these contexts creates the overarching framework in which the roles of whitebark pine become defined. There are three fundamental elements of this framework: symbolic meanings, aesthetic attractiveness, and instrumental utility.

The Elements of Whitebark Pine Transactions

Symbolic meaning. Many of the meanings associated with our natural world are strongly guided by the way they have emerged as symbols of our culture and heritage. The aesthetic value of the bald eagle (*Haliaeetus leucocephalus*), for example, is strongly complemented by its symbolic association with freedom, our nation, and the ability to recover from endangered status.

Trees are an historically powerful symbol within the fundamental institutions of social organization. Schroeder (1991) describes the prominent role that trees take within many of the world's religions, myths, and cultural rituals. Schroeder points to examples such as the mythical "World Trees," which stand at the center of the universe within many mythical traditions, and the dedication of sacred groves such as the Garden of Eden to illustrate a legacy of reverence for nature that has been dominated by trees as symbols. While we often use terms like "Mother Nature" loosely, they are actually quite indicative of the personal and symbolic value that humans have associated with landscapes. Dismissing this level of importance can be a critical managerial mistake. Schroeder argues that "Many managers view the public as victims of misinformation, and assume that if correct information about resource management could be effectively communicated, then public protest would be greatly diminished" (28). He further suggests that "Beneath the surface of natural resource conflicts, such as the spotted owl (*Strix occidentalis*) and old growth controversies, there may be powerful unconscious archetypes that do not respond to logical argument or rational persuasion" (24). So it may be for maintaining and restoring whitebark pine forests.

A role of trees in American heritage can be viewed in the paintings of Thomas Cole of the Hudson River School. It is the tree that often serves as the gatekeeper to the world, which is wild, mysterious, and sublime. Cole's painting *The Oxbow* (Figure 13-1) illustrates the desirability of having the safe and beautiful pastoral landscape juxtaposed with the sublime and foreboding wilderness within the same image. As Cole's painting suggests, the American aesthetic ideal relies heavily on a balance of the challenging, unpredictable, and foreboding qualities of landscapes that are unaltered by humans, with an organized, softened, pastoral, and safe landscape that can be labeled "the beautiful." *The Oxbow* also illustrates the unique role of trees in many of these depictions. Trees often serve the role of gatekeeper between what is civilized and what is wild.

There are several implications of the symbolic character of whitebark pine

Figure 13-1. *The Oxbow*, by Thomas Cole. The Metropolitan Museum of Art, gift of Mrs. Russell Sage, 1908. (08.228). See text for discussion of the symbolism of trees, particularly as the "gatekeeper."

ecosystems that are important to their management. First, since Western society tends to be dominated by scientific empiricism, people with interests in whitebark systems may not be fluent in the language of symbolism and may feel uncomfortable expressing symbolic meanings in processes dominated by scientists and technically trained planners. Further, people who hold strong feelings about landscapes may not find themselves with the vocabulary necessary to express these feelings, thus choosing to express themselves with such clichés as "Mother Nature." Since current paradigms of management rely on systematic empiricism, there is often little motivation to explore one's personal (and often unconscious) attitudinal dimensions that may underlie resistance to a proposed action. At longer time frames, the dynamic nature of landscapes must be acknowledged, and it may be this key quality, expressed empirically or emotionally, that forms the basis for constructive social discourse.

Aesthetic attractiveness. Trees that serve in the gatekeeper role must have certain qualities. Those qualities may function as an indication of what is to come or what has just been experienced. The trees themselves are often gnarled, twisted, and weathered. They demonstrate an enduring aspect of the setting in which they live and the challenging conditions by which these areas must be experienced. This aspect of attractiveness is clearly in line with the picturesque notion of attractiveness about which much of American landscape preference has been organized. A picturesque landscape includes a combination of "beautiful" features such as soft vegetation, water, and good spatial visual penetration with the "sublime" features such as jagged rocks, gnarled trees, and hazardous travel conditions.

The picturesque aesthetic model has been criticized for its romantic and static orientation (Gobster 1993). Gobster points to Aldo Leopold as an early proponent of an ecologically based aesthetic. Leopold's aesthetic appreciation is based on ecological function as opposed to the form of the object. From this orientation, a whitebark pine stand that has been intentionally burned may be perceived as beautiful, if the viewer understands and appreciates the role of fire in maintaining ecosystem health. While this perspective has a growing following, the cautions provided by Schroeder (1991; see page 268, this volume) should qualify the potential to simply educate aesthetic conflicts from existence. Whitebark pine declines pose a real dilemma for both forms of aesthetic appreciation. On the one hand, the presence of snags is often the basis for photography and wonder; yet those snags may have resulted from "unnatural" biophysical processes.

Instrumental utility. Whitebark pine stands serve a functional role in the experience of wildlands. Mature trees provide visitors with shade, a place to hang food, shelter from inclement weather, and contribute to the physical character of travel corridors and campsites. Whitebark pine stands also tend to occur in areas where opportunities to experience solitude, closeness to nature, tranquillity, and self-reliance are maximized. These unique and scarce experiences are often difficult to acquire outside of subalpine settings. Active manipulation of environmental attributes is frequently viewed as inconsistent with these motivations for the experience. These are also settings in which on-site educational resources (e.g., signs) are typically deemed inappropriate.

As Cole (1990) points out, not only is whitebark pine present along trails in wilderness, but also in some wildernesses as many as 78 percent of the campsites are located within these ecosystems. Removing whitebark pine from such a setting (e.g., the decline) is likely to change the character of the visit, and be visually and experientially disturbing. On-site education is unlikely to mitigate such effects. Yet, active manipulation of vegetation and associated activities is disruptive, at least in the short run, leading to a major management dilemma. In summary, trees and whitebark pine forests are valued not only for their aesthetic character and contribution to a recreational experience, but also because they symbolize certain environments and social commitments.

The Transaction with Whitebark Pine

When people transact with the whitebark pine, therefore, it is usually during a visit to remote backcountry areas that are often inaccessible except by foot or horseback. The type of person who is motivated for a wilderness experience is often characterized as one who seeks escape from the day-to-day world, self-sufficiency, solitude, and an unconfined setting. Whitebark pine plays a critical role in the development and characterization of that setting for recreationists. It has a particular relationship with the aesthetic and recreational components of the transaction. Whitebark pine also plays an important symbolic role in the transition from modern society to untrammeled wildness that extends to audiences far beyond the recreationist. The symbolic significance of these settings provides an emotional overlay to responding to declines, which are largely induced by human activity (fire suppression, exotic disease) and which are largely invisible to the visitor. Activities to restore such systems, although perhaps leading to a more natural system in the long run, may have immediate, apparent effects on the experience.

Acceptability of Restoration Actions

The near unanimous agreement among scientists about whitebark pine declines and needed restoration poses perplexing dilemmas. Clearly, while the problem of whitebark pine losses has been posed as a biotechnical one (the ultimate cause is human acts), the institutional context, definitions of wilderness and meanings assigned to nature and the places where whitebark pine systems occur complicate how the problem will be addressed. To find a favorite campsite closed for restoration, for example, may alter the individual perception of the entire area. The place simply will not seem the same if the site of specific memories, meanings, and assigned values is removed from the experience. Restoration is an immediate, definable consequence of active management, while losses of whitebark pine are more gradual, occurring over long time spans and may not be immediately visible to the visitor, even long-term ones. Both lead to important visual, recreational, and symbolic effects.

Although there appears to be a consensus among scientists about causes of the declines and potential remedies, there may be no such similar agreement among the various publics that retain interest in wilderness, a condition that suggests that whitebark pine restoration—at least in wilderness and similar areas—is at best a "wicked problem" (Allen and Gould 1986). Wicked problems, briefly, are those for which scientists may agree about cause-effect relationships, but goals are contested. Wicked problems are addressed in significantly different ways than in situations where scientists agree and there is a consensus on meanings and goals. Wicked problems are not solved, but resolved, because the problem is settled with compromises over goals. We could also argue that the situation is messy, because not all scientists may eventually agree on restoration prescriptions and their effects.

Two goals may be in conflict here: maintenance of whitebark pine (the appearance of naturalness) and the requirement that wilderness be untrammeled. One cannot necessarily argue that the goal of "untrammeled" has been abandoned because an important ecological factor has been excluded for a relatively short period of time in the life of an ecosystem. Fire exclusion policies may have been inappropriately applied to designated wilderness during a period when the knowledge of effects was limited. In order to develop acceptable management actions, one of these goals must be identified as an ultimately constraining goal, and achieving the other requires that the constraining goal be initially compromised to an acceptable limit, and then the other be compromised as needed (Cole 1995). Scientists proposing restoration of whitebark pine in wilderness, including both the use of prescribed fire and the planting of blister rust–resistant seedlings, have implicitly assumed that maintaining naturalness (whitebark pine) is the ultimately constraining goal, so that the "untrammeled" goal will be compromised in order to maintain whitebark pine conditions. The public, for which we manage public lands, may not agree.

We understand this controversy. Some would argue that the goal of untrammeled conditions has already been compromised with a fire exclusion policy and the introduction of an exotic disease, white pine blister rust. But it does not follow that active manipulation reduces trammeling in the long run. We do agree, however, that little pushes in the right direction may be more acceptable.

The primary question is one of how much compromise in each of the two goals is acceptable. The concept of acceptability in a political/social sense is relevant here. Public lands exist because of a social commitment to protection and management to meet broad conservation and economic development goals. Public land management occurs within the context of extensive, continuing debate about goals of management and management techniques. Since the public in the broad sense provides the funding for management, its perceptions and concepts of acceptable actions are critical components of restoration planning. Within this context, the potential for long, drawn-out planning processes, which focus on the wrong questions and end in elevated levels of frustration, is very high. Exploring the symbolic dimensions of whitebark pine early in development of a restoration strategy would help identify what questions are relevant and appropriate. Processes that encourage broad discussions, emphasize multiple perspectives and meanings, and invite deliberation may have a higher potential for success.

Yet, successful restoration requires that scientists and managers understand the social and political acceptability of management actions. Without a fundamental consensus on goals and management responses, managers have little support for actions, and such actions may interfere with the human-environment transaction. Will this transaction have an authentic base, or will it be viewed, in Gunn's (1991) terms, as a "fake." While such language may seem strong, wilderness experiences are constructed based on transactions with authentic attributes of the environment. Genetically engineered whitebark pine and wildernesses that are now trammeled are simply not authentic. On the other hand, wildernesses missing their native population of whitebark pine are not authentic either.

Brunson (1992) suggests a number of working propositions that would be useful in developing greater understanding of the acceptability of restoration activities, assuming that maintenance of whitebark pine is the ultimate constraining goal. Brunson suggests these more as hypotheses than definitive conclusions. They form a good framework for discussing the acceptability of treatments among affected publics and recreationists but should be considered relative to one another. There are situations in which individual propositions may conflict with one another.

Acceptability may apply to conditions, but is a function of causes—doing things that imitate nature generally tends to be more acceptable than more artificial interventions. In this sense, use of fire as a restoration tool may be more acceptable than other techniques. Several studies support fire as a means of management. For example, in the early 1990s, the Interior Columbia Basin Ecosystem Management Project commissioned a study of attitudes toward forest management (Brunson et al. 1994; McCool et al. 1997). The survey informed respondents that "Some people favor the introduction of fire in federal forest lands to control disease, insects, and excessive fuel levels. Others suggest this use of fire is unnecessary and dangerous." Respondents were then asked to choose their preference regarding fire from five statements. The top choice by a wide margin was, "We should suppress wildfires in federal forests managed for timber; however, controlled fire may be used to protect forest health." The second choice was, "We should suppress wildfires in federal forests only if they threaten human lives or property; otherwise we should allow fire to resume its natural role in forests."

A study conducted by McCool and Stankey (1986) showed that wilderness users had become more supportive of naturally ignited fires in wilderness over a ten-year period. However, when it comes to more active intervention—such as the use of roads to access areas in need of mechanical treatment, such support withers. For example, a national opinion poll conducted in 1994 (Frederick/Schneider, Inc.) asked respondents to choose between two statements regarding managing forest areas where roads have not been built. Fifty-five percent said they do not favor building roads, because roads create soil erosion and destroy the wild character of the forest; whereas 40 percent chose a statement that favored building more roads to increase access for fighting fires, allowing the forest to be thinned, and for recreational uses. Pacific Northwest residents slightly favored building more roads (48 percent to 46 percent).

Conditions that arise from natural "causes" are virtually always thought of as acceptable. Perceptions of the public concerning the decline of whitebark pine and its causes have not been reported in the literature. Scientists attribute such declines to a combination of fire exclusion and blister rust, both results of human decisions and actions. However, because such losses may *appear* to the public to be a result of natural processes (succession, disease), the public may view them as being more acceptable than if such losses were the direct result of active interventions. If this admittedly speculative conclusion is reasonably correct, scientists have a considerable job ahead—that of convincing the public that declines are not natural. Appropriate, interactive venues for deliberation among interested publics will be required to build the fundamental understanding leading to notions of acceptable practices.

Acceptability of a condition (such as the whitebark pine decline) is not an issue, unless there is a feasible alternative to that condition. Conditions caused by nature are viewed as acceptable because often they are thought of as "unpreventable" or "unforeseeable." However, such views may change if a feasible alternative to this condition is presented. In the case of whitebark pine decline, if the public sees no feasible alternative to this condition (and the condition is a result of natural causes), it may be difficult to generate support for restoration. However, if restored ecosystems are portrayed as feasible alternatives to the replacement of whitebark pine (and the public accepts that depiction), there may be greater public acceptability for restoration activities. It is unclear at this point, because of lack of research, how the public views the acceptability of the loss of whitebark pine. Without such data, it then is difficult to assess the social and political acceptability of restoration, particularly in wilderness settings.

In the presence of feasible alternatives, acceptability is a function of the preferability, probability, and propriety of those alternatives. Once a set of feasible alternatives is generated, acceptability is linked to public assessment of the likelihood that management will lead to restored systems. Judgments concerning the acceptability of alternative restorative actions are based on many variables, including beliefs about probabilities, trust in the institution initiating the restorative action, attitudes toward environment and human interactions, and beliefs that restoration should occur. This means that gaining public interest in restoration and consent to initiate actions requires not only that scientists must guarantee a fairly high level of confidence that actions will restore the systems, but also that managers understand the public's perceptions, attitudes, and concerns. Problematically, the knowledge to provide this high level of confidence occurs over time frames of biological significance that are much longer than budgetary cycles. This scientific knowledge will be tentative as well for many years as we seek and gain greater understanding of how whitebark pine ecosystems function and how they respond to varying management treatments. Any information that the scientific community can provide on the probable time frame for necessary restoration would be highly valuable in this regard. A fundamental objective is to reduce the need for long-term trammeling, making achieving the objective within a specific time frame critical (Shoemaker 1984).

Acceptability is a function of perceived risk associated with the condition or restoration activity. Here we are dealing with two fundamentally important questions. First, does the public see whitebark pine declines leading to a negative impact? In this sense, does the decline represent, as we argued earlier, the risk of a loss in a particular dimension of wilderness and backcountry recreational experiences or in the untrammeled character of wilderness? Do whitebark pine losses negatively impact the symbolic meanings or ecological attractiveness of subalpine settings? If so, there would be greater support for restoration activities. How are these risks valued in relation to other risks, such as losses in picturesque aesthetics or a sense of wildness that would accompany active intervention? Second, there are risks of unintended consequences from restoration—such as a manager-ignited prescribed fire escaping the identified boundaries. How does the public perceive these risks? How are such risks balanced—risk of wildfire against restoring the ecosystem—in the public's mind? How are the risks of whitebark pine ecosystem declines considered against the possibilities—perhaps temporarily—of disrupting notions of wildness in visitors' experiences?

Checkland and Scholes (1990) argue that planners need to consider three criteria in evaluating proposed actions:

1. Efficacy (Will the proposed action work?)—for example, will the restoration action lead to restored whitebark pine systems?
2. Efficiency (Does the proposed action give the "biggest bang for the buck"?)
3. Effectiveness (Does the proposed action meet the longer-term goal?)—for example, will restoration lead to ecosystems that are self-regulating, natural, and untrammeled?

An action meeting one of these criteria well may not meet others particularly well, thus complicating the analysis. In a sense, then, the risk analysis is multidimensional and broader than a limited examination of biophysical consequences.

Acceptability depends on the local context. We noted earlier that values and roles of whitebark pine forests are strongly dependent on both the biophysical context and the social context. Restoration activities acceptable in one location may not be acceptable in another. For some places that scientists feel restoration is needed and possible, the public may feel unwilling to engage in such activities—because of the special character of such places, risk of negative consequences, or the general landscape context. This suggests that restoration actions are uniquely local in character, integrating scientific findings with locally derived meanings of whitebark pine ecosystems.

Acceptability is a function of social influences. Individual judgments about the acceptability of restoration activities occur within the context of reference group norms. Such reference groups may develop positions for or against restoration, and thus sway the attitudes of individuals. Reference groups may articulate particular meanings of wild places that are at odds with the manager's or scientist's definition. Scientists may define a whitebark pine forest in relatively abstract terms, such as a place where ecological functions can occur relatively unhindered, and thus these places may serve useful benchmark purposes. A local backcountry user group may define the same location as an important area for backcountry recreation, whereas a nearby Native American tribe may view the area as a culturally significant spiritual site. These varying definitions of place really mean that the same physical location is several different places. The acceptability of a management action thus varies by meaning. Understanding what meanings are held, how they are distributed among the affected population, and who in the institutions hold decision-making authority is essential to implementing restoration actions.

Acceptability can refer to a pleasing condition or a "barely tolerable" one. There is a difference between conditions that are judged as *acceptable* and those that are *preferred.* Preferred conditions are those that are desirable, whereas acceptable conditions are tolerable ones. Thus, people may feel that in light of the alternatives, declines in whitebark pine are acceptable but not desired. The public may prefer an ecosystem to remain in both its natural and untrammeled condition but, given the costs to these values and the risks associated with restoration, may feel that losses in whitebark pine are acceptable. This is an important issue because it reflect the types of trade-off considerations that managers and the public will be making.

Institutional Issues in Managing Whitebark Pine Communities

Scientists are nearly unanimous in their assessment that whitebark pine ecosystems require immediate restoration activity to counteract the results of decades of fire exclusion and the introduction of white pine blister rust (see especially Chapter 1). Such restoration activities occur within an institutional context that is as complex as whitebark pine ecology is intricate. Three aspects of this institutional context are noteworthy here: (1) questions about the acceptability of restoration in areas protected by the Wilderness Act, (2) the scientific capacity to understand and predict consequences of restoration activities, and (3) policy constraints dealing with the use of natural fire in wilderness.

There is a significant overlap between wilderness and whitebark pine ecosystems (about 49 percent of whitebark pine is located in designated wilderness and in national parks, and about 48 percent in national forests [R. E. Keane, unpublished data]). Whitebark pine ecosystems within designated wilderness must be managed in accordance with Section 2 (c) of the act, which defines wilderness in part as a place "untrammeled by man" and which "generally appears to have been affected by the forces of nature, with the imprint of man's work substantially unnoticeable." Section 4 (b) requires agencies that manage wilderness to do so in such a manner as to preserve the wilderness character of the area. These are powerful philosophical statements of what wilderness is and should be. The words were carefully chosen during eight years of debate and fifty-six revisions. The notion of wilderness as self-willed and untrammeled is an important philosophical foundation to any management regime. These concepts represent a significant symbolic step in protecting from human development and intervention the remaining vestiges of nature-dominated landscapes.

Cole (1996) argues that managers must achieve three somewhat incompatible goals simultaneously: (1) ensuring that wilderness landscapes remain in a natural or pristine condition, (2) protecting lands from "human control" so that they remain untrammeled and are not consciously and intentionally manipulated, and (3) providing a variety of direct public benefits, such as for recreation, science, and education. The definitions of "natural," "wild," and "untrammeled" are critical to understanding the effects of restoration activities and their appropriateness in wilderness. Aplet (1999) proposes that natural has at least two meanings: (1) the absence of human modification, such as structures and roads, and (2) an intact biota. Restoration would seem to be an activity that would restore naturalness in at least the second meaning. Untrammeled communicates the idea of freedom—in the case of wilderness, land that is "self-willed." Wildness consists of these two dimensions, which are orthogonal. Aplet (1999) argues ". . . it is possible to conceive of wildness as increasing in two dimensions: from the controlled to the 'self-willed' along a gradient of freedom, and from the artificial to the pristine along a gradient of naturalness . . . where freedom and naturalness are highest is the wilderness" (355).

Each of the three goals is somewhat compatible and yet often in conflict with the other. For example, providing opportunities for the "primitive and unconfined" recreation called for in the act means that some level of impact on naturalness will occur and is acceptable. There is also a question about whether current wilderness management policies (such as fire exclusion) are achieving these goals.

Suppression of fires and the presence of blister rust in designated wilderness is a trammeling, because such a policy essentially "controls" the character of the landscape: conditions are not the result of natural processes but rather of human intervention, if only indirect. While fire suppression and exotic biological agents may trammel the wilderness (it is not self-willed), some may still view the resulting conditions as "natural." In the case of whitebark pine, such effects go far beyond the trees themselves and include loss of biodiversity (Chapter 12).

If wilderness is established in part to protect natural conditions, some operational definition of natural backgrounds must exist to describe baseline conditions and processes. Yet, the question of what is natural is one that is a function of social norms, preferences, and ethical positions. Just as Greider and Garkovich (1994) argue that every river is more than one river, we maintain that the transactions people have with nature in wilderness settings vary from one person to another. To some, what is natural is simply defined as the absence of human intrusions. To others, natural is synonymous with the notion of ecological integrity (the presence of all "original" pieces, see Callicott and Mumford 1997), whereas for still others it means maintenance of ecological function. In terms of restoration activity, such definitions are crucial to designing processes that sustain the type of nature people want.

A number of scientists have suggested the use of prescribed fire and various mechanical treatments to restore whitebark pine ecosystems (e.g., Chapter 18). Intervention in the way of prescribed, manager-ignited fires would maintain an area's natural character but may be problematic in terms of the institutional constraint on wilderness being "untrammeled"—a term deliberately, thoughtfully chosen during the congressional debate over the Wilderness Act. Thus, those proposing restoration are engaged in a highly philosophical and symbolically significant trade-off between natural and untrammeled, yet may not have the ethical framework needed to guide the decisions. Restoration decisions are complicated by the trammeling (e.g., fire suppression) that has been permitted to occur in the past in the sense that naturally occurring fire has been deliberately excluded in many wildernesses. Such decisions about trade-offs in achieving conflicting goals are intrinsically political ones that are informed, but not determined, by science (Cole and McCool 1997; Krumpe and McCool 1997). And, a substantial debate—which we cannot address here because of space limitations—rages about the appropriateness and ethical foundations for landscape restoration activities (see especially Cowell 1993; Katz 1991; Light and Higgs 1996), which are all the more relevant for wilderness landscapes; that is, is a landscape a natural one if restored by humans?

These questions are complicated by the presence of modified whitebark pine ecosystems in wilderness, systems that have been impacted by human activity, particularly through fire exclusion and blister rust. Thus, we argue that these ecosystems are currently unnatural (biologically) and trammeled. Restoration activities thus would impact a system that is legally natural, but biologically a result of human intervention (if only indirect). Would such "mucking around" achieve the intent of the Wilderness Act? Clearly, there are long-term and short-term viewpoints here. In the short-term, restoration may trammel the wilderness; in the long-term, such a "nudge" may allow the area to return to its "self-willed" goal. Would restoration be ongoing or a one-time event? Would restoration achieve its goal or leave us with a modified system that has been even more modified?

Third, the effects of restoration activities on an ecosystem that is unnatural are unpredictable, given our current understanding of fire regimes in subalpine settings and institutional constraints on the use of fire. Science is limited in its capacity to fully understand and therefore predict the consequences of all actions, particularly at larger spatial and longer temporal scales. Fire in whitebark pine systems is a complex phenomenon, with varying fire intensities, frequencies, and scales occurring in different types of whitebark pine ecosystems. There may be a limit to the human capacity to fine-tune fire as a tool to emulate the differences among these ecoystems.

Manager-ignited fires are unlikely to replicate the natural range of variation in fire because of a variety of institutionally imposed limitations—such as regional preparedness guidelines (when fire danger is extreme), risk of fires "escaping" wilderness boundaries, threats from fires to lives and property, fuel loadings, and so on (Mutch 1995). Whitebark pine exists primarily in settings where fires often burned under extreme environmental conditions. Because agencies, in the words of one reviewer of this manuscript, are limited to burning under optimal conditions—with prime attention to the safety of fire management personnel—it is unlikely that restoration fire and its consequent effects can closely emulate natural processes. This poses the dilemma to wilderness managers: Restoration efforts as currently construed are, from an institutional perspective, unnatural, and are counter to the idea of wilderness.

Policies that allow natural ignitions (lightning-caused fires) to burn are also unlikely to fully replicate the natural range of variation for similar reasons: The extreme fire weather that is required to burn in subalpine settings means that many other fires are also burning, and agency preparedness guidelines require full suppression on natural ignitions. In the Selway-Bitterroot Wilderness astride the Montana-Idaho border, which has the longest-standing and arguably the most soundly based natural fire ignition policy in the country, data show that the spatial distribution, frequency, and intensity of fires is dissimilar from what occurred prior to Euro-American settlement (Brown et al. 1995). As with manager-ignited fires, a variety of institutional factors constrain full implementation of a natural fire policy.

The public is also deeply sensitive to the presence of smoke in the skies. Many people live near whitebark pine systems because of the forest and mountain amenities associated with them. They want to see those amenities. Again, fire management institutions tend to be risk averse; prescribed fire cannot be precisely controlled, and there is always the probability of fires moving away from their intended targets. The extent to which these issues are present in any given situation would have a bearing on restoration using fire, inside or outside designated wilderness.

Although we have outlined the considerable institutional and knowledge-based barriers to fire as a restoration tool, we are well aware that a continued fire exclusion policy has significant negative effects as well. The chapters in this volume certainly alert readers to the very high probability that an important plant community and the wildlife species that depend on it could be lost, and lost quickly. Such losses would be irreversible within scales of human significance. It could be that prescribed fires would have fewer negative biophysical and symbolic consequences than maintaining a policy of fire exclusion.

In summary, defining the problem of whitebark pine losses and restoration as solely biological-technical issues would be a serious miscasting of the puzzle. It would marginalize the powerful emotional images of wilderness users and enthusiasts. Active manipulation of vegetation, including use of fire, was probably never even considered during the debates over the Wilderness Act as it wended its way through Congress. It is a significant question about how such restoration activities would be interpreted in terms of the Wilderness Act. Whereas manipulation may appear to maintain natural conditions (if we could agree on what we meant by natural), it would lead to trammeling of the wilderness. Wildernesses contain a rich set of cultural, spiritual, and individual meanings, the values of which are strongly related to its untrammeled character. Certainly, such manipulation, while itself a technical procedure, is a political/ethical decision, complicated by multiple definitions of nature and constrained by institutional policies with respect to the use of fire—either manager-caused or natural ignitions. Complete discussion of the ramifications of these questions is essential prior to implementation of restoration. Testing of fire as a restoration technique in similar adjacent areas may provide important scientific information about its efficacy, and allow the use of wilderness as an important baseline (control) for experiments. Yet, even this policy would yield little to inform a potentially politically volatile debate about the appropriateness of restoration activities in wilderness.

Conclusions

We contend that whitebark pine ecosystems serve important symbolic, aesthetic, and instrumental functions in recreational experiences. Whitebark pine forests undoubtedly also serve important scientific purposes as we seek to expand our understanding of the natural world, and their retention is critical to understanding ecological function and process. Declines in whitebark pine abundance and distribution have important managerial, scientific, aesthetic, and symbolic implications as to how, what, and where restoration activities take place.

First, the importance of aesthetic, symbolic, and instrumental values of whitebark pine to recreationists and others interested in wild landscapes are not well known. How much and what kind of value recreationists and others concerned place on these dimensions is uncertain, yet is clearly fundamental to understanding the meaning, and therefore the significance, of whitebark pine declines and to the potential consequences of restoration. Scientific meanings are well articulated in the technical literature, and losses can be accounted for in terms of knowledge and science. "Educating" the public about these values will not necessarily create agreement about restoration activity because other meanings may be more important. Unfortunately, the current lack of research on such meanings leaves us with a large degree of uncertainty.

But since these other meanings have yet to be identified for backcountry recreationists, we cannot identify the values, meanings, and representations that are perhaps being irreversibly forfeited through the decline in whitebark pine. Support for restoration is founded on understanding of what is being lost. We suggest that continued declines in whitebark pine populations and distribution could have major consequences to the experiences people demand from wilderness areas. Whitebark pine communicates a sense of ruggedness, remoteness, and wildness to people visiting these ecosystems. The change in vegetation and biodiversity accompanying whitebark pine declines may no longer communicate such increasingly scarce impressions that are likely important to wilderness experiences. Since settings for wilderness experiences are themselves relatively scarce, changes in whitebark pine systems are particularly significant. Yet, without studying the meanings recreationists attach to these systems, it is difficult to document potential losses.

Second, the acceptability of restoration activities in designated wilderness is unknown. The literature suggests that recreationists are willing to accept management interventions that address the negative consequences of recreational use on biophysical attributes. But little research reports the acceptability of management actions taken to restore backcountry situations to some previous, nature-dominated setting. Since many types of restoration activities could conceivably be proposed for designated wilderness, acceptability is clearly an important question, as is the legal framework for management action. The Wilderness Act requires that federal agencies take only the "minimal" action needed to manage for the values for which the wilderness was designated. What restoration activities meet this requirement? Acceptability implies trade-offs. Can the trade-offs of restoration, or of nonrestoration, be identified for recreationists and others interested in the untrammeled character of wilderness?

Restoration occurs at the interface of science, policy, and public preferences. While many scientists and managers may be ready to implement restoration, such activities must be acceptable, even desirable to the larger, politically alert and active population. A fundamental issue in developing a restoration program is understanding acceptability and the values upon which it is based. Particularly important are restoration plans that detail the objectives, time frames, methods, and monitoring. These are needed to ensure informed decision making.

Trade-offs between the concepts of natural and untrammeled are not yet identified. Aggressive use of manager-ignited fire to restore a scientist's definition of a natural system may be less acceptable than continuing to trammel wilderness ecosystems through fire exclusion. Such actions tend to be viewed as precedent-setting, and are therefore much more controversial: What's next, genetically altered grizzly bears that are not aggressive to humans? Understanding the basics of what would make restoration activity acceptable to scientists, managers, and members of the public requires that we consider perceptions, social influences, and attributions of the causes of the problem. Factors affecting the acceptability of trade-offs need uncovering if restoration is to be implemented with popular support.

Third, the scientific and institutional capacity to address whitebark pine declines is challenged not only by the legislative foundation (e.g., Wilderness Act) for managing many whitebark pine ecosystems, but also by a variety of policy and technical fire management dictates. These may not only be at odds with the aesthetic and symbolic values of these ecosystems but also must be carried out in an environment of uncertainty. Can land managers ignite and manage fires that emulate those found in nature? Fire is natural. Are manager-ignited fires natural or simply an acceptable surrogate for nature? What if scientific predictions about consequences are wrong? What happens then?

More manipulation? These questions may make restoration of whitebark pine systems more a fancy of scientists than a reality for managers.

Fourth, we are confronted with a perplexing dilemma in designated wilderness: On one hand, lack of action to deal with whitebark pine losses may lead to additional and potentially irreversible systemic changes in subalpine ecosystems. These losses not only have notable ecological effects, they may also lead to major aesthetic and symbolic outcomes as well. Doing nothing preserves what may be an apparent but unreal untrammeled character of wilderness, but leads to lots of negative effects. On the other hand, intervention, through active vegetative manipulation, may save whitebark pine and protect its role in recreational experiences but lead to areas being "trammeled" and thus a loss, at least temporarily, of their wilderness character, a value of tremendous importance to Americans. Intervention may also negatively impact both picturesque (e.g., burnt or logged landscapes) and ecological aesthetics (the landscape is an artificial one). Dealing with this dilemma involves not only gaining better understanding of whitebark pine ecology, but also in establishing venues for increased public dialogue of the values at stake.

For whitebark pine in designated wilderness, we cannot overemphasize the sensitivity of intervening in the sense of active manipulation. Restoration occurs within the context of a widening debate over the meanings of wilderness (Callicott and Nelson 1998). If we modify wilderness to protect whitebark pine, why not do so to enhance habitat for grizzly bears? If we enhance habitat for grizzly bears, why not engage in riparian management to increase the salmon population upon which they would feed? If we do these things, what makes wilderness different from other publicly managed wildlands? De facto interventions in the case of fire exclusion may be interpreted as errors of omission rather than errors of commission. Such errors are often viewed more tolerably, but their consequences may be just as—or more—insidious.

In writing this chapter, we gained a great deal of humility for the complexity of the whitebark pine dilemma, especially when placed within wilderness—the land-use designation most designed to facilitate managerial humility. It is clear that the dearth of research about restoration acceptability in wilderness is a severely limiting factor. It is similarly clear that the philosophical debate about restoration is lagging behind our technical capabilities (Light and Higgs 1996; Barry 1998). It is our opinion that there is a need to clarify the objectives of wilderness restoration in terms of naturalness and trammeling. Aided by clarified objectives, time frames of trammeling as a *means* can be established and impacts to human experiences and philosophies can be better understood, while wild wilderness can be considered as an *end*.

During our visit to that high-mountain lake, we never would have thought that those gnarled and twisted trees, those gatekeepers of wildness that reflected our visions of wilderness, would be so problematic to manage. In that whitebark pine forest, the problem seemed pretty simple, if intricate. Do those crooked trunks provide the answer to the dilemma we have created? Will they continue to provide the inspiration for thousands of recreationists as they seek an experience of wildness? Will our next transaction with a whitebark pine be with one that was planted there, or genetically altered? If those trees have survived through dozens if not hundreds of winters, then maybe they hold the clue to these complex questions. Do we have the patience to find out?

Acknowledgments

David Parsons, Aldo Leopold Wilderness Research Institute, Missoula, Montana, and James Burchfield, Bolle Center for People and Forests, the University of Montana, provided thought-provoking reviews of an earlier version of this manuscript. We thank the Metropolitan Museum of Art, New York, New York, for permission to use *The Oxbow* by Thomas Cole.

Literature Cited

Allen, G. M., and E. M. Gould Jr. 1986. Complexity, wickedness, and public forests. Journal of Forestry 84:21–26.

Anderson D. H., W. F. Freimund, D. W. Lime, L. H. McAvoy, D. G. Pitt, and J. L. Thompson. 1992. Recreation and aesthetic resources: A technical paper for a generic environmental impact statement on timber harvesting and forest management in Minnesota. Draft Report to the Minnesota Environmental Quality Board, St. Paul, Minnesota.

Aplet, G. 1999. On the nature of wildness: Exploring what wilderness really means. Denver University Law Review 76:347–357.

Barry, D. 1998. Toward reconciling the cultures of wilderness and restoration. Restoration and Management Notes 16:125–127.

Borrie, W. T. 1995. Measuring the multiple, deep and unfolding aspects of the wilderness experience using the experience sampling method. Ph.D. dissertation, Virginia Polytechnic Institute and State University, Blacksburg.

Brown, P. J., B. J. Driver, and C. McConnell. 1978. The opportunity spectrum concept and behavioral information in outdoor recreation resource supply inventories: Background and information. Pages 74–83 *in* H. G Lund, V. J. LaBau, P. F. Folliott, and D. W. Robinson, editors. Integrated inventories of renewable natural resources: Proceedings of the workshop. USDA Forest Service, Rocky Mountain Forest and Range Experiment Station, General Technical Report RM-55, Fort Collins, Colorado.

Brown, J. K., S. F. Arno, L. S. Bradshaw, and J. P. Menakis. 1995. Comparing the Selway-Bitterroot fire program with presettlement fires. Pages 48–54 *in* J. K. Brown, R. W. Mutch, C. W. Spoon, and R. H. Wakimoto, editors. Proceedings: Symposium on fire in wilderness and park management. USDA Forest Service, Intermountain Research Station, General Technical Report INT-GTR-320, Ogden, Utah.

Brunson, M. B. 1992. Social acceptability of new perspectives, practices, and conditions. Final Project Report, Department of Forest Resources, Oregon State University, Corvallis.

Brunson, M., B. Shindler, and W. D. Schreskhise. 1994. Mail survey of natural resource issues on public lands in the West: Results from a public survey. Report on file with the USDA Forest Service, Interior Columbia Basin Ecosystem Management Project, Walla Walla, Washington.

Callicott, J. B., and K. Mumford. 1997. Ecological sustainability as a conservation concept. Conservation Biology 11:32–40.

Callicott, J. B., and M. P. Nelson, eds. 1998. The great new wilderness debate. Athens: University of Georgia Press.

Checkland, P., and J. Scholes. 1990. Soft systems methodology in action. John Wiley and Sons, New York.

Cole, D. N. 1990. Recreation in whitebark pine ecosystems: Demand, problems, and management strategies. Pages 305–309 *in* W. C. Schmidt and K. J. McDonald, compilers. Proceedings—Symposium on whitebark pine ecosystems: Ecology and management of a high-mountain resource. USDA Forest Service, Intermountain Research Station, General Technical Report INT-270, Ogden, Utah.

———. 1995. Defining fire and wilderness objectives: Applying limits of acceptable change. Pages 42–47 *in* J. K. Brown, R. W. Mutch, C. W. Spoon, and R. H. Wakimoto, editors. Proceedings: Symposium on fire in wilderness and park management. USDA Forest Service, Intermountain Research Station, General Technical Report INT-GTR-320, Ogden, Utah.

———. 1996. Ecological manipulation in wilderness—An emerging management dilemma. International Journal of Wilderness 2:15–19.

Cole, D. N., and S. F. McCool. 1997. The limits of acceptable change process: Modifications and clarifications. Pages 61–68 *in* S. F. McCool and D. N. Cole, compilers. Proceedings—Limits of acceptable change and related planning processes: Progress and future direction. USDA Forest Service, Intermountain Research Station, General Technical Report INT-GTR-371, Ogden, Utah.

Cowell, C. M. 1993. Ecological restoration and environmental ethics. Environmental Ethics 15:19–32.

Driver, B. L., P. J. Brown, G. H. Stankey, and T. G. Gregoire. 1987. The ROS planning system: Evolution, basic concepts and research needed. Leisure Sciences 9:201–212.

Frederick/Schneider, Inc. 1994. Results from a nationwide survey on forest management. Prepared for American Forests.

Galliano, S. J., and G. M. Loeffler. 1995. Place assessment: How people define ecosystems. Technical report on file with USDA Forest Service, Interior Columbia Basin Ecosystem Management Project, Walla Walla, Washington.

Gobster, P. H. 1993. The aesthetic experience of sustainable forest ecosystems. Paper presented at the Conference for Sustainable Ecological Systems, Flagstaff, Arizona, July 12–15.

Gunn, A. S. 1991. The restoration of species and natural environments. Environmental Ethics 13:291–310.

Greider, T., and L. Garkovich. 1994. Landscapes: The social construction of nature and the environment. Rural Sociology 57:1–24.

Ittelson, W. H. 1973. Environment and cognition. Seminar Press, New York.

Katz, E. 1991. The ethical significance of human intervention in nature. Restoration and Management Notes 9:90–96.

Krumpe, E., and S. F. McCool. 1997. Role of public involvement in the Limits of Acceptable Change wilderness planning system. Pages 16–20 *in* S. F. McCool and D. N. Cole, compilers. Proceedings—Limits of acceptable change and related planning processes: Progress and future directions. USDA Forest Service, Intermountain Research Station, General Technical Report INT-GTR-371, Ogden, Utah.

Light, A., and E. S. Higgs. 1996. The politics of ecological restoration. Environmental Ethics 18:227–247.

Litton, R. B. Jr. 1972. The aesthetic dimensions of landscape. Pages 262–291 *in* J. V. Krutilla, editor. Natural environments: Studies in theoretical and applied analysis. Johns Hopkins University Press, Baltimore, Maryland.

———. 1974. Visual vulnerability of forest landscapes. Journal of Forestry 72:392–397.

McCool, S. F., and G. H. Stankey. 1986. Visitor attitudes toward wilderness fire management policy—1971–84. USDA Forest Service, Intermountain Research Station, Research Paper INT-357, Ogden, Utah.

McCool, S. F., J. A. Burchfield, and S. Allen. 1997. Social assessment. Pages 1871–2009 *in* T. M. Quigley and S. J. Arbelbide, editors. An assessment of ecosystem components in the Interior Columbia Basin and portions of the Klamath and Great Basins. USDA

Forest Service, Pacific Northwest Research Station, General Technical Report PNW-GTR-408, Volume IV, Portland, Oregon.

Mutch, R. W. 1995. Prescribed fires in wilderness: How successful? Pages 38–41 *in* J. K. Brown, R. W. Mutch, C. W. Spoon, and R. H. Wakimoto, editors. Proceedings: Symposium on fire in wilderness and park management. USDA Forest Service, Intermountain Research Station, General Technical Report INT-GTR-320, Ogden, Utah.

Nash, R. 1982. Wilderness and the American mind, 3d edition. Yale University Press, New Haven, Connecticut.

Pitt, D. G., and E. H. Zube. 1987. Management of natural environments. Pages 1009–1041 *in* D. Stokols and I. Altman, editors. Handbook of environmental psychology, Vol. 2. John Wiley and Sons, New York.

Schroeder, H. W. 1991. The spiritual aspect of nature: A perspective from depth psychology. Pages 25–30 *in* G. A. Vanderstoep, editor. Proceedings of the northeastern recreation research symposium. USDA Forest Service, Northeastern Research Station, General Technical Report NE-160, Saratoga Springs, New York.

Shoemaker, J. H. 1984. Writing quantifiable river recreation management objectives. Pages 249–253 *in* J. S. Popadic, D. I. Butterfield, D. H. Anderson, M. R. Popadic, editors. Proceedings: National river recreation management symposium. Louisiana State University, Baton Rouge.

Stankey, G. H., D. N. Cole, R. C. Lucas, M. E. Peterson, and S. S. Frissell. 1985. The limits of acceptable change (LAC) system for wilderness planning. USDA Forest Service, Intermountain Research Station, General Technical Report INT-178, Ogden, Utah.

Worf, W. 1997. The pathway to an enduring resource of wilderness. Presented at the national wilderness management conference for line officers, June 3. Arthur Carhart National Wilderness Training Center, Missoula, Montana.

Part IV

Restoring Whitebark Pine Communities

The chapters in this final section focus on the restoration and management of whitebark pine communities impacted by fire exclusion, white pine blister rust, and increasing fragmentation. Collectively, the authors provide the strategies and hands-on methodology for maintaining whitebark pine on the landscape, and explore the feasibility of restoration within the context of environmental change and today's social and economic institutions.

Chapter 14, "Landscape Ecology and Isolation: Implications for Conservation of Whitebark Pine," by Penelope Morgan and Michael P. Murray, examines the patterns of occurrence of whitebark pine communities across the landscape and particularly the issue of progressive fragmentation and isolation. The authors explore the regional dynamics of whitebark pine distribution by examining data from the northern Rocky Mountains of the United States and describing variation temporally, spatially, and at different scales. They discuss how spatial variation in fire frequency, size, and severity creates landscape-level differences in communities, and how altered fire regimes and the introduction of blister rust have changed this pattern. Many populations are becoming smaller and more isolated from other populations through time, increasing their risk of extinction. These communities should be a high priority for restoration action.

The chapter by Donna Dekker-Robertson and Leo P. Bruederle, "Management Implications of Genetic Structure," explains why we need to understand how genetic variation is distributed within and among whitebark pine populations. The authors review the evolutionary forces that shape population genetic structure, the methodology used to determine genetic structure in pines, and the general patterns of genetic structure found in conifers and their implications for forest management. Most important, they discuss the population genetic structure of whitebark pine with respect to the development of seed transfer guidelines for the species—that is, guidelines about how far whitebark pine seeds (as seedlings, usually) may be moved from collection locations to planting sites, and how to increase the likelihood of obtaining seeds with blister rust–resistant genotypes in infected whitebark pine stands. Information from Chapter 15 directly pertains to Chapter 16, "Growing Whitebark Pine Seedlings for Restoration," by Karen E. Burr, Aram Eramian, and Kent Eggleston. These authors begin with a review of the guidelines for seed transfer and seed collecting from rust-infected stands, followed by the steps involved in growing healthy seedlings for restoration plantings: Identifying areas with potential cone crops a year in advance, protecting cones from Clark's nutcrackers, collecting and transporting cones, extracting seeds, assessing seed quality, storing seeds, stratifying seeds, germinating seeds, planting germinated seeds, growing seedlings in the greenhouse, and packing seedlings for transportation to planting areas. Whitebark pine seeds require a complex protocol for successful germination and seedling production because of their natural dormancy patterns. The level of methodological detail provided by these authors is invaluable, given that growing seedlings for planting must be an important part of whitebark pine restoration efforts where seed sources have been diminished or lost and where blister rust resistance must be enhanced. The USDA Coeur d'Alene Tree Nursery, where all three authors work, has developed the current procedure for germinating whitebark pine seeds through trial-and-error effort and grows the majority of whitebark pine seedlings that are being planted.

Chapter 17, by Raymond J. Hoff, Dennis E. Ferguson, Geral I. McDonald, and Robert E. Keane, "Strategies for Managing Whitebark Pine in the Presence of White Pine Blister Rust," applies information learned from the intensive effort to breed rust-resistant western white pine (*Pinus monticola*) to the blister rust epidemic in whitebark pine. First, the issues of blister rust susceptibility and genetic resistance to blister rust are examined for both western white pine and whitebark pine. Tests indicate that seedlings grown from surviving whitebark pine trees in stands with high mortality rates from blister rust have good resistance to blister rust, and thus surviving trees have "usable levels of heritable resistance." Next, the authors discuss different aspects of management strategies; they revisit the recently developed seed transfer guidelines for whitebark pine, but explain in detail the justification for very broad seed transfer rules. They urge the establishment of seed banks, the use of prescribed fire to promote natural regeneration, and the planting of rust-resistant seedlings. Finally, they present an integrated management approach to controlling the blister rust infection, which applies epidemiological principles, including monitoring infection rates, together with assessing local "hazard" levels in whitebark pine stands, enhancing rust resistance both naturally and through plantings and prescribed fire, and controlling *Ribes* populations through eradication programs.

Chapter 18, "Restoration Concepts and Techniques," by Robert E. Keane and Stephen F. Arno, presents a detailed, stepwise procedure for identifying whitebark pine communities that are candidates for restoration, prioritizing stands for treatment, devising restoration treatments, implementing the treatments, and monitoring the success of the treatments. Much of the authors' experience comes from work in five research sites in western Montana and eastern Idaho, where restoration efforts are in whitebark pine communities that are successionally advanced and infected by blister rust. Historical patterns of whitebark pine occurrence have been reconstructed for these areas. The treatment options discussed and evaluated include prescribed mixed-severity fire and/or silvicultural cuttings that mimic this effect, including creating small, open patches ("nutcracker openings") as caching areas for nutcrackers; prescribed natural or ignited stand-replacement fire; and possibly planting blister rust–resistant seedlings where seed sources are no longer available. Monitoring treated stands through time, especially through remeasurement of permanent plots, provides feedback on the outcome of different treatment approaches. The general restoration procedures outlined in the chapter are applicable to any fire-dependent ecosystem worldwide.

Hal Salwasser and Dan E. Huff, in Chapter 19, "Social and Environmental Challenges to Restoring White Pine Ecosystems," discuss the barriers to "timely and effective intervention" in the decline of whitebark pine and other white pine communities, and the complex forces that affect forest communities. They begin by describing how traditional forestry values—that is, managing for forest products—have been replaced by the concepts of sustainability and ecosystem management, particularly with respect to biodiversity. Next, they present five principles for ecosystem management, which comprise a reality check on idealistic goals, and discuss ongoing changes in society and our institutions that further challenge ecosystem management. The chapter then focuses on white pine ecosystems and the forces of change in these communities, which will complicate management efforts. The chapter is a sobering reminder that whitebark pine restoration is impossible without societal, institutional, and economic support, and that the goals of restoration must be realistic in a complex, changing world.

Finally, the brief concluding chapter, "Whitebark Pine Restoration: A Model for Wildland Communities," by Stephen F. Arno, Diana F. Tomback, and Robert E. Keane, presents several generalizations extrapolated from the experience with whitebark pine communities:

1. Protected "natural" areas do not avoid the human influences that perturb ecosystems.
2. The problem of fire suppression disrupting natural ecosystems occurs worldwide; many natural communities are changing and disappearing because fire has been eliminated from the landscape.
3. Invasive exotic species are also a worldwide problem, altering community composition and function.
4. Ecological restoration is the only hope for returning some semblance of naturalness to these communities, but even then the success may be limited by social or institutional constraints.
5. The approach used by whitebark pine researchers—collaborative, focused research including simulation modeling to predict the effects of different management actions—is recommended for other disturbed ecosystems as well.

The authors point out that pro-active management is more realistic than attempting to eliminate human influences from communities.

Chapter 14

Landscape Ecology and Isolation: Implications for Conservation of Whitebark Pine

Penelope Morgan and Michael P. Murray

Whitebark pine (*Pinus albicaulis*) is a keystone species of subalpine forest ecosystems, providing watershed protection, scenic value, and food for wildlife, including the Clark's nutcracker (*Nucifraga columbiana*) and black (*Ursus americanus*) and grizzly (*Ursus arctos horribilis*) bears (Arno and Hoff 1989; Chapter 1). Although whitebark pine occurs predominantly in wilderness areas, national forests, and national parks, and exclusively in high-mountain landscapes not usually disturbed by people, whitebark pine has declined dramatically in abundance in parts of its range (Arno 1986; Keane and Arno 1993; Kendall et al. 1996; Chapter 11). In the northern Rocky Mountains of the United States, 42 percent of whitebark pine trees on remeasured plots died in the twenty years from 1971 to 1992 (Keane and Arno 1993). Keane et al. (1994) determined that there was moderate or high whitebark pine mortality across 61 percent of the subalpine forest landscapes in the 600,000-hectare Bob Marshall Wilderness Complex in western Montana (Keane et al. 1994). The combined effects of the exotic disease white pine blister rust (*Cronartium ribicola*) and mountain pine beetle (*Dendroctonus ponderosae*) have created "ghost forests" over extensive areas at high elevations in Idaho (Ciesla and Furniss 1975); these disturbance factors, coupled with eighty years of fire suppression in the northern Rocky Mountains, have placed whitebark pine in serious jeopardy.

Fire maintains whitebark pine dominance in northern Rocky Mountain landscapes, especially in communities where whitebark pine is seral. Fire creates regeneration opportunities for whitebark pine by providing open patches in the landscape. Clark's nutcrackers disperse the seeds of whitebark pine, preferentially caching them in open and burned areas—conditions where whitebark pine thrives (Tomback et al. 1990; Chapter 5). Thanks to the Clark's nutcracker, whitebark pine has a distinct advantage in regenerating large, burned areas relative to conifers that depend on windblown seeds (Tomback et al. 1990; Chapter 5). In the northern Rocky Mountains of the United States, adjacent areas of Canada, and the Intermountain West, fire exclusion in the greater part of the twentieth century has led to increases in subalpine fir (*Abies lasiocarpa*) and Engelmann spruce (*Picea engelmannii*) from advancing succession, except on the harshest sites where these aggressive competitors cannot grow. Even within areas that once supported extensive whitebark pine forests, openings suitable for whitebark regeneration are increasingly fewer, farther apart, and more distant from whitebark pine seed sources that are themselves declining in abundance from the combined effects of blister rust, mountain pine beetle, and fire exclusion. Most of the openings created by fire and other disturbance are rapidly filled with other tree species or with other vegetation (Chapter 9).

The pattern of occurrence of whitebark pine communities across forest landscapes varies greatly over the geographic range of whitebark pine (Figure 14-1). Many whitebark pine populations are in environments isolated at the top of mountain ranges separated by low valleys where whitebark pine cannot grow, such as, those of the Great Basin (Figure 14-1). Elsewhere, such as in the Greater Yellowstone Area, suitable environments are extensive and well connected. Within a mountain range, the topographic, soil, and microclimatic conditions suitable for whitebark pine establishment are scattered and patchy, and often separated by lakes, streams, meadows, scree, talus, erosion faces, bedrock, and other forest types. Meadows or exposed rocks often surround upper subalpine communities of whitebark pine, isolating them from the closest neighboring whitebark pine communities.

Isolated populations of whitebark pine may be highly vulnerable to extinction. The threats to long-term survival of the trees in small populations of whitebark pine are numerous. They include fire, disease, and advancing succession. If small, isolated whitebark pine communities become locally extinct through disturbance, disease, or beetle infestation, the probability that the sites will be recolonized by whitebark pine is lower the farther they are from a seed source.

Whitebark pine forests have become more isolated where they have declined in abundance. In much of the northern Rocky Mountains, the decline in whitebark pine is great (Arno 1986; Keane and Arno 1993); however, the magnitude of the decline varies greatly from one place to another (Keane et al. 1994; Kendall et al. 1996; Smith and Hoffman 1998). In some places, whitebark pine communities have increased in abundance and extent over the last century, as described later in this chapter. The rates of decline, the threats to continued dominance by whitebark pine, and the probability of disturbance, including fire, mountain pine beetle, and blister rust, depend on location within a landscape and the characteristics of the adjacent communities from which disturbances can spread. The degree to which isolated populations are at risk, or lost and less likely to be recolonized, depends not only on the size of the pop-

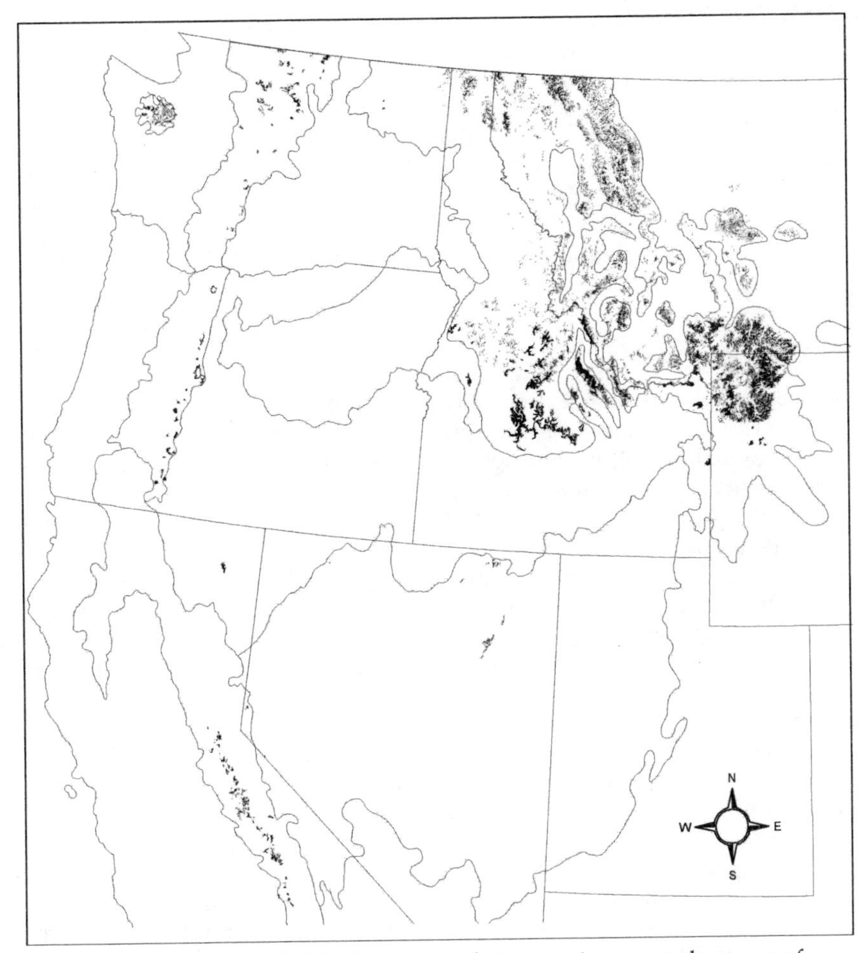

Figure 14-1. Isolation of whitebark pine populations varies across the range of whitebark pine within the western United States. Black areas indicate presence of whitebark pine and solid lines, other than state borders, are ecoregion boundaries. Source: GAP analysis (Scott et al. 1993).

ulation and the distance from adequate seed sources, but also on the biophysical setting, landscape configuration, and the composition of adjacent forests.

Our objectives for this chapter are to describe (1) the landscape ecology of whitebark pine, including landscape dynamics; (2) the degree to which whitebark pine communities are isolated; and (3) the implications of both for conservation and restoration of whitebark pine ecosystems. We argue that if conservation and restoration efforts are to be successful, we must understand the patterns of whitebark pine at local, landscape, and regional scales, and how these patterns have changed through time. Where whitebark pine is becoming more isolated, our conservation efforts should focus on reversing that trend.

Landscape Ecology

Landscape analyses are useful for characterizing ecological conditions, spatial patterns, and trends in disturbance and succession. Comparisons among landscapes and across multiple spatial and temporal scales help us to understand and evaluate change in the relative abundance, structure, and pattern of occurrence of whitebark pine forests (Swetnam et al. 1999). Such understanding is crucial to formulating effective conservation strategies (Holling and Meffe 1996). This also provides a reference for developing and evaluating conservation and restoration priorities and success (Morgan et al. 1994a; Landres et al. 1999).

Spatial Dynamics

Forested landscapes change over time, depending on site conditions (temperature, moisture, and soils, often integrated as potential vegetation type), plant species (growth, mortality, and species life history characteristics), and disturbances (type, frequency, severity, size, spatial pattern, and interactions). These forces interact to create variable and dynamic spatial patterns across landscapes (Krummel et al. 1987).

The abundance and spatial pattern of whitebark pine communities vary greatly from one watershed to the next. This is evident from the comparison of the oldest (circa 1940) and most recent (circa 1990) digitized aerial photographs for 228 randomly selected watersheds, each about 10,000 hectares in size, across the Interior Columbia River Basin (Quigley and Arbelbide 1997; Hessburg et al. 1999). This area encompasses all of Idaho, western Montana, eastern Oregon, eastern Washington, and adjacent parts of Nevada, Utah, and Wyoming. Because field data were not available to provide ground-truth, the accuracy of the maps was not assessed.

Historically, only 70 of the 228 mapped watersheds supported whitebark pine as the dominant cover type on suitable sites (subalpine fir and whitebark pine potential vegetation types). Whitebark pine and alpine larch (*Larix lyallii*) cover types occupied more than 10 percent of the area either historically or currently in only 22 of these 228 mapped watersheds; in 16 of these watersheds, whitebark pine forests are currently less extensive than they were historically. Despite the dramatic decline in the extent of whitebark pine in some areas (Keane and Arno 1993; Chapter 11), there is such high variability from one mapped watershed to another that the relative abundance and landscape pattern of whitebark pine communities have not changed significantly in the last fifty years when evaluated at this scale (Table 14-1). Thus, the average changes in areal extent (percent of the watershed, PCT), the size of the largest patch of whitebark pine (largest patch index, LPI), patch density (PD), mean patch size (MPS), or distance between patches (MNN, mean nearest neighbor) were not statistically different (Table 14-1). To understand these overall trends for all whitebark pine regardless of structure, we looked at the data for old and young whitebark pine forests (these data are not shown in Table 14-1). Neither the extent nor pattern of old whitebark pine forests has changed significantly ($P>0.05$). However, the mean size of young whitebark pine communities has more than doubled, from 35 to 71 hectares, and the largest patch index has

Table 14-1. Changes in spatial patterns of whitebark pine communities dominated by whitebark pine (*Pinus albicaulis*) or alpine larch (*Larix lyallii*) on subalpine fir (*Abies lasiocarpa*), and whitebark pine potential vegetation types for watersheds in the Interior Columbia River Basin. Statistical comparisons (P-values) are based on a paired t-test, n = 84, for PCT,[1] and on t-tests with unequal variances for LPI,[2] PD,[3] MPS,[4] and MNN[5] (n = 76 historical and 70 current). Watersheds are about 10,000 hectares in size. Data provided by Interior Columbia River Basin Ecosystem Management Project (Quigley and Arbelbide 1997; Hessburg et al. 1999).

	PCT	LPI	PD	MPS	MNN
Historical	7.6 ± 12.6	4.4 ± 12.0	0.1 ± 0.1	166 ± 779	740 ± 1,393
Current	7.0 ± 8.3	3.4 ± 4.4	0.1 ± 0.1	90 ± 114	1,073 ± 1,730
P-value	0.334	0.252	0.274	0.209	0.117

[1] PCT = Percent of the landscape area.
[2]LPI = Largest patch index, a measure of the size of the largest patch.
[3]PD = Patch density (Number/km^2).
[4]MPS = Mean patch size (km^2).
[5]MNN = Mean nearest neighbor distance (km).

also doubled from 0.7 to 1.4 hectares (P = 0.006), even though the mean percent of landscapes occupied did not change significantly. Young whitebark pine forests occupied about 2 percent of the landscapes both historically and currently. Thus, in the watersheds mapped in this sample, there are currently more large patches of young forests dominated by whitebark pine forests than there were historically. Many of these young whitebark pine communities probably established following extensive fires that occurred in the Interior Columbia River Basin in the late 1800s and early 1900s.

Whitebark pine abundance declined dramatically (by more than 10 percent of the historical extent) in 41 of the 70 mapped watersheds in which whitebark pine occurred historically. For instance, in the five watersheds of the Blackfoot River in Montana, the areal extent of forests dominated by whitebark pine declined significantly from 4.8 ± 3.7 percent in 1935 to 1.1 ± 1.1 percent of the watershed area in 1994 (P = 0.032, paired t-test, n = 5). Patch density also decreased significantly from 0.11 ± 0.11 historically to 0.05 ± 0.03 patches per square kilometer (P = 0.012, one-tailed t-test, unequal variance, n = 5). However, mean patch size was highly variable (43.5 ± 24.1 hectares historically and 29.8 ± 3.4 hectares currently) and did not change significantly (P = 0.152, one-tailed t-test, unequal variance, n = 5). Thus, in these watersheds, there are fewer communities dominated by whitebark pine, and fewer large ones.

Elsewhere, whitebark pine increased in areal extent. In 10 of the 76 watersheds that currently support whitebark pine communities in the Interior Columbia Basin, whitebark pine was abundant (>10 percent of the watershed area) either historically or currently. In these ten watersheds, the average extent of whitebark pine communities increased significantly from 12.5 ± 8.1 to 20.7 ± 8.8 percent of the watershed area ($P<0.0000$, paired t-test and $n = 10$ for this and all other tests in this paragraph). The average patch size increased significantly from 78.1 ± 6.9 hectares historically to 213.4 ± 182.5 hectares currently ($P = 0.002$), as did the size of the largest patch (LPI was 4.7 ± 3.9 hectares historically and is 10.9 ± 6.3 hectares currently, $P = 0.010$). Patch density did not change significantly, suggesting that the patches dominated by whitebark pine simply got bigger. Distances between patches were so variable that the differences from 380.8 ± 232.0 meters historically to 262.4 ± 131.6 meters currently was not statistically significant ($P = 0.103$). In these and many other areas, the number and size of trees, including whitebark pine, have increased, as forests regenerated following extensive fires in the late 1800s and early 1900s and as fire suppression became increasingly effective through the 1900s. For instance, in the West Big Hole mountain range of Idaho and Montana, Murray et al. (1998) found that during the last 120 years, the area burned had decreased by 87 percent relative to the previous 120 years, and that the basal area of whitebark pine has increased by 12 percent.

Overall, then, the spatial pattern of whitebark pine communities is highly variable from one place to the next and scale-dependent. It is possible that the data are in error or at a scale that is too coarse. However, the analysis suggests that we exercise caution in generalizing trends from one place to another and that field assessments be conducted to evaluate the structure, composition, and trends in local whitebark pine communities. In more than half of the watersheds sampled across the Interior Columbia River Basin, whitebark pine communities are much less abundant now than they were 50 years ago. In those watersheds where whitebark pine has increased in abundance, this does not imply that whitebark pine is thriving. Where whitebark pine is infected with blister rust, future declines are very likely, unless disturbances create many regeneration opportunities.

Fire Ecology at the Landscape Scale

Fire is a major recurrent disturbance process in the northern and central Rocky Mountains and adjacent Canadian Rocky Mountains. There, fire has shaped subalpine landscapes since the last Pleistocene glaciation (Mehringer et al. 1977; Tande 1979; Agee 1993; Morgan et al. 1994b). Because fire is key to the structure, composition, and function of whitebark pine communities in these regions, changes in ecosystems in this century are a reflection of the change in fire patterns.

Fire regime refers to the "nature of fires occurring over an extended period of time" (Brown 1995). Fire regimes in upper subalpine settings are complex, because the frequency, severity, and size of fires vary greatly (Arno and Petersen 1983; Despain 1991; Morgan and Bunting 1990; Barrett 1994). For whitebark pine forests, the mean number of years between fires historically varied from 29 to 144 years for mixed-severity fires, and from 80 to 500 years for stand-replacing fires (Arno 1986; Morgan and Bunting 1990; Morgan et al. 1994b; Keane et al. 1994; Murray et al. 1998; Chapter 4). These numbers are an approximation of the average return interval of fires at a particular point or within a small area (typically less than 4 hectares).

Fire frequency at the landscape scale is expressed as fire rotation or area frequency. Fire rotation is the number of years between fires divided by the proportion of the area burned, and thus expresses the number of years it would take to burn an area the size of the landscape (Agee 1993). Area frequency is the number of fires per 100 square kilometers per year (Agee 1993). Both require data on fire size as well as frequency. Fire rotations and area frequency for whitebark pine forests are available for only a few locations within the northern Rocky Mountains that support whitebark pine (Table 14-2 and Figure 14-2). Fire rotations have lengthened considerably (from seven to seventeen times) with fire exclusion in upper subalpine forests (Rollins et al. 2000). These limited data suggest that the subalpine forest landscapes in which whitebark pine communities are found are burning far less often than they once did.

The ecological effects of fires depend not only on fire frequency but also on fire severity, size, and spatial pattern. Most historical fires in whitebark pine forests were patchy and of mixed severity (Morgan and Bunting 1990; Morgan et al. 1996). For those sites supporting whitebark pine communities, infrequent (76 to 150 years between fire events) mixed-severity fire regimes predominated historically (circa 1900) on the relatively cool-dry, warm-moist, and warm-dry biophysical settings that could support the species. Very infrequent (151 to 300 years) stand-replacing fire regimes historically predominated on cool-moist sites supporting whitebark pine communities (Morgan et al. 1996). Stand-replacing fires have been defined as those that kill more than 70 percent of the extant tree basal area, and mixed-severity fires as those that kill between 20 percent and 70 percent of the tree basal area (Agee 1993; Morgan et al. 1996).

Where stand-replacing fires are the norm, burned patches tend to be relatively large—usually hundreds of hectares in size (Agee 1998). The rela-

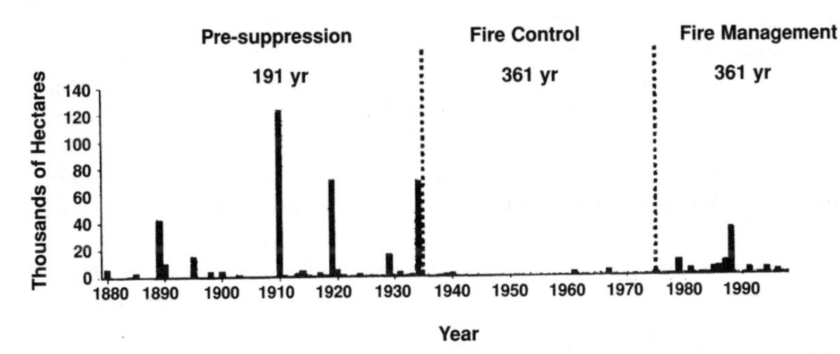

Figure 14-2. Area burned by fires, 1880 to 1996, in the Selway-Bitterroot Wilderness, Idaho and Montana (Rollins et al. 2000). A total of 545,321 hectares burned within the 547,370-hectare wilderness area within those 117 years. Dotted lines separate three time periods. We report fire rotations (years, printed in large numbers at the top of the graph) calculated by Rollins et al. (2000) for each time period for the upper subalpine forests that support whitebark pine communities, but not for the wilderness area as a whole.

Table 14-2. Landscape-level fire history of whitebark pine forests of the West Big Hole Range in Idaho and Montana, an area of 75,170 hectares (Murray et al. 1998).

	1754–1873	1874–1993
Mean interval between fires (yr)	51–119	52–112
Fire frequency within the study area (number of fires)	2	9
Area frequency (Number of fires per 100 km^2 per year)	0.167	0.047
Fire rotation (Years to burn area equal in size to study area)	184	1364

tively few and distinct tree age classes, especially of whitebark and lodgepole pine (*Pinus contorta*), reflect the long interval between fires and the paucity of residual live trees in burned areas. Although many different whitebark pine communities established from the same fire event, they likely vary greatly in structure and composition depending on site conditions, proximity to seed sources, and fire effects. Between 1979 and 1990, 70 percent of the fires occurring in whitebark pine forests of the Selway-Bitterroot Wilderness area in northern Idaho and Montana were stand-replacing (Brown et al. 1994). Of these stand-replacing fires, 18 percent were crown fires, whereas most (52 percent) were lethal surface fires. To the degree that the relative proportion of subalpine fir has increased and downed woody fuel has accumulated, these fire effects are probably more severe than historically occurred (Morgan and Bunting 1990; Morgan et al. 1994b). Where such conditions predominate over extensive areas, landscapes are vulnerable to large, stand-replacing fires.

Mixed-severity fire regimes create complex mosaics, with patches of diverse structure and composition. Patches are typically 1 to 300 hectares in size, and highly variable with many small and medium-sized patches (Agee 1998). Edges are many and varied between the forest communities of different structure and composition that result from different residual densities of trees. This reflects the mixture of fire effects from nonlethal to stand-replacing often occurring from the same fire, but certainly also among the different fires occurring through time. Low-severity fires maintain local pockets of mature whitebark pine that are more likely to survive subsequent fires and provide a seed source for regenerating burned patches in the surrounding landscape (Morgan and Bunting 1990). However, both mixed-severity and stand-replacing fires are patchy, and the sequence of successional development follows multiple pathways (Chapter 9).

Changes in the spatial pattern of whitebark pine communities through time are consistent with changes documented in fire frequency, size, extent, and rotation. The current spatial pattern of whitebark pine communities reflects past fire occurrence, and provides information needed for reintroducing a semblance of the native fire regime (Chapter 18).

Fire Size

Both small and large fires are important to whitebark pine regeneration, especially where whitebark pine is seral to subalpine fir and Engelmann spruce. Fire creates regeneration opportunities, recycles biomass, and maintains whitebark pine on the landscape in the face of advancing succession on sites where it is seral.

Both historically and currently, the majority of fires are small (Morgan and Bunting 1990; Morgan et al. 1994b). For instance, during 12 years in the Selway-Bitterroot Wilderness area, 84 percent (340 of 403 fires recorded) of all fires occurring at upper elevations were smaller than 4 hectares in size (Brown et al. 1994). When large fires occurred historically, they often lasted for weeks or months and spread across the landscape through multiple forested communities including whitebark pine. Few large fires are restricted to upper subalpine settings.

Fires burn extensive areas of whitebark pine communities only in years of extreme drought (Rollins et al. 2000). For instance, in the 547,370-hectare Selway-Bitterroot Wilderness area, 6 of the 117 years between 1880 and 1996 account for 72 percent of the area burned: 1889, 1910, 1919, 1929, 1934, and 1988 (Figure 14-2). In most years (111 of 117), less than 1 percent of the wilderness burned. Only 29 percent of the 114,064 hectares of upper subalpine forests that support whitebark pine communities burned in the 117 years between 1880 and 1996. Of the 30 years in which more than 1,000 hectares burned in the Selway-Bitterroot Wilderness area, 80 percent were drier than average (Rollins et al. 2000).

Fire sizes among isolated whitebark pine communities may be restricted due to the small patches of fuel and the acute variations in topography, microclimate, and fuels. In contrast, fires burning in topography of moderate to low relief are often large, because the forests are contiguous, and particularly when they are driven by winds in extreme drought years. For example, in Yellowstone National Park one of the 1988 fires burned nearly 215,000 hectares (Rothermel et al. 1994).

Fire Exclusion

Euro-American settlement and subsequent development of the western landscapes changed fire regimes dramatically through fire suppression, settlement in mountain valleys, grazing, and road construction (Agee 1993). Today, even the most remote populations of whitebark pine in wilderness areas are affected by fire exclusion (Chapters 9 and 11). Although many modern fires have the potential of spreading into whitebark pine forests, they are usually extinguished at lower elevations. Subalpine fuels are usually too moist to support extensive fires, except for late summer during unusually dry years, although some fires, mostly small, occur nearly every year. During dry years, the lower elevations are usually even drier, and managers are unwilling to risk fire spread.

The consequences of fire suppression are clearly evident in statistics for area burned per year in the Selway-Bitterroot Wilderness area (Figure 14-2), although climate and weather also influenced fire patterns (Rollins et al. 2000). Most (87 percent) of the total area burned from 1880 to 1994 did so prior to 1935 (Rollins et al. 2000). Whereas these data do not represent all fires, and fires were not consistently detected and mapped through time, the largest fires are included, making this a reasonable representation of area burned through time (Rollins et al. 2000). Brown et al. (1994) estimated that fire intervals for whitebark pine forests have extended to more than twice their natural return time despite a relatively liberal natural burn policy during the last 25 years.

Fire exclusion has had greater consequences for subalpine forests, especially those supporting whitebark pine, than for the wilderness as a whole (Brown et al. 1994). In the upper subalpine forests that support whitebark pine communities, fire rotations during 1935 to 1975, an era of fire control, were more than seventeen times that experienced during the early part of this century under relatively natural (but also drier) conditions (1880 to 1934), and more than nine times that experienced since 1975 (Figure 14-2) (Rollins et al. 2000). Fire rotations during the latter period of "fire management" (1975–1996), when some lightning-ignited fires were allowed to burn naturally, were still almost double that experienced before 1935 (Figure 14-2) (Rollins et al. 2000).

For whitebark pine communities occurring in an isolated mountain range, the West Big Hole in Montana, the effects of fire exclusion are even more extreme (Table 14-2). There, the area burned was reduced by 87 percent in the last 120 years relative to the previous 120 years (Murray et al. 1998). This probably reflects the proportionately greater reduction in fire frequency in the surrounding settled landscapes (Murray et al. 1998), but it could also be due in part to climate differences.

Natural resource management agencies are more successful at suppressing fires when they are small, so we have fewer small patches created now than historically. Small fires increase landscape heterogeneity by increasing the number of small areas dominated by young forest, open parklike groves of relatively pure whitebark pine, or other vegetation. Large fires also increase landscape heterogeneity, but at a different scale. Where fires have been excluded from subalpine landscapes, landscape homogeneity increases as advancing succession leads to more and larger areas dominated by late-seral composition and structure (Keane et al. 1996, 1999; Chapter 9). Romme (1982), Baker (1992, 1993), and Keane (Chapter 9) documented simplified shape, diversity, and richness of forested patches under the influence of fire suppression.

Thus, in the subalpine forest landscapes of the northern Rocky Mountains, whitebark pine communities have burned less often recently compared to the late 1800s and early 1900s (e.g., Murray et al. 1998), especially in the last 60 years (Brown et al. 1994; Rollins et al. 2000). Managers are inclined to allow some natural fires to burn in large wilderness areas but not usually in small ones. Fires burning in small wilderness areas are more likely to threaten adjacent private lands. Although fires ignited by lightning have been allowed to burn in some wilderness areas under prescribed conditions, in those wilderness areas without approved wildland fire management plans, all fires have been suppressed. The recently revised policy for fire management on federal lands, however, allows more flexible decisions, particularly when fires benefit resources (USDA 1998).

Where fire frequency declined over the last century, advancing succession has created more continuous conditions conducive to the development and spread of crown fires. In many landscapes, the forests surrounding whitebark pine communities have changed under fire exclusion; the accumulated fuels, multistrata forest structure, and composition are now more likely to elevate fires to stand-replacing intensities (Hessburg et al. 1999). In such landscapes fires are less likely to occur, but when they occur, fires are more likely to be extensive and severe.

Landscape Dynamics

Subalpine forest landscapes change with succession and disturbance. Disturbances in addition to fire alter the abundance and pattern of whitebark pine communities in high-mountain landscapes. The current composition and diversity of subalpine forest landscapes is a legacy of past disturbances, including fires, bark beetles, and blister rust, as well as topographic and site differences.

On sites that support subalpine fir, bark beetles and blister rust accelerate succession by killing whitebark pine more often than competing conifers (Keane et al. 1990; Chapter 9). Mountain pine beetles cause infrequent (Perkins and Swetnam 1996) but high mortality of whitebark pine. Mountain pine beetle infestations killed many whitebark pine trees in the Selway-Bitterroot Wilderness in the late 1870s, 1930s, and late 1980s (Kipfmueller et al. 1999). Where fires have been suppressed, old lodgepole pine communities have increased in extent (Kendall and Arno 1990) and may attract and harbor large populations of beetles that subsequently spread to adjacent whitebark pine communities (McGregor and Cole 1985).

Landscape dynamics have changed greatly with fire exclusion and the introduction of blister rust. Blister rust mortality is increasing in intensity within whitebark pine communities and spreading across the landscapes in which whitebark pine communities are found (Keane et al. 1994; Chapter 11). Because blister rust reduces whitebark cone and seed production (Chapter 6) long before it kills trees, the fungal disease has greatly reduced the potential for whitebark pine to resume its historical importance in the landscape, and has altered the function and dynamics of many subalpine landscapes (Keane et al. 1990, 1994, 1999).

Human-induced climate change may threaten whitebark pine communities as well. Subalpine forests have changed in response to climate variability throughout the Holocene (Baker 1990). Whitebark pine has been present in the Rocky Mountains for more than 10,000 years. Although it has declined before (Baker 1990), the current decline may be more severe than those of the past. Bartlein et al. (1997) predicted that the range of high-elevation species, including whitebark pine, will shrink as global temperature increases in the Greater Yellowstone Area. In other words, whitebark pine could be lost from Yellowstone National Park if the area experiences the warmer temperatures and drier summers predicted with a doubling of the carbon dioxide content of the atmosphere (Bartlein et al. 1997).

The effects of future climate change will likely be mediated through fire (Baker 1995). Climate is an important determinant of fire occurrence at regional and subregional scales (Swetnam and Betancourt 1998), and lightning fires are predicted to occur more frequently if average temperature increases (Price and Rind 1994). Furthermore, predicted warmer temperatures due to increasing greenhouse gas concentrations may increase the frequency, extent, and severity of fires in subalpine forests (Ryan 1991; Price and Rind 1994; Keane et al. 1999). Whitebark pine communities in landscapes that have experienced relatively few fires recently and those already damaged by blister rust and under competitive pressure of advancing succession are likely to be vulnerable, as are those whitebark pine communities growing on harsh sites close to timberline.

Fire exclusion, blister rust, and mountain pine beetle infestations in synergism have led to more homogeneous subalpine forest landscapes. Landscape dynamics have changed substantially from historical patterns (Keane et al. 1990, 1994), and will be further altered by climate change. Future patterns of occurrence and intensities of disturbances, including fire, bark beetles, and blister rust, will determine whether or in what form whitebark pine communities will persist on the landscape.

Isolation

One of the characteristics of the landscapes in which whitebark pine communities are found is their isolation. One of the most isolated metapopulations (i.e., groups of small populations) of whitebark pine occurs on the Pine Forest Range of northeastern Nevada, roughly 130 kilometers from the nearest metapopulation in the Warner Range in northeastern California. However, most whitebark pine forests are much less isolated than this. Whitebark pine communities typically occur within 10 kilometers of each other, except in the northern Basin and Range ecoregion of northern Nevada and south-central Oregon, where the mean distance between whitebark pine communities is 14 kilometers (Table 14-3). Within individual landscapes where whitebark pine communities have declined in abundance (e.g., Table 14-1), the remaining

Table 14-3. Isolation of whitebark pine communities within ecoregions. See text for explanation.

Ecoregion	Nearest Neighbor Distance (Mean ± standard deviation)	Mean Proximity Index (22 km)
Blue Mountains	2.3 ± 3.7 km	0.6
Cascades[1]	2.8 ± 5.6 km	175.9
Middle Rocky Mountains	0.3 ± 0.3 km	1,685.0
Northern Basin and Range	14.1 ± 58.1 km	95.8
Northern Rocky Mountains	0.4 ± 7.0 km	57.9
Olympic Range	0.4 ± 0.5 km	113.3
Sierra Nevada	1.8 ± 4.9 km	92.6

[1]Olympic Range analyzed separately.

whitebark pine communities are smaller and more distant from one another. Where whitebark pine communities are isolated and becoming more so, those communities are at risk of extinction.

In the following section, we discuss the theoretical effects of isolation. We follow with a documented example of extinction of isolated montane and subalpine pines (including but not limited to whitebark pine) in the Great Basin, and then evaluate the degree of isolation of whitebark pine in six different ecoregions in the United States. These analyses emphasize isolation at regional scales, but we also discuss the implications of isolation at the landscape scale, relating this back to the spatial pattern of whitebark pine communities within landscapes.

Theoretical Implications of Isolation

There are three ways, in theory, that isolation affects whitebark pine: increased vulnerability to disturbance, decreased regeneration success, and decreased genetic diversity (MacArthur and Wilson 1967; Hanski and Gilpin 1997). Apparently, whitebark pine communities are not sufficiently isolated to threaten their genetic diversity (Chapter 8), but the other factors could lead to the extinction of isolated whitebark pine communities.

Genetic isolation occurs when the distance between whitebark pine communities becomes too great for seed or pollen exchange. Metapopulations are groups of closely aggregated populations with intermittent seed or pollen exchange, resulting in gene flow (Hanski and Gilpin 1997). If distances are too great between the component populations, then no new genetic material is imported, and there is no chance of new genes being incorporated. Genetic isolation can lead to inbreeding and genetic drift with accompanied lower fitness, thereby resulting in gradual extinction (Brussard 1990). Brussard (1990) recognized the potential genetic consequences of isolation for whitebark pine, but subsequent work has shown that pollen and even seed exchange may occur over reasonably long distances (see Chapters 8 and 15). Bruederle et al. (1998) and Jorgensen and Hamrick (1997), for example, reported new evidence that the majority of the observed variation in genetic diversity in whitebark pine was due to differences among individuals within populations. The populations Bruederle et al. (1998) studied in the Greater Yellowstone Area are not as isolated as many elsewhere within the range of whitebark pine (Figure 14-1). However, their data confirm that pollen exchange and perhaps seed dispersal by Clark's nutcracker limits many of the genetic risks associated with isolation for most populations (Bruederle et al. 1998; Chapter 8). But extremely small, isolated populations may conceivably experience genetic risks, including inbreeding and drift (Chapter 12).

Clark's nutcrackers can disperse seed over distances of 10 kilometers or more (Vander Wall and Balda 1977; Tomback 1978). However, because birds are less likely to cache seeds in areas that are far from seed sources (Tomback et al. 1990), those openings suitable for whitebark pine establishment that are far from cone-producing trees are less likely to be populated by young whitebark pine. Thus, where landscapes have become more continuously forested and dominated by late-seral forests of subalpine fir and Engelmann spruce, whitebark pine is less likely to replace whitebark pine communities that have been lost to disturbance.

Smaller whitebark pine communities are more vulnerable to disturbance than are large ones, at least in theory (Nilsson and Grelsson 1995). When relatively few individuals occur in a small area, they may all be killed when disturbances such as fire, disease, or insects spread from nearby forests. Because small whitebark pine communities have a high ratio of edge to core area, disturbance spreading from adjacent forests can permeate them entirely. Local extinction may also result when trees die before new recruits are established. Perhaps the greatest problem for a small, isolated whitebark population is the damage caused by blister rust. The combined effects of blister rust reducing seed production and fewer fire disturbances creating openings for regeneration are particularly threatening to isolated whitebark pine communities. If they are also adversely affected by climate change, the outlook is bleak without implementation of a restoration effort.

Local Extinction of Isolated Pines in the Great Basin

We do not know whether any isolated whitebark pine populations have gone extinct, but local extinction of other montane pines has evidently occurred on the smaller mountain islands in the Great Basin (Wells 1983; Thompson 1990). Paleoecological evidence indicates that during the late Pleistocene, Great Basin bristlecone pine (*Pinus longaeva*), limber pine (*P. flexilis*), and other montane conifers in the region occurred at lower elevations than they do today (Wells 1983; Thompson 1984). A general climatic shift to warmer and drier conditions during the Holocene (from circa 10,000 years ago to present) (Thompson 1984) probably restricted conifer ranges to the smaller expanses of higher-elevation habitat (Wells 1983; Spaulding 1990). According to Wells (1983) and Thompson (1990), this could have led to the local extinction of pines on smaller mountain islands, as predicted by the theory of island biogeography (MacArthur and Wilson 1967). At least seven Great Basin mountain ranges have lost their pine populations (Wells 1983; Thompson 1984, 1990; Wigand and Nowak 1992; Charlet 1995, 1996).

If whitebark pine communities isolated on small islands of habitat are more vulnerable to extinction, we should find whitebark pine more often on large than on small mountaintops. We examined the current distribution of whitebark, bristlecone, and limber pines in the northern Basin and Range ecoregion (Omernik 1987). Indeed, the modern-day distribution of these three pines at high elevations is significantly correlated with the areal extent of individual mountain islands ($P<0.000$, Mann-Whitney U-test). For our analysis, we combined digital elevation data and vegetation maps (Scott et al. 1993) with more detailed field data (Charlet 1995) in a geographic information system (GIS). We defined a "mountain island" as any area greater than 10 hectares above 2,300 meters elevation. Of the 982 smallest mountain islands, only 10 (1 percent) support pine populations. Conversely, of the 48 largest mountain islands, 32 (67 percent) support pine populations. Mountain islands with these three pines averaged 34,665 hectares in size, whereas islands with no pines averaged 616 hectares in size. Although climate, available habitat, bedrock or soil type, prior occurrence of the pines, and interspecific competition are also important, these data suggest that small, isolated pine populations may be more vulnerable to extinction. It is of course possible that some of these islands did not previously support pines.

From the historical changes in pine distribution in the Great Basin, we can infer that the extreme small size and isolation of some whitebark pine communities currently and in the future make them vulnerable to extinction.

Isolation Rangewide

To evaluate the degree of isolation of whitebark pine communities throughout the range of the species in the western United States, we performed a coarse-scale assessment using the mapped distribution of the whitebark pine cover type (Scott et al. 1993). To compare among regions, we partitioned the U.S. range of whitebark pine by Omernik's (1987) ecoregions (Figure 14-1). Using FRAGSTATS (McGarigal and Marks 1995), we calculated the mean and standard deviation of nearest neighbor distances, which measures the shortest distance from the edge of one whitebark pine community to another. The smallest patches mapped were 4 hectares in area. We also used the proximity index (Gustafson and Parker 1992) as another measure of isolation within a given distance, in this case 22 kilometers, the greatest known dispersal distance for Clark's nutcracker (Vander Wall and Balda 1977). A low proximity index value indicates that whitebark pine communities are isolated, because neighboring communities are small, distant, or both.

Most whitebark pine forest communities are found within an average of less than 14 kilometers of the nearest neighboring whitebark pine community (Table 14-3). In the Blue Mountains and northern Rocky Mountain ecoregion of the United States, there is a low proximity index not because the communities are distant, but because they are small. Specifically, even though patches in the Blue Mountains are within an average of 2.3 ± 3.7 kilometers of one another, the very low proximity index is a reflection of the very small size of the average whitebark pine community in this region (Table 14-3). It is likely that the currently low proximity index of whitebark pine communities in the northern Rocky Mountains of the United States was higher historically (Keane and Arno 1993; Chapter 11).

In summary, many whitebark pine communities are isolated, and although there is little evidence that they are genetically at risk, isolated whitebark pine communities are more vulnerable to disturbance. When whitebark pine trees die in isolated whitebark pine communities, they are less likely to be replaced by regenerating whitebark pine.

Implications for Conservation and Restoration

In the northwestern United States, where whitebark pine is an important seral forest component and blister rust damage is the greatest (Chapter 11), we must rethink our current management strategy. We are not managing subalpine forest landscapes in a way that allows nature to create refugia for whitebark pine—those burned areas where whitebark pine dominates in the early stages of succession. Refugia are places (the location of which will vary with time) where species persist for longer than in the surrounding landscape. Where whitebark pine is an early successional forest component, refugia for whitebark pine *depend* on fire, in contrast to the common belief that refugia are areas not burned. Within a landscape and through time, refugia are more likely in particular topographic or physiographic locations where the disturbance regimes, rates of successional change, and environmental conditions are favorable (Camp et al. 1997).

For whitebark pine, refugia are regeneration sites usually created by fires near mature, seed-producing communities of whitebark pine. With fewer fires, sites suitable for regeneration are fewer and increasingly far from seed sources. At the same time, successional replacement, blister rust, and bark beetles contribute to the continued decline in whitebark pine seed sources. More fires across the landscape not only result in more whitebark pine regeneration but also expand the scale of mass selection for blister rust resistance (Chapter 17).

How much whitebark pine will remain on landscapes and where it will be found will depend upon where we choose to take action—allowing more fires to burn. Unfortunately, even if restoration programs grow substantially, efforts will remain dwarfed in comparison to the need. Equally important is where we choose not to act. Ironically, the fire exclusion policy is practically universal. Whether or not we choose active management, the combination of successional advancement, disturbance, and climate will continue to modify whitebark pine forests.

Prioritizing areas for action is crucial to a long-term conservation strategy. Human ingenuity and thoughtful analysis will help identify those landscapes and the sites within them where whitebark pine can best be maintained. The costs of inaction are great—the local extinction of whitebark pine, particularly for blister rust–infected stands. Keane and Arno (Chapter 18, this volume) suggest prioritizing locations for conservation and restoration based on successional status, severity of blister rust infection, the presence of a seed source within 10 kilometers, whether the stand is in grizzly bear habitat, and road or trail access.

Prioritizing restoration efforts should also consider the degree of isolation and the dynamics of the landscapes in which whitebark pine is found. The function and ecological integrity of whitebark pine communities depend on the surrounding landscapes. High-priority landscapes would be those where the relative abundance and spatial pattern of whitebark pine–dominated communities substantially depart from patterns expected under native fire regime and within the ecoregion. Where whitebark pine communities are becoming more isolated through human action (or inaction), those trends should be at least halted and preferably reversed. The size distribution of whitebark pine communities, their structure and species composition, and their distance from other whitebark pine communities are all measures of landscape pattern that can be determined from historical and current aerial photographs (Hessburg et al. 1999). To be effective, restoration and conservation must address the species both across landscapes and within individual forest communities. The great variability in the landscape pattern of whitebark pine communities requires that we must be careful not to apply a "one size fits all" approach to management. Restoration efforts must account for the variable degree of decline, and the differences in past, current, and probable future extent, spatial pattern, disturbance regimes, and climate in whitebark pine–dominated landscapes.

Whitebark pine has declined more in the northern Rocky Mountains of the United States than elsewhere in its range, but only in some watersheds. The greatest loss has occurred on productive sites where whitebark pine is seral to subalpine fir and Engelmann spruce, and where blister rust infection rates are high. These sites are the most important for many wildlife species. Reintroducing the native fire regime at the landscape scale is crucial if whitebark pine communities are to be a part of the legacy we leave for future generations.

Acknowledgments

We thank Calvin Farris and Mike Scott of the Cooperative Fish and Wildlife Research Unit, University of Idaho, for support for the Great Basin landscape analysis. Paul Hessburg and Anne Black kindly made data available from historical and current aerial photograph comparisons. Miles Hemstrom, Steve Arno, Bob Keane, Diana Tomback, Steve Bunting, Matt Rollins, and Kurt Kipfmueller provided helpful reviews that greatly improved the manuscript. This material is based on work supported by the National Science Foundation under Grant No. SBR-9619410, the USDA Forest Service under RJVA INT-94913, and the Idaho Forest, Wildlife and Range Experiment Station.

Literature Cited

Agee, J. K. 1993. Fire ecology of Pacific Northwest forests. Island Press, Washington, D.C.

———. 1998. The landscape ecology of western forest fire regimes. Northwest Science 72:4–34.

Arno, S. F. 1986. Whitebark pine cone crops: A diminishing source of wildlife food? Western Journal of Applied Forestry 1:92–94.

Arno, S. F., and R. J. Hoff. 1989. Silvics of whitebark pine (*Pinus albicaulis*). USDA Forest Service, Intermountain Research Station, General Technical Report INT-253, Ogden, Utah.

Arno, S. F., and T. D. Petersen. 1983. Variation in estimates of fire intervals: A closer look at fire history on the Bitterroot National Forest. USDA Forest Service, Intermountain Forest and Range Experiment Station, General Technical Report INT-301, Ogden, Utah.

Baker, R. G. 1990. Late quaternary history of whitebark pine in the Rocky Mountains. Pages 40–48 *in* W. C. Schmidt and K. J. McDonald, compilers. Proceedings—Symposium on whitebark pine ecosystems: Ecology and management of a high-mountain resource. USDA Forest Service, Intermountain Research Station, General Technical Report INT-270, Ogden, Utah.

Baker, W. L. 1992. Effects of settlement and fire suppression on landscape structure. Ecology 73:1897–1887.

———. 1993. Spatially heterogeneous multi-scale response of landscapes to fire suppression. Oikos 66:66–71.

———. 1995. Long-term response of disturbance landscapes to human intervention and global change. Landscape Ecology 10:143–159.

Barrett, S. W. 1994. Fire regimes on andesitic mountain terrain in northeastern Yellowstone National Park, Wyoming. International Journal of Wildland Fire 4:65–76.

Bartlein, P. J., C. Whitlock, and S. L. Shafer. 1997. Future climate in the Yellowstone National Park region and its potential impact on vegetation. Conservation Biology 11:782–792.

Brown, J. K. 1995. Fire regimes and their relevance to ecosystem management. Pages 171–178 *in* Proceedings of the Society of American Foresters Annual Meeting. Society of American Foresters, Bethesda, Maryland.

Brown, J. K., S. F. Arno, S. W. Barrett, and J. P. Menakis. 1994. Comparing the prescribed natural fire program with presettlement fires in the Selway-Bitterroot wilderness. International Journal of Wildland Fire 4:157–168.

Bruederle, L. P., D. F. Tomback, K. K. Kelly, and R. C. Hardwick. 1998. Population genetic structure in a bird-dispersed pine, *Pinus albicaulis* (Pinaceae). Canadian Journal of Botany 76:83–90.

Brussard, P. 1990. Pages 315–318 *in* W. C. Schmidt and K. J. McDonald, compilers. Proceedings—Symposium on whitebark pine ecosystems: Ecology and management of a high-mountain resource. USDA Forest Service, Intermountain Research Station, General Technical Report INT-270, Ogden, Utah.

Camp, A., C. Oliver, P. Hessburg, and R. Everett. 1997. Predicting late-successional fire refugia pre-dating European settlement in the Wenatchee Mountains. Forest Ecology and Management 95:63–77.

Charlet, D. A. 1995. Great Basin montane and subalpine conifer diversity: Dispersal or extinction pattern? Ph.D. dissertation, University of Nevada, Reno.

———. 1996. Atlas of Nevada conifers. University of Nevada Press, Reno.

Ciesla, W. M., and M. M. Furniss. 1975. Idaho's haunted forests. American Forests 81:32–35.

Despain, D. G. 1991. Yellowstone vegetation: Consequences of environment and history. Roberts Rinehart Publishing, Boulder, Colorado.

Gustafson, E. J., and G. R. Parker. 1992. Relationships between landcover proportions and indices of landscape spatial pattern. Landscape Ecology 7:101–110.

Hanski, I., and M. E. Gilpin. 1997. Metapopulation biology: Ecology, genetics and evolution. Academic Press, New York.

Hessburg, P. F., B. G. Smith, S. G. Kreiter, C. A. Miller, R. B. Salter, C. H. McNicoll, and W. J. Hann. 1999. Historical and current forest and range landscapes in the Interior Columbia River Basin and portions of the Klamath and Great Basins. Part II: Linking vegetation patterns and landscape vulnerability to potential insect and pathogen disturbances. USDA Forest Service, Pacific Northwest Research Station, General Technical Report PNW-GTR-458, Portland, Oregon.

Holling, C. S., and G. K. Meffe. 1996. Command and control and the pathology of natural resource management. Conservation Biology 10:328–337.

Jorgensen, S. M., and J. L. Hamrick. 1997. Biogeography and population genetics of whitebark pine. Canadian Journal of Forest Research 27:1574–1585.

Keane, R. E., and S. F. Arno. 1993. Rapid decline of whitebark pine in western Montana: Evidence from 20-year remeasurements. Western Journal of Applied Forestry 8:44–47.

Keane, R. E., S. F. Arno, J. K. Brown, and D. F. Tomback. 1990. Modeling stand dynamics in whitebark pine (*Pinus albicaulis*) forests. Ecological Modeling 51:73–95.

Keane, R. E., P. Morgan, and J. P. Menakis. 1994. Landscape assessment of the decline of whitebark pine (*Pinus albicaulis*) in the Bob Marshall Wilderness Complex, Montana, USA. Northwest Science 68:213–229.

Keane, R. E., P. Morgan, and J. D. White. 1999. Temporal patterns of ecosystem processes on simulated landscapes in Glacier National Park, Montana, USA. Landscape Ecology 14:311–329.

Kendall, K. C., and S. F. Arno. 1990. Whitebark pine—An important but endangered wildlife resource. Pages 264–273 *in* W. C. Schmidt and K. J. McDonald, compilers.

Proceedings—Symposium on whitebark pine ecosystems: Ecology and management of a high-mountain resource. USDA Forest Service, Intermountain Research Station, General Technical Report INT-270, Ogden, Utah.

Kendall, K., D. Schirokauer, E. Shanahan, R. Watt, D. Reinhart, R. Renkin, S. Cain, and G. Green. 1996. Whitebark pine health in northern Rockies national park ecosystems: A preliminary report. Nutcracker Notes 7:16. http://www.mesc.usgs.gov/glacier/nutnotes.htm

Kipfmueller, K. F., T. W. Swetnam, and P. Morgan. 1999. Timing of whitebark pine mortality due to mountain pine beetle in the Selway-Bitterroot wilderness in Montana and Idaho. Unpublished funded research proposal on file at Laboratory of Tree-Ring Research, University of Arizona, Tucson.

Krummel, J., R. Gardner, G. Sugihara, R. O'Neill, and P. Coleman. 1987. Landscape patterns in a disturbed environment. Oikos 48:321–324.

Landres, P., P. Morgan, and F. J. Swanson. 1999. Overview of the use of natural variability concepts in managing ecological systems. Ecological Applications 9:1179–1188.

MacArthur, R. H., and E. O. Wilson. 1967. The theory of island biogeography. Princeton University Press, Princeton, New Jersey.

McGarigal, K., and B. J. Marks. 1995. FRAGSTATS: Spatial pattern analysis program for quantifying landscape structure. USDA Forest Service, Pacific Northwest Research Station, General Technical Report PNW-GTR-351, Corvallis, Oregon.

McGregor, M. D., and D. M. Cole. 1985. Practices and considerations for noncommercial forests. Pages 56–57 *in* M. D. McGregor and D. M. Cole, editors. Integrating management strategies for the mountain pine beetle with multiple-resource management of lodgepole pine forests. USDA Forest Service, Intermountain Research Station, General Technical Report INT-174, Ogden, Utah.

Mehringer, P. J. Jr., S. F. Arno, and K. L. Petersen. 1977. Postglacial history of Lost Trail Pass bog, Bitterroot Mountains, Montana. Arctic and Alpine Research 9:345–368.

Morgan, P., and S. C. Bunting. 1990. Fire effects in whitebark pine forests. Pages 136–141 *in* W. C. Schmidt and K. J. McDonald, compilers. Proceedings—Symposium on whitebark pine ecosystems: Ecology and management of a high-mountain resource. USDA Forest Service, Intermountain Research Station, General Technical Report INT-270, Ogden, Utah.

Morgan, P., G. H. Aplet, J. B. Haufler, H. C. Humphries, M. M. Moore, and W. D. Wilson. 1994a. Historical range of variability: A useful tool for evaluating ecosystem change. Journal of Sustainable Forestry 2:87–111.

Morgan, P., S. C. Bunting, R. E. Keane, and S. F. Arno. 1994b. Fire ecology of whitebark pine forests of the northern Rocky Mountains, U.S.A. Pages 136–141 *in* W. C. Schmidt and F.-K. Holtmeier, compilers. Proceedings—International workshop on subalpine stone pines and their environment: The status of our knowledge. USDA Forest Service, Intermountain Research Station, General Technical Report INT-GTR-309, Ogden, Utah.

Morgan, P., S. C. Bunting, A. E. Black, T. Merrill, and S. Barrett. 1996. Fire regimes in the Interior Columbia River Basin: Past and present. Unpublished final report for RJVA INT-94913. On file at Fire Sciences Laboratory, USDA Forest Service, Intermountain Research Station, Missoula, Montana.

Murray, M. P., S. C. Bunting, and P. Morgan. 1998. Fire history of an isolated subalpine mountain range of the Intermountain Region, United States. Journal of Biogeography 25:1071–1080.

Nilsson, C., and G. Grelsson. 1995. The fragility of ecosystems: A review. Journal of Applied Ecology 32:677–692.

Omernik, J. M. 1987. Ecoregions of the conterminous United States. Annals of the Association of American Geographers 77:118–125.

Perkins, D. L, and T. W. Swetnam. 1996. A dendroecological assessment of whitebark pine in the Sawtooth-Salmon River region, Idaho. Canadian Journal of Forest Research 26:2123–2133.

Price, C., and D. Rind. 1994. The impact of a $2XCO_2$ climate on lightning-caused fires. Journal of Climate 7:1484–1494.

Quigley, T. M., and S. J. Arbelbide, technical editors. 1997. An assessment of ecosystem components in the Interior Columbia Basin and portions of the Klamath and Great Basins. USDA Forest Service, Pacific Northwest Research Station, General Technical Report PNW-GTR-405, Portland, Oregon.

Rollins, M. G., T. W. Swetnam, and P. Morgan. 2000. Twentieth-century fire patterns in the Gila/Aldo Leopold wilderness Complex New Mexico and the Selway-Bitterroot wilderness area Idaho/Montana. In D. N. Cole and S. F. McCool, editors. Proceedings: wilderness science in a time of change. USDA Forest Service, Rocky Mountain Research Station, Proceedings RMRS-P-000, Fort Collins, Colorado.

Romme, W. H. 1982. Fire and landscape diversity in subalpine forests of Yellowstone National Park. Ecological Monographs 52:199–221.

Rothermel, R. C., R. A. Hartford, and C. H. Chase. 1994. Fire growth maps for the 1988 Greater Yellowstone Area fires. USDA Forest Service, Intermountain Research Station, General Technical Report INT-304, Ogden, Utah.

Ryan, K. C. 1991. Vegetation and wildland fire: Implications for global change. Environmental Management 17:169–178.

Scott, J. M., F. Davis, B. Csuti, R. Noss, B. Butterfield, C. Goves, H. Anderson, S. Caicco, F. D'Erchia, T. C. Edwards Jr., J. Ulliman, and R. G. Wright. 1993. Gap Analysis: A geographic approach to protection of biological diversity. Wildlife Monographs 123:1–41. http://www.gap.uidaho.edu/

Smith, J., and J. Hoffman. 1998. Status of white pine blister rust in Intermountain Region white pines. USDA Forest Service, Intermountain Region, State and Private Forestry, Forest Health Protection, Report No. R4-98-02, Ogden, Utah.

Spaulding, W. G. 1990. Vegetation dynamics during the last deglaciation, southeastern Great Basin, U.S.A. Quaternary Research 33:188–203.

Swetnam, T. W., and J. L. Betancourt. 1998. Mesoscale disturbance and ecological response to decadal climatic variability in the American Southwest. Journal of Climate 11:3128–3147.

Swetnam, T. W., C. D. Allen, and J. L. Betancourt. 1999. Applied historical ecology: Using the past to manage for the future. Ecological Applications 9:1189–1206.

Tande, G. F. 1979. Fire history and vegetation pattern of coniferous forests in Jasper National Park, Alberta. Canadian Journal of Botany 57:1912–1931.

Thompson, R. S. 1984. Late Pleistocene and Holocene environments in the Great Basin. Ph.D. dissertation, University of Arizona, Tucson.

———. 1990. Late quaternary vegetation and climate in the Great Basin. Pages 200–239 *in* J. L. Betancourt, T. R. Van Devender, and P. S. Martin, editors. Packrat middens: The last 40,000 years of biotic change. University of Arizona Press, Tucson.

Tomback, D. F. 1978. Foraging strategies of Clark's nutcracker. The Living Bird 16:124–161.

Tomback, D. F., L. A. Hoffman, and S. K. Sund. 1990. Coevolution of whitebark pine and nutcrackers: Implications for forest regeneration. Pages 118–129 *in* W. C. Schmidt and K. J. McDonald, compilers. Proceedings—Symposium on whitebark pine ecosystems: Ecology and management of a high-mountain resource. USDA Forest Service, Intermountain Research Station, General Technical Report Int-270, Ogden, Utah.

U.S. Department of Agriculture (USDA). 1998. World Wide Web site: http://www.fs.fed.us/land/wdfire5.htm. October 20, 1998. USDA Forest Service, Washington, D.C.

Vander Wall, S. B., and R. B. Balda. 1977. Coadaptations of Clark's nutcracker and the piñon pine for efficient seed harvest and dispersal. Ecological Monographs. 47:89–111.

Wells, P. V. 1983. Paleobiogeography of montane islands in the Great Basin since the last glaciopluvial. Ecological Monographs 53:341–382.

Wigand, P. E., and C. L. Nowak. 1992. Dynamics of northwest Nevada plant communities during the last 30,000 years. Pages 40–61 *in* C. A. Hall Jr., V. Doyle-Jones, and B. Widawski, editors. The history of water: Eastern Sierra Nevada, Owens Valley, White-Inyo Mountains. White Mountain Research Station Symposium, California.

Chapter 15

Management Implications of Genetic Structure

Donna Dekker-Robertson and Leo P. Bruederle

All species possess some genetic variation. Genetic structure is defined as the way in which this genetic variation is distributed within a species. Managers must have a good understanding of the genetic structure of whitebark pine (*Pinus albicaulis*) populations in order to make decisions about conservation or any type of supplemental reforestation. In the absence of such information, problems can range from maladaptation of planting stock on the one hand to unnecessary expenses for seed collection and storage on the other.

Genetic variation accumulates slowly over evolutionary history and provides the raw material upon which natural selection acts, thereby allowing adaptation to a dynamic environment. Over time, genetic structure results from the joint actions of evolutionary forces, including gene mutation, limited gene flow, random drift, and selection. The magnitude of the effects of these evolutionary forces are themselves determined by recent evolutionary events, as well as by the natural history and biogeography of a species. This chapter discusses why an understanding of genetic structure is important, what forces determine genetic structure, what is known about genetic structure in whitebark pine, and what is being done with this knowledge.

Why Study Genetic Structure?

Understanding the genetic structure of whitebark pine populations is important for two reasons: to direct genetic conservation efforts, and to guide restoration projects. This is a time of relatively rapid change in whitebark pine stands, which stems from the profound influence of humans in this century (Chapter

1). Two major threats to genetic diversity could severely impact whitebark pine: white pine blister rust (*Cronartium ribicola*) and fire exclusion. A fungal pathogen that was introduced into North America early in the twentieth century, blister rust has since spread through the native range of nearly all of the continent's five-needled white pines (Chapter 10). Fire exclusion has altered the successional mix in stands historically dominated by whitebark pine (Chapter 18). At this time, whitebark pine populations are experiencing high levels of mortality from the combined effects of nearly a century of fire suppression and blister rust. Some populations may be lost entirely, and a loss of genetic diversity is likely over the whole species.

Blister rust is slowly spreading throughout the range of whitebark pine. The disease impacts whitebark reproduction in two ways: by killing trees outright and by killing the upper, cone-bearing portion of the tree (Chapter 17). Studies indicate that there is some genetic resistance in the species (Chapter 17). However, neither the basis of this resistance nor the distribution of the genes controlling it is known. There are two possibilities: Either blister rust resistance genes are found throughout the range of whitebark pine at varying frequencies, or these genes are unusual and limited in their distribution.

Managers have a distinct advantage if blister rust resistance genes are found throughout the range of the species. Under this scenario, the danger of losing whole populations to rust is reduced, because there should be survivors scattered across the range. By encouraging as much regeneration as possible, natural selection by the rust will eliminate those trees without rust resistance genes. Over time, the frequency of rust resistance genes should increase, as will the proportion of rust-resistant individuals in a stand (Chapter 17).

If blister rust resistance is unusual and limited, it will be important to identify and collect from as many resistant parent trees as possible. It will also be critical to understand the genetic structure of the species to be certain that rust-resistant seeds are correctly deployed. It is possible that managers may be presented with a difficult dilemma: rust resistance or local adaptation, but not both.

Fires open up the subalpine environment, making sites attractive to Clark's nutcrackers (*Nucifraga columbiana*) for caching whitebark seeds (Chapter 5). At the same time, fires also eliminate competing vegetation such as subalpine fir (*Abies lasiocarpa*). Both of these effects enhance whitebark pine regeneration. However, fire exclusion ultimately influences genetic structure, as it may lead to reduced reproduction and the loss of alleles or even populations. Blister rust will compound this situation by reducing the number of potential parent trees.

Awareness that the genetic diversity of a species is both important and irreplaceable is growing. Once genetic structure is deduced, genetic conservation entails saving representative genotypes from enough populations of a species to preserve the majority of its genetic diversity. These genotypes may be saved by collecting and storing seeds and/or establishing orchards, or by setting aside portions of the native range as genetic reserves (Millar and Libby 1991). A combination of both approaches best ensures that genotypes will be maintained for the future, but requires a strong administrative commitment in the form of facilities and funds. In an era of tight budgets, this type of conservation project is difficult to justify to managers with limited resources. However, in the meanwhile, the genetic base of whitebark pine is becoming narrower, which may make the species less resilient to environmental change.

The restoration of lost or diminished populations of whitebark pine has attracted both interest and funding. Following a perturbation, insufficient seeds may remain for Clark's nutcrackers to cache. It may be necessary or desirable to plant additional whitebark pine, supplementing the natural regeneration, to maintain its presence in the stand. Seed transfer guidelines ensure that after a catastrophic event, new seedlings for reforestation will be well adapted to the site in which they are planted. In general, the guidelines are based on duplicating to some extent the elevation, latitude, and longitude of the stands of origin, and are developed by examining patterns of genetic differentiation among provenance samples planted together in a common environment, called a common-garden test (Westfall 1992). In the absence of data from a common-garden test, seed transfer guidelines tend to be very conservative to prevent maladaptation. In general, it is desirable to minimize the distance between the stand of origin and the outplant site, both latitudinally and elevationally.

Whitebark pine is a keystone species of the subalpine ecosystem (Chapter 1), and as such, it may be possible to justify a program to breed for blister rust resistance. Should such a breeding program be developed, seed transfer guidelines are required to ensure that seedlings from this program are planted in environments where they can be expected to thrive.

Forces That Determine Genetic Structure

In order to understand the role of genetic structure, it is useful to discuss the concept of genetic variation in greater detail. When one examines the range of morphology found in species, it becomes obvious that there is a great deal of variation. What one sees, called the *phenotype,* is determined by genes, the environment, and the interaction of the two. The genes are encompassed within DNA, an elegant molecule comprising variable sequences of four different nucleotides. These, in turn, make up genes that produce proteins or enzymes.

Mutations in a gene can lead to minor or major alterations in the resulting protein. In some cases proteins lose their ability to function in the cell, while in others some change occurs to the functional protein. The variant forms of one gene created by mutation are called *alleles,* and the resulting different proteins or enzymes are referred to as *allozymes*. Hundreds of alleles may exist for a single gene, each producing slightly different yet functional forms of a protein.

Mutations can be selectively beneficial, neutral, or detrimental, depending on their effect on the protein, as well as the role of that protein in the cell. Natural selection favors the reproduction of individuals possessing beneficial alleles, whereas those with detrimental mutations will leave fewer offspring. Although the frequency of mutation is low, new mutations constantly arise. Some, if not most, are selectively neutral, conferring neither benefit nor detriment. Others are quickly eliminated by natural selection. In most cases, those mutations that affect the structure or function of a protein are detrimental.

Natural selection, however, may not work against these detrimental mutations, or choose beneficial mutations, until environmental conditions become extreme.

Because most mutations (nonsomatic) are inherited, they can accumulate over evolutionary time. These mutations are the basis for all variation in the genome. Two other forces that influence genetic structure are gene flow and random genetic drift. Whereas mutation is the ultimate source of genetic variation, gene flow and drift influence the manner in which this variation is distributed within and among populations.

Gene flow follows from the movement of individuals among populations. In pines, as with other plants, both seeds and pollen are involved. For example, a Clark's nutcracker harvesting seeds from one mountaintop population of whitebark pine and caching them at another site illustrates gene flow involving seeds. Similarly, pollen grains being blown by the wind from one mountaintop and pollinating developing female whitebark pine cones at another site is also gene flow. Regardless of the mechanism, gene flow among populations continuously introduces and reintroduces alleles from one population into another. This also allows the introduction of mutations into populations distant from the one in which the mutation arose.

Random genetic drift, on the other hand, refers to chance fluctuations in the frequency of alleles that can also lead to allele loss. Isolated stands of whitebark pine may show the effects of drift, which are magnified in small populations. Drift may occur when a population passes through a genetic bottleneck in which the size of the population is reduced. It is quite possible that whitebark pine is passing through such a bottleneck as a result of the mortality caused by blister rust coupled with the lack of regeneration resulting from fire exclusion.

It is useful to think of the genetic variation comprising a species in two ways: the amount due to differences among individuals within a population, and the amount due to differences among populations. In other words, how different is one whitebark pine from another growing 2 meters away relative to another whitebark pine growing 200 meters away, relative to still another whitebark pine growing on a mountaintop 200 kilometers away? In species in which most of the genetic variation is found among populations, spatially separated populations are genetically different from each other. When variation is found primarily within populations, populations that may be kilometers apart are not genetically different from each other.

Genetic changes in traits that occur gradually over a geographic area, rather than abruptly, are termed *clinal.* Genetic variation often corresponds to changes in geography, such as increasing elevation or latitude, generally as a result of a selection gradient that corresponds to this change (e.g., shorter growing seasons). Clinal variation is most common in conifer species with large, continuous distributions over which geography and climate gradually change (Millar and Libby 1991). Many adaptive traits in conifers exhibit clinal variation; for example, the difference in growing season from south to north has a profound impact on such traits as date of budbreak, date of budset, and period of shoot elongation in many species. Other examples of this clinal variation include height growth of white fir (*Abies concolor*) (Hamrick and Libby 1972) and cold hardiness in lodgepole pine (*Pinus contorta*) (Rehfeldt 1988).

Variation can also be distributed discontinuously. When abrupt genetic differences are found, natural selection and/or genetic drift may be the reason (Millar and Libby 1991). Whitebark pine may have such a pattern of variation. It is typically found only at high elevations, and as such, populations are disjunct and insular. These island-like whitebark pine populations are separated by kilometers of lower-elevation forest that are populated by other conifer species. This isolation can be expected to reduce gene flow with a concomitant increase in the variation among populations, although this tendency may be countered by nutcrackers' long-distance caching behavior and long-distance pollen flow. Since whitebark pine is found in small, discontinuous stands throughout much of its range, the lack of gene flow coupled with drift may together play a more important role in determining population structure than in other species with more continuous distributions.

Many variables have been demonstrated to influence significantly the amount and distribution of variation in a species. In pines, these include breeding system; mode of seed dispersal, geographic range, and regional distribution; successional status; and population size, density, and spatial distribution (Loveless and Hamrick 1984). Others can be expected to influence genetic structure in whitebark pine specifically, such as seed harvesting and caching by Clark's nutcrackers (Chapter 8).

The degree to which populations are inbred or outcrossed influences genetic structure by increasing or decreasing the variation among populations. Self-pollination is the most extreme form of inbreeding; however, even outcrossing species may have some degree of inbreeding as a result of pollen exchange between related individuals. Inbreeding tends to increase among-population variation and results in genetic drift; over time, individuals within populations become more closely related, while becoming more distantly related to individuals in other populations. This is termed population *divergence*. Outcrossing, on the other hand, tends to decrease the tendency toward divergence by countering genetic drift and retaining gene flow among populations. Whitebark pine, like other conifers, is primarily an outcrossing species. By promoting the recombination of parental genes, outcrossing increases the genetic variation present within populations, as well as within the species as a whole (Loveless and Hamrick 1984).

Conifers are wind-pollinated, which promotes a high degree of within-population variation due to outcrossing. Wind-dispersed pollen may travel a long distance, which may act to reduce divergence among populations. Most conifers also have wind-dispersed seeds. Short-distance seed dispersal leads to the clustering of relatives in a stand, and promotes inbreeding and thus divergence among populations (Loveless and Hamrick 1984). In contrast, long-distance dispersal tends to prevent populations from diverging (Loveless and Hamrick 1984). Whitebark pine presents an interesting contrast because of its bird-mediated dispersal of seeds. Normally, long-distance animal dispersal of seeds leads to homogeneity among populations and may reduce family structure (relatedness of adjoining individuals) within stands. However, nutcrackers cache seeds together that have a high probability of coming from the same tree

(Tomback 1988; Chapter 5). Several studies have shown a high degree of family structure within the clusters of whitebark pine trees arising from a single cache (e.g., Furnier et al. 1987; Rogers et al. 1999; Chapter 8). Over an entire stand, in contrast, individuals in neighboring clusters are not more similar than individuals in distant clusters (Furnier et al. 1987; Rogers et al. 1999; Chapter 8). Whitebark pine has been shown to have a higher degree of inbreeding than many other conifers, which may result from pollination within tree clusters (Jorgenson and Hamrick 1997). Yet, a cache may be located at a considerable distance or a different elevation from the parent tree, depending on the distance of the nutcracker's flight; this would tend to decrease population divergence.

Species such as whitebark pine that flower seasonally and synchronously have the potential for extensive gene flow, which both inhibits subdivision within populations and keeps populations from diverging. Also, whitebark pine in the northern Rockies varies in the number of female cones produced from year to year (Weaver and Forcella 1986; Morgan and Bunting 1992). Although the precise factors that determine a good cone year in whitebark pine are unknown (Weaver and Forcella 1986), the phenomenon extends over a large geographic area, indicating the potential for extensive gene flow during good cone years.

Methods for Determining Genetic Structure

Three principal methods have been used to assess genetic structure in pines: allozyme analysis; DNA-based approaches, including the RAPD (Randomly Amplified Polymorphic DNA) technique; and common-garden tests. Each has advantages and disadvantages, and each describes variation somewhat differently. For that reason, the technique that is chosen by the investigator depends on the specific objectives of the proposed research, as well as on the available resources.

Allozyme Analysis

Until recently, allozyme analysis was inarguably the most commonly used tool applied by geneticists to assess genetic diversity. Several investigators have used allozyme analysis to describe population genetic structure in whitebark pine (Yandell 1992; Bruederle et al. 1998; Jorgenson and Hamrick 1997; Rogers et al. 1999). Allozyme analysis involves separation of varying forms of proteins derived from a single gene. Tissue samples are collected from a presumably representative sample of individuals comprising a population, and from populations selected to represent the presumed range of variation present in the species under consideration. Typically, geographic, ecological, and taxonomic variation are taken into consideration when selecting populations for these purposes.

The tissue samples are treated to release the allozymes, which are then separated on an electrophoretic gel, that is, by different allozyme migration rates when a current is applied. Based on the mobility of the stained allozymes, the genotype of each individual can be deduced. These data are then analyzed to obtain a variety of descriptive statistics, thereby characterizing genetic diversity and the manner in which it is organized within and among populations. These statistics reveal the portion of genetic diversity that is attributable to differences among individuals within and among populations.

DNA-based Techniques

Recent advances in molecular biology have yielded additional techniques that involve the polymerase chain reaction (PCR) to analyze genes and other DNA sequences from chromosomes in the nucleus, as well as those present in mitochondria and chloroplasts. In the RAPD technique, short nucleotide primers are used to amplify random DNA fragments, which are then separated based upon molecular weight using acrylamide or, more commonly, agarose gel electrophoresis. In other techniques used for gene flow studies, a primer is used to amplify a sequence that is known to contain a genetic marker. In any case, the fragments are visualized using a variety of stains, although ethidium bromide viewed under ultraviolet wavelengths is the most commonly employed. As before, the data may be analyzed to obtain a variety of descriptive statistics.

The inheritance and expression of the aforementioned molecules vary, providing powerful population genetic tools. Whereas allozymes and RAPDs are bi-parental in pines, mtDNA (mitochondrial DNA) is maternal in inheritance and cpDNA (chloroplast DNA) is paternal in inheritance. As such, it is possible to obtain very different impressions of genetic structure. For example, in a study by Latta and Mitton (1997), mtDNA showed much higher among-population differentiation than cpDNA in populations of limber pine (*Pinus flexilis*), which, like whitebark, has wind-dispersed pollen and bird-dispersed seeds. The large difference between mtDNA and cpDNA genetic structure shows that pollen movement is contributing the bulk of interpopulation gene flow in limber pine (Latta and Mitton 1997).

Although many informative exceptions exist (e.g., Stutz and Mitton 1988), it has been generally assumed that natural selection does not operate on allozymes and other molecular markers. Adaptive traits, on the other hand, are generally expected to be under strong selection pressure, because organisms that are poorly adapted to their environment will be quickly selected against. Allozyme analysis and DNA-based techniques may therefore provide little insight into the manner in which genetic variation for adaptive traits is apportioned.

Common-Garden Tests

In the future, DNA-based techniques may make it possible to assess adaptive variation by screening for the genes that control adaptation. It is clear, however, that such traits are governed by the effects of many genes working in concert (quantitative traits). Molecular mapping of most quantitative traits in conifers is well in the future. However, patterns of adaptive variation in plants may still be assessed using common-garden tests. In this type of test, seeds from many different locations are grown together in one or more locations and adaptive traits are measured (e.g., Rehfeldt et al. 1984). In conifers, some of the traits commonly measured include date of budbreak, date of budset, duration of shoot elongation, height, diameter, and cold hardiness. Other traits such as blister rust resistance can also be directly assessed.

Adaptation may involve a plant's tolerance of such environmental factors as drought, poor soils, shading, herbivory, or cold. The length of the growing season (frost-free period) is a selection factor in some regions (Rehfeldt 1989). Shoot elongation occurs more quickly when species have a short growing season, as is typical for those like whitebark pine at high elevations or latitudes. Both height and diameter growth are slower in species with short growing seasons than in species with longer growing seasons. An example is lodgepole pine, where adaptation reflects a balance between selection for traits that provide a high growth potential in mild environments and traits that provide high tolerance to early autumn frost in severe environments (Rehfeldt 1988). Common-garden tests of conifers allow the researcher to assess these traits and to predict the environmental conditions under which each population could be expected to thrive. Further, common-garden tests that are replicated at more than one location may give some indication of genotype by environment interactions, which both molecular techniques and allozyme analysis are incapable of detecting. As such, common-garden tests determine whether a species is a genetic generalist or specialist.

When adaptive variation is predominantly among populations, these species are assumed to be genetic specialists. Variation among individuals from different populations is controlled by genes with steep clines (Rehfeldt 1989). On the other hand, when the genetic variation in a species occurs predominantly within populations rather than among populations, these species are assumed to be genetic generalists. Individuals from different populations may be more similar to each other genetically than to other individuals in the same population. Genetic specialists may not be moved far without the seedlings being at an adaptational disadvantage (Rehfeldt 1989), whereas in species that are genetic generalists, individuals from hundreds of kilometers apart may be quite similar in their adaptative traits and fully capable of thriving in either location.

Common-garden tests do not require special laboratory equipment or techniques, but they do require long-term cultivated field or greenhouse space. Results from common-garden tests are slower to obtain than those from laboratory studies. In fact, a common-garden test of ponderosa pine (*Pinus ponderosa*) was established at Priest River Experimental Forest in northern Idaho near the beginning of the twentieth century and is still yielding information about which provenances are capable of long-term survival in that region.

Pros and Cons

Laboratory studies based on selectively neutral biochemical markers and common-garden tests for adaptational traits may yield similar, or substantially different results pertaining to genetic diversity and the manner in which it is apportioned within and among populations (Karhu et al. 1996). At a recent meeting, there was unanimous agreement among the sixty-eight scientists and forest managers present that characterizing a species' genetic diversity exclusively by allozymes is inadequate for developing genetic conservation studies, deploying pedigreed material to plantation sites, or restoring keystone, threatened or endangered species (Libby et al. 1997). Nevertheless, laboratory studies yield results more quickly, which will allow decisions to be made if the species is in peril.

Since each method for determining genetic variation provides a different type of information, the best approach to assessing genetic structure is to start common-garden and laboratory studies simultaneously. Data from laboratory studies will provide insight into the amount and apportionment of selectively neutral variation in the species, patterns of gene flow, and breeding system. This information may help develop preliminary genetic conservation guidelines. They are not, however, currently capable of assessing genotype by environment interaction, blister rust resistance, or variation in adaptive traits. Common-garden tests will take longer, but can address the inheritance of quantitative traits, describe a species as a genetic generalist or specialist, and provide guidance for decisions pertaining to seed transfer and genetic conservation. When both types of study are conducted, a complete picture of genetic structure emerges.

Implications of Known Patterns of Genetic Variation

Several general patterns have emerged from studies of genetic structure in conifers: (1) genetic homogeneity, (2) genetic heterogeneity with no geographic correlates, (3) genetic heterogeneity with clinal variation over long distances, and (4) genetic heterogeneity with clinal variation over short distances. Each of these patterns has somewhat different implications for management.

Genetic homogeneity. Red pine (*Pinus resinosa*), which is found in the north-central and northeastern portions of the United States and in Canada, is the best-known member of this small group. Both adaptive variation and molecular studies have revealed little genetic variation either within or among populations (Fowler and Morris 1977; Mosseler et al. 1992). This lack of genetic variation has been attributed to the pine's recent evolutionary history, including the effect of glaciation upon its distribution. It is thought that the species was restricted to small glacial refugia, resulting in an intense genetic bottleneck and a marked reduction in genetic variation (Fowler and Morris 1977).

Species with little genetic variation require minimal intervention to manage their genetic resources. Seed transfer guidelines are not necessary, and selective breeding programs need not be concerned with adaptive traits. Genetic conservation efforts may be limited to collecting germplasm from a small number of individuals or populations. This pattern is quite unlikely for whitebark pine, based on existing allozyme data.

Genetic heterogeneity with no geographic correlates. In these species, populations do not show adaptive differences or clinal variation. Nevertheless, they are genetically distinct from other conspecific populations. An example of this is Torrey pine (*Pinus torreyana*), a rare species found in only two small populations in California. There is no within-population variation detectable by allozyme studies; each population is genetically identical at each of fifty-nine allozyme loci (Ledig and Conkle 1983). There are, however, some minor differences between the two populations at two allozyme loci. This variation is presumably caused by founder effects (Millar and Libby 1991). There are no obvious adaptive differences between the two populations.

In this situation, it is not necessary to save material from across the range to ensure adapted planting stock, nor are seed transfer guidelines necessary. Genetic conservation may be limited to preservation of a few genotypes from each population for the sake of allelic diversity. In whitebark pine, variation that is not clinal is a possibility for two reasons: the species grows in disjunct stands through much of its range, and the environment in which it grows is uniformly harsh (Chapter 2 and 3), which may suppress much variation in adaptive traits. This pattern is supported by results from Howard's (1999) adaptational study of whitebark pine, which showed significant differences in height and diameter growth among populations with no clear geographic or elevational patterns.

Genetic heterogeneity with continuous clinal variation over long distance. Genetic changes often occur gradually over space rather than abruptly (Millar and Libby 1991). An example of this is Scots pine (*Pinus sylvestris*) in Europe, where the native range of the species stretches across Europe and Asia from the Atlantic to the Pacific and from Finland to the southern portion of Spain. Variation corresponds to geography, with growth beginning and ceasing earlier in populations derived from higher altitudes and latitudes (Giertych 1991; Oleksyn et al. 1998). Populations differ from each other in dry matter production (Langlet 1959) and height growth (Wright and Baldwin 1957; Shutyaev and Giertych 1997) over long distances. One might consider the species in this group as "specialized generalists" in that, although there are adaptive differences among populations, seeds must be moved a long distance to be maladapted.

With long-distance clinal variation, management requirements are intermediate. Seed transfer zones would be fairly large, and seeds moved freely over long distances. However, breeders would have to be cautious about using parents from distantly located populations as breeding stock. For effective genetic conservation, reserves and seed collections should occur across the range to capture most or all of the genetic variation. This pattern of variation is quite possible for whitebark pine with regard to adaptive traits, because the range of the species extends from the central Sierra Nevada north into Canada and eastward to Wyoming.

Genetic heterogeneity with clinal variation over short distances. In this group, clinal variation is correlated with geography and elevation of the seed source; thus, populations located relatively close to each other may be genetically different. Douglas-fir (*Pseudotsuga menziesii*) shows clinal genetic variation that is correlated with both geography and elevation of the seed source (Rehfeldt 1989). Variation corresponds directly to differences in number of frost-free days per growing season, and population adaptation may be viewed as a balance between selection for cold hardiness and height growth (Rehfeldt 1989). Seeds from low-elevation sources may not be moved more than 240 meters in elevation, whereas those from higher-elevation stands may be moved somewhat further (350 meters). Those populations adapted to a short frost-free period must break bud later, set bud sooner, and thus protect themselves from frost. Those populations adapted to a milder climate must break bud sooner and set bud later in order to compete with others for height growth.

The resources required to manage genetic specialists are considerable. Selective breeders must maintain separate orchards to ensure that pollen contamination does not produce maladapted seed. Seed transfer zones and guidelines must be very conservative for the same reason. To capture most or all of the genetic diversity, a series of reserves and seed collections must be made. A larger sample of populations are needed for these reserves and collections than would be required for a species with less among-population variability. It seems less likely, however, that whitebark pine would show this pattern of extreme genetic specialization, because there is much less diversity in the environments in which whitebark pine is found. Existing allozyme data also do not support this pattern.

Genetic Structure of Whitebark Pine

With respect to the genetic structure of whitebark pine, several studies using allozyme analysis have been completed, and an adaptive variation study is currently underway (Chapter 8). Enough data exist to make preliminary and informed decisions regarding conservation of genetic resources in this species, and to develop the first set of guidelines for seed transfer.

Allozyme data collected at a regional level and rangewide level have revealed low levels of genetic variation at both the population and the species level (Chapter 8). What variation exists is found largely within populations, which is typical for most pines (Jorgenson and Hamrick 1997). What this means is that two individuals from populations many miles apart may be as similar to each other as two individuals from the same population. While significant regional differentiation exists, it represents a small amount of the total genetic diversity that has been described for whitebark pine (Jorgenson and Hamrick 1997).

Allozyme studies by a number of authors have revealed much about gene flow in whitebark and other bird-dispersed pines. Schuster et al. (1989) showed that most populations of limber pine separated by more than about 400 meters in elevation do not have overlapping pollination periods and thus are incapable of pollen-mediated gene flow. Nevertheless, migration was clearly occurring between high- and low-elevation populations, implying that either stepping-stone pollen transfer between intermediate populations or seed transfer must be occurring. However, Latta and Mitton (1997) have demonstrated that gene flow in limber pine is largely attributable to pollen movement, which would imply that nutcrackers are not as important as agents of gene flow in this species. On the other hand, Bruederle et al. (1998) report that populations of bird-dispersed pines, including whitebark and limber pine, are less divergent than populations of wind-dispersed pines, suggesting that bird-facilitated gene flow is more effective than wind dispersal in preventing population differentiation. A study to determine whether pollen or seed transfer is more important in whitebark pine genetic structure is presently underway at the University of Idaho (S. J. Brunsfeld, personal communication).

Seed Transfer Guidelines

Preliminary seed transfer data from a study of whitebark pine conducted across a latitudinal-moisture gradient in Idaho showed little variation observed among populations for water-use efficiency, whereas significant variation was observed for height and diameter growth (Howard 1999). There was no clear geographic pattern to the variation relative to region of seedling origin. However, whitebark pine seedlings from all populations studied adapted phenotypically to site, allocating biomass to height growth and aboveground biomass on wet sites and to belowground biomass and stems on dry sites. Most important, Howard (1999) found that seedlings taken from stands with high blister rust mortality (and thus expected to have greater than average resistance to blister rust) did not show important differences in growth, biomass allocation, and water-use efficiency compared to native populations when moved to other climatic regimes. This indicates that these potentially rust-resistant seedlings may be capable of broad adaptation, which may be necessary if widespread planting is required to stabilize losses to blister rust.

In order to develop sound seed transfer guidelines, a full-scale adaptive variation study is needed. Such a study for whitebark pine using a common-garden test was begun in the early 1990s with seed collections from stands in northern Idaho and western Montana (D. Dekker-Robertson, unpublished data). Seeds were collected and stored from more than sixty sites scattered throughout the region, and sown in spring 1998. In spring 2001, the seedlings will be measured and phenological data collected; this will continue for one to two subsequent growing seasons. Eventually, the seedlings will be outplanted at Priest River Experimental Forest, Idaho, for long-term measurement and observation. Subsequent studies for other regions are planned.

As the final results of this adaptive variation study are still years away, the USDA Forest Service has developed a set of preliminary seed transfer guidelines (Mahalovich and Hoff, in press; see also Chapter 17) that may be modified and refined at the conclusion of this test, or over the longer term as necessary. These guidelines are based on the results from allozyme studies, which indicate that there is little variation within the species and little differentiation of populations. This points toward a generalist strategy.

The guidelines encourage cone collections from stands with high blister rust mortality, because the surviving trees have a high probability of carrying blister rust resistance genes. In addition, managers should refrain from moving seeds from sites with low blister rust infection to sites with high blister rust infection. It is unlikely that seeds collected from low-infection sites will carry high levels of resistance, since natural selection by the rust has not yet weeded out the highly susceptible trees.

The Forest Service guidelines suggest eight seed zones for whitebark pine in the West, based on mountain range (Chapter 16). The authors of the guidelines expressed concern that cone collectors not be too absolute about seed zone boundaries (Mahalovich and Hoff, in press). This is supported by Howard's

(1999) study that indicates that whitebark pine seedlings may be capable of phenotypically adapting to sites much different climatically from those in which the seeds originated.

Decision Making in a Time of Crisis

Time is of the essence. Our ability to carry out genetic conservation and stand restoration projects will be diminished as time passes and the existing stands continue to lose individuals to blister rust and advancing succession. In those portions of the range where blister rust infection levels are highest, most or all of the whitebark pine regeneration may become infected (Chapter 1), which means that as older trees die, few to no young trees will take their place. The genetic implications of failing to take action are therefore serious.

As managers plan restoration projects, the data available, along with our understanding of the ecology of whitebark pine, indicate that seeds may be moved within a mountain range without fear of maladaptation. Bruederle et al. (1998) conclude that although local genetic variation should be considered, it is probably of less concern in whitebark pine than in wind-dispersed pine species. Howard (1999) is more cautious, advocating supplemental planting only where natural regeneration after fire is likely to be sparse. Both agree that seed sources may derive from larger geographic areas. Until results from adaptive variation tests are available, managers planting whitebark pine seedlings should avoid moving seedlings interregionally.

Enough is already known about whitebark pine genetic structure to develop a coherent plan for genetic conservation. Seed collections should be made soon to preserve the remaining genetic diversity. From the allozyme data, we know that there is not a great deal of genetic variation in the species as a whole, and what there is tends to be concentrated within populations. For this reason, fewer populations will need to be sampled to capture the genetic diversity fully. As there is some interregional differentiation, the populations sampled should be scattered through the range of whitebark pine.

In a way, genetic variation is something like topsoil. It is created very slowly over time, and the process cannot be duplicated or hurried. Genetic conservation, like soil conservation, is therefore a long-term goal that requires commitment from long-term institutions like governments. With the present threats to whitebark pine, this process cannot begin too soon.

Acknowledgments

The authors thank Steven J. Brunsfeld, University of Idaho, for his helpful comments on this manuscript.

Literature Cited

Bruederle, L. P., D. F. Tomback, K. K. Kelly, and R. C. Hardwick. 1998. Population genetic structure in a bird-dispersed pine, *Pinus albicaulis* (Pinaceae). Canadian Journal of Botany 76:83–90.

Fowler, D. P., and R. W. Morris. 1977. Genetic diversity in red pine: Evidence for low genetic heterogeneity. Canadian Journal of Forest Research 7:343–347.

Furnier, G. R., P. Knowles, M. A. Clyde, and B. P. Dancik. 1987. Effects of avian seed dispersal on the genetic structure of whitebark pine populations. Evolution 41:607–612.

Giertych, M. 1991. Provenance variation in growth and phenology. Pages 87–101 *in* M. Giertych, and C. S. Matyas, editors. Genetics of Scots pine. Elsevier, Amsterdam, The Netherlands.

Hamrick, J. L., and W. J. Libby. 1972. Variation and selection in western U.S. montane species. I. White fir. Silvae Genetica 21:29–35.

Howard, J. 1999. Transplanted whitebark pine regeneration: The response of different populations to variation in climate in field experiments. Master's thesis, Department of Biological Sciences, University of Montana, Missoula.

Jorgenson, S. M., and J. L. Hamrick. 1997. Biogeography and population genetics of whitebark pine, *Pinus albicaulis*. Canadian Journal of Forest Research 27:1574–1585.

Karhu, A., P. Hurme, M. Karjalainen, P. Karvonen, K. Karkkainen, D. Neale, and O. Savolainen. 1996. Do molecular markers reflect patterns of differentiation in adaptive traits of conifers? Theoretical and Applied Genetics 93:215–221.

Langlet, O. 1959. A cline or not a cline—A question of Scots pine. Silvae Genetica 8:13–22.

Latta, R. G., and J. B. Mitton. 1997. A comparison of population differentiation across four classes of gene marker in limber pine (*Pinus flexilis* James). Genetics 146:1153–1163.

Ledig, F. T., and M. T. Conkle. 1983. Gene diversity and genetic structure in a narrow endemic, Torrey pine (*Pinus torreyana* Parry ex Carr.). Evolution 37:79–86.

Libby, W. J., F. Bridgwater, C. Lantz, and T. White. 1997. Genetic diversity in commercial forest tree plantations: Introductory comments to the 1994 SRIEG meeting papers. Canadian Journal of Forest Research 27:397–400.

Loveless, M. D., and J. L. Hamrick. 1984. Ecological determinants of genetic structure in plant populations. Annual Review of Ecology and Systematics 15:65–95.

Mahalovich, M. F., and R. J. Hoff. In press. Whitebark pine operational cone collection instructions and seed transfer guidelines. *In* Seed handbook for regions 1–4. USDA Forest Service, Forest Service Handbook 2409.26f.

Millar, C. I., and W. J. Libby. 1991. Strategies for conserving clinal, ecotypic and disjunct population diversity in widespread species. Pages 149–170 *in* D. A. Falk and K. E. Holsinger, editors. Genetics and conservation of rare plants. Oxford University Press, New York.

Morgan, P., and S. C. Bunting. 1992. Using cone scars to estimate past cone crops of whitebark pine. Western Journal of Applied Forestry 7:71–73.

Mosseler, A. J., K. N. Egger, and G. A. Hughes. 1992. Low levels of genetic diversity in red pine confirmed by random amplified polymorphic DNA markers. Canadian Journal of Forest Research 22:1332–1337.

Oleksyn, J., M. G. Tjoelker, and P. B. Reich. 1998. Adaptation to changing environments in Scots pine populations across a latitudinal gradient. Silva Fennica 32:129–140.

Rehfeldt, G. E., R. J. Hoff, and R. J. Steinhoff. 1984. Geographic patterns of genetic variation in *Pinus monticola*. Botanical Gazette 145:229–239.

Rehfeldt, G. E. 1988. Ecological genetics of *Pinus contorta* from the Rocky Mountains (USA): A synthesis. Silvae Genetica 37:131–135.

———. 1989. Ecological adaptations in Douglas-fir (*Pseudotsuga menziesii* var. *glauca*): A synthesis. Forest Ecology and Management 28:203–215.

Rogers, D. L., C. I. Millar, and R. D. Westfall. 1999. Fine-scale genetic structure of whitebark pine (*Pinus albicaulis*): Associations with watershed and growth form. Evolution 53:74–90.

Schuster, W. S., D. L. Alles, and J. B. Mitton. 1989. Gene flow in limber pine: Evidence from pollination phenology and genetic differentiation along an elevational transect. American Journal of Botany 76:1395–1403.

Shutyaev, A. M., and M. Giertych. 1997. Height growth variation in a comprehensive Eurasian provenance experiment of *Pinus sylvestris* L. Silvae Genetica 46:332–349.

Stutz, H. P., and J. B. Mitton. 1988. Genetic variation in Engelmann spruce associated with variation in soil moisture. Arctic and Alpine Research 20:461–465.

Tomback, D. F. 1988. Nutcracker-pine mutualisms: Multi-trunk trees and seed size. Pages 518–527 *in* H. Ouellet, editor. Acta XIX Congressus Internationalis Ornitholgici. Vol. 1. University of Ottawa Press, Ottawa, Ontario, Canada.

Weaver, T., and F. Forcella. 1986. Cone production in *Pinus albicaulis* forests. Pages 68–76 *in* R. C. Shearer, compiler. Proceedings—Conifer tree seed in the inland mountain West Symposium. USDA Forest Service, Intermountain Research Station, General Technical Report INT–203, Ogden, Utah.

Westfall, R. D. 1992. Developing seed transfer zones. Pages 313–398 *in* L. Fins, S. T. Friedman, and J. V. Brotschol, editors. Handbook of quantitative forest genetics. Kluwer Academic Publishers, Dordrecht, The Netherlands.

Wright, J. W., and H. I. Baldwin. 1957. The 1938 International Union Scotch pine provenance test in New Hampshire. Silvae Genetica 6:2–14.

Yandell, U. G. 1992. An allozyme analysis of whiteback pine (*Pinus albicaulis* Engelm.). M.S. thesis, University of Nevada at Reno.

Chapter 16

Growing Whitebark Pine Seedlings for Restoration

Karen E. Burr, Aram Eramian, and Kent Eggleston

There have been major losses of mature whitebark pine (*Pinus albicaulis* Engelm.) trees during the past several decades, and ecologists forecast continued mortality without an earnest effort to reverse this trend. Restoration work aimed at reducing mortality, particularly that caused by white pine blister rust (*Cronartium ribicola*), has begun by identifying and reproducing rust-resistant genotypes. This has resulted in a demand for greenhouse-grown seedlings propagated with seeds from resistant trees. It has been estimated that as few as 10 to 15 percent of whitebark pine seeds germinate in the first year under natural conditions, whereas nearly 90 percent germinate in the laboratory if treated (McCaughey 1993; Chapter 6). A successful nursery production program that combines germination under laboratory conditions with protected and accelerated growth of germinants under greenhouse conditions offers an opportunity to maximize the potential of valuable seed resources. Successful nursery production, however, does not depend solely on events and practices occurring at the nursery. Timely cone collection and safe delivery of high-quality seeds is a critical first step in the process. This chapter discusses nursery production of whitebark pine seedlings from cone collecting to delivery of healthy seedlings at the planting site.

Cone Collection

Cone Crop Assessments

Successful cone collection requires considerable planning, good timing, and conscientious protection of the cones every step of the way. Financial, personnel, and equipment resources must first be made available. Then cones must be located; assessed for genetic sustainability; monitored for quality; protected from animals; gathered without injury to collector, tree, or seeds; and transported to a seed extraction site. An understanding of cone biology is an invaluable asset to have in the planning stage.

The purpose of making cone crop assessments is to locate trees with cones and determine their suitability for collection well in advance of the actual cone-collecting activity. Once sites for restoration have been identified, the question of where to look for cone-bearing trees is very important. It is generally assumed that "local" sources of seeds will produce seedlings maximally adapted to "local" sites. However, recent allozyme loci studies (e.g., Jorgensen and Hamrick 1997) indicate that whitebark pine has very little genetic differentiation among stands; it appears to be similar to western white pine (*Pinus monticola*) in this regard (USDA Forest Service 1999a) (see Chapter 15). Thus, the word "local" takes on a broad meaning. The USDA Forest Service (1999a) concludes that whitebark pine seeds can be safely moved among sites within several conservative seed transfer boundaries based on mountain ranges (Table 16-1). For example, seeds collected in the Sierra Nevada may be safely sown (or planted as seedlings) anywhere within that mountain range, but should not be moved to the Cascade Range without some risk of maladaption. A slightly broader whitebark pine seed transfer rule based on the point of seed origin may also be considered for improving the disease resistance of the population: 1° latitude +/– 80 kilometers, 1° longitude +/– 80 kilometers, and no restriction on elevation transfer (USDA Forest Service 1999a; see also Chapter 17). In British Columbia, seed planning zones to guide seed transfer have been delineated based on broad climatic regions for most conifers on an individual species basis (Eremko et al. 1989; Portlock 1996).

Once possible cone collection sites have been located within the designated region, choosing which trees to collect from must be done carefully (USDA Forest Service 1999b). A primary goal is to enhance the resistance of trees in the next generation to white pine blister rust. In whitebark pine stands with blister rust mortality of 90 percent or more, selecting apparently resistant trees as a

Table 16-1. Conservative geographic seed transfer boundaries for whitebark pine (*Pinus albicaulis*) based on mountain ranges (USDA Forest Service 1999a).

Number	Seed Transfer Zone
1	Sierra Nevada Range
2	Cascade Range (California to British Columbia)
3	Selkirk and Cabinet Ranges
4	Lake Pend Oreille to Lolo Pass, Bitterroot Range
5	Mission Range east to Glacier National Park
6	Mountain areas east of Glacier National Park, except Yellowstone National Park
7	Greater Yellowstone National Park and the Teton Range
8	Lolo Pass, Bitterroot Range to the Idaho Plateau
9	Great Basin Ranges, Nevada (and other isolated populations)

cone source is not difficult. All living trees, although infected, are likely to have some resistance. Any individual trees without cankers are the optimal cone sources, but if none can be found, trees with five or fewer cankers are chosen. In stands with 50 to 90 percent mortality, individuals with five or fewer cankers are also chosen. It is recommended that no cones be collected from stands with less than 50 percent mortality from blister rust. Natural selection has not progressed far enough to identify resistant trees with confidence. Field guides are available to aid in recognizing blister rust infection on whitebark pine (Hoff 1992).

The first assessment of the cone crop on the selected trees may be conducted the year before the cones are collectible. At this time, the purple seed cones are approximately 2 or 3 centimeters long (McCaughey 1994; Chapter 6) and can usually be counted from the ground with binoculars. A subsequent survey is made in the year of collection following spring fertilization to update estimates of the quantities of cones available and the numbers of seeds ripening within the cones. This second estimate is lower than the first, because winter losses and the impacts of disease and insects are subtracted. The information gathered determines the feasibility of collecting. Cone collecting may not be possible in a given year, because the frequency of large whitebark pine cone crops is once every three to five years (Krugman and Jenkinson 1974). Once the decision is made to collect, it is necessary to determine the numbers of cones to be collected, and to plan financial, staffing, and equipment needs.

Monitoring Cone and Seed Maturity

Cones must be periodically monitored in the weeks just prior to collection to evaluate maturity or ripeness. The degree of maturity achieved before collection affects seed vulnerability to handling damage, and seed yield, viability, germination percentage, speed of germination, and longevity in storage (Eremko et al. 1989). Thus, the importance of this step cannot be emphasized too strongly.

Three to five two-year-old cones can be sampled from the crowns of half a dozen trees distributed about the collection site at intervals in the weeks prior to maturation (Eremko et al. 1989; Portlock 1996; USDA Forest Service 1999b). To ensure a reliable assessment, it is better to collect fewer cones per tree than to decrease the number of trees sampled. Ripeness is determined by cutting cones lengthwise through the center of the cone with a cone cutter (manufacturing plans, MTDC 1990) and inspecting the exposed seeds (Figure 16-1a). Cone features that can be observed without cutting, such as cone length, width, and specific gravity, are not reliable indicators of maturity (McCaughey 1995).

There should be six to eight filled seeds exposed on the cut surface of a typical cone, if the crop is developing well. When the seeds are ripe, the megagametophyte, or storage tissue surrounding the central embryo cavity (Kolotelo 1997), is white, firm, and fills the seed. The megagametophyte should not shrink away from the seed coat after being exposed for a few hours, and should certainly not have a viscous milky texture, which occurs in very immature seeds. Also, embryos are considered fully mature when they occupy 90 percent or more of the length of the embryo cavity (Eremko et al. 1989; Portlock 1996; USDA Forest Service 1999b). Kolotelo (1997) presents photographs of imma-

Figure 16-1. Shoshone National Forest cone-collecting operation at Union Pass, Wind River Range, Wyoming, September 1999. Cones were collected from whitebark pine (*Pinus albicaulis*) trees that appeared either to be blister rust–free (*Cronartium ribicola*) or had very few cankers. Seeds were extracted and the seedlings grown at the USDA Forest Service, Coeur d'Alene Tree Nursery, Coeur d'Alene, Idaho, for planting in Shoshone National Forest in stands with high mortality from blister rust or high infection levels. a. Cone cutter used to check the ripeness of seeds. b. Harvesting cones with extendable pruning poles from the bucket of a cherrypicker on loan from the local telephone company. c. The use of a cherrypicker for cone collecting simplifies access to the upper tree canopy, where most of the female cones are borne. Photo credits: Norma Williamson, *The Dubois Frontier.* d. Typical morphological variation among dormant, nine-month-old whitebark pine seedlings. These seedlings have completed their first growing cycle, including bud set and meeting chilling requirements, in the greenhouse at the Coeur d'Alene Tree Nursery. Photo credit: Karen E. Burr.

ture embryos to illustrate the correct method for calculating the length of the embryo relative to the length of the embryo cavity. Numerous developmental problems that can be detected by cutting the seeds are also presented in this reference, with a decision-making chart to estimate germinability based on visual inspection.

Whitebark pine seeds are usually not ripe in the U.S. Rocky Mountains until at least August 15 (USDA Forest Service 1999b). Weaver and Forcella (1986) reported that maturation of undisturbed cones continues as late as mid-September or early October. Variability in annual weather patterns can affect maturation dates by weeks (Halstrom 1994). Although the date of maturation varies with location and weather, cones need protection from harvesting by animals to remain on the trees through September, which permits embryo growth to approach the desired 90 percent of the embryo cavity. Collection of viable seeds from cones unprotected from birds or squirrels requires frequent monitoring of animal activity as well as seed development. This helps ensure that cone collection starts at the latest possible date before the animals harvest a significant portion of the crop, but rarely ensures that seeds are fully mature prior to harvest.

Cone Crop Protection

Protecting cones to minimize animal damage and harvesting is becoming standard practice when seeds are in short supply or of high enough value to warrant the time and expense. Cone enclosures made of aluminum foil, paper bags, and nylon mesh are ineffective (VonBonin 1994), but hardware cloth or wire cone enclosures are not likely to be destroyed or removed by Clark's nutcrackers (*Nucifraga columbiana*) and pine squirrels (*Tamiasciurus hudsonicus* or *T. douglasii*). Fabrication of cone cages made of hardware cloth with 0.64-centimeter (1/4-inch) squares is described by the USDA Forest Service (1999b). A piece of hardware cloth approximately 60 × 90 centimeters is needed for each cage; this is folded in half lengthwise, and the open 90-centimeter seam is crimped closed. The tube is slipped over a cone-bearing branch, and the corners nearest the branch tip are folded over so birds cannot get to the cones from the top. The bottom edge of the tube is fastened to the branch to prevent slipping or blowing off in the wind. Cage dimensions should be adjusted to fit branch size and allow room for cones to grow and mature.

Cages must be placed over the second-year cones as early as June in northern Idaho. In other parts of the Rocky Mountains, cages may be installed in July to protect adequately from nutcracker activity. The precise timing of cage installation at each site should be determined by cone crop monitoring. Protected cones remain attached to the trees and can be collected until the sites become inaccessible because of snow (McCaughey and Schmidt 1990). Cones cannot be left on the tree to be gathered the following spring, however, because the seeds do not remain viable (Lanner 1982).

Collection Methods and Techniques

Three methods are commonly used to collect cones from whitebark pine trees: climbing, pole-pruning, and shooting (Eremko et al. 1989; Portlock 1996; USDA Forest Service 1999c). The most cost-effective method varies depending on site accessibility, tree form, size of the cone crop, and numbers of cones to be collected, as well as the skills and training of the collectors (Figure 16-1b, c).

Trees best suited to climbing are those with well-spaced branches capable of supporting the climber (MTDC 1996). Climbing spurs and tree-climbing ladders, if tree form permits, are used to reach the live crown of those trees without branches close enough to the ground to permit free climbing. Spurs do wound the tree and are not recommended for repeated climbing of the same trees. The cones should be within the climber's reach, while the climber remains belted to the main trunk of the tree. Pruning poles and cone hooks, used to pull branches toward the climber, are raised and lowered from the crown by rope. The climber works from the top of the crown down, placing the cones in sacks suspended from branches. Cone sacks are lowered by rope from the crown, rather than dropped to the ground, to minimize cone damage.

The use of an extendable pruning pole from the ground is effective for collecting from shorter trees. A 9-meter (30-foot) pole is manageable by most adults, but a fully extended 12-meter (40-foot) pole may require two people to maneuver. Clippers or saws are mounted on pole ends. Pruning is done judiciously to minimize defoliation and wounding, which might negatively impact future cone crops. Cones should not be collected from the lower crown because a high percentage of self-pollination (inbreeding) is likely.

Branches and cones can be shot out of trees easily and quickly using a .22-caliber rifle with a variable 3×9 scope. Alternatively, Portlock (1996) reports that stems up to 15 centimeters in diameter can be felled with a .30-30–caliber rifle and soft-nose shells. Shooting may be practical for smaller collections of cones from otherwise unreachable upper crowns. The shooter must be a marksman, and if on U.S. federal forest land, certified to do this work. Extreme care is taken to ensure the safety of surrounding property and persons, because bullets striking tree limbs and trunks are often deflected in unexpected directions.

If it is deemed reasonable to forgo individual tree collections (e.g., all remaining trees in a stand have blister rust resistance), collecting from pine squirrel middens, which are piles of cones, is a possible option (Eremko et al. 1989; Portlock 1996; USDA Forest Service 1999c). However, collecting from middens is generally discouraged when collecting by other methods is possible, because the parent trees of cached cones are unknown, and the cones are more likely to be infected with seedborne fungi that can adversely impact germination. When using cached cones, the USDA Forest Service (1999b) recommends collecting from no fewer than three middens in a collection area, yielding 200 to 300 cones, to increase the odds of adequately representing the gene pool. Middens are not difficult to locate, because squirrels usually use the same ones for several years. Squirrels do start cutting cones prior to seed maturity, so cones must be opened and examined as previously described. Cones are removed only from the clean, drier, upper layers of middens to reduce infection by seedborne fungi. Because cone middens are usually located in cool, shady, damp areas, cones collected from them are wetter than those collected by other methods (Portlock 1996). Thus, proper field care to promote drying is particularly important to maintain the quality of cones from middens.

Field Care of Cones

Seeds are perishable and may become damaged if cones are not properly handled after collection, even if they are collected when fully mature (Eremko et al. 1989). Cones are picked from branches, cleaned free of debris and damaged cones, and bagged in burlap sacks as quickly as possible. Sacks are half-filled and tied with twine at the top of the sack so that the cones are spread in a layer about two cones thick when the sack is stored horizontally. Sacks are clearly identified with a cone tag attached at the tie and with a duplicate tag inside. The tag indicates ownership, lot identification, and numbers of sacks in the lot. Sacked cones are kept dry, cool, and well ventilated at the collection site, and are transported to temporary storage facilities daily. To prevent seed injury, sacked cones must not be thrown or dropped in the process of moving them from one location to the next.

Once the cones are at the interim storage location, the primary concerns are to promote drying and to minimize overheating (Eremko et al. 1989; Portlock 1996). Cones that remain damp will be susceptible to mold. Respiration of living tissues within the cones generates heat that must be dissipated. Storage in cone sheds protects cones from radiant heating and precipitation. If cones arrive wet at the temporary storage facility—that is, with surface moisture and/or in wet sacks—they are removed from sacks, spread out, air-dried for several days, and then re-bagged in clean, dry sacks. Separation of sacks, laid on their sides on racks allowing complete air circulation, assures that temperatures inside the sacks do not exceed 25°C. Heat damage to seeds is often evident from yellowing and withering of the megagametophyte (USDA Forest Service 1999c).

Transportation of Cones to the Extractory

Extractories, or facilities where seeds are extracted from cones, are often co-located with nurseries and long-term storage units for cleaned seeds: they are rarely located near field sites where cones are collected. The primary concern during lengthy transport of large quantities of cones to extractories is overheating (Eremko et al. 1989; Portlock 1996). Sacked cones are loaded onto trucks just prior to transport and are kept dry, cool, and well ventilated on racks as necessary. Parking of loaded trucks in the sun, as well as open transport during rainy weather, must be avoided. The shortest route to the extractory is used to minimize transport duration. The arrival time at the extractory is coordinated with the extractory staff to facilitate efficient transfer of the cones and related documentation. Completed collection reports are delivered with the cones so that tracking seed identity and ownership through the extraction process is error-free. The report details information about the collection date and method, the site location and elevation, the condition of the cones and seeds upon arrival at the extractory, as well as the information on the cone tag.

Nursery Techniques

Seed Extraction and Processing

A full-service nursery, such as the U.S. Forest Service's Coeur d'Alene Tree Nursery, can remove seeds from the cones, test the quality of seeds, store seeds for future use, prepare seeds for germination, produce seedlings, prepare seedlings for field planting, store seedling for future planting, and deliver seedlings to planting sites.

Seeds are extracted from mature whitebark pine cones quickly and with little effort, once the cones have dried to the point where they can be crumbled apart by hand. The cones of other conifers are often dried by placing them, unsacked, on a kiln at 40°C for about a week (Krugman and Jenkinson 1974; USDA Forest Service 1999d). Whitebark pine cones, however, are best dried more slowly over one or two months at room temperature on drying racks with good air circulation. The hope is that slower drying permits further embryo maturation. Once dry, cone scales separate and can be broken apart from the central cone axis, allowing the seeds to fall away. Portions of the scales may not separate easily even when dry. Examination of the seeds removed from these scales by hammering them apart indicates that such seeds are not fully developed and can be discarded.

The extraction process results in a mix of seeds and cone pieces that are both larger and smaller than the seeds. Automated cleaning machines, such as scalpers (model 36-S-2, Hance Corp., Westerville, Ohio) and clippers, (model M-2B, A.T. Ferrell and Co., Saginaw, Michigan) are available for use with large seed lots. These are two-screen cleaners that separate the seeds from debris, based on size differences. A screen with holes larger than the seeds allows seeds to fall through while trapping the larger cone scales. A screen with holes smaller than the seeds allows finer particles to fall through while trapping the seeds. With small lots of whitebark pine cones, the clipper screens are used by hand to achieve the same results.

Next, debris the same size as the seeds and hollow seeds are removed with pneumatic separators. The Rainbow, the Barnes Tree Seed Separator (Barnes Tree Improvement Co., Cottage Grove, Oregon), and the Dakota Blower (E.L. Erickson Products, Brookings, South Dakota) all have a vertical air column that blows up from beneath the seeds. Separation occurs when lighter, empty seeds and debris are lifted up and away, while heavier, full seeds fall. The Rainbow quickly separates the majority of the filled seeds from a mix of hollow and full seeds. The small fraction with hollow seeds is then cleaned with one of the other two smaller machines to recover full seeds that were removed with the hollow seeds.

Seed Quality Testing

The first and most informative test performed on the cleaned seeds is an X ray. A nondestructive X ray of 50 or 100 randomly sampled seeds of a seed lot can be taken in several minutes. This procedure clearly reveals hollow, partially filled, and completely filled seeds, embryo presence and size within the embryo cavity, substantial mechanical damage such as seed coat or megagametophyte cracking, and insect larvae within seeds. X rays illustrating all of these conditions are presented in Leadem (1981).

First, the X ray is used to assess the quality of seed cleaning. If hollow seeds are present, the lot is re-cleaned. Hollow seeds are a ready source of fungi that will form visible mycelia after only a few weeks when fresh seeds are stored at

2°C. A small fraction of partially filled seeds is usually accepted because of the difficulty in separating it from small, completely filled seeds of similar weight, and the need to maximize the potential of small whitebark pine seed lots.

Next, the X ray is used to estimate the potential germination percentage of the newly collected whitebark pine seed lot. Once the obvious deductions have been made for partially filled seeds, seeds without embryos, and mechanical and insect damage, the size of the embryo relative to the embryo cavity is the critical feature. When embryos fill 25 to 40 percent of the embryo cavity, cone-scale tissue is often attached to the seed coat, and little or no germination is possible. When embryos fill at least 60 percent of the embryo cavity, reasonable germination of fresh seeds is possible, although successful storage of the seeds is doubtful. Better germination and storage are achieved when embryos fill at least 75 percent of the embryo cavity, possibly because ripening continues after the cones have been collected (Eremko et al. 1989). Most whitebark pine seeds that arrived at the Coeur d'Alene Tree Nursery were in these varying stages of immaturity; thus, no whitebark pine seed lot in our experience germinated or stored as well as fully mature (embryo length 90 percent of the embryo cavity) seeds of the other pines we grow, e.g., western white, lodgepole (*P. contorta*), and ponderosa (*P. ponderosa*).

Estimation of potential germination from an X ray does have limitations. The damaging effects of excessive drying or frozen storage at high moisture content cannot be detected. Similarly, complex dormancy mechanisms are undetectable in the radiographs. This is particularly problematic after frozen storage. For example, a review of X rays of 100 individually numbered, previously frozen seeds revealed no feature that could distinguish the 50 percent of them that germinated in a standard test. All looked capable of germinating.

Additional information on a seed lot is gathered once the X ray indicates that the seeds are of sufficient quality to proceed. Purity analysis determines the percentage of pure seed (versus debris) in the lot. Minimum purity standards exist for most conifers (USDA Forest Service 1999d); for example, 97 percent is the minimum acceptable purity for ponderosa pine. No standard is specified for whitebark pine, but achieving greater than 99 percent pure seed is not difficult for cleaned seeds of this size.

Seed moisture content is measured quickly, using 5 grams of seeds and a Mettler balance with an infrared dryer attachment (model PM400 with LP16M, Mettler Instrument Corp., Hightstown, New Jersey). The test is destructive, which is problematic for very small lots, but it is essential for lots that will be frozen for future use. Fresh seeds often have a moisture content of 12 to 15 percent, but freezer storage at this moisture level has been lethal in our experience. Seeds with a moisture content higher than 9 percent are air-dried and retested until they reach 9 percent, if the seeds are to be stored frozen. Once seed moisture is at an acceptable level, seed weight is measured as seeds per pound or as grams per 100 seeds. Both purity and weight are needed to determine the bulk weight of seeds to sow.

Standard operating procedure dictates that a germination test be conducted on all new seed lots. However, this may or may not occur with whitebark pine. If a seed lot is to be completely sown shortly after collection, a germination test will be incorporated into the operational sow. Germination testing may not be valuable for lots that are partially sown shortly after collection due to the 110-day wait for results, especially if the X ray determines roughly the quantity of seeds to sow. Owners of small lots to be stored for later use may not want to sacrifice 200 seeds for an estimate of the germination percentage, but may instead choose to rely on the information provided by the X ray. Owners of large lots to be stored for later use may choose to take a 200-seed sample to estimate pre-storage germination percentage. However, germination percentages of lots in storage at the Coeur d'Alene Tree Nursery have been observed to drop 30 to 65 percent after one or two years of frozen storage. Data gathered prior to storage are thus of limited value for estimating the quantity of seeds to sow after storage.

When a germination test is conducted at the Coeur d'Alene Tree Nursery, the protocol is the same as that used for operational germination of seed, presented in the sections "Seed Stratification" (p. 335) and "Germination" (p. 338) of this chapter. The protocol has been substantially changed from that previously reported (Eggleston and Meyer 1990; McCaughey and Schmidt 1990; McCaughey 1994). Standardized environmental conditions and pretreatments for germinating seeds of countless species have been developed by the Association of Official Seed Analysts (1993) and the International Seed Testing Association (1985). The purpose of standardization is to assure that germination test results reported by many testing facilities are comparable. However, the treatments and conditions specified for whitebark pine produce very poor results in our experience, and we have chosen not to use them. Reports of whitebark pine germination in the literature also reflect a consistent divergence from the standard procedures of the seed testing associations (e.g., Pitel and Wang 1980). In reports prior to the 1980s, whitebark pine germination percentages were characterized as poor and variable. This is probably the result of working with seeds having varying levels of embryo immaturity, as well as a general failure to understand the requirements of the species. Recent reports in the literature, as well as our own experiences germinating whitebark pine seeds, are more encouraging.

Seed Storage

The ability to store seeds is essential if seeds are to be available when cone crops are not. Extracted seeds are prepared for storage as quickly as possible to minimize deterioration. Seed lots to be stored for sowing in future years are moisture tested and weighed, and then sealed in 6-mil plastic and stored within covered barrels at –15°C until needed. When seeds are removed from these lots, they are brought to room temperature and then sampled quickly to maintain the initial seed moisture content. The practice is the same for other conifer seeds stored at the Coeur d'Alene Tree Nursery. Mature seeds of many species of *Pinus* may be stored for fifteen to twenty years or more under these conditions (Krugman and Jenkinson 1974). Immature *Pinus* seeds generally do not have this level of longevity.

Storage longevity of whitebark pine seeds has not been determined; we have had difficulty germinating seeds after frozen storage. For example, a seed lot with embryos 60 percent of the length of the embryo cavity had 80 percent germination success when sown prior to frozen storage, but only 55 percent germination success after two years of storage. The seeds may not have stored well because they were immature. It may be necessary to increase embryo development prior to frozen storage, as is done with Swiss stone pine (*Pinus cembra*) (Frehner and Schönenberger 1994). The dormancy-breaking requirements may have changed once the seeds were dried and frozen, and we were less successful because we used the same germination pretreatment before and after storage. A similar situation has been observed in lodgepole pine: Seeds from the previous year's cones, which overwintered on the tree, responded differently to stratification than seeds of current-year cones collected before winter (Hellum and Dymock 1986).

For temporary storage prior to stratification and planting, fresh seeds are sealed in 6-mil plastic with the moisture content they had following extraction. They are then stored at 2°C for about one month, depending upon extraction date, until it is time to begin stratification to ready the seeds for germination in synchrony with the early April greenhouse sowing schedule.

Seed Stratification

Stratification is the practice of placing seeds in a moist environment to hasten afterripening or to overcome dormancy (Bonner 1984). Afterripening includes the physiological processes of seed maturation, which occur after cone collection, that may be necessary for germination, for example, embryo development as evidenced by continued elongation. Dormancy refers to the condition of a seed when it is mature, viable, and imbibed, but it does not germinate, even in the presence of favorable environmental conditions (Bonner 1984). Dormancy assures that seeds do not germinate as winter approaches. Most mature whitebark pine seeds are dormant and will not germinate when simply planted, watered, and kept warm. The choice of stratification treatment depends on the dormancy mechanisms involved. These mechanisms are poorly understood in whitebark pine, making this one of the more challenging aspects of growing this species. Physiological dormancy within the embryo itself, as well as seed coat dormancy resulting from impermeability to gases or moisture or mechanical restrictions both appear to be involved (Leadem 1986; Pitel and Wang 1990; Chapter 6). An operational stratification process has been selected at the Coeur d'Alene Tree Nursery that addresses many of the possible afterripening needs and dormancy mechanisms.

First, whitebark pine seeds to be sown, whether just collected or thawed after frozen storage, are placed in mesh bags and soaked for forty-eight hours in running tap water so they will take up water while being aerated. This is a common practice for conifer seeds, which reduces the levels of seedborne pathogenic fungi relative to standing water soaks (Axelrood et al. 1995). Seed lots expected to have high germination success are kept separated from seed lots of poor quality to reduce fungal contamination. The amount of seedborne fungi, both saprophytic (e.g., *Penicillium* molds, James 1995) and pathogenic (e.g., *Fusarium* spp., James 1986), is strongly related to seed quality and can vary dramatically among lots.

Poor-quality seeds (e.g., with incomplete development, cracked seed coats, etc.) can develop very high levels of fungal growth during stratification and germination. However, we are not currently applying surface sterilization treatments prior to soaking to reduce the levels of seedborne fungi, although this is frequently recommended (James 1986; Landis 1989a; James 1995; Littke 1996). Landis (1989a) reviewed the mixed results achieved with hydrogen peroxide and sodium hypochlorite on species other than whitebark pine. The correct combination of chemical concentration, soaking time, and rinsing protocol to match the particular seed coat thickness and surface texture to reduce pathogens and increase germination has been reported by some workers, while others report doing more harm than good with these treatments. Our results have not been promising. For example, soaking in 30 percent hydrogen peroxide for varying lengths of time (see Barnett 1976) drastically reduced saprophytic molds but did not improve germination success or noticeably reduce the percentage of germinants infected by seedborne *Fusarium*. Soaking in 1 percent bleach for five or ten minutes did not reduce seedborne *Fusarium*, but it did reduce 7-day germination by 35 to 70 percent in some lots. Although further testing of sterilant concentrations and treatment duration is needed, *Fusarium* propagules may survive these types of treatments because they are carried inside the seeds or are trapped in seed coat cracks (Landis 1989a). Since surface sterilants are not 100 percent effective, and the lengthy stratification process promotes fungal development and provides ample opportunity for recontamination, achieving greater numbers of surviving seedlings is difficult with heavily contaminated lots. Removing hollow seeds, sanitizing nursery equipment (Neumann et al. 1997), keeping different lots of seed separated, and using the running water soak are all important. But we believe the action that makes the greatest impact on seedborne pathogen levels is to begin with good-quality, mature seeds.

Following the forty-eight-hour running water soak, whitebark pine seeds receive a 28-day warm stratification treatment to improve embryo development. The seeds, still in mesh bags, are placed in 1-mil plastic bags (to slow water loss while permitting oxygen and carbon dioxide exchange) and are then spread out one or two seeds deep on shelves in germinators. The germinators, resembling large household refrigerators with excessive shelves, lights, and fans, are maintained at 20°C night/22°C day (68/72°F) with a twelve-hour photoperiod. The mesh bags of seed are removed from the plastic bags weekly for a one-hour running water soak to maintain complete imbibition, leach out inhibitory substances if present, and keep fungi in check. We tested 0-, 14-, and 28-day warm stratification periods on fresh seeds of one lot with embryos averaging 60 percent of the length of the embryo cavity at collection. Germination was 14 percent with no warm stratification, 53 percent with 14 days, and 80 percent with 28 days, when followed with the rest of the stratification and germination procedures described below. It was possible to watch the embryos enlarge by taking X rays before and during warm stratification. Similar before and after X rays are presented in Leadem (1986). At the end of 28 days, nearly all embryos in this lot had more than doubled in diameter, becoming as wide as the embryo cavity and 90 percent as long as the cavity. This evidence for embryo development during warm stratification suggests that the seed coat dormancy is not an issue of impermeability to oxygen or water preventing embryo growth, but rather one of mechanical resistance to radical emergence, assuming externally supplied oxygen and water are necessary for embryo growth.

Some seeds do germinate during warm stratification. They are planted when radical emergence is noted during inspection at weekly soaks. The percentage of these seeds without dormancy mechanisms varies by lot but is typically less than 1 percent. It may be a bit of a nuisance from an operational perspective to care for such a small percentage of the seedlings so far ahead of the main crop. However, it is very important to maintain the diversity of the gene pool with regard to the seed dormancy trait from an ecological perspective. Whereas it is established that genetic diversity is essential in general for natural selection in whitebark pine, diversity in the dormancy trait may provide food for Clark's nutcrackers more than one year after a cone crop, thereby strengthening the mutualistic relationship (Lanner and Gilbert 1994). Seeds of western white pine have a wide range of dormancy mechanisms (Hoff 1986), and perhaps the same will be demonstrated for whitebark pine seeds with further investigation.

After the 28-day warm stratification period and a one-hour running water soak, the seeds are spread out, still in mesh bags, and returned to new 1-mil plastic bags, and placed in a dark stratification room at 2°C for approximately 60 days. The combination of 28 days warm and 60 days cold stratification was better than 28 days warm and 0, 14, 28, or 45 days cold, but extending the cold stratification to 90 and 120 days produced similarly good results. Others have found 60 days cold stratification to produce better results than 30 days (Leadem 1986; Pitel and Wang 1990). The weekly soaks are continued during cold stratification to leach out inhibitory substances that may be present and to continue to remove seedborne fungi.

Once cold stratification is completed, seed coat dormancy is addressed. The seed coats are nicked with a scalpel on the line visible on the seed coat that marks the junction of the two halves where the splitting will occur. The nick, approximately 2 millimeters long, is made at the radical end of the seed, 1 millimeter from the pointed tip of the seed coat so as not to break the tip of the seed coat and damage the end of the radical. Cutting into the megagametophyte tissue is avoided, because that damaged tissue appears to be an entry point for fungal infection. With this approach, water and oxygen are not limited by the membrane surrounding the megagametophyte. Impermeability as a dormancy mechanism, if initially present, is probably removed during the 90 days of stratification (Hoff 1986). Seeds that crack open during warm or cold stratification are not nicked. Cracked seeds generally comprise only 5 to 20 percent of the total number of seeds in a lot. Nicking is quite time-consuming, but an alternative method useful on a large scale that similarly results in high, uniform germination (i.e., most seeds germinate, and at the same time) necessary for operational production has not been found. Without nicking, the other 95 to 80 percent of the seeds will not germinate, given the stratification treatments described. Others have reached the same conclusion (Leadem 1986; Pitel and Wang 1990). Completely removing the seed coat does not improve germination beyond that achieved with nicking (Leadem 1986).

Germination

Seeds are placed nicked edge down and not touching each other (Pitel and Wang 1980) on wet 20-ply Kimpak germination papers in covered germination boxes (14 × 14 × 4-centimeter box of clear rigid plastic). Seed water balance is maintained by uptake from the paper, and seeds that develop excessively moldy seed coats can be removed with minimal contamination of neighboring seeds. Seeds that develop excessive mold in the first few days under germinating conditions are generally not viable.

The germination boxes are kept in germinators at 20°C night/22°C day with a twelve-hour photoperiod. The day temperature is cool relative to the standard 30°C germination test conditions recommended for many conifer species (International Seed Testing Association 1985; Association of Official Seed Analysts 1993). However, this approach results in greater germination success apparently by reducing the rate of seedborne fungal development while not noticeably impacting the speed of germination. Most seeds capable of germinating do so within the first two weeks, but germination will often continue for six to eight weeks. The later germinants have low vigor and tend to succumb to diseases once reaching the greenhouse. It is probably cost-effective from an operational perspective to discontinue the germination process at five weeks. If speed of germination is only a function of the physiological quality of the seeds and not regulated by genetic differences, the practice should have no ecological implication.

Planting Germinants

When radicals emerge 3 or 4 millimeters from the seed coats, the germinants are planted by hand, one per cell, into 164-cm^3 (10-in^3) Ray Leach Super Cells (Stuewe and Sons, Inc., Corvallis, Oregon) (Landis 1990a). Seeds are planted one per cavity, because about 90 percent of the seeds planted with radicals emerging will continue to grow if maintained in a low-stress environment. The growing medium is a sphagnum peat-Douglas-fir wood chip blend (Landis 1990b). Once emerged, a radical can easily grow 1 or 2 centimeters in a twenty-four-hour period in a germination box, which is similar to growth rates observed under experimental conditions by Jacobs and Weaver (1990). Thus, the seeds must be planted promptly to ensure straight taproot development and to minimize planting time. Seeds are planted so that the tops of the seeds are at the surface of dampened medium. Floating of the seeds up and out of the medium with future watering is reduced by pre-watering the containers and surrounding the seeds with medium. Conversely, planting the seeds any deeper delays the start of photosynthesis and can drastically reduce the survival of seedlings with initially low vigor levels. The medium is topped with a 0.5-cm layer of perlite (Landis 1990b) to reduce evaporative water losses.

Initially, containers at the Coeur d'Alene Tree Nursery are kept in the extractory at a constant 21°C air temperature. Light watering by hand is sufficient to keep the perlite damp, since evaporative losses are minimal. The containers remain there for about a week until germinants begin to come up through the perlite. Only then are they moved to a greenhouse where environmental conditions fluctuate more.

This entire process is quite different from standard practice for conifer container production. Typically, conifer seeds are mechanically sown (without nicking) after cold stratification onto the surface of dry medium with multiple seeds per cell, topped with perlite, and moved directly to a greenhouse with daily high temperatures reaching 24°C. We did try the standard practice, altered only by hand sowing after nicking, but germination was reduced by half. Many seeds died with radicals penetrating the medium prior to shoot growth. These potential germinants were likely killed by seedborne pathogens that were at high levels due to sowing multiple infected seeds per cell at higher medium temperatures. Some seeds floated up into the perlite with watering and dried out. Drying may have been a problem in general for the seeds, because they were sown on the surface of the medium, covered only with the light, porous perlite. Although most conifers naturally germinate on the surface, whitebark pine seeds apparently have evolved to germinate when buried (McCaughey and Weaver 1990). Another possible reason for the decrease in germination is that optimal germination temperatures are simply lower and fluctuate less for whitebark pine than for other conifers that inhabit lower elevational zones.

Seedling Greenhouse Production

At this stage in the process at the Coeur d'Alene Tree Nursery, responsibility for the care of the whitebark pine is transferred from the seed horticulturist to the greenhouse seedling horticulturist. Information about the crop, such as germinant numbers, vigor, and disease issues, is transferred as the newly planted germinants are physically moved to the greenhouses.

First growing season and overwintering. The emerging whitebark pine seedlings are placed in the greenhouse in early April when other conifer seedling crops are at the same stage of development. The photoperiod or perceived daylength is typically increased to eighteen hours to promote growth in most conifers this early in the year. We are trying a twenty-four-hour photoperiod with whitebark pine, which is often necessary with very high elevation ecotypes of other conifer species (Landis et al. 1992). Day and night temperatures are kept around 21°C, which is possible at our location in April. This is a challenge on sunny days later in the year, making shade cloth or other greenhouse covering a must (Landis et al. 1992), if the seeds do not start the stratification process by the previous December. Minimally fluctuating growing temperatures during this first *establishment phase* provide an advantage to whitebark pine rather than to seedborne fungi.

Taproot extension, shedding of the seed coat, and opening of the cotyledons occurs in that order within the first week or two in the greenhouse. Low-vigor seedlings have difficulty shedding the megagametophyte tissue before it dries and strangles the cotyledons, preventing them from opening. This tissue is often removed by hand without damage to the seedling to increase survival potential. Once the seed coats are shed, fertilizer is applied with irrigation. There are many complete conifer fertilizers available which promote rapid growth (Landis 1989b). At the Coeur d'Alene Tree Nursery, the same fertilizer is used for whitebark pine as for other conifers in production.

During the next six or seven weeks, growth occurs predominantly in the root system and primary needles, with some increase in stem diameter resulting from lignification. Lignification is an important step in developing resistance to *Fusarium* infection of stems. Controlling *Fusarium* and other disease organisms continues to be critical during this period (Landis 1989a). The best approach to minimize seedborne *Fusarium* infection of germinants is to collect mature seeds that will germinate rapidly and grow vigorously. However, an equally vigorous sanitation and fungicide application program can be essential. The primary cause of seedling mortality throughout the entire greenhouse process is infection by pathogenic seedborne fungi within the first several weeks. A 25 percent mortality rate is estimated for first-year losses in high-vigor seedlings. This mortality rate can double or triple in low-vigor seedlings from heavily infected seeds.

At three months, mature needle buds form and lateral shoot development begins. This second stage of development is the *rapid growth phase,* and is defined as the period when seedlings grow in height and weight at an exponential rate (Landis et al. 1992). In reference to whitebark pine, "rapid" is a bit of a misnomer. Seedlings increase in height very little. The mature needles are often longer than the seedling is tall. However, seedling weight does increase substantially due to lateral branching and an increase in stem diameter. Because disease problems are less threatening, day temperatures are permitted to fluctuate more than during the establishment phase, although spikes above 30°C are reduced with misting.

Whitebark pine seedlings enter the third growth stage known as the *hardening phase* at about five months of age when shoot buds set, which is apparently timed by internal cues. In contrast, most greenhouse-grown conifers are induced to set buds when seedlings approach target sizes by treatments, such as decreasing temperatures, daylength, and/or nitrogen levels, and increasing water stress. However, with whitebark pine, we have been forced to follow its lead, not vice versa, as we would like. Whitebark pine seedling shoots are rarely longer than 5 centimeters at five months. Mature needle fascicles are evident on 80 percent of the seedlings. All seedlings, whether only with primary needles or with a combination of primary and secondary needles, have huge (7 to 10 millimeters long), elongated buds. Many seedlings are shrubby in form with lateral buds similar in size to their terminal bud. Root and stem diameter growth continues for another month or two, depending on the sowing date. Stem diameter can approach 5 millimeters, with the average over 3.5 millimeters. For comparison, western white pine shoot height and stem diameter average 15 centimeters and 3.0 millimeters, respectively, at this time.

A combination of changing from high- to low-nitrogen fertilizer, declining day and night temperatures, and a shortening natural daylength promotes dormancy induction. A period of chilling is then required to overcome seedling bud dormancy and resume shoot growth.

The seedlings are gradually hardened in a greenhouse or shelterhouse beginning in September or October. The hardening process is associated with an increase in stress tolerance. Irrigation is tapered off, resulting in moderate water stress. Air temperatures drop to near freezing, and shoot tissues develop an ability to withstand freezing temperatures without injury. Fertilizer is withheld when day and night temperatures no longer permit root growth. Seedlings are protected from below-freezing temperatures in either location, however. Roots do not cold harden to the extent that shoots do, and this eliminates the necessity to mulch the containers to protect root systems from low-temperature injury.

Second growing season. Seedlings are induced to begin shoot growth with warm temperatures and a twenty-four-hour photoperiod in a greenhouse in early February to increase the length of the growing season (Figure 16-1d). A single flush of shoot growth typically lasts six to eight weeks. A second flush of shoot growth can be promoted in most seedlings if high-nitrogen fertilizer is applied until the terminal and lateral buds to be overwintered appear in June. The nutrient regime is then changed to high phosphorus and potassium to promote root growth and stem diameter development. At this time, average shoot height is 15 centimeters, although some seedlings reach 25 centimeters before setting buds. Stem diameter can approach 8 millimeters with an average of 5 millimeters. Extensive lateral branching is common. In general, disease problems are not as severe in the second season. However, whitebark pine seedlings require more treatment for root diseases, for example, *Cylindrocarpon* spp. (James 1991), *Fusarium,* and root zone insects—particularly fungus gnats (Sciaridae, *Bradysia* spp.) and springtails (Order Collembola)—than other container-grown conifer seedlings. Losses in the second season from all causes typically do not exceed 25 percent.

Seedling Storage and Transport to the Planting Site

Seedlings are prepared for summer (July), or fall (September) outplanting after the second growing season, as well as for spring planting in the following year. Seedlings are developing terminal and lateral buds in July. Seedlings for fall planting have fully developed buds with dormancy induced to eliminate the possibility of flushing until the following year. Seedlings for spring planting have chilling requirements met and are ready to flush when temperatures warm at the planting site. White root tips should be present on the seedlings delivered in July and September to ensure good root growth upon planting.

Seedlings for summer or fall planting are extracted from the containers immediately before scheduled pick-up or delivery. Whitebark pine seedlings develop extensive root systems during two growing seasons and easily hold the soil together in a firm plug. They are placed ten per plastic bag, covering just the root systems to minimize drying of roots. The bags of seedlings are then packed with minimal air space in waxed boxes, which are put into refrigerated (2°C) storage. Seedlings are delivered by refrigerated (2°C) truck for immediate planting.

Seedlings for planting the following spring are extracted from the containers only after the buds are fully developed, root and stem diameter growth have stopped, and shoots and roots are cold hardy. The seedlings are then bagged and boxed as above for the summer and fall crops. Boxed seedlings are stored just below freezing until they are shipped to the planting sites in refrigerated trucks.

Table 16-2. Planning timeline for whitebark pine (*Pinus albicaulis*) seedling production activities.

Year	Season	Activity
0	Summer	Assess developing female conelets
1	Spring	Re-assess cone crop after fertilization
1	Summer	Install cone cages
1	Fall	Collect cones and ship to extractory
1	Winter	Dry cones; clean seeds; X-ray seeds; determine seeds to sow/store; begin 30-day warm/60-day cold stratification
2	Spring	Nick and sow seeds; plant germinants and move to greenhouse
2	Summer	Initiate buds; shoot height averages 5 centimeters
2	Fall	Develop buds and induce dormancy
2	Winter	Meet dormancy requirements
3	Spring	Break overwintering buds and start second flush
3	Summer	Initiate buds; shoot height averages 15 centimeters; box and ship seedlings
3	Fall	Develop buds and induce dormancy; box and ship seedlings
3	Winter	Harden, box, and store for spring planting
4	Spring	Ship seedlings

Final Comments

There may be increasing demand for whitebark pine seedlings as restoration projects become more widespread and as mortality from blister rust reduces seed availability. Considerable time and planning are required between the decision to restore a whitebark pine community using nursery-grown seedlings and the actual planting of a site. The timeline of events for whitebark pine seedling production is summarized to guide that planning process (Table 16-2). Regardless of the species being restored, similar steps are taken based on the reproductive characteristics of the species of interest. A nursery, with knowledge of those reproductive characteristics and the facilities to maximize that reproductive potential, can be a valuable asset in the restoration process.

Acknowledgments

The authors thank Robert P. Karrfalt, USDA Forest Service, Dry Branch, Georgia, and Mary F. Mahalovich, USDA Forest Service, Moscow, Idaho, for technical reviews of the manuscript.

Literature Cited

Association of Official Seed Analysts. 1993. Rules for testing seeds. Journal of Seed Technology 16:1–113.

Axelrood, P. E., M. Neumann, D. Trotter, R. Radley, G. Shrimpton, and J. Dennis. 1995. Seedborne *Fusarium* on Douglas-fir: Pathogenicity and seed stratification method to decrease *Fusarium* contamination. New Forests 9:35–51.

Barnett, J. P. 1976. Sterilizing southern pine seeds with hydrogen peroxide. Tree Planters' Notes 27:17–19, 24.

Bonner, F. T. 1984. Glossary of seed germination terms for tree seed workers. USDA Forest Service, Southern Forest Experiment Station, General Technical Report SO-49, New Orleans, Louisiana.

Eggleston, K., and J. Meyer. 1990. Containerized whitebark pine nursery production in the Forest Service northern region. Pages 366–367 *in* W. C. Schmidt and K. J. McDonald, compilers. Proceedings—Symposium on whitebark pine ecosystems: Ecology and management of a high-mountain resource. USDA Forest Service, Intermountain Research Station, General Technical Report INT-270, Ogden, Utah.

Eremko, R. D., D. G. W. Edwards, and D. Wallinger. 1989. A guide to collecting cones of British Columbia conifers. Forestry Canada—British Columbia Ministry of Forests, Forestry Canada, FRDA Report 055, Victoria, British Columbia.

Frehner, E., and W. Schönenberger 1994. Experiences with reproduction of cembra pine. Pages 52–55 *in* W. C. Schmidt and F. -K. Holtmeier, compilers. Proceedings—International workshop on subalpine stone pines and their environment: The status of our knowledge. USDA Forest Service, Intermountain Research Station, General Technical Report INT-GTR-309, Ogden, Utah.

Halstrom, L. K. 1994. Whitebark pine cone development and collections in 1993. Pages 102–107 *in* K. C. Kendall and B. Coen, compilers. Workshop proceedings: Research and management in whitebark pine ecosystems. Glacier National Park, West Glacier, Montana.

Hellum, A. K., and I. Dymock. 1986. Cold stratification for lodgepole pine seed. Pages 107–111 *in* R. C. Shearer, compiler. Proceedings—Conifer tree seed in the Inland Mountain West symposium. USDA Forest Service, Intermountain Research Station, General Technical Report INT-203, Ogden, Utah.

Hoff, R. J. 1986. Effect of stratification time and seed treatment on germination of western white pine. Pages 112–116 *in* R. C. Shearer, compiler. Proceedings—Conifer tree seed in the Inland Mountain West symposium. USDA Forest Service, Intermountain Research Station, General Technical Report INT-203, Ogden, Utah.

———. 1992. How to recognize blister rust infection on whitebark pine. USDA Forest Service, Intermountain Research Station, Research Note INT-406, Ogden, Utah.

International Seed Testing Association. 1985. International rules for seed testing: Rules and annexes. Seed Science and Technology 13:299–520.

Jacobs, J., and T. Weaver. 1990. Effects of temperature and temperature preconditioning on seedling performance of whitebark pine. Pages 134–139 *in* W. C. Schmidt and K. J. McDonald, compilers. Proceedings—Symposium on whitebark pine ecosystems: Ecology and management of a high-mountain resource. USDA Forest Service, Intermountain Research Station, General Technical Report INT-270, Ogden, Utah.

James, R. L. 1986. Diseases of conifer seedlings caused by seedborne *Fusarium* species. Pages 267–271 *in* R. C. Shearer, compiler. Proceedings—Conifer tree seed in the Inland Mountain West symposium. USDA Forest Service, Intermountain Research Station, General Technical Report INT-203, Ogden, Utah.

———. 1991. *Cylindrocarpon* root disease of container-grown whitebark pine seedlings, USDA Forest Service Nursery, Coeur d'Alene, Idaho. USDA Forest Service, Northern Region, Forest Pest Management Report 91-8, Missoula, Montana.

———. 1995. Fungi on Douglas-fir and ponderosa pine cones from the USDA Forest Service Nursery, Coeur d'Alene, Idaho. USDA Forest Service, Northern Region, Forest Pest Management Report 95-5, Missoula, Montana.

Jorgensen, S. M., and J. L. Hamrick. 1997. Biogeography and population genetics of whitebark pine, *Pinus albicaulis*. Canadian Journal of Forest Research 27:1574–1585.

Kolotelo, D. 1997. Anatomy and morphology of conifer tree seed. British Columbia

Ministry of Forests, Nursery and Seed Operations Branch, Forest Nursery Technical Series 1.1. British Columbia, Canada.

Krugman, S. L., and J. L. Jenkinson. 1974. *Pinus* L. Pine. Pages 598–638 *in* C. S. Schopmeyer, technical coordinator. Seeds of woody plants of the United States. USDA Forest Service, Agriculture Handbook No. 450, Washington, D.C.

Landis, T. D. 1989a. Disease and pest management. Pages 1–99 *in* T. D. Landis, R. W. Tinus, S. E. McDonald, and J. P. Barnett. The container tree nursery manual, Vol 5. USDA Forest Service, Agriculture Handbook 674, Washington, D.C.

———. 1989b. Mineral nutrients and fertilization. Pages 1–67 *in* T. D. Landis, R. W. Tinus, S. E. McDonald, and J. P. Barnett. The container tree nursery manual, Vol. 4. USDA Forest Service, Agriculture Handbook 674, Washington, D.C.

———. 1990a. Containers: Types and functions. Pages 1–39 *in* T. D. Landis, R. W. Tinus, S. E. McDonald, and J. P. Barnett. The container tree nursery manual, Vol. 2. USDA Forest Service, Agriculture Handbook 674, Washington, D.C.

———. 1990b. Growing media. Pages 41–85 *in* T. D. Landis, R. W. Tinus, S. E. McDonald, and J. P. Barnett. The container tree nursery manual, Vol. 2. USDA Forest Service, Agriculture Handbook 674, Washington, D.C.

Landis, T. D., R. W. Tinus, S. E. McDonald, and J. P. Barnett. 1992. Atmospheric environment. The container tree nursery manual, Vol. 3. USDA Forest Service, Agriculture Handbook 674, Washington, D.C.

Lanner, R. M. 1982. Adaptations of whitebark pine for seed dispersal by Clark's nutcracker. Canadian Journal of Forest Research 12:391–402.

Lanner, R. M., and B. K. Gilbert. 1994. Nutritive value of whitebark pine seeds, the question of their variable dormancy. Pages 206–211 *in* W. C. Schmidt and F. -K. Holtmeier, compilers. Proceedings—International workshop on subalpine stone pines and their environment: The status of our knowledge. USDA Forest Service, Intermountain Research Station, General Technical Report INT-GTR-309, Ogden, Utah.

Leadem, C. L. 1981. Quick methods for determining seed quality in tree seeds. Pages 64–72 *in* R. F. Huber, compiler. High-quality collection and production of conifer seed. Canadian Forestry Service, Information Report NOR-X-235, Edmonton, Alberta, Canada.

———. 1986. Seed dormancy in three *Pinus* species of the Inland Mountain West. Pages 117–124 *in* R. C. Shearer, compiler. Proceedings—Conifer tree seed in the Inland Mountain West symposium. USDA Forest Service, Intermountain Research Station, General Technical Report INT-203, Ogden, Utah.

Littke, W. 1996. Seed pathogens and seed treatments. Pages 187–191 *in* T. D. Landis and D. B. South, technical coordinators. National proceedings, Forest and Conservation Nursery Associations. USDA Forest Service, Pacific Northwest Research Station, General Technical Report PNW-GTR-389, Portland, Oregon.

McCaughey, W. W. 1993. Whitebark pine regeneration: Results from a study on the Gallatin National Forest. Nutcracker Notes 1:5–6, <http://www.mesc.usgs.gov/glacier/nutnotes.htm>.

———. 1994. The regeneration process of whitebark pine. Pages 179–187 *in* W. C. Schmidt and F. -K. Holtmeier, compilers. Proceedings—International workshop on subalpine stone pines and their environment: The status of our knowledge. USDA Forest Service, Intermountain Research Station, General Technical Report INT-GTR-309, Ogden, Utah, USA.

———. 1995. Seasonal maturation of whitebark pine seed in the greater Yellowstone ecosystem. Nutcracker Notes 5:7–10, <http://www.mesc.usgs.gov/glacier/nutnotes.htm>.

McCaughey, W. W., and W. C. Schmidt. 1990. Autecology of whitebark pine. Pages 85–96 *in* W. C. Schmidt, and K. J. McDonald, compilers. Proceedings—Symposium on whitebark pine ecosystems: Ecology and management of a high-mountain resource. USDA Forest Service, Intermountain Research Station, General Technical Report INT-270, Ogden, Utah.

McCaughey, W. W., and T. Weaver. 1990. Biotic and microsite factors affecting whitebark pine establishment. Pages 140–150 *in* W. C. Schmidt and K. J. McDonald, compilers. Proceedings—Symposium on whitebark pine ecosystems: Ecology and management of a high-mountain resource. USDA Forest Service, Intermountain Research Station, General Technical Report INT-270, Ogden, Utah.

[MTDC] Missoula Technology Development Center. 1990. Sandia cone cutter. Drawing Number MTDC-841, Missoula, Montana.

———. 1996. National tree climbing field guide. USDA Forest Service, Technology Development Program Publication 9624-2819-MTDC, Missoula, Montana.

Neumann, M., D. Trotter, and D. Kolotelo. 1997. Seed soaking tank sanitation methods to reduce risk of contamination of seedlots by *Fusarium*. Page 17 *in* E. van Steenis, editor. Seed and seedling extension topics, Vol. 10, Numbers 1 & 2. British Columbia Ministry of Forests, Victoria, British Columbia, Canada.

Pitel, J. A., and B. S. P. Wang. 1980. A preliminary study of dormancy in *Pinus albicaulis* seeds. Bi-monthly Research Notes 36:4–6.

———. 1990. Physical and chemical treatments to improve germination of whitebark pine seeds. Pages 130–133 *in* W. C. Schmidt and K. J. McDonald, compilers. Proceedings—Symposium on whitebark pine ecosystems: Ecology and management of a high-mountain resource. USDA Forest Service, Intermountain Research Station, General Technical Report INT-270, Ogden, Utah.

Portlock, F. T. 1996. A field guide to collecting British Columbia conifers. Canadian Forest Service—British Columbia Ministry of Forests, Canadian Forest Service, FRDA II Publication Fo42-258, Victoria, British Columbia, Canada.

USDA Forest Service. 1999a. Transfer rules. Chapter 4 *in* Seed handbook. Draft Forest Service Handbook 2409.26f (Regions 1–4), Missoula, Montana.

———. 1999b. Cone collection strategy. Chapter 5 *in* Seed handbook. Draft Forest Service Handbook 2409.26f (Regions 1–4), Missoula, Montana.

———. 1999c. Collecting and handling the cones. Chapter 7 *in* Seed handbook. Draft Forest Service Handbook 2409.26f (Regions 1–4), Missoula, Montana.

———. 1999d. Nursery cone and seed operations. Chapter 8 *in* Seed handbook. Draft Forest Service Handbook 2409.26f (Regions 1–4), Missoula, Montana.

VonBonin, F. 1994. Green Mountain whitebark pine cone collection effort. Nutcracker Notes 3:7, <http://www.mesc.usgs.gov/glacier/nutnotes.htm>.

Weaver, T., and F. Forcella. 1986. Cone production in *Pinus albicaulis* forests. Pages 68–76 *in* R. C. Shearer, compiler. Proceedings—Conifer tree seed in the Inland Mountain West symposium. USDA Forest Service, Intermountain Research Station, General Technical Report INT-203, Ogden, Utah.

Chapter 17

Strategies for Managing Whitebark Pine in the Presence of White Pine Blister Rust

Raymond J. Hoff, Dennis E. Ferguson, Geral I. McDonald, and Robert E. Keane

Whitebark pine (*Pinus albicaulis*) is one of many North American white pine species (*Pinus* subgenus *Strobus*) susceptible to the fungal disease white pine blister rust (*Cronartium ribicola*) (Chapter 10). Blister rust has caused severe mortality (often reaching nearly 100 percent) in many stands of whitebark pine north of 45° latitude in western North America. The rust is slowly moving south through the range of whitebark pine and other white pine species (Chapters 10 and 11).

In whitebark pine, the rust typically kills the upper, cone-bearing branches long before the tree dies, thus reducing or ending seed production and, consequently, future regeneration. Whitebark pine is a keystone species that increases biodiversity in the subalpine zone in a multitude of ways, especially by providing seeds as a wildlife food source (Chapters 1 and 12). The loss of whitebark pine will lower the environmental carrying capacity for many forest animals as well as alter forest composition and distribution in the upper elevations.

One of the few options to reverse severe losses of whitebark pine to blister rust is to increase the level of genetic resistance. Previous efforts to increase blister rust resistance in western white pine (*Pinus monticola*) provide some guidance for the process. This chapter presents information on the presence of resistance in whitebark pine to white pine blister rust. We will use this knowledge, together with knowledge about the ecology of whitebark pine and the blister rust fungus, to propose integrated rust management strategies for restoring whitebark pine communities.

Blister Rust Susceptibility in Whitebark Pine

White pine blister rust entered North America from Europe early in the twentieth century at two locations (Chapter 10). One entry point was through New York State before 1906, when several million three-year-old seedlings of eastern white pine (*Pinus strobus*) were imported from nurseries in Europe and outplanted at many forest sites (Spaulding 1911). Only 1 to 3 percent of these seedlings were infected with blister rust; nonetheless, the infection quickly took hold in the native stands of eastern white pine (Chapter 10).

The second entry point was Vancouver, British Columbia, Canada, in 1910, when a few hundred seedlings of eastern white pine were imported from a nursery in France and were planted at Point Grey near Vancouver. However, blister rust in the West was not observed until September 1921, when it was discovered on European black currant (*Ribes nigrum*), one of the many species of the genus *Ribes* that are obligate alternate hosts in the life cycle of blister rust (Mielke 1943; Chapter 10). A hurried survey completed before winter arrived revealed infection on European black currant and a few exotic white pines throughout the lower Fraser River Valley of British Columbia. Surveys in 1922 showed that many native western white pine trees in western British Columbia were infected, and infection of European black currant was observed in western Washington.

The first infected whitebark pine was discovered in 1922 in the arboretum of the University of British Columbia, Vancouver (Bedwell and Childs 1943). In 1926, the rust was discovered on native whitebark pine near the Birkenhead River in the Coast Range of British Columbia, 160 kilometers north of Vancouver (Lachmund 1926). There, whitebark pine occurred in association with western white pine, and Lachmund (1928) observed that whitebark pine appeared to be seven to ten times more susceptible to blister rust than was western white pine.

To further determine the relative susceptibility of western white pine and whitebark pine to blister rust, Bedwell and Childs (1943) established study plots of young whitebark and western white pine in natural and nursery settings. The nursery plot was in the same general area where Lachmund (1928) made his observations, but the natural stands were located in Washington, Oregon, and Idaho. In the natural stands, where blister rust had been present for about ten years, infection level for western white pine ranged from 0 to 70 percent, averaging 28 percent, with 0.06 cankers per 1,000 needles. Whitebark pine infection ranged from 59 percent to 100 percent, averaging 83 percent, with 1.22 cankers per 1,000 needles. In the nursery test, the number of cankers per 1,000 needles for western white pine was 0.18, and for whitebark pine 1.42. Percent infection for the nursery test was not given. These results certainly confirmed Lachmund's (1928) observation.

Furthermore, Bedwell and Childs (1943) observed that whitebark pine trees were dying faster than western white pine because of extremely high numbers of branch cankers. They concluded that the greater susceptibility of whitebark pine was due in part to the longer retention time of needles (5.3 years for whitebark versus 3.8 years for western white pine) and to the higher susceptibility of current year's needles. Estimation of the absolute infection rate (see Chapter 10 for definition of r) from published data (Bedwell and Childs 1943) indicates that whitebark pine was 4.8 times more susceptible than western white pine west of the crest of the Cascade Mountains, but was 72.5 times more susceptible east of the Cascade crest. If whitebark pine is several times more susceptible than western white pine—as estimated by Lachmund (1926), Bedwell and Childs (1943), and McDonald and Hoff (Chapter 10)—the survival of this species appears bleak. However, recent data indicate that western white pine and whitebark pine may have about equal susceptibility when density of *Ribes* plants per hectare are equal (Tomback et al. 1995; Chapter 10). Although the relative susceptibility of these two species is still unclear, it is clear that they are both very susceptible to blister rust.

A large amount of data collected by foresters, pathologists, and geneticists paint a bleak picture for whitebark pine (Chapter 11). The data confirm high levels of mortality by blister rust on many sites north of 45°N latitude and an increasing level of infection south of that line. But, there is hope that human intervention can help save whitebark pine and restore these ecosystems.

Blister Rust Resistance

It is unusual to find 100 percent mortality from blister rust in stands of either whitebark or western white pine (Hoff et al. 1994; Chapters 10 and 11). In areas where blister rust has infected and killed most of the whitebark and western white pine, often one or more trees have no visible cankers, which indicates the possibility of genetically controlled resistance to the rust. Bingham (1983) estimated that 1 in 10,000 western white pine trees was canker-free in high-infection areas. Then, too, even the surviving but cankered trees may have genes for resistance to blister rust. In western white pine, the number of resistant trees becomes apparent in blister rust–infected stands with an increasing level of mortality (Hoff et al. 1976). On average, the most susceptible trees die first. The last to die, if indeed they die, would be the most resistant. Alternatively, because of nonrandom distribution of spores and/or infection microclimate, blister rust epidemics may not achieve 100 percent rust incidence (Chapter 10). Therefore, healthy trees may be "escapes" rather than phenotypically resistant.

Tests for Resistance: Western White Pine

Methods for determining the level of resistance in western white pine were worked out by Bingham et al. (1960) and later modified by Hoff and McDonald (1980). Recently, a new breeding and seed orchard plan has been prepared by Mahalovich and Eramian (2000). Methods for determining resistance levels are reviewed by McDonald and Hoff (Chapter 10), and since they are germane to the determination of resistance in whitebark pine, they will be briefly described.

Breeding for resistance in western white pine started in 1950 (Bingham 1983; Chapter 10). The first major objective was to determine if the few rust-free western white pines, growing among neighbors supporting hundreds of cankers, had heritable rust resistance. Seeds were collected from canker-free trees, called candidates, which were cross-bred with other candidates. Bingham's tests included candidates from the inland range of western white pine, that is, eastern Washington, northern Idaho, and western Montana. Seeds of the first four progeny tests in 1952 to 1955 were sown, grown, and inoculated with blister rust in a nursery near Spokane, Washington. The next tests in 1960 to 1970 were completed in Moscow, Idaho. Two-year-old seedlings were inoculated by suspending blister rust–infected *Ribes* leaves over them in the fall. Experimental control was established by using seeds from infected trees (called "comparison trees") located in the same stands that contained the candidates. The 1952 to 1955 progeny tests were divided into three units and outplanted at three different forest sites. The 1960 to 1970 tests remained in Moscow for the duration of the tests.

The first data tallied were the presence and the number of blister rust needle spots on the secondary needles. The needle spots were easily visible by June, 9 months after inoculation. The second data tallied were the presence of blister rust needle spots, cankers, and/or bark reactions (easily visible bark lesions caused by the seedling attempting to kill the rust) 12 months after inoculation. The third, fourth, and fifth data tallied were for cankers and/or bark reactions and mortality due to blister rust 24, 36, and 48 months after inoculation.

The 1952 progeny test showed that blister rust resistance in western white pine was inherited. Seedlings from the candidates were 17.9 percent canker-free, whereas the comparison or control trees were 5.3 percent canker free. The data also indicated the existence of additional underlying resistance mechanisms.

Additional information from the western white pine tests changed the way we view blister rust resistance; seedlings from surviving, cankered trees in high-mortality stands also have resistance. This insight came from comparing the level of resistance of the seedlings from the comparison trees in the early 1950s, when the level of mortality of the parental stands ranged from 0 to about 15 percent, to the level of resistance of seedlings from comparison trees in the early 1960s, when the level of mortality of parental stands had increased to 80 to 90 percent. About 5.3 percent of the seedlings from the comparison trees for the 1952 to 1955 progeny tests showed blister rust resistance (Bingham et al. 1960), but for the 1964 progeny tests, it was 22.8 percent. Most important, the resistance in the candidate trees increased to 39.3 percent (Hoff et al. 1976). The higher resistance of the candidate trees is due, in part, to the fact that four of the twelve pollen parents used in the breeding design had already been selected (in the 1952 progeny test) for high resistance. It is also probable that better candidates had been selected. Future selection of candidate trees was relaxed to include trees with cankers, depending on the level of stand mortality from blister rust (Mahalovich and Eramian 2000). This exemplifies the high resiliency of western white pine, probably the result of its high within-stand genetic variation (Rehfeldt et al. 1984). In just one generation of selection by blister rust, resistance has already achieved a useful level.

One of the cankered seedlings in the 1954 progeny test illustrates the longevity of infected trees that must have some resistance to blister rust. This exceptional tree was artificially inoculated in 1956 and had a canker by 1958. In 1981 the tree was 14.7 meters tall and was 90 percent girdled by blister rust (Hoff 1984). The tree died in 1991, having lived with a very large canker for

33 years. The average height for trees in this same plot in 1981 was 14.3 meters for cankered trees (12 trees remained) and 14.2 meters for canker-free trees (14 trees remained). Most of the 953 trees planted at this site had died within a few years after inoculation, and most of the canker-free trees were moved to an arboretum near Moscow, Idaho, by 1965.

Tests for Resistance: Whitebark Pine

Methodology developed to determine the level of resistance in western white pine was used for two whitebark pine tests. The purpose of the first test was to compare the levels of resistance to blister rust among nineteen white pine species (subsections *Cembrae* and *Strobi*) from Asia, Europe, and North America. In spring 1970, seeds were sown in pots arranged in randomized blocks (replications). Two-year-old seedlings were inoculated with blister rust and inspected annually for 3 years (Hoff et al. 1980). In prior tests when seeds of whitebark pine were collected from the general population, whitebark pine was ranked as the most susceptible to blister rust among white pine species (Bingham 1972). But in the 1970 test, the seeds of whitebark pine came from trees with no visible cankers in high-mortality stands (>90 percent by blister rust), and the seeds of western white pine came from a mix of resistant candidates. The level of canker-free seedlings of whitebark pine ranked fourth at 46 percent. To our surprise, western white pine ranked fifth with 36 percent canker-free seedlings, which showed less resistance than whitebark pine.

The second resistance test for whitebark pine was established in 1989 by R. J. Hoff (unpublished data). The purpose of this test was to relate the level of resistance to blister rust in whitebark pine to varying levels of mortality caused by blister rust. Seeds from three high-mortality stands (>90 percent), three moderate-mortality stands (40–60 percent), and three low-mortality stands (<10 percent) were included in the test. Three years after inoculation, 44.4 percent of the seedlings from the high-mortality stands were canker-free, 11.9 percent from moderate-mortality stands, and 0.9% from low-mortality stands (Table 17-1). These results together suggest that surviving whitebark pine trees from high-mortality stands possess usable levels of heritable resistance.

Table 17-1. Rust resistance of whitebark pine (*Pinus albicaulis*) seedlings three years after inoculation with white pine blister rust (*Cronartium ribicola*). The seedlings were grown from parent stands having three levels of mortality caused by blister rust.

Parent Stand Mortality	Number of Seedlings	Spots Per Meter of Needles	Percent Not Cankered	Number of Seedlings by Resistance Mechanism		
				Needle Shed	Short Shoot	Bark Reaction
> 90%	304	8.0	44.4	50	47	38
40–60%	134	10.2	11.9	4	7	5
<10%	226	5.4	0.9	1	0	1

Resistance Mechanisms

Many mechanisms of resistance have been observed and measured in western white pine (Table 17-2). Mahalovich and Eramian (2000) have included most of these mechanisms in their western white pine–blister rust breeding and seed orchard plans. At this time, tests of resistance of whitebark pine to blister rust have not been detailed enough to determine whether all the resistance mechanisms in western white pine are present. However, three main resistance mechanisms have been observed in whitebark pine:

1. Needle shed—characterized by the premature shedding of the needles that have rust infections, that is, needle spots.
2. Short-shoot—seedlings with this mechanism have needle spots, but when the fungus grows down the needle, it dies as it enters the short shoot. The short shoot is the small, stemlike appendage at the base of the needles that holds the five needles of a fascicle together.
3. Bark reaction—here the fungus can grow into the stem, but it soon dies after being walled off by special cortex cells in the bark, which becomes visible on the surface of the stem as lesions.

In one test of whitebark pine, needle shed was the most common resistance mechanism observed (Hoff et al. 1980). Although this trait also occurred in all

Table 17-2. Defense mechanisms against white pine blister rust (*Cronartium ribicola*) observed in western white pine (*Pinus monticola*).

Defense Symptom	Possible Inheritance
Resistance in secondary needles to a yellow-spot forming race.	Recessive gene
Resistance in secondary needles to a red-spot forming race.	Dominant gene
Resistance in secondary needles to yellow-green-island spot forming race.	Dominant gene
Resistance in secondary needles to red-green-island spot forming race.	Dominant gene
Resistance in secondary needles that prevents spot formation.	Unknown
Reduced frequency of secondary needle infections.	Nondominant gene
Slow fungus growth in secondary needles.	Polygenic
Premature shedding of infected secondary needles	Recessive gene
Fungicidal reaction in the short-shoot.	Recessive gene
Fungicidal reaction in the stem.	Polygenic
Slow fungus growth in the stem.	Polygenic
Tolerance.	Polygenic

other pines showing resistance, it was highest in frequency in whitebark pine. The short-shoot and bark reactions that were the most prevalent resistance mechanisms in the other pines occurred only at a moderate level in whitebark pine. In a 1989 test (R. J. Hoff, unpublished data), the three main resistance mechanisms were about equal in number (Table 17-1).

There are other noteworthy observations from the whitebark pine resistance tests (Hoff et al. 1980; R. J. Hoff, unpublished data). First, some seedlings had huge cankers but did not die until the fourth year after inoculation. Second, cankers in some seedlings grew very slowly, so that these seedlings were still alive five years after inoculation. These observations are signs of two additional resistance mechanisms—"tolerance" and "slow canker growth." If we could breed for these and other resistance mechanisms, the resulting new variety of whitebark pine should provide the level of resistance needed to maintain itself in the presence of blister rust.

McDonald and Dekker-Robertson (1998) and McDonald and Hoff (Chapter 10) argue that all resistance mechanisms thus far observed in Idaho western white pine and whitebark pine fall into the category of horizontal resistance, so-called because when a variety of pine is tested against several different races of the rust, the histograms for proportion infected are equal, although they can vary from near 0 to 1 (Zadoks 1972). This terminology comes from agriculture (Van der Plank 1968; Simmonds 1991), where horizontal resistance factors are usually controlled by minor genes that can be overcome by more aggressive races of rust, although this may take a long time. Mortality typically remains low. Horizontal resistance contrasts sharply with vertical resistance, because when a vertically resistant variety is tested against various races of the rust, the histograms for proportion infected are either 1 or 0. Vertical resistance is nearly always controlled by major dominant genes and is usually (almost invariably) overcome by new races of rust, often within five years or so, rendering the varieties completely susceptible. Mortality is rapid. For example, a dominant gene for resistance to blister rust was found in sugar pine (*Pinus lambertiana*) (Kinloch et al. 1970), but it was soon overcome by a new race of blister rust (Kinloch and Comstock 1981; Chapter 10).

Management Strategies

It appears that whitebark pine may have stable horizontal resistance to blister rust. The next step is to use one or more strategies to incorporate the resistance genes into future generations of whitebark pine. The discussion that follows provides ideas for managing whitebark pine in restoration efforts. Appropriate management actions will depend on factors such as site conditions, history of blister rust infection, potential for insect outbreaks, landscape considerations, and landowner objectives. Most of these management activities are doubly beneficial because they can increase rust resistance while restoring whitebark pine communities.

Seed Collection

Whitebark pine seeds must be collected for gene conservation, planting, and for use in any rust-resistance breeding program. Protection of maturing cones is a major consideration because of seed losses to pine squirrels (*Tamiasciurus* spp.) and Clark's nutcrackers (*Nucifraga columbiana*). Squirrels and nutcrackers can harvest every cone and seed from whitebark pine trees, especially in stands where cone crops are reduced by blister rust. The best strategy is to collect cones whenever there is a good seed crop, and harvest cones from many stands so they can be stored and used when needed. Details of cone and seed collection are summarized by Burr et al. (Chapter 16).

Blister rust–resistance levels in seeds will be higher if cones are collected only from remnant trees that have no cankers, because they probably have the highest resistance in the stand (although some of these isolated trees could be escapes rather than resistant [Chapter 10]). These remnant trees are probably widely scattered and would definitely require the use of wire mesh bags to protect maturing cones from squirrels and nutcrackers. Because of the expected higher levels of resistance, this alternative may be worth the effort.

A better alternative, in terms of levels of resistance, would be to collect pollen from many canker-free or lightly cankered trees (i.e., trees that are apparently rust resistant found in stands having more than 90 percent rust-caused mortality) and artificially pollinate other canker-free or lightly cankered trees. This would require that pollination bags be placed over the strobili (female cones) before they become receptive, artificially pollinating them, removing the pollination bag, and using wire mesh bags to protect developing cones. Each tree would need to be climbed five times to complete the process. Whereas this is a good procedure for research purposes, the process is expensive and probably not very practical.

Seed Transfer

Most plant species vary genetically across environmental gradients. If plant material (seedlings, pollen, seeds, or clones) is transported too far, there is danger of plants being maladapted to their new environment. Plans to transfer whitebark pine seeds or seedlings from one site to another must avoid making transfers that are genetically maladapted. That is, we must know if there are genetic differences within whitebark pine from one stand to another, or from one mountain range to another, that control adaptive traits. If there are no differences, plant materials could be transferred throughout the range of whitebark pine. If there are differences, where do we draw the boundaries between populations within species?

Several researchers have evaluated genetic variation among populations of whitebark pine using allozymes (Jorgenson and Hamrick 1997; Bruederle et al. 1998; Chapter 8). So far, the results indicate that most of the variation is within populations and that the variation among populations is small and not limiting. More useful are tests that relate survival traits—such as growth, frost hardiness, drought hardiness, and pest resistance—with various aspects of the environment such as latitude, longitude, elevation, and habitat. Whereas the Rocky Mountain Research Station has initiated seed transfer studies for whitebark pine (Chapter 15), we cannot wait for conclusions, which are some years off, and instead must consider alternatives based on what we now know.

Recent summaries and reworking of genetic data indicate that more liberal seed transfer guidelines can be established, because conifers in general have high amounts of genetic variation (see also Chapter 15). Genetic variation among populations is nearly always detected, but genetic variation within populations is nearly always greater (Hamrick et al. 1994; Rehfeldt 1994; Mitton 1995; Bruederle et al. 1998). The degree of differentiation among conifer populations has turned out to be quite variable. For example, significant genetic differences occur over short elevational gradients among populations of interior Douglas-fir (*Pseudotsuga menziesii*) (Rehfeldt 1989), lodgepole pine (*Pinus contorta*) (Rehfeldt 1988), ponderosa pine (*Pinus ponderosa*) (Rehfeldt 1991), and western larch (*Larix occidentalis*) (Rehfeldt 1982), but not among populations of western white pine (Rehfeldt et al. 1984). Genetic variation within populations of most western conifers is so high that, even if many of the seedlings are not adapted to the site, there will be seedlings within the mix that are adapted. Conversely, some of the seedlings from an on-site population will not be adapted to that specific site.

Another argument for liberal seed transfer guidelines is the short growing season at high elevations. Frost-free periods decrease by ninety days for every 1,000 meters in elevation gain (Baker 1944). For several conifers, increasing elevation increases the size of population differentiation intervals. For example, for Douglas-fir below 1,000 meters elevation, the population boundary is 200 meters; between 1,000 meters and 1,525 meters, it is 350 meters; and above 2,000, meters differentiation is not detected (Rehfeldt 1989). The conclusion for high-elevation Douglas-fir is that either genetic variability has been exhausted (Douglas-fir is genetically homogeneous at high elevations) or the environment is homogeneously severe (the frost-free period is very short, less than twenty days at timberline) and much further decrease in the growing season would end the range of Douglas-fir.

Recent research results for lodgepole pine show that populations within this species grow in suboptimal environments because adjacent populations of lodgepole pine outcompete them for space (Rehfeldt et al. 1999). The amount of space supplanted depends upon the breadth of lodgepole pine's fundamental niche (Hutchinson 1958). Therefore, if survival and growth are the desired traits for seed transfer, the populations with the fastest growth coupled with adequate survival traits would be selected. On the other hand, if survival is the only trait that is selected, the area where populations would be adapted is much broader. We think that the populations of whitebark pine are not strongly differentiated because selection has been mainly for survival and seed production.

Four main factors support broad transfer rules for whitebark pine. First, the environment over the range of whitebark pine is so uniformly severe that the genetic structure among populations, even populations separated by long distances, will likely be similar. The species is restricted at its upper elevations by the most severe climatic conditions and at the lower elevations by competition from other tree species (Chapter 3). Second, many whitebark pine germinants survive the hot, dry conditions resulting from a site that has been burned. Third, whitebark pine trees can tolerate summer frost that would kill or severely damage other tree species (Chapters 1 and 3). Fourth, whitebark pine seeds are bird-dispersed. Gene flow of bird-dispersed seeds is faster and farther than that of wind-dispersed seed; Clark's nutcrackers have been observed transporting pine seeds up to 22 kilometers and 12 kilometers or farther for whitebark pine specifically (Tomback and Linhart 1990; Bruederle et al. 1998; Chapter 5). Seeds dispersed by wind for species such as western white pine and Douglas-fir travel only 100 meters or so, depending upon wind speed (Wright 1976). Estimated gene flow or populations of limber pine (*Pinus flexilis*), a bird-dispersed pine, separated by 1,700 meters, averaged 11.1 migrants (genes) per generation (Schuster et al. 1989). The gene pools of bird-dispersed seed, even at long distances, are more likely to be mixed than are gene pools of wind-dispersed seed.

Therefore, we propose relatively broad transfer rules for whitebark pine with certain limitations (USDA Forest Service 1999; Mahalovich and Hoff 2000). The most important limit is to restrict transfers with respect to blister rust mortality. First, transfer should be only among high-mortality stands or from high mortality to moderate or low mortality, and never from low to high mortality. Second, transfer should only be within those sites for which whitebark pine restoration is possible or needed, that is, no need to consider krummholz environments. We may be able to grow whitebark pine at low elevations with protection from insects or disease and with periodic cleaning and weeding. This would be useful in speeding up growth and seed production.

We propose the following transfer rules: In the inland mountains of the United States and Canada, land managers should transfer whitebark pine plant material (seed, seedlings, pollen, or clones) no farther than 80 kilometers from the point of origin (USDA Forest Service 1999; Mahalovich and Hoff 2000). At this time we propose no restrictions on elevation transfer.

Gene Conservation

There are a number of important reasons to consider gene conservation as an integral part of managing whitebark pine. If whitebark pine populations became locally extinct or severely depleted, there should be alternative ways to reestablish populations. Establishment of seed banks should be strongly considered. Whitebark pine seeds can be collected from several different regions and stored in freezers, following recommendations from Burr et al. (Chapter 16). However, fresh seeds have higher germination success than stored seeds; in fact, seeds stored in freezers lose viability over time (McCaughey 1994; Chapter 16).

Establishment of gene banks is also important. Seedlings from various whitebark pine sites could be planted in low fire- or rust-hazard areas. Then, genetically adapted seeds, pollen, or clones are available to reforest sites. An advantage of gene banks over stored seeds is that genetic material can be produced indefinitely from gene banks.

Another management treatment to consider is protecting individual whitebark pines that are apparently resistant to blister rust. Many of these existing trees are in danger of being crowded out by succession to more shade-tolerant Engelmann spruce (*Picea engelmannii*) and subalpine fir (*Abies lasiocarpa*) in the Inland Northwest. Wildfires are also a threat. Individual trees can be protected by thinning back encroaching competition and moving the debris away from trees. Reduced competition should allow whitebark pine to live longer with less stress, grow faster, have fuller crowns, and produce larger cone crops.

Protection of whitebark pine forests from wildfire and mountain pine beetle (*Dendroctonus ponderosae*) epidemics that originate in lower-elevation forests requires a landscape perspective. Both wildfire and beetles can sweep upward and kill the few remaining rust-resistant whitebark pine. Therefore, management of whitebark pine forests needs to be considered at the landscape level (see Chapter 14).

Openings in the Forest Canopy

Historically, periodic wildfires in the interior northwestern United States and southwestern Canada helped maintain whitebark pine communities by creating openings in the forest canopy and returning successionally advanced communities to earlier seral stages, where whitebark pine was an important component (Chapter 4). Clark's nutcrackers cache seeds in newly created openings, and caches that are not later retrieved for food by nutcrackers germinate and establish whitebark pine seedlings (Hutchins and Lanner 1982; Tomback 1982; Chapter 5). Openings must be large enough to allow enough sunlight to reach the forest floor, so that whitebark pine germinants can survive and grow.

Fire exclusion practices in the twentieth century have drastically reduced both the number of fires and the area burned, resulting in replacement of whitebark pine with shade-tolerant conifers (Chapters 1, 4, and 9). Today, we need to use fire as a restoration tool (see also Chapter 18).

Restoration may be accomplished through controlled prescribed burns, although wildfires could also become controlled burns, especially in national parks and wilderness areas. However, the use of fire in whitebark pine communities has limitations because of abruptly changing weather, scattered fuels, and short growing seasons (Chapter 18). In addition, increasing human populations near forested lands and undesirable smoke pollution may place limits on the use of fire (Chapter 1). The advantage of fire in whitebark pine communities is that it removes competing trees and shrubs at a reasonable cost (Keane et al. 1989; Chapter 18); the risk is that rust-resistant trees will be inadvertently killed. Rare whitebark pine trees that remain free of blister rust should be protected.

An alternative to fire is to create openings in the forest canopy, primarily by harvesting trees. Results obtained by manual, silvicultural methods are more precise than fire, but are more costly and may require access for equipment. Proceeds from harvesting can pay for treatments in some situations. Openings created by harvesting, whether burned or not, are used by nutcrackers for caching whitebark pine seeds (Chapter 5).

Natural Regeneration

The easiest and least costly method to increase the proportion of rust-resistant genes in whitebark pine populations would be to apply a seedtree silvicultural system that utilizes apparently rust-resistant whitebark pine trees as the seed source. Clark's nutcrackers cache seeds in openings, effectively "planting" a new stand of whitebark pine seedlings (Tomback 1982; Chapter 5). Natural selection will favor the survival of rust-resistant trees for both seed-bearing trees and their progeny. The process is initiated by selecting sites that have high mortality caused by blister rust, because the seeds of surviving trees would likely contain rust-resistant genes. Site preparation can reduce competition and attract nutcrackers.

A major consideration with natural regeneration is inbreeding depression. In most conifers, perhaps all, trees that are crossed with themselves or close relatives produce seedlings that grow slower, are less hardy, and often exhibit lethal genes (Wright 1976). Stands of whitebark pine that have only a few trees left will produce a high number of selfed seedlings. These trees will likely be slow growing with even less competitive ability than seedlings from outcrossed trees.

Planting Seedlings and Tending Regeneration

Sites that do not have enough surviving whitebark pine to provide an adequate seed supply should be planted with nursery-grown seedlings (see Chapter 16). Planting may be a good choice where an increased certainty of regeneration establishment is desired, new stands need to be established quickly, local populations have gone extinct, whitebark pine populations are so small that inbreeding depression is a strong possibility, or rust-resistant seedlings need to be established in areas not yet heavily impacted by blister rust. Planting sites must be open enough for adequate sunlight to reach the forest floor, which means it may be necessary to create openings in the forest canopy.

Seedlings should be grown from seeds collected in stands having high mortality (>90%) caused by blister rust, because the surviving trees may have some resistance to the rust. About twice the number of seedlings needed to meet management objectives should be planted, because many trees will be killed by blister rust (depending on the local rust hazard) and other causes. Planted seedlings should be large (e.g., three years old), since they have higher survival rates than planted smaller/younger trees. Good root development of planted whitebark pine is important. Natural whitebark pine regeneration quickly puts down a taproot to aid in survival in hot, dry conditions that kill other conifer seedlings (Chapter 6).

Competing vegetation—mostly spruce, subalpine fir, and lodgepole pine, but also large shrubs or other vegetation—should be removed if open planting sites cannot be found. Although there is little doubt that fire would enhance survival and growth on most sites, successful planting may be achieved without fire once openings in the forest canopy are created.

The need to monitor and manage whitebark pine continues after seedlings become established. Young stands may be overly dense and need thinning. Whitebark pine should not be thinned, because blister rust will kill nonresistant trees. Caution should be used in thinning competing conifer species and shrubs, because open areas allow establishment and growth of *Ribes,* the obligate alternate host of blister rust. Below 80 to 90 percent full sunlight, the number of *Ribes* shrubs decreases rapidly (McDonald et al. 1981), so trade-offs must be considered between keeping high stand densities that shade out *Ribes* shrubs and keeping low stand densities that help survival and growth of whitebark pine. Thinning alone is not a desirable option for western white pine, because it can increase the amount of cankering (Hungerford et al. 1982).

A management technique developed for western white pine may be useful for whitebark pine as well. Pruning lower live branches on western white pine regeneration helped reduce infections and mortality from blister rust (Schwandt et al. 1994; Barth 1994; Hunt 1998). All lower branches can be pruned as a preventative measure, or infected branches can be pruned before blister rust reaches the main stem. If rust hazard is too high, pruning will not provide any benefit. A management model designed to assist in making decisions about pruning western white pine is under development (G. I. McDonald, in preparation), and this model could be adapted for whitebark pine.

Ribes Management

Management of local *Ribes* populations can increase the survival of whitebark pine. Research on western white pine shows that high rates of infection are associated with trees close to *Ribes* shrubs. The number of cankers on western white pine usually drops to negligible amounts if pines are 300 meters or more from a *Ribes* shrub (Chapter 10). Although blister rust spores can travel farther to infect pines, reducing *Ribes* populations should lower the local blister-rust hazard level (Chapter 10). Thinning of overly dense stands becomes a viable option if *Ribes* populations are also reduced. *Ribes* management has diminished the blister rust hazard in the northeastern United States (Ostrofsky et al. 1988; Martin 1944), even though it was not deemed successful under western conditions (Toko et al. 1967; Chapter 10).

Chemical Control of Blister Rust

In the 1950s and 1960s, actidione and phytoactin appeared to be promising chemicals for controlling blister rust, but they were not effective enough, and this part of the blister rust control program was terminated in 1966 (Chapter 10). Recently, two new chemicals proved effective against blister rust infection over a short time period (Johnson et al. 1992). Three-year-old sugar pine seedlings were treated with foliar sprays of triadimefon and benodanil and inoculated with blister rust three weeks later. After six months, the seedlings were inoculated again, and this time only triadimefon was effective. Berube (1996) inoculated eastern white pine seedlings with blister rust two weeks after foliar spray treatment with triadimefon. He found that only 3.8 percent of the treated seedlings but 70.8 percent of the untreated seedlings developed blister rust symptoms. Kelly and Williams (1985) used triadimenol, and a compound closely related to triadimefon, as a dressing on loblolly pine (*Pinus taeda*) seeds, and then inoculated the seedlings with fusiform rust (*Cronartium quercuum* f. sp *fusiforme*). They found that both compounds were effective in decreasing fusiform infection for up to thirty-six days and that triadimefon was the most effective of the two chemicals.

Slow-release fertilizer plugs containing triadimefon have been produced, and studies are underway by G. I. McDonald to see how much, and for how long, these protect against blister rust. A slow-release fertilizer plug impregnated with triadimefon would be useful both for protecting seedlings and for imparting faster growth to planted whitebark pine seedlings.

Establishing a Rust-Resistance Breeding Program

The most complex and costly approach to increasing blister rust resistance in whitebark pine is to conduct a traditional breeding program (Wright 1976). This approach, patterned after the western white pine breeding program (Bingham 1983; Mahalovich and Eramian 2000), would provide the highest gain in resistance. A variety of resistance-gene mechanisms that may not occur together in nature for many years could be packaged into individual trees. Trees that appear to have high resistance, good growth, and good seed production would be selected and cross-bred. The seedlings would be artificially inoculated, and the best families and individuals within families selected and grown in seed orchards. Seed orchards are usually planted on the best growing sites, often good agricultural land. For most conifers, large amounts of seeds are produced in fifteen to twenty years. For whitebark pine, it may take thirty to forty years or more to produce comparable amounts of seeds.

Recently, seed orchards have been moved into greenhouses. This provides opportunities to accelerate growth and stimulate pollen and seed production. Consequently, several conifers have been made to produce seed within five to seven years. While five to seven years seems too optimistic for whitebark pine, perhaps the number of years to seed production could be greatly reduced.

Integrated Management

Conifer populations at risk for disease, like those of whitebark pine, can be managed more efficiently by integrating information from a variety of sources, including what is known as hazard assessment. Hazard assessment involves estimation of risk stand by stand or over other geographic scales. Some preliminary attempts at an integrated management approach (Ostrofsky et al. 1988; Geils et al. 1999) were not completely successful. A fully integrated approach should include principles of comparative epidemiology, computer modeling, satellite imagery, and GIS. One key to successful deployment of resistance genes in whitebark pine is matching levels of resistance to degree of hazard (McDonald 1979).

Hazard in higher-elevation forests varies across the landscape. Hoff (unpublished data) sampled nineteen stands of whitebark regeneration for rust incidence and classified stands as moist or dry (see also below). Three of the seven stands classed as dry were rust-free after about twenty years' exposure to blister rust. A recent survey of blister rust in 100 whitebark pine communities in southern Idaho, Utah, and western Wyoming determined that 41 percent of the stands surveyed were rust-free (Smith and Hoffman 1998, 2000). Possible explanations for rust-free stands are lack of humidity and few to no *Ribes* shrubs (Chapter 10). Mapping of *Ribes* hazard could facilitate the management of whitebark pine.

Van der Plank (1963), in his theoretical treatment of diseases, emphasizes that any control measure that reduces absolute infection rate, including horizontal resistance, is additive for "simple interest" diseases such as blister rust (also see Chapter 10). So horizontal resistance to blister rust in whitebark pine, *Ribes* eradication, pruning of cankers, or reduction of incidence by chemicals should be additive in their effects. Thus, blister rust impact expected under various combinations of blister rust hazard and mix of controls can be predicted.

How much success can we expect? The variable r in models describing disease spread refers to absolute infection rate in incidence/year (Chapter 10). The average r-value of twelve stands of susceptible whitebark pine was 0.176 (Chapter 10). There was one extremely high-hazard stand (0.691), four high-hazard stands (0.253, 0.221, 0.212, 0.281), four showing moderate hazard (0.124, 0.124, 0.08, 0.089), and three with low hazard (0.018, 0.013, 0.007). The extreme-, high-, and moderate-hazard stands were located in relatively moist environments, whereas the three low-hazard stands were located in more dry environments.

Whitebark pine mostly grows in three types of ecological communities: seral, climax, and tree-line (krummholz form) (Arno and Hoff 1989). Seral whitebark trees are found on sites moist enough to support subalpine fir and/or mountain hemlock (*Tsuga mertensiana*) (Cooper et al. 1991). Climax communities are found on drier, harsher sites, generally south of 47° N latitude (Arno and Hoff 1989). This distribution leads to a natural grouping of whitebark stands into moist (seral) and dry (climax) classes. A series of nineteen plots were established in whitebark pine regeneration (fifteen to twenty-five years old) in 1992 to survey blister rust damage (R. J. Hoff, unpublished data). We classified twelve of these plots as "moist" and seven as "dry" on the basis of their plant associations, and then computed infection rate r (Chapter 10). The average r-value for the seven dry stands was 0.007 and for the twelve moist stands 0.046, indicating that moister aspects and areas are a hazard to whitebark pine.

An outplanting of western white pine in a high-rust-hazard area gives us insight to possibilities for resistance. This plantation contained susceptible, resistant first-generation, and resistant second-generation western white pine (Chapter 10). The number of *Ribes* shrubs was 3,700 per hectare. After six years, blister rust incidence had reached nearly 100 percent in susceptible trees. Absolute infection rate (r) for this susceptible lot was 0.504. The infection rate for the first generation of rust-resistant western white pine was three times less, and for the second generation, six times less (McDonald and Dekker-Robertson 1998). After twelve years of exposure, all susceptible trees were dead, and at twenty-six years almost 100 percent of the resistant first generation and 93 percent of the resistant second generation were infected. Mortality from blister rust after twenty-six years for the first generation was 78 percent, and for the second generation 56 percent. So, even at this high-hazard site, many of these western white pine trees will likely survive to maturity. The resistance from high-mortality stands of whitebark pine (Table 17-1) is comparable to the resistance of the first generation of rust-resistant western white pine, and should reduce the r-value by three times and second generation by six times. Blister rust–resistant populations of whitebark pine should be outplanted at extreme hazard sites to verify their r-values.

Over a half-billion *Ribes* plants have been removed from thousands of hectares in North American white pine forests (Chapter 10). Because of the high cost versus effectiveness, the *Ribes* eradication program was stopped in 1966 (Chapter 10). *Ribes* eradication in eastern white pine forests of Maine has been more successful. Ostrofsky et al. (1988) evaluated a *Ribes* eradication program after seventy years of effort in Maine. The average absolute infection rate (r-value) for no *Ribes* eradication was 0.091. For stands with *Ribes* eradication, the average r-value was 0.038. Therefore, trees were becoming infected at a rate 2.5 times less for treated than for untreated stands. Other studies with similar results in eastern white pine were reported by Martin (1944) and Robbins et al. (1988). Removal of *Ribes* shrubs in whitebark pine stands should reduce the r-value by 2.5 times.

In order to estimate mortality rates, the ratio of the number of dead trees to number of infected trees (i), was computed for the "moist" and "dry" whitebark pine stands (R. J. Hoff, unpublished data). A higher i-value means more mortality relative to infection. This ratio was also computed for young western white pine stands (McDonald 1982). Results for i were whitebark-dry = 0.22, whitebark-moist = 0.25, and western white pine = 0.31. The i ratio could vary considerably with interaction among wave years, tree age, and any other factors that would influence relative rust growth rate. Nevertheless, i facilitates comparisons of predicted performance, so we will use a conservative value of 0.35. Absolute infection rate (r in Equation 1, Chapter 10) was multiplied by i to predict blister rust mortality. The multipliers we used to adjust absolute infection rate r for the effects of first-generation resistance and removal of *Ribes* shrubs are 0.333 and 0.4, respectively. Then, mortality due to blister rust was estimated by multiplying r, adjusted or not, by the incidence-mortality ratio, 0.35.

Comparisons of predicted performance are then computed for low ($r < 0.05$), moderate ($r = 0.05$ to 0.2), high ($r = 0.2$ to 0.4), and extreme ($r > 0.5$) hazard sites. The whitebark pine stand at Sawtell Peak (see Chapter 10) represents a low-hazard site. Computed and actual values after forty years of exposure were absolute infection rate, r = 0.018; actual incidence at time t, y = 0.51; actual mortality = 0.24, and predicted mortality = 0.22. If selected rust-resistant whitebark pine trees from high-mortality stands (Table 17-1) were planted (absolute infection rate, r reduced by 0.33) at Sawtell Peak, the expectation is r = 0.006, with actual incidence of infection at time t, y = 0.26, and mortality = 0.1 after fifty years. The manager could probably live with this amount of impact. Adding *Ribes* control would give r = 0.007, then y = 0.30 and mortality = 0.12.

A stand at Mt. Brundage, located in central Idaho, can serve as the moderate hazard example. Here, r was 0.123, so after fifty years, y = 1.00 and mortality = 0.88. This level of unacceptable damage would be reduced to only 0.87 infection and 0.51 mortality by planting resistant seedlings. The addition of *Ribes* management would lead to a combined effect of 0.56 and 0.25, respectively, which is within acceptable limits. Our high-hazard example is based on a stand (r = 0.253) located in the Olympic Mountains of Washington (see Chapter 10). After fifty years, mortality under these high-hazard conditions would be about 0.98. Application of the combination of first-generation resistance and *Ribes* management would decrease mortality to 0.45. *Ribes* management and second-generation resistance should reduce damage to an acceptable level of 0.26 in fifty years.

Our extreme-hazard stand is Wasco County 2, located on the southeastern flank of Mt. Hood in Oregon (see Chapter 10). Application of second-generation resistance would result in 0.87 mortality after fifty years. Adding *Ribes* management reduces this mortality to a relatively high 0.55. Management in extreme hazard situations will probably require the development of chemical controls and/or advanced generation populations of resistant whitebark pine.

The efficacy of *Ribes* eradication also depends on the source of inoculum. If inoculum is produced locally (within 300 meters), eradication would be successful at lowering the infection rate. If the inoculum comes from off-site (>1 km), for example from lower elevations where *Ribes* is mixed with western white pine stands, then eradication would be less feasible. We do not currently have technology that will reliably identify sources of inoculum. There are stands, however, where the inoculum is definitely local. The forests below are too dry, do not contain *Ribes,* or there are no white pines present.

Concluding Comments

There is sufficient research to show that natural resistance in whitebark pine to white pine blister rust exists, and it is passed to the next generation. Several resistance mechanisms have been documented, and these mechanisms are likely to be stable (horizontal resistance). We can use knowledge gained from the western white pine breeding program to package multiple resistance mechanisms to provide even more stability.

Several approaches will assist restoration efforts through integrated management. They include seed collection, seed transfer, gene conservation, use of fire, planting, natural regeneration, tending of regenerated stands, blister rust hazard mapping, *Ribes* management, chemical control of blister rust infection, and development of a blister rust–resistance breeding program with its supporting field outplantings. We must also keep in mind that all five-needled white pines behave in a similar fashion regarding blister rust. This means that most of the cost of developing integrated management tools can be spread over all the species. Most approaches that help restore whitebark pine can also be used to increase the level of rust resistance in future generations. These approaches are additive, in that each will increase the level of surviving trees.

We need to develop and implement strategies that will restore whitebark pine ecosystems as quickly as possible. The need is urgent, because the major ongoing threats to whitebark pine—blister rust, mountain pine needle, and secondary succession to shade-tolerant species—continue to decrease the size of whitebark pine populations. Delays in implementing management are actually a decision to accept even greater declines in whitebark pine populations.

Humans are responsible for the introduction of white pine blister rust to North America, which has dramatically reduced populations of five-needled white pines, including whitebark pine. We have accelerated the succession of whitebark pine ecosystems to shade-tolerant conifers through fire suppression activities. The good news is that we can help reverse declines in whitebark pine populations and restore whitebark pine ecosystems. Whitebark pine is going through an extreme evolutionary bottleneck at present; the species and the ecosystems it occupies can benefit greatly from our assistance.

Acknowledgment

We thank Eugene Van Arsdel, Texas A & M University, for reviewing an earlier draft of this chapter.

Literature Cited

Arno, S. F., and R. J. Hoff. 1989. Silvics of whitebark pine (*Pinus albicaulis*). USDA Forest Service, Intermountain Research Station, General Technical Report INT-253, Ogden, Utah.

Baker, F. S. 1944. Mountain climates of the western United States. Ecological Monographs 14:223–254.

Barth, R. S. 1994. Pruning and excising white pine on the Wallace Ranger District as a blister rust control measure. Pages 333–334 *in* D. M. Baumgartner, J. E. Lotan, and J. R. Tonn, editors. Proceedings, Interior cedar-hemlock-white pine forests: Ecology and management. Cooperative Extension, Washington State University, Pullman, Washington.

Bedwell, J. L., and T. W. Childs. 1943. Susceptibility of whitebark pine to blister rust in the Pacific Northwest. Journal of Forestry 41:904–912.

Berube, J. A. 1996. Use of triadimefon to control white pine blister rust. Forestry Chronicle 72:637–638.

Bingham, R. T. 1972. Taxonomy, crossability, and relative blister rust resistance of 5-needle white pines. Pages 271–280 *in* R. T. Bingham, R. J. Hoff, and G. I. McDonald, editors. Biology of rust resistance in forest trees: Proceedings of a NATO-IUFRO advanced study institute. USDA Forest Service, Miscellaneous Publication 1221, Washington, D.C.

———. 1983. Blister rust resistant western white pine for the Inland Empire: The story of the first 25 years of the research and development program. USDA Forest Service, Intermountain Research Station, General Technical Report INT-146, Ogden, Utah.

Bingham, R. T., A. E. Squillace, and J. W. Wright. 1960. Breeding blister rust resistant western white pine. II. First results of progeny tests including preliminary estimates of heritability and rate of improvement. Silvae Genetica 9:33–41.

Bruederle, L. P., D. F. Tomback, K. K. Kelly, and R. C. Hardwick. 1998. Population genetic structure in bird-dispersed pine, *Pinus albicaulis* (Pinaceae). Canadian Journal of Botany 76:83–90.

Cooper, S. V., K. E. Neiman, and D. W. Roberts. 1991. Forest habitat types of northern Idaho: A second approximation. USDA Forest Service, Intermountain Research Station, General Technical Report INT-236, Ogden, Utah.

Geils, B. W., D. A. Conklin, and E. P. Van Arsdel. 1999. A preliminary hazard model of white pine blister rust for the Sacramento Ranger District, Lincoln National Forest. USDA Forest Service, Rocky Mountain Research Station, Research Note RMRS-RN6, Fort Collins, Colorado.

Hamrick, J. L., A. F. Schnabel, and P. V. Wells. 1994. Distribution of genetic diversity among populations of Great Basin conifers. Pages 147–161 *in* K. T. Harper, L. L. St. Clair, K. H. Thorne, and W. M. Hess, editors. Natural history of the Colorado Plateau and Great Basin. University of Colorado Press, Boulder.

Hoff, R. J. 1984. Resistance to *Cronartium ribicola* in *Pinus monticola:* Higher survival of infected trees. USDA Forest Service, Intermountain Research Station, Research Note INT-343, Ogden, Utah.

Hoff, R. J., and G. I. McDonald. 1980. Improving rust resistant strains of inland western white pine. USDA Forest Service, Intermountain Research Station, Research Paper INT-245, Ogden, Utah.

Hoff, R. J., G. I. McDonald, and R. T. Bingham. 1976. Mass selection for blister rust resistance: A method for natural regeneration of western white pine. USDA Forest Service, Intermountain Forest and Range Experiment Station, Research Note INT–202, Ogden, Utah.

Hoff, R. J., R. T. Bingham, and G. I. McDonald. 1980. Relative blister rust resistance of white pines. European Journal of Forest Pathology 10:307–316.

Hoff, R. J., S. K. Hagle, and R. G. Krebill. 1994. Genetic consequences and research challenges of blister rust in whitebark pine forests. Pages 118–128 *in* W. C. Schmidt and F.-K. Holtmeier, compilers. Proceedings—International workshop on subalpine stone pines and their environment: The status of our knowledge. USDA Forest Service, Intermountain Research Station, General Technical Report INT–306, Ogden, Utah.

Hungerford, R. D., R. E. Williams, and M. A. Marsden. 1982. Thinning and pruning western white pine: A potential for reducing mortality due to blister rust. USDA Forest Service, Intermountain Forest and Range Experiment Station, Research Note INT-322, Ogden, Utah.

Hunt, R. S. 1998. Pruning western white pine in British Columbia to reduce white pine blister rust losses: 10-year results. Western Journal of Applied Forestry 13:60–63.

Hutchins, H. E., and R. M. Lanner. 1982. The central role of Clark's nutcracker in the dispersal and establishment of whitebark pine. Oecologia 55:192–201.

Hutchinson, G. E. 1958. Concluding remarks. Cold Spring Harbor Symposium on Quantitative Biology 22:415–427.

Johnson, D. R., B. B. Kinloch Jr., and A. H. McCain. 1992. Triadimefon controls white pine blister rust on sugar pine in a greenhouse test. Tree Planters Notes 43:7–10.

Jorgenson, S. M., and J. L. Hamrick. 1997. Biogeography and population genetics of whitebark pine, *Pinus albicaulis*. Canadian Journal of Forest Research 27:1574–1585.

Keane, R. E., S. F. Arno, and J. K. Brown. 1989. FIRESUM—An ecological process model for fire succession in western conifer forests. USDA Forest Service, Intermountain Research Station, General Technical Report INT-266, Ogden, Utah.

Kelley, W. D., and J. C. Williams. 1985. Effects of triadimefon and triadimenol as seed dressings on incidence of fusiform rust on loblolly pine seedlings. Plant Disease 69:147–148.

Kinloch, B. B., and M. Comstock. 1981. Race of *Cronartium ribicola* virulent to major gene resistance in sugar pine. Plant Disease 65:604–605.

Kinloch, B. B., G. K. Parks, and C. W. Fowler. 1970. White pine blister rust: Simply inherited resistance in sugar pine. Science 167:193–195.

Lachmund, H. G. 1926. Western blister rust investigations, 1926. Pages 19–23 *in* Report of Proceedings, Western white pine blister rust conference, Portland, Oregon.

———. 1928. Investigative work, 1928. Pages 12–19 *in* Report of Proceedings, Western white pine blister rust conference, Portland, Oregon.

Mahalovich, M. F., and A. Eramian. 2000. Breeding, seed orchard, and restoration plan for the development of blister rust resistant white pine for the northern Rockies. USDA Forest Service, Northern Region and Inland Empire Tree Improvement Cooperative, Missoula, Montana.

Mahalovich, M. F., and R. J. Hoff. 2000. Whitebark pine operational cone collection instructions and seed transfer guidelines. Nutcracker Notes 11:10–13. http://www.mesc.usgs.gov/glacier/nutnotes.htm

Martin, J. F. 1944. *Ribes* eradication effectively controls white pine blister rust. Journal of Forestry 42:255–260.

McCaughey, W. W. 1994. The regeneration process of whitebark pine. Pages 179–187 *in* W. C. Schmidt and F.-K. Holtmeier, compilers. Proceedings—International workshop on subalpine stone pines and their environment: The status of our knowledge. USDA Forest Service, Intermountain Research Station, General Technical Report INT-306, Ogden, Utah.

McDonald, G. I. 1979. Resistance of western white pine to blister rust: A foundation for integrated control. USDA Forest Service, Intermountain Forest and Range Experiment Station, Research Note INT-252, Ogden, Utah.

———. 1982. Resistance-hazard alignment: A blister rust management philosophy. Pages 355–277 *in* D. Miller, compiler. Proceedings: Breeding insect and disease resistant forest trees. USDA Forest Service, Timber Management Staff, Washington, D.C.

McDonald, G. I., and D. L. Dekker-Robertson. 1998. Long-term differential expression of blister rust resistance in western white pine. Pages 285–295 *in* R. Jalhaner, editor. Proceedings of first IUFRO rusts of forest trees, Saariselka, Finland. Finnish Forest Research Institute, Research Papers 712.

McDonald, G. I., R. J. Hoff, and W. R. Wykoff. 1981. Computer simulation of white pine blister rust epidemics: 1. Model formulation. USDA Forest Service, Intermountain Forest and Range Experiment Station, Research Paper INT-258, Ogden, Utah.

Mielke, J. L. 1943. White pine blister rust in western North America. Bulletin 52, School of Forestry, Yale University, New Haven, Connecticut.

Mitton, J. B. 1995. Genetics and the physiological ecology of conifers. Pages 1–36 *in* W. K. Smith and T. M. Hinckley, editors. Ecophysiology of coniferous forests. Academic Press, New York.

Ostrofsky, W. D., T. Rumpf, D. Struble, and R. Bradbury. 1988. Incidence of white pine blister rust in Maine after 70 years of a *Ribes* eradication program. Plant Disease 72:967–970.

Rehfeldt, G. E. 1982. Differentiation of *Larix occidentalis* populations from the northern Rocky Mountains. Silvae Genetica 31:13–19.

———. 1988. Ecological genetics of *Pinus contorta* from the Rocky Mountains (U.S.A.): A synthesis. Forest Ecology and Management 28:203–215.

———. 1989. Ecological adaptations in Douglas-fir (*Pseudotsuga menziesii* var. *glauca*): A synthesis. Silvae Genetica 37:131–135.

———. 1991. Models of genetic variation for *Pinus ponderosa* in the Inland Northwest (U.S.A.). Canadian Journal of Forest Research 21:1491–1500.

———. 1994. Evolutionary genetics, the biological species, and the ecology of the interior cedar-hemlock forests. Pages 91–100 *in* D. M. Baumgartner, J. E. Lotan, and J. R. Tonn, editors. Proceedings, Interior cedar-hemlock-white pine forests: Ecology and management. Cooperative Extension, Washington State University, Pullman, Washington.

Rehfeldt, G. E., R. J. Hoff, and R. J. Steinhoff. 1984. Geographic patterns of genetic variation in *Pinus monticola*. Botanical Gazette 145:229–239.

Rehfeldt, G. E., C. C. Ying, D. L. Spittlehouse, and D. A. Hamilton Jr. 1999. Genetic responses to climate for *Pinus contorta:* Niche breadth, climate change and reforestation. Ecological Monographs 69:375–407.

Robbins, K., W. A. Jackson, and R. E. McRoberts. 1988. White pine blister rust in the eastern upper peninsula of Michigan. Northern Journal of Applied Forestry 5:263–264.

Schuster, W. S., D. L. Alles, and J. B. Mitton. 1989. Gene flow in limber pine: Evidence from pollination phenology and genetic differentiation along an elevational transect. American Journal of Botany 76:1395–1403.

Schwandt, J. W., M. A. Marsden, and G. I. McDonald. 1994. Pruning and thinning effects on white pine survival and volume in northern Idaho. Pages 167–172 *in* D. M. Baumgartner, J. E. Lotan, and J. R. Tonn, editors. Proceedings, Interior cedar-hemlock-white pine forests: Ecology and management. Cooperative Extension, Washington State University, Pullman, Washington.

Simmonds, N. W. 1991. Genetics of horizontal resistance to diseases of crops. Biological Review 66:189–241.

Smith, J. P., and J. T. Hoffman. 1998. Status of white pine blister rust in Intermountain Region white pines. USDA Forest Service, Intermountain Region, State and Private Forestry, Forest Health Protection Report No. R4-98-02, Ogden, Utah.

———. 2000. Status of white pine blister rust in the Intermountain West. Western North American Naturalist 60:165–179.

Spaulding, P. C. 1911. The blister rust of white pine. USDA Bulletin Number 206. Washington, D.C.

Toko, H. V., D. A. Graham, C. E. Carlson, and D. E. Ketchum. 1967. Effects of past *Ribes* eradication on controlling white pine blister rust in northern Idaho. Phytopathology 57:1010.

Tomback, D. F. 1982. Dispersal of whitebark pine seeds by Clark's nutcracker: A mutualism hypothesis. Journal of Animal Ecology 51:451–467.

Tomback, D. F., and Y. B. Linhart. 1990. The evolution of bird-dispersed pines. Evolutionary Ecology 4:185–219.

Tomback, D. F., J. K. Clary, J. Koehler, R. J. Hoff, and S. F. Arno. 1995. The effects of blister rust on post-fire regeneration of whitebark pine: The Sundance burn of northern Idaho (U.S.A.). Conservation Biology 9:654–664.

USDA Forest Service. 1999. Transfer rules. Chapter 4 *in* Seed handbook. Draft Forest Service Handbook 2409.26f (Regions 1–4), Missoula, Montana.

Van der Plank, J. E. 1963. Plant disease—Epidemics and control. Academic Press, New York.

———. 1968. Disease resistance in plants. Academic Press, New York.

Wright, J. 1976. Introduction to forest genetics. Academic Press, New York.

Zadoks, J. C. 1972. Reflections on disease resistance in annual crops. Pages 43–63 *in* R. T. Bingham, R. J. Hoff, and G. I. McDonald, editors. Biology of rust resistance in forest trees: Proceedings of a NATO-IUFRO advanced study institute. USDA Forest Service, Miscellaneous Publication 1221, Washington, D.C.

Chapter 18

Restoration Concepts and Techniques

Robert E. Keane and Stephen F. Arno

Innovative techniques are needed to restore the health of whitebark pine (*Pinus albicaulis*) communities in the northern Rocky Mountains of the United States, inland West, and adjacent areas of Canada, because of the detrimental effects of the exotic disease white pine blister rust (*Cronartium ribicola*) coupled with fire exclusion policies and recent mountain pine beetle (*Dendroctonous ponderosae*) epidemics (Chapters 1 and 11). Land managers need guidelines to return native patterns, processes, compositions, and structures to declining whitebark pine communities in these regions. Even in community types where whitebark pine is less fire dependent (Chapter 4), these techniques may be useful for facilitating the establishment of blister-rust-resistant trees.

This chapter outlines an approach for planning, implementing, and monitoring whitebark pine restoration projects to accomplish this goal. Specific restoration treatments and techniques are discussed in the context of the ecosystem processes being restored. This approach is presented more as a guide to restoration efforts rather than as a step-by-step procedure. All steps discussed in this chapter need not be completed if time, money, or personnel are lacking. Many restoration projects can be implemented in small areas as demonstrations (i.e., prototypes) without the multiple spatial analyses and detailed planning processes presented here. Most of the principles and procedures presented in this chapter apply to the restoration of other fire-dependent ecosystems as well.

Whitebark pine ecosystem restoration, as discussed here, does not imply that historical stand structures should be re-created using silviculture or prescribed fire, as is commonly the case (Bonnicksen and Stone 1985; Apfelbaum and Chapman 1997; Bradshaw 1997). We suggest that ecosystem restoration must emphasize the return of ecosystem processes rather than historical stand characteristics to succeed over the long term (Parsons et al. 1986; Chambers 1997; Crow and Gustafson 1997; Michener 1997). Historical disturbance regimes, stand structures, and landscape patterns should be used as guides rather than goals in restoration efforts. It is important that restored processes complement prevailing environmental conditions for long-term success of restoration activities (Apfelbaum and Chapman 1997; Parker and Pickett 1997). We assume that when important processes, such as the fire regime, are restored to an ecosystem, suitable stand and landscape structures and compositions will follow (Parsons et al. 1986; Parker and Pickett 1997). This approach becomes complicated by widespread white pine blister rust infection and mortality that is not native to the ecosystem. However, we believe that the maintenance of native fire regimes is the single most important management action to ensure conservation of whitebark pine. Native fire regimes create favorable habitat for seed caching by Clark's nutcrackers (*Nucifraga columbiana*) that will effectively regenerate rust-resistant whitebark pine, because most seeds will come from trees that survived blister rust infections (Chapter 17).

Restoration Research Projects

Many ideas and methods discussed in this chapter are from an ongoing research project investigating methods for restoring whitebark pine to the landscape in and around the Bitterroot Mountains of western Montana and east-central Idaho (Figure 18-1) (Keane and Arno 1996; Keane et al. 1996a). This project, Restoring Whitebark Pine Ecosystems (RWPE), began in 1993 and involves five research sites in whitebark pine forests that differ in biophysical environment, rust infection levels, and stand structure. Prescribed burning and silvicultural cuttings are being used to reestablish or maintain whitebark pine in areas where the species is declining because of blister rust and advancing succession. Experience gained from the planning, design, and implementation of this effort have been incorporated into this chapter. We know of no other restoration research projects in whitebark pine ecosystems. Since the RWPE project is ongoing, and some of the conclusions are based on limited data, new results gathered from these study sites will lead to refinements in the proposed restoration methods.

Four spatial scales were used in planning the RWPE project—coarse, mid, watershed, and stand. Their boundaries were dictated by the availability of comprehensive spatial data layers (Keane et al. 1996a). The coarse scale encompassed the 89-million-hectare Interior Columbia River Basin (Figure 18-1) at a 1 km^2 pixel resolution and a map scale of 1:250,000 (Quigley and Arbelbide 1997). Many vegetation and biophysical data layers were created for this area as part of the Interior Columbia Basin Ecosystem Management Project (Quigley et al. 1996). The 2.2-million-hectare Selway-Bitterroot Wilderness Complex defined the midscale analysis area with data layers resolved at 900 m^2 pixel resolution and 1:24,000 map scale (Figure 18-1). Keane et al. (1998) used climate, vegetation, topography, and satellite image classification data layers (Redmond and Prather 1996) to map vegetation and fuels on this landscape. Individual watersheds containing the RWPE research sites defined

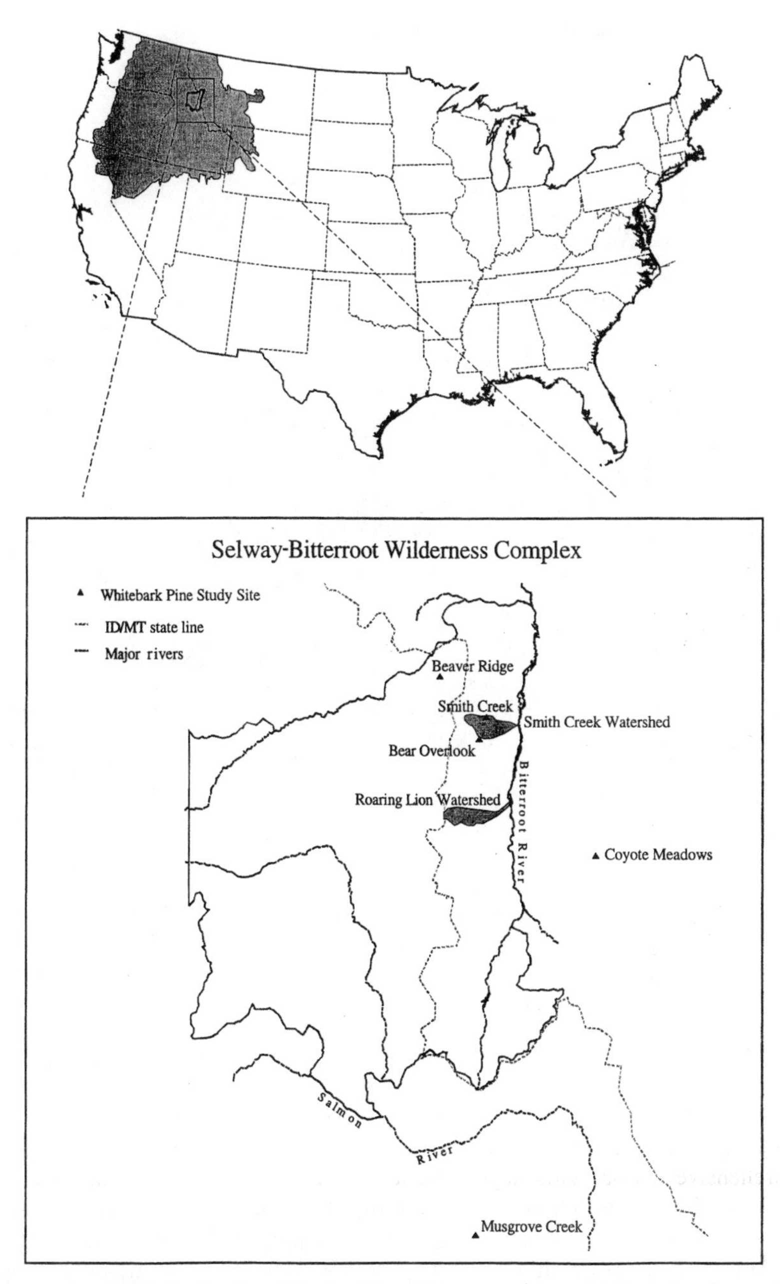

Figure 18-1. Whitebark pine (*Pinus albicaulis*) research restoration sites in west-central Montana. The larger boundary delineates the Interior Columbia River Basin analysis area used for coarse-scale analysis. The interior boundary is of the Selway-Bitterroot Wilderness Complex used to analyze midscale data. Triangles indicate the locations of research sites in the restoration project Restoring Whitebark Pine Ecosystems (RWPE). Smith Creek Watershed is shown as an example of a fine-scale current landscape, and Roaring Lion Watershed is shown as an example of fine-scale historical landscape.

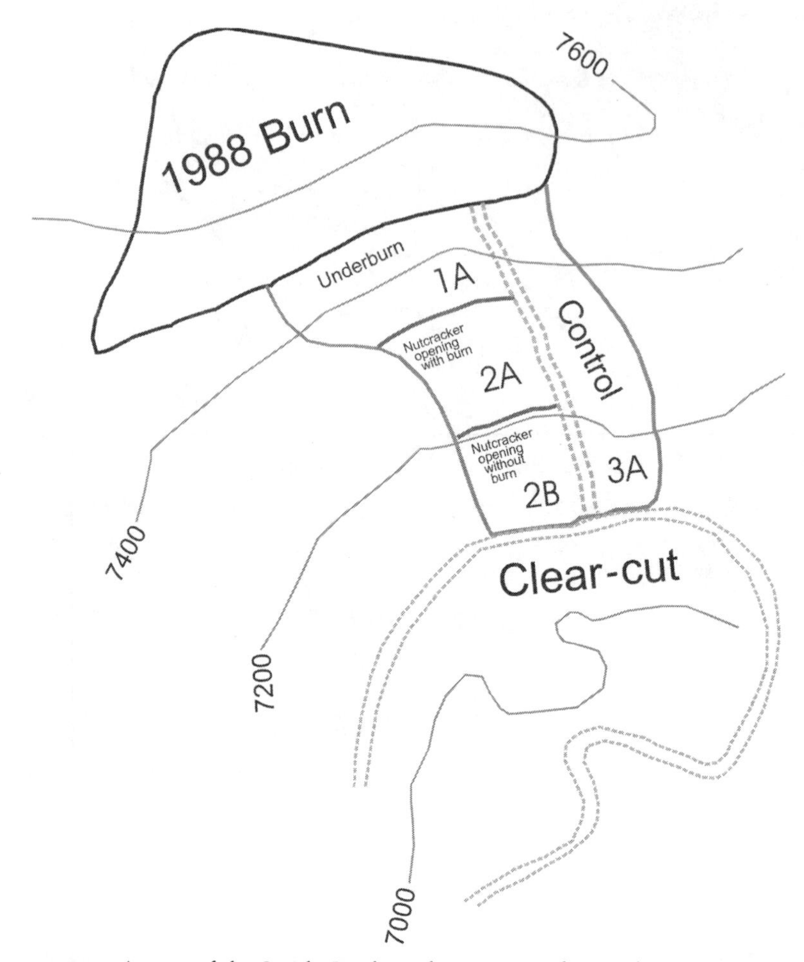

Figure 18-2. A map of the Smith Creek study site, one of several research sites in the Restoring Whitebark Pine Ecosystems (RWPE) project. Smith Creek has three treatment units (1A, 2A, 2B) and a control (3A).

the landscapes at resolutions similar to the midscale. The Roaring Lion Watershed (Figure 18-1) was the only watershed with historical vegetation data layers created from the interpretation of 1930s aerial photographs (circa 1930) (Hessburg et al. 1999a). Although interpretation of historical photography can be suspect because of the lack of reference conditions, these images provide the only spatially explicit data of historical characteristics. Again, these tools should be used as guides, not targets. Individual research sites are composed of several treatment units, as illustrated in Figure 18-2 for the Smith Creek study site. These RWPE treatments areas were mapped at 90 m^2 resolution and 1:4,000 map scale and are referred to extensively in this chapter (Keane and Arno 1996).

The Restoration Process

The most important task in any restoration effort is to define project objectives clearly (Eggers 1990; Michener 1997). However, a concise statement of objectives may require information about the restoration area that may not be available. Obviously, it is necessary to know that whitebark pine stands are declining before a project is initiated to restore their health. Field reconnaissance and literature searches often provide key information to craft a succinct initial objective statement, and this objective can be amended as more information becomes available during restoration planning and implementation. The primary objective of any ecosystem restoration effort should be to reintroduce, to the landscape, the ecological processes and characteristics that were modified by human activities so that the processes are in harmony with environmental conditions (Apfelbaum and Chapman 1997). This, of course, is rather difficult, because future climate, environmental, and social trends are nearly impossible to assess, predict, and manage (Bartlein et al. 1997).

This chapter is organized into seven steps that are important in whitebark pine restoration efforts once a comprehensive objective statement has been developed:

1. An *inventory* of existing landscape and stand characteristics is conducted at multiple scales.
2. Important ecosystem processes of landscapes and stands are *described* to provide background for the design of restoration activities.
3. Landscapes and stands are then *prioritized* for restoration treatment.
4. These prioritized landscapes and stands are *selected* based on the inventory, description, and prioritization.
5. Treatments are then *designed* for each selected stand or landscape based on the inventory and description.
6. These treatments are *implemented* in the most efficient means possible.
7. Treatment sites are *monitored* to evaluate restoration success.

Restoration Planning

The planning phase of a restoration project is composed of the steps needed to craft appropriate treatments. This critical phase involves the inventory, description, prioritization, selection, and design steps.

Inventory

A comprehensive, spatially explicit ecosystem inventory is essential for designing a defensible and successful whitebark pine restoration project (Apfelbaum and Chapman 1997). This inventory can be conducted across multiple time and space scales, albeit extensive spatial data sets may be lacking for all scales (Bell et al. 1997; Crow and Gustafson 1997; Michener 1997; Ziemer 1997). The inventory should emphasize fire characteristics so that treatments can be designed based on fire histories collected at the stand, landscape, and regional levels (Swanson et al. 1990). A thorough review of the literature and interviews with local and regional experts can augment sampled inventory information or can be used as a substitute when spatial and field data are unavailable for coarse- and midscales.

Coarse-scale data layers describing past and present distribution, processes, and health of whitebark pine communities across a large geographical area provide a regional context for restoration activities (Keane et al. 1997b). For example, it might be important to initiate restoration activities in an area that contains 90 percent of whitebark pine stands for that entire region to maximize efficiency. Potential vegetation type (PVT), cover type, and structural stage data layers provide the basic information to link local restoration efforts into a regional framework. This vegetation triplet (PVT, cover type, structural stage) was used to characterize the Interior Columbia River Basin (Quigley and Arbelbide 1997; Keane et al. 1996d), Selway-Bitterroot Wilderness Complex (Keane et al. 1998), Smith Creek Watershed, and many RWPE study stands (Keane and Arno 1996). Lands having the potential to support whitebark pine can be identified from the PVT layer based on the PVT classification (e.g., species coverage tables in Pfister et al. 1977), whitebark pine community composition can be described from the cover type, and the vertical stand structure can be determined from structural stage layers (Eyre 1980; O'Hara et al. 1996). Classification categories for the vegetation triplet are often different for each spatial scale, because inherent classification and aggregation errors in coarse data layers prevent their extrapolation to finer scales (Rastetter 1996). However, Keane and Arno (1996) and Keane et al. (1998) designed classification categories and cross-reference tables to link each vegetation classification category hierarchically across three spatial scales (fine: Smith Creek; mid: Selway-Bitterroot Wilderness; and coarse: Interior Columbia River Basin) for restoration planning.

Midscale data can quantify the diversity and character of local landscapes within a region (Swanson et al. 1990; Crow and Gustafson 1997; Ziemer 1997). The size, spatial distribution, and composition (cover type and structural stage) of stands mapped across midscale areas describe landscape composition and structure (i.e., patch dynamics) that can be used to prioritize landscapes for treatment and also to design the type, size, intensity, and location of restoration treatments. For example, stand-replacement fire regimes common in high-elevation forests of the Bob Marshall Wilderness Complex might indicate that large, prescribed fires might be appropriate.

The structure and composition of past, present, and future landscapes at the midscale provide a more accurate means to identify and prioritize whitebark pine stands for restoration activities (Swanson et al. 1997). For example, since subalpine fir (*Abies lasiocarpa*) becomes dominant during the late-seral stages, the proportion of the landscape characterized by subalpine fir cover types may provide a means to assess some of the landscapes in need of fire (Keane et al. 1996d). Proximity of possible restoration sites to roads, trails, and rocklands, which might provide access for treatment and possible fire control structures, can also be described using midscale layers. This is accomplished by assessing the ecosystem characteristics of the stands that comprise the analysis landscape. Field data gathered for each stand can be summarized to descriptive attributes meaningful to restoration, or vegetation classification categories, if only a subset of stands were sampled. For example, Keane et al. (1998)

sampled canopy closure and other fuel attributes for about 3 percent of the Selway-Bitterroot Wilderness and then generalized the summarized data to all combinations of categories in the PVT, cover type, and structural stage classifications. These results were then extrapolated to the unsampled area using existing satellite imagery and terrain modeling created by Redmond and Prather (1996).

A comprehensive inventory of every stand on the restoration landscape is useful for identification of those areas that need treatment, but this detailed inventory is often unrealistic. The spatial and size distribution of stands, along with the fire regime, will dictate the frequency, size, and location of possible restoration treatments (Pickett and White 1985; Turner and Gardner 1991; Hessburg et al. 1999b). For instance, landscapes with small patches (i.e., stands) may indicate that low-intensity, nonlethal surface fires are an appropriate restoration treatment for stands with abundant fire-tolerant whitebark pine (Turner et al. 1994).

Fine-scale, stand-level inventories are often needed to provide detail for the design and implementation of restoration activities. Ecosystem information needed at this scale depends primarily on restoration objectives and usually includes stand structure (composition, age, and size), fuels, and plant composition (Keane and Arno 1996). Although it is probably not feasible to measure these characteristics for every stand on the target landscape, it may be necessary to measure them for each stand to be restored, especially if monitoring treatment effects are important.

The ECODATA system developed by the USDA Forest Service Northern Region is an effective inventory tool that provides methods, databases, and programs for the description, analysis, and evaluation of ecosystems (Keane et al. 1990; Jensen et al. 1993). Although there are many standardized inventory procedures available to measure stands, it is important that the information collected match the restoration objectives. For example, if the return of fire is an important restoration goal, then fire scars should be sampled and dated, and stand age structure should be assessed to characterize past fire regimes and successional cycles (Arno and Sneck 1977; Barrett and Arno 1988). If fuel reduction is specified in the fire prescription, then fuel loadings should be measured. A detailed plot sampling design, such as ECODATA, provides baseline conditions for detecting change and assessing treatment success.

We modified ECODATA forms and methods to inventory treatment stands for the RWPE study. Ten plots were systematically located in each treatment unit, with the first plot established randomly. Stand structure information was obtained by measuring diameter at breast height (DBH, centimeters), height (meters), age (years), and crown height (meters) on all trees above a threshold diameter (10 centimeters for this study) inside a 0.04-hectare fixed plot. Seedling and sapling tree data were measured on a smaller 0.001-hectare fixed plot for sampling efficiency (Hann et al. 1990). We gathered fuels information along two 15.2-meter transects at each plot, using methods described by Brown (1970). Cover (percent) and average height (meters) of all vascular plant species were estimated within each fixed area plot using four 1-square-meter microplots. We also estimated the percent crown killed by blister rust for each whitebark pine tree inside plot boundaries (Keane et al. 1994; Keane and Arno

1996). The sampling intensity used in this research effort is probably too high for most management applications. A possible alternative is to measure the same information on fewer plots.

Description

Planning scientifically based restoration activities requires a complete description of the ecosystem processes and characteristics of the landscape to be treated. Information gathered from the literature can be augmented with information gained from the preceding inventory to describe the restoration site. This description will provide the background needed to devise comprehensive restoration treatments. Published whitebark pine research is relatively meager when compared to other tree species, especially in the area of ecosystem restoration. McCaughey and Weaver (1990b) developed an annotated bibliography that may be useful. The Fire Effects Information System contains current citations and summarized ecological literature useful for whitebark pine management activities (Fischer et al. 1996). Issues of *Nutcracker Notes* (http://www.mesc.usgs.gov/glac/nutnotes.htm), a news digest of current research and management activities in whitebark pine, also provide useful and current information for restoration planning. However, the inventory data gathered from the project area will provide the most useful and germane information about the target ecosystem. Spatial data layers may also be summarized during this step.

The data layers prepared for the Interior Columbia Basin Ecosystem Management Project were used to describe historical and current status of whitebark pine in the Pacific Northwest for the RWPE project (Quigley and Arbelbide 1997). Keane et al. (1996d) used these coarse-scale layers to compare whitebark pine status in the Bob Marshall Wilderness Complex with its status across the Interior Columbia River Basin. A large proportion (approximately 10 percent) of the region's whitebark pine cover types was within the Bob Marshall Wilderness Complex, indicating the importance of maintaining healthy whitebark pine in this reserve. About 11 percent of the Interior Columbia River Basin lands that can potentially support whitebark pine are found in the Selway-Bitterroot Wilderness Complex, but a large portion of this percentage (over 15 percent) were in the late-successional spruce-fir (Engelmann spruce [*Picea engelmannii*] and subalpine fir) cover type (Table 18-1). The Smith Creek Watershed had similar current proportions of spruce-fir and whitebark pine cover types compared with the entire Selway-Bitterroot Wilderness Complex (Table 18-1). In addition, it appears that the conditions and incidence of blister rust at the Smith Creek study site were representative of the high elevations of the watershed and of the Selway-Bitterroot. This information, interpreted across scales, indicates that the Selway-Bitterroot is an excellent area for restoration efforts in the Interior Columbia Basin because of the high percentage of lands in potential whitebark pine habitat, coupled with the apparent decline of these forests due to rust and advanced succession (Kendall and Arno 1990; Keane and Arno 1993) (Table 18-1). The Smith Creek watershed appears to be a good restoration landscape, because it is representative of the Selway-Bitterroot Wilderness Complex (SBWC) (Table 18-1).

Table 18-1. Characteristics of historical and current whitebark pine (*Pinus albicaulis*) landscapes across multiple scales. Only the Smith Creek study site is presented at the fine scale in this example. The Roaring Lion Watershed was used to quantify historical conditions from 1930s aerial photos. Interior Columbia River Basin (ICRB) historical conditions were simulated from data recorded circa 1900. SBWC = Selway-Bitterroot Wilderness Complex.

	Coarse Scale (ICRB)		Midscale (SBWC)		Smith Creek Watershed Scale		Smith Creek Research Site	
Classification Category	Historical	Current	Historical	Current	Historical	Current	Historical	Current
Area (km^2)	890,000	890,000	11,540	11,540	66	66	0.2	0.2
Whitebark pine potential (km^2)	49,967	49,967	873	873	10	12	0.2	0.2
Whitebark cover type (km^2)	17,116	9,451	—	48	8	2	0.2	0.0
Spruce-fir cover type (km^2)	11,414	19,813	—	423	1	6	0.0	0.2
Predominant structural stage[1]	Pole	Pole	Pole	Mature	Pole	Seedling	Pole	Mature
Fire severity class[2]	Mixed	Lethal	Mixed	Lethal	Mixed	Mixed	Mixed	Mixed
Fire return interval (years)	76–150	151–300	80–120	—	80–130	170	80–120	180
Average annual precipitation (mm yr^{-1})	990	990	1,200	1,200	—	—	—	—
Rust-caused mortality (%)	0	50	0	60	0	85	0	91
Data source	Quigley & Arbelbide (1997)	Quigley & Arbelbide (1997)	Brown et al. (1994)	Keane et al. (1998)	Roaring Lion watershed	Keane et al. (1998), Keane and Arno (1996)	Keane and Arno (1996)	Keane and Arno (1996)

[1]Structural Stages—Seedling: 0–10 cm diameter; Pole: 10–25 cm diameter; Mature: 25–50 cm diameter.
[2]Fire Regimes—Mixed: Mixed-severity fire regime; Lethal: Stand-replacement fire regime.

Stand and landscape descriptions should also include a summary of the important processes that shaped the structure and composition of the restoration area, such as historical and current fire, insect, disease, and climate regimes. This information provides the guidelines for the restoration project. For example, field data revealed that the Smith Creek upper subalpine areas historically experienced mixed-severity fires with fire return intervals of about 80 to 250 years (Keane and Arno 1996; Hartwell 1997). Prior to 1900, these landscapes were dominated by whitebark pine, but whitebark pine is being replaced by subalpine fir because of mortality from the 1930s mountain pine beetle epidemic and, more recently, from white pine blister rust (Keane and Arno 1993; Hartwell 1997). Around 1900, the Interior Columbia River Basin had nearly twice the area containing whitebark pine cover types, and around 1930, the Roaring Lion Watershed had eight times more whitebark pine area than subalpine fir area (Table 18-1). Therefore, restoration treatments should simulate the effect of mixed-severity fire to maintain whitebark pine on the landscape, and additional restoration activities may be necessary to mitigate the effects of blister rust. The size of the treated areas can be determined from the patch dynamics of the historical and current landscapes.

Many fire regimes are found on Selway-Bitterroot landscapes because of diverse topography and ecosystems, but stand-replacement fires are probably the most common in the upper subalpine ecosystems (Barrett and Arno 1992; Brown et al. 1994), which is similar to the fire regime predicted for whitebark pine forests of the Interior Columbia River Basin (Table 18-1) (Keane et al. 1994). However, mixed-severity fire regimes were common in the Smith Creek and Roaring Lion Watershed analysis areas (Table 18-1). The mixed-severity regime has components of both stand-replacement fires and nonlethal surface fires mixed in space and time, creating small to large patches where either all trees are burned and killed (stand-replacement) or only the fire-tolerant species survived (underburn). At the Beaver Ridge study site, patches of the 1910 stand-replacement fire were intermixed with patches from a low-severity underburn in 1833. The Musgrove Creek project area experienced a stand-replacement fire more than 250 years ago. An old-growth stand in the Coyote Meadows study area experienced at least one nonlethal underburn in 1855, but trees that survived the fire probably originated from a stand-replacement event during the 1600s.

Landscape fire history and fire locations can be assessed from intensive fire history sampling (Barrett and Arno 1988), or from fire atlases, stand ages, or burn areas from photos to determine the proportion of the landscape that historically burned every year or decade. Average fire size can also be quantified from the fire atlases or landscape patch characteristics (Hessburg et al. 1999b). These two statistics can then be used to design the type and size of fire treatment.

Descriptions of historical landscape and stand conditions provide information on the compositional and structural conditions that would be created by a restored fire regime. Landscape pattern (i.e., structure) and composition are important in ecosystem restoration, because appropriate spatial configurations facilitate recruitment of native flora and fauna (Bell et al. 1997). Whitebark pine cover types were more extensive on the historical landscapes of the Interior Columbia Basin, Selway-Bitterroot Wilderness Complex, Roaring Lion Watershed, and Smith Creek treatment area compared with conditions today (Table 18-1). Some stands were in structural stages indicative of occasional, nonlethal surface fires, but most whitebark pine stands tended to be in early- to midseral structural stages. In the 1930s, subalpine fir encompassed about 5 percent of the Roaring Lion Watershed, compared with approximately 20 percent today. The historical range of subalpine fir landscape coverage can be coarsely determined from the historical fire return interval, or more accurately from historical landscape data and successional simulation modeling (Chapter 9). Prescribed fire and cutting guidelines can be designed to create stand conditions that represent the effects of the historical fire regime.

Landscape structure and composition can be described from landscape metrics that are calculated from software packages including FRAGSTATS (McGarigal and Marks 1995), R.LE (Baker and Cai 1990), and UTOOLS (Ager and McGaughey 1997). These programs calculate indices that describe relative landscape differences, but these indices cannot be tested for statistical significance (McGarigal and Marks 1995). However, landscape metrics do provide a means to summarize, compare, and contrast landscape characteristics. The following landscape indices, at a minimum, may be suitable for landscape pattern assessment: The mean patch size, its range and variability, patch density, mean nearest neighbor, and edge density provide measures of landscape diversity and fragmentation, which help quantify treatment size and location (Turner and Gardner 1991). Largest patch index (LPI) quantifies the maximum size of possible landscape treatments. Contagion (the higher the index, the greater the aggregation) and mean nearest neighbor indices describe how landscapes are aggregated or clumped, and this can guide decisions about the spatial configuration of treatment blocks (McGarigal and Marks 1995). Other indices can be included in the landscape analysis, depending on restoration objectives and future landscape metric research (Turner et al. 1994). For example, if the shape of the treatment area is important, there are many indices that can quantify shape differences (McGarigal and Marks 1995). Hann (1990) used polygon shape indices to describe three whitebark pine landscapes in Montana and Idaho. A major difficulty in interpreting landscape metrics for upper subalpine landscapes is the presence of nonforest types (e.g., rock, glacier, alpine) that are not seral to whitebark pine types. These can be excluded from the analysis or treated as a different PVT.

Landscape metrics were compared across scales for the Smith Creek treatment site (Table 18-2). Historically, the Roaring Lion drainage contained proportionately more whitebark pine, which were killed by the 1930s mountain pine beetle epidemic and blister rust (Keane and Arno 1996; Hartwell 1997). The Smith Creek study area treatments were too small to influence landscape metrics and had negligible effects on landscape structure and composition (Table 18-2). The Roaring Lion Watershed has larger patches and lower patch density, because it was mapped from 1930s aerial photographs that tended to be small-scale and fuzzy, whereas the Smith Creek Watershed was mapped from satellite imagery, which typically has higher resolution. This illustrates the

Table 18-2. Summary of landscape metrics for three reference landscapes (see Figure 18-1 for geographic location). Metrics are defined in McGarigal and Marks (1995). Roaring Lion Watershed was included because it was mapped using 1930s aerial photos that provide a historical context (Quigley and Arbelbide 1997).

Landscape Metric	All Selway-Bitterroot Landscapes	Smith Creek Watershed Pre-treatment	Smith Creek Watershed Post-treatment	Roaring Lion Watershed Current	Roaring Lion Watershed Historical
Area (km^2)	2616[1]	66	66	66	66
Patch density (patches 100 ha^{-1})	5.66	6.72	6.72	1.35	1.31
Mean Patch Size (ha)	17.7	14.9	14.9	65.1	76.4
Contagion Index	50.8	45.3	45.3	56.3	55.1
Interspersion-Juxtaposition Index	68.5	73.0	73.1	63.3	68.4
Shape Index	4.82	3.49	3.49	3.24	2.71
Fractal Dimension	1.46	1.42	1.42	1.51	1.43
Edge Density (m/ha)	93.0	93.6	93.5	46.7	40.9
Mean Nearest Neighbor (m)	47.1	23.2	23.1	93.6	113.2

[1]The Selway-Bitterroot layer was clipped to 2,616 km^2 to include both Smith Creek and Roaring Lion Watersheds.

importance of consistent mapping strategies in comparing and assessing landscape patterns. Smith Creek metrics compare well with those averaged across the SBWC, but Smith Creek patches tended to be smaller and less contagious (farther apart).

Special attention must be given to interpreting landscape metrics in the context of the processes that control patch dynamics. Hann (1990) mentions that patch size, shape, and juxtaposition are highly correlated with landform, topography, and soils as well as historical disturbance patterns; it is important to identify the most important factor that governs landscape structure when designing restoration activities.

An estimate of the historical range of landscape indices provides important thresholds for designing treatment sizes and shapes. Estimates can be computed from averages of several similar landscapes mapped for the same time period (Hessburg et al. 1999a) or computed for one landscape across long time spans using simulation modeling (Keane et al. 1999). Hessburg et al. (1999b) found patches in the whitebark pine zone that were 15 to 41 hectares in size, based on a collection of 120 landscapes on the east slope of the Cascade Range. He mentions that restoration treatments should be designed to create future landscapes with characteristics that fall within 80 percent of the variation of historical conditions. Statistics for historical conditions can be computed from chronological sequences of historical imagery, summarized across many landscapes for one or more time periods, or simulated using spatial succession models (Hessburg et al. 1999a; Keane et al. 1999; Chapter 9).

Prioritization

At a minimum, three important elements should be used to prioritize whitebark pine restoration areas: stand condition, distance to seed source, and management issues. An assessment of current structure and composition of whitebark pine forests can be used to identify those stands with the greatest need for treatment. Stands that experienced severe blister rust mortality (greater than 80 percent dead) should be treated first, because seeds planted by the nutcracker after treatment may have a higher probability of being rust resistant (Chapter 17). Successional status could also be evaluated; if a major portion of a stand or landscape comprises subalpine fir (i.e., late-seral communities), then those stands need treatment to return to early-seral conditions. Second, the proximity and relative abundance of whitebark pine and subalpine fir seed sources need to be quantified to evaluate the potential for successful whitebark pine regeneration. Those rust-infected stands that are distant (greater than 10 kilometers, see Chapter 9) from any whitebark pine seed source should be high priority for treatment, because their further decline might result in local extinction, and their treatment might require planting of rust-resistant seedlings. Last, key management issues require individual evaluation to tailor prioritization criteria to local areas. For example, grizzly bear (*Ursus arctos*) habitat and watershed protection issues may be considered during prioritization.

A weighted matrix approach is useful for prioritizing restoration sites, although the process is highly subjective. Prioritization elements for potential stands or landscapes are rated on a simple, relative scale, and then each element is weighted to reflect its relative importance to land management and ecosystem restoration (see Table 18-3, for example). It is important that ratings all use the same ordinal scale, and that the scales increase in magnitude with the level of concern. A scale of 1 (low) to 3 (high) was used in Table 18-3, because absolute measurements of prioritization elements were costly and not as important as the direct comparison of all elements across all stands or landscapes. Successional status was based on cover type as classified from satellite imagery (Keane et al. 1998). Simple measures of rust severity were determined for each stand, whereas proximity to seed sources was computed with geographical information system (GIS) software, using a buffer of 10 kilometers around all whitebark pine stands in the Selway-Bitterroot (Keane et al. 1998). Grizzly bear habitat was determined from the Bitterroot Grizzly Bear Environmental Impact Statement (U.S. Fish and Wildlife 1997) and Eggers (1986). A 1-kilometer-wide buffer around all roads and trails defined lands accessible for treatment. Values assigned to each element were then multiplied by the assigned weight to determine the prioritization score for each stand (Table 18-3). The

Table 18-3. An example of a weighted matrix for prioritizing stands for restoration in the Smith Creek Watershed. See text for explanation.

Element	Weight	Stand 1	Stand 2	Stand 3	Stand 4	Stand 5
Successional status[1]	10	3	2	3	1	3
Rust severity[2]	10	2	3	3	3	3
Seed source[3]	5	1	2	2	2	1
Grizzly habitat[4]	1	1	1	1	1	1
Accessiblity[5]	10	3	3	1	1	3
Score (weight × rating)	—	86	91	81	61	96

[1]Succession status: 1 = early seral (whitebark pine), 2 = mid-seral (mixed whitebark pine/fir), 3 = late seral (subalpine fir)
[2]Rust severity: 1 = low severity, 2 = moderate severity, 3 = high severity
[3]Seed source: 1 = adjacent, 2 = within 10 km, 3 = more than 10 km
[4]Grizzly habitat: 1 = not important, 2 = potentially important, 3 = important
[5]Accessiblity: 1 = no roads or trails near area, 2 = roads or trails near area, 3 = area has road
Weight: 1 = not important, 10 = most important

restoration objective and current land management plan will help quantify weights for each element.

Selection

The selection of treatment areas dictates the details of treatment design and implementation. The prioritization analysis forms a good foundation for stand or landscape selection, but ultimately, the final selection must also depend on opinions of others involved in the project. For example, it makes no sense to plan a prescribed burn for high-priority stands when the fire manager feels that any prescribed fire would be difficult to control. Moreover, because of ambiguity, some factors important to prioritization efforts cannot be included as elements in the above approach. For example, it would be difficult to rate the potential success of a restoration project or the value of treatment to landscape diversity. So, the selection process should account for all unevaluated factors, as well as the results of the prioritization analysis.

Generally, there are four questions germane to selecting a stand or landscape for restoration. First, is the stand in the late stages of succession? This can be evaluated by the coverage of the shade-tolerant species (e.g., subalpine fir) and the age of the stand (greater than 90 percent of the average fire-return interval). Climax whitebark pine sites where subalpine fir rarely dominates the overstory will probably not need restoration because of the long fire-return intervals and slow succession rates. Next, are many of the stands on a landscape late seral or near climax? The proportion of subalpine fir cover types on the Interior Columbia River Basin historical upper subalpine landscape was probably within the range of 3 to 20 percent, assuming the entire landscape can support whitebark pine. Keane et al. (1994) computed this threshold at 8 percent for the Bob Marshall Wilderness Complex. We estimate it is roughly 12 percent for the Selway-Bitterroot Wilderness, based on fire regime studies, computer simulations, and local knowledge (Barrett and Arno 1992; Brown et al. 1994). This, of course, depends on the landforms that comprise the landscape (e.g., presence of rock fields, alpine). Third, are the stands experiencing heavy blister rust mortality? Treatment of rust-infected stands to create caching habitat for nutcrackers is perhaps the only way to regenerate naturally rust-resistant whitebark pine seedlings (Chapter 17). Last, can restoration treatments be successful?

Design

Restoration treatments should be designed to match the characteristics of the important natural disturbance processes prevalent on the project landscape. Since fire is the keystone disturbance that shaped most historical whitebark pine landscapes (Chapter 4), treatments should emulate fire's effect. For example, properly designed silvicultural thinning could simulate the effect of a nonlethal surface fire in whitebark pine stands. Designing the prescribed fire or silvicultural treatment for the project area is a complex task that requires comprehensive descriptions of the landscape and forest communities. The following discussion presents possible kinds of treatments to reproduce the effect of the two major whitebark pine fire regimes: mixed-severity and stand-replacement.

Nonlethal underburns are not discussed in detail here because they are either part of the mixed-severity regime or they occur in high, dry whitebark pine stands (i.e., climax whitebark pine sites, Chapter 9) that only represent a small percent of the whitebark pine range (Chapter 4). Many of these climax whitebark pine stands are still somewhat healthy because of the inhospitable conditions for rust infection and the long fire cycles, which minimize the adverse effects of fire exclusion, so they probably do not need immediate restorative action (Chapter 11). Some older stands may be experiencing subalpine fir encroachment because of recent favorable weather conditions. In these cases, a low-severity fire may be appropriate to kill subalpine fir seedlings and saplings to maintain the health of overstory whitebark pine. Thinning treatments to eliminate subalpine fir might simulate natural fire, but many smaller seedlings would probably persist after the cutting, resulting in the need for future thinnings at frequencies shorter than prescribed fire intervals.

Mixed-severity fires. A mixed-severity fire regime is characterized by fires of different intensities in space and time, creating complex patterns of tree survival and mortality. Individual fires can be nonlethal (i.e., underburns), stand-replacement, or contain elements of both (Morgan et al. 1994). Mixed-severity fires can create two types of stand structures, depending on pre-burn species composition and fire intensity. Sometimes the fire burns in sparse ground fuels at low severities, killing the smallest trees and the most fire-susceptible overstory species, particularly subalpine fir. Severities increase if the fire enters areas with high fuel loads or if the fire gains entrance into tree crowns due to higher winds (Lasko 1990). These lethal fires can kill most trees across the landscape (Keane et al. 1994).

The patchy nature of mixed-severity fire regimes would indicate that their effect is best evaluated at landscape scales rather than at the stand level. Mixed-severity burn patches are often 1 to 30 hectares in size, depending on topography and fuels, and these openings provide important caching habitat for the Clark's nutcracker (Hutchins and Lanner 1982; Tomback et al. 1990; Norment 1991; Hessburg et al. 1999b). Patch sizes may be estimated from historical landscape patch distributions or the average size of a fire event (Arno et al. 1993; Hessburg et al. 1999b). Design treatment parameters that mimic mixed-severity fires can be very liberal, depending on the restoration objective and the current site conditions, because of the wide range of severities and patch sizes created by this fire regime. For example, a prescribed fire that kills all trees in a stand may be appropriate when the restoration objective is to enhance whitebark pine regeneration and when fuel loadings or rust mortality are high.

The Smith Creek, Beaver Ridge, Coyote Meadows, and Bear Overlook study areas historically experienced mixed-severity surface fires at intervals of 60 to 200 years; two types of treatments for these areas were designed to simulate the effect of fire in this mixed-severity regime. All four areas had an overstory of mature subalpine fir and spruce with abundant understory fir regeneration. The primary objective of the treatments was to kill all subalpine fir and spruce and spare as many mature whitebark pine trees as possible. First, a low- to moderate-severity fire (i.e., underburn) was planned for one treatment unit at each of the four study areas (Figures 18-1 and 18-2). The fire prescriptions, which were designed by the national forest fire managers for the Smith Creek and Coyote Meadows units, minimized fire spread and intensity while still achieving the stated objective. Based on output from the BEHAVE fire behavior model (Andrews 1986), these prescriptions were designed to generate flame lengths between 0.2 and 1.5 meters and spread rates less than 0.2 km h^{-1}. Preliminary tree mortality predictions from the FOFEM model (Reinhardt et al. 1997) indicated that these target flame lengths would kill over 80 percent of the subalpine fir trees and only 20 percent of the whitebark pine trees.

Parts of the Musgrove Creek, Beaver Ridge, Coyote Meadows, and Bear Overlook treatment units did not have sufficient fuels to carry the fire to all parts of the stand, so we created an adjacent treatment unit where some standing subalpine fir were cut and left on the ground (i.e., slashed) to improve continuity of the fuel bed. Fuel augmentation by slashing does not require the felling of every fir, but just enough to enhance the fuel bed so that the fire will visit more parts of the unit. This treatment will also widen the prescribed burning window, because the same target fire behavior and effects can achieved with higher fuel moistures. At Musgrove Creek, about 5 hectares were slashed in less than two days by a crew of twelve people. The mixed-severity fire treatment for the Musgrove site was more severe than the other sites because of the low density of mature whitebark pine and high-severity fire regime evident from landscape structure. We designed this burn to kill most trees but possibly leave some large, cone-bearing whitebark pine. At the Beaver Ridge study site, over 20 hectares were slashed, because the 1910 fire had created a patchy fuel bed across some portions of the site. The Coyote Meadows underburn unit (about 3 hectares) was slashed in about four hours by two people who cut only subalpine fir trees and felled them in areas with sparse fine fuels. At the Bear

Overlook site, about 20 hectares were heavily slashed to generate high fire intensities and allow a larger burning window. Because there were few live whitebark pine trees on the site, we were less concerned about postfire whitebark pine survival.

A wide variety of silvicultural cutting methods can be used with and without postharvest burning to mimic the effects of mixed-severity fire regimes in successionally advanced whitebark pine forests (Eggers 1990). Controlling composition by cutting all subalpine fir, spruce, lodgepole pine (*Pinus contorta*), and dying whitebark pine (i.e., species selection cuts) would probably simulate nonlethal and lethal surface fire effects. Group and individual tree selection methods are probably appropriate for emulating patchy mixed-severity fire regimes, where either all trees are removed in small patches or all understory trees and overstory fir and spruce are cut, leaving only cone-producing whitebark pine (Eggers 1990; DeBell et al. 1997). Also, stand structure may be controlled by removing certain age and size classes (Eggers 1990). Cutting small trees and fire-susceptible trees may recreate stand structures that were common under a nonlethal or mixed-severity fire regime and will also eventually improve tree vigor and cone production (Keane et al. 1990; DeBell et al. 1997).

Silvicultural cuttings at the Smith Creek study site were designed to mimic patchy mixed-severity burns (Figure 18-2). Since the site is accessible by road, fire patches were simulated by cutting all trees, except healthy, cone-bearing whitebark pine, in small, 0.1-hectare circular units called nutcracker openings to entice nutcrackers to cache whitebark pine seeds in that vicinity (Tomback 1998) (Figure 18-2). Norment (1991) found that nutcrackers were most abundant in 0.1- to 2-hectare disturbed or nonforest patches. All subalpine fir and spruce, some lodgepole pine, and dying whitebark pine were removed from the forested areas between the nutcracker openings. We did this to limit the seed rain from the wind-dispersed, shade-tolerant tree species that could outcompete nutcracker-cached whitebark pine regeneration after treatment. Nutcracker openings were created in two treatment units, with one unit burned at the same time as the underburn mentioned above and the other unit left unburned with minimal slash on the ground. The units were circular to maximize the distance to subalpine fir seed sources, even though the shape indices indicate that an irregular-shaped cutting unit is more ecologically appropriate.

A similar noncommercial treatment created nutcracker openings at the Beaver Ridge site. All fir, spruce, lodgepole, and dying whitebark pine were cut and left on-site in large 1- to 2-hectare patches (i.e., nutcracker openings). Branches were cut from downed trees and piled both inside and outside the cleared opening to allow nutcrackers access to the ground for future seed caching. Half of the 30-hectare harvest area was burned, and half of the nutcracker openings (burned and unburned) were planted with whitebark pine seedlings. A Coyote Meadows treatment unit was thinned to remove all fir and spruce, and half of that unit will receive a low-severity burn. A treatment unit at the Bear Overlook site was thinned to remove all lodgepole pine, fir, and spruce trees to enhance cone production on the healthy whitebark pine that was left uncut.

Stand-replacement fires. Many whitebark pine forests in northwestern Montana and northern Idaho originated from large, stand-replacement fires that killed over 90 percent of the trees (Ayers 1901; Gabriel 1976; Arno 1980; Keane et al. 1994; Morgan et al. 1994). Large, stand-replacement fires also occurred in mixed-severity fire regimes but at intervals greater than 400 years. Whitebark pine has a unique advantage in colonizing the large stand-replacement burns, because the Clark's nutcracker can disperse whitebark pine seeds farther into a disturbed area than wind can disperse seeds of competing tree species (Tomback 1989, Tomback et al. 1990; Sund et al. 1991; Tomback 1998; Chapter 5). Therefore, the restoration of forests characterized by this type of fire regime may require the creation of large burned areas using prescribed fire (Keane and Arno 1996).

The target size and shape of a prescribed, stand-replacement burn can be estimated from historical landscape patterns, the spatial distribution of seed-bearing trees, prevailing winds, and natural fire control structures, such as talus slopes or alpine meadows (Hann 1990; Keane et al. 1990; Keane and Arno 1996). Stand-replacement burns can be small if subalpine fir seed sources are downwind and downhill, or confined to a topographic position where wind alone would have a difficult time carrying subalpine fir or spruce seeds into the burn. Seed dispersal maps for each tree species can be created using GIS based on the dispersal curves for wind-dispersed seeds (McCaughey et al. 1985) and bird-dispersed seeds (Tomback et al. 1990). As an example, a 100-hectare stand-replacement fire is planned for the Beaver Ridge study area to create nutcracker caching habitat where wind-dispersed tree species are at a colonization disadvantage. We programmed the seed dispersal equations into a GIS and estimated that about 50 percent of the proposed Beaver Ridge burn area has a less than 1 percent chance of receiving subalpine fir seeds from surrounding stands, assuming no wind and topographic effects and that no subalpine fir survived the fire. That proportion increases to 75 percent when the predominant wind is from the southwest (normal for this site), and downhill fir seed sources are considered.

Very few silvicultural tools can be used in upper subalpine forests to simulate stand-replacement fire regimes. Clear-cutting or clear-felling (trees cut and left on-site) is seldom used because of possible site degradation in these fragile, high-mountain environments. Moreover, the large areas created by stand-replacement fires may preclude using clear-cutting to replicate fire effects.

Probably the most practical tools for creating large, burned openings, especially in wilderness and roadless settings, are prescribed natural fires, recently termed "wildland fire use for resource benefit" (Keane and Arno 1996). A prescribed natural fire results from a lightning ignition that is allowed to burn within a given set of weather and fuel conditions (i.e., prescription) without prior fuel-break construction. A management-ignited prescribed fire is a fire started by fire managers in a roadless or wilderness setting that is allowed to burn within a fire prescription. Most whitebark pine forests are found in remote roadless or wilderness settings with little road access (Keane 2000), so fire-control structures used in conventional prescribed fire, such as firelines and dozer lines, are difficult to construct. Silvicultural cuttings are difficult, expensive, and often not permitted. Therefore, prescribed natural fires may be the only management tool to treat declining whitebark stands in roadless areas.

There are many advantages to using prescribed natural fires as a restoration tool. First, they are more similar to the natural process than any other treatment. Ignitions usually occur during the summer, the season when most whitebark pine forests burned historically. Second, a summer ignition can be allowed to burn over many weeks, creating the mosaic of low- to high-severity fire that was common historically (Morgan et al. 1994). This makes prescribed natural fire useful for both mixed-severity and stand-replacement fire regimes. Third, more area can be treated with lower cost than with conventional prescribed fire, because fire control structures are minimal and there are usually fewer people managing the fire. Last, it may be easier to implement a stand-replacement fire as a prescribed natural fire, because crown fires would be difficult to control using conventional prescribed fire techniques. The major disadvantage of prescribed natural fires is that they can grow too large and become wildfires that can be dangerous to human life and property.

Blister rust considerations. The health of a stand will also dictate treatment design. If all whitebark pine trees in the project area are heavily infected with rust (greater than 50 percent crown kill), it makes little sense to leave these trees unless some show outward signs of rust resistance. Rust-infected whitebark pine trees with more than 20 to 30 percent of the crown killed will probably never produce cone crops again. Provided that other seed sources are available within a 10-kilometer distance, it may be easier and more efficient to burn the entire site and kill all trees to create favorable caching habitat for the nutcracker. Many whitebark pine trees will continue to die from blister rust regardless of restoration treatment (Tomback et al. 1995). Thus, it is essential that native disturbance processes remain intact so that caching habitat for nutcrackers will continually be created. In this way, seeds from rust-resistant whitebark pine can be planted across the landscape, thus ensuring the survival of the species.

Elimination of rust from infected stands is not a viable or preferred restoration alternative, because it is expensive, ineffective, and ecologically unsound. Application of aerial sprays of antibiotic solutions did not stop rust infection or mortality in western white pine (*Pinus monticola*), although new chemicals show some promise (Brown 1969; Chapters 10 and 17). Moreover, these sprays are costly to apply and have the potential to harm other ecosystem components (Chapter 17). Removal of the alternate host *Ribes* spp. by mechanical or herbicide treatments also does not seem to be a practical means of preventing rust infections, considering the limited success of the *Ribes* eradication effort during the 1930s to the 1960s (Carlson 1978). *Ribes* will sprout from rhizomes after cutting, and the seeds lie dormant in the soil until fire occurs. Some *Ribes* control, however, in conjunction with rust resistance in whitebark pine is being reconsidered as an integrated management strategy (Chapter 17). Pruning rust-infected pine branches may delay tree mortality, but ultimately, it is highly probable that future infections will eventually kill the tree (Dooling 1974; Hoff and Hagle 1990; Hunt 1998). Infected whitebark pine saplings were pruned at the Coyote Meadows study site to evaluate the efficacy of this treatment.

The design of restoration activities should anticipate the future outlook for blister rust infection, since the rust is continually expanding its range and intensifying locally (Chapters 10 and 11). At a minimum, whitebark pine stands should be evaluated for possible rust-resistant, mature, seed-producing individuals. Planting rust-resistant seedlings may be warranted in special cases, such as in critical grizzly bear habitat or in areas where there are no other potential whitebark pine seed sources within 10 km. Planting may be unnecessary if there are enough rust-resistant individuals (greater than 10 per hectare) within the 10-kilometer dispersal shadow (Krebill and Hoff 1995). Alternatives to planting rust-resistant seedlings, such as broadcast seeding, are ineffective because of heavy seed predation by mammals (McCaughey and Weaver 1990a). As a general restoration tool, planting rust-resistant seedlings is expensive and time-consuming (Chapter 16). It may not be necessary in many cases because of the efficiency of nutcracker caching and the high genetic gain in rust resistance of future progeny regenerated naturally from seeds of rust-resistant trees (Krebill and Hoff 1995).

Other design criteria. Many other factors, including accessibility, tree value, and stand health, must be evaluated in the design of restoration treatments. Timber cutting and other manipulative mechanical treatments might be too costly if there are few roads and limited access to the project area. Prescribed fires and prescribed natural fires might be the only cost-effective restoration tool for remote whitebark pine stands. Although most whitebark pine stands have little timber value, local market conditions might favor certain fuelwood or cutting treatments to generate income while still accomplishing restoration goals. For example, it might be cost-effective to haul subalpine fir, spruce, and lodgepole pine logs to the road, which we did at the Smith Creek site.

Since fire and climate regimes are highly variable in time and space, scheduling treatments and designing future landscapes may be a futile task. Climate will continue to change over the next 10 to 1,000 years, and the rate of structural and compositional development across a landscape will also be highly variable (Mehringer et al. 1977; Baker 1990; Bartlein et al. 1997; Ferguson 1997). In some cases, the status and pattern of low-elevation stands on the landscape will directly affect whitebark pine ecosystem health. It is also highly probable that political, social, and economic conditions will change during the century-long successional developmental periods common in whitebark pine forests (see Chapter 19). Moreover, advances in research and technology may foster development of better restoration tools and techniques. So, instead of conventional treatment schedules, we propose an adaptive management approach where landscapes are evaluated every ten to twenty years to assess their need for restoration.

Simulation modeling. Simulating future landscapes and stand structures with and without restoration treatments and wildfire can be an integral part of restoration design (Keane et al. 1990; Parker 1997). Since succession modeling is discussed in detail in Chapter 9, only the design of simulation scenarios will be discussed here. Any succession model can be used to simulate treatment effects, but the best models have the ability to simulate landscape and stand-level succession and disturbance dynamics. Examples include the SIMPPLLE model (Chew 1997), Fire-BGC (Keane et al. 1996c), LANDSUM (Keane et al. 1997a), VDDT (Beukema and Kurz 1998), and FVS (Wykoff et al. 1982). We did not simulate the future impact of our whitebark pine restoration activities on the landscape because our study sites are probably too small to influence most landscape processes (Table 18-2).

It is important to select a succession model with a simulation resolution that matches treatment effects. For example, the LANDSUM model simulates succession as a change in cover type or structural stage at yearly time intervals, so that any treatment or disturbance effect that is not manifest at the cover type and structural stage levels, and at the annual time step, probably should not be simulated. Chew (1990) used the FVS model to create target whitebark pine stand and landscape conditions. The Interior Columbia Basin Ecosystem Management Project simulated landscape and disturbance dynamics in a nonspatial domain with the VDDT model (Beukema and Kurz 1998), and then used the simulation results to parameterize the CRBSUM spatial model (Keane et al. 1996d; Quigley and Arbelbide 1997). Keane et al. (1996b) also used CRBSUM to simulate coarse-scale restoration effects on the Interior Columbia River Basin and the Bob Marshall Wilderness Complex. Detailed successional pathway information that can be entered into the VDDT or LANDSUM models to simulate landscape dynamics at multiple scales is available in Chapter 9.

The best set of simulation scenarios bracket the entire range of ecosystem responses, given the important processes prevalent in the system and the objective of the simulation exercise. For example, Keane et al. (1997a) simulated ecosystem dynamics on a Glacier National Park landscape to monitor vegetation change with and without fire under current and future weather conditions (i.e., four scenarios). Landscape-level simulations allow quantification of future landscape composition and structure that would be impossible to reproduce with field sampling.

The following four scenarios might describe the range of impacts for restoration activities. One scenario would be a "*status quo*" simulation, where all fire disturbances are suppressed and no restoration activities are implemented to represent the effects of passive management on declining whitebark pine landscapes. This is especially important for stands experiencing blister rust mortality. A complementary option would be the "*natural*" scenario, where native fire regimes are simulated but no restoration activities are scheduled. A "*restoration only*" scenario and a "*restoration and wildfire*" scenario would depict the effects of restoration activities on landscapes where wildfires are either excluded or allowed to burn. The suppression of all wildfires from real-life whitebark pine landscapes is unrealistic in most cases, so 80 to 90 percent reduction is often used (Keane et al. 1996d). Simulation scenarios should be crafted in the context of the restoration objective (Parker 1997). For example, harvesting scenarios should be simulated if a silvicultural restoration treatment is proposed.

Restoration Implementation

Implementing Treatments

The implementation phase is the execution of the planning phase. It is composed of the implementation of the planned treatments and the monitoring of the effects of those treatments on the ecosystems.

Prescribed fire. Because whitebark pine ecosystems are remote, they are difficult to treat with prescribed fire. High-elevation whitebark pine forests experience late snowmelt and abundant summer precipitation, and thus are rarely dry enough in the summer to carry a prescribed fire (see Chapter 4). Moreover, in those occasional years where the high country is dry enough for a summer fire, the rest of the landscape is usually in extreme fire danger; spotting from high-elevation fires may start severe fires in low-elevation forests (Brown et al. 1994). Autumn is the best season for prescribed fire in whitebark pine forests. However, fine herbaceous and woody fuels are rarely cured by the beginning of fall because of high summer precipitation. Therefore, it is desirable that an early, hard frost (below –4°C) kill most herbaceous plants and shrub foliage so that they will dry quickly to provide cured fine fuels for fire propagation. Drying usually occurs during the ensuing warm, early autumn conditions, common in the western United States.

The Smith Creek prescribed burn was one of the first successful prescribed fires in a mature whitebark pine stand in the United States. This prescribed fire burned two adjacent units (harvest/burn and underburn) comprising 2 hectares each within a mature whitebark pine stand (Figure 18-2). The first unit had been treated to create nutcracker openings with some slash left on the site, while the second unit was left undisturbed. Temperatures the day of the burn ranged from 10°C to 16°C, and relative humidities ranged from 21 percent to 31 percent. Log moisture contents ranged from 16 percent to 28 percent, and fine woody fuel (less than 2.54 centimeters in diameter) moisture contents ranged from 14 percent to 33 percent. Both units were burned using 3 meter-wide strip head fires to keep intensities low. The fire burned approximately 52 percent of the mature underburn stand (unit 1A, Table 18-4), given the complex fine-scale mosaic of fuel moistures, fuel loadings, and shading. The fire smoldered for about nine days until a snowfall extinguished it. Low subalpine fir mortality and fuel consumption in the underburn unit 1A (Table 18-4) illustrate the need to (1) enhance the fuel bed by slashing, (2) conduct more fire treatments in the following years, or (3) burn under drier conditions.

The outcome of the Smith Creek burn related directly to the amount of fuel on the ground. The harvested unit had the highest fuel loadings because of the cut slash, which created a more continuous fuel bed throughout the stand (80 percent fire coverage, Table 18-4). As a consequence, this slash unit also had higher fire intensities, where flame lengths averaged 1 to 2 meters compared to less than 1 meter in the unharvested mature stand. Scorch heights, scorched crown volume, and bole char height were also greater in the harvested unit. This resulted in greater tree mortality (especially in whitebark pine), greater fuel consumption, and more mineral soil exposed (Table 18-4). In retrospect, it was unnecessary to leave whitebark pine trees in the burned nutcracker openings, because most were killed by fire, and it would have been difficult to

Table 18-4. Harvest and burn effects on the tree and fuels characteristics one year after the Smith Creek prescribed burn. See Fig. 18-2 and text for explanation. Note: mortality estimates for treatment 2A are for trees that remained after the harvest. It is expected that mortality estimates will double by 5 years after burn.

Postburn Characteristics	Mature Stand (1A) Underburn	Harvest and Burn Unit (2A)
TREE		
Tree mortality (%)	19.5	24.6
Whitebark pine mortality (%)	16.8	20.0
Subalpine fir mortality (%)	29.9	36.7
Ave scorch height (m)	6.3	9.1
Ave crown volume scorch (%)	32.9	59.9
Ave bole char height (m)	3.4	3.2
FUEL		
Fire coverage (%)	56.0	81.2
Fuel consumption (%)	32.0	44.6
Fuel consumption (kg m-2)	1.4	2.9
Log consumption (%)	28.1	33.3
GROUND COVER		
Soil cover (%)	18.4	38.7
Rock cover (%)	2.4	5.4
Wood cover (%)	10.0	11.0

reduce fire intensity to minimize whitebark pine mortality and still achieve desired effects.

Some statistically significant changes in tree structure and fuel loading occurred as a result of the restoration treatments. All sapling- and pole-sized subalpine fir were removed by the harvest treatment (Figure 18-2, units 2A, 2B), but only 30 percent of the same-sized fir were removed by the underburn treatment (unit 1A), indicating that a low-intensity surface fire alone might not entirely restore whitebark pine stands in mixed-fire regimes. Another fire or cutting may be needed to achieve prescription goals. Approximately 40 percent of the downed logs were consumed, with the greatest consumption in the harvest and burn treatment (unit 2A), presumably because the additional small woody fuels created a more intense fire. This compares well to the effects of natural mixed-severity burns (Chapter 4).

The most important concern in the RWPE treated areas is that subalpine fir will regenerate more quickly than whitebark pine. There are abundant subalpine fir seed sources within 100 meters of treated areas, and the subalpine fir have a good cone crop every two to three years (Alexander et al. 1990). Good whitebark pine seed crops only occur once every three to five years in healthy trees and are rarer in diseased trees (Weaver and Forcella 1986; Arno and Hoff 1990; McCaughey and Schmidt 1990). Because the majority of whitebark pine trees in this area are infected with blister rust, which kills the cone-bearing branches first, it is unlikely that bountiful cones crops will occur while the treated site remains attractive to nutcracker caching. All stands in this study will require another burn or cutting in ten to twenty years to kill new and surviving subalpine fir.

Silvicultural cutting. Commercial harvests of whitebark pine stands for restoration purposes are difficult to implement because of inaccessibility, low timber value, and rugged topography (Keenan et al. 1970; Kipfer 1992; Keane and Arno 1996). Consequently, the felled trees from restoration cuttings in unhealthy whitebark pine stands will usually be left either in piles or scattered throughout the stand. A prescribed burn should follow any cutting treatment to eliminate small fir and spruce seedlings and to reduce slash to allow Clark's nutcrackers access to the ground surface for seed caching (Tomback et al. 1990; Keane and Arno 1996). If prescribed burning is not feasible, we recommend that slash be removed, piled, or lopped and scattered to eliminate any obstructions for nutcracker caching and to improve decomposition rates. Decomposition rates in these areas are slow, and the longer the slash remains on the ground, the greater the chance for an intense stand-replacement fire that could kill the remaining cone-bearing whitebark pine trees.

Nutcrackers tend to cache seeds near visual cues such as stumps, rocks, and logs (Tomback et al. 1993; Tomback 1998). These protected microsites also seem to ameliorate severe site conditions and improve seedling survival (McCaughey and Weaver 1990a). Restoration treatments that create openings in harvested whitebark stands should improve nutcracker caching habitat by increasing the amount of ground a nutcracker can see from the surrounding canopy and provide visual cues for nutcracker spatial pattern recognition.

We observed a group of seven nutcrackers caching whitebark pine seeds in the Smith Creek harvested openings over a two-week period in the fall of 1996. Even though slash covered 20 to 30 percent of ground in both unit 2A and unit 2B (Figure 18-2), these nutcrackers found open spots for caching seeds. The Musgrove and Beaver Ridge sites were burned in October 1999, also a good cone crop year, and extensive nutcracker caching was observed within five days after burn.

Monitoring

Monitoring the effects of restoration treatments provides critical information to evaluate treatment efficacy (Michener 1997). More important, monitoring is vital for building comprehensive databases for others to use as reference in their restoration projects. Monitoring information can be shared via newsletters, journals, databases, and periodic workshops or symposia. This is especially important for these little-studied, rust-ravaged ecosystems, because there are so few examples of restoration treatments. Monitoring information also provides the educational material that land management agencies need to educate the public about these complex restoration techniques.

Monitoring is best accomplished by remeasurement of permanent plots. Monitoring design may be intensive where many variables are measured on numerous plots (Keane and Arno 1996), or less rigorous where a limited set of measurements are taken on only a few representative plots (Michener 1997). Monitoring can be at the landscape level, where historical and current vegetation layers developed from satellite imagery or aerial photo interpretation are compared to evaluate successional change (Kratz et al. 1994), or at fine scales, where individual stands are evaluated with plot-based sampling.

Sampling design should accurately describe all variables needed to evaluate the restoration objective, while accounting for differences in site conditions. Ideally, several plots should be randomly established in each site type (i.e., habitat type) within the treatment area to characterize treatment impacts comprehensively. However, many land management agencies have limited funds or personnel for such an intensive monitoring effort for ecosystem restoration, which argues for some compromise that still meets project objectives. At least three randomly established plots per site type (i.e., potential vegetation type) and treatment type are recommended as a minimum for monitoring efforts. For projects where only one plot in each treatment-site combination is possible, we recommend that the plot be located in a representative portion of the treatment unit (Mueller-Dombois and Ellenburg 1974). Repeated photography from fixed points is also a valuable, low-cost tool to complement these measurements.

The objectives of the restoration treatment will define the variables to measure at each monitoring plot. For example, assessment of postfire tree survival requires measurements of tree density, size, and species composition, especially if removing subalpine fir was the primary purpose of the treatment. Suggested methods for measuring many ecosystem characteristics are available in the ECODATA system (Keane et al. 1990; Jensen et al. 1993). We recommend, at a minimum, that fuels, tree structure, and ground cover be measured on every monitoring plot and that the size of the plot be at least 0.04 hectare. Surface fuels can be measured using the Brown (1970) transect techniques; trees can be tallied by diameter class and species on a fixed-area circular plot; and ground cover can be assessed using microplot or line intersect techniques (Jensen et al. 1993). Plot centers are permanently marked by a metal stake, and the center coordinates are recorded using a geographic positioning system and bearing and distance to selected trees. Detailed maps, location descriptions, and copious photos should document monitoring plot locations so that they can be relocated. Monitoring plots should be established before the treatment so that subsequent effects can be interpreted using predisturbance data as reference. Since succession is slow in most whitebark pine ecosystems, remeasurement may be done at ten-year intervals after the posttreatment measurements.

Monitoring landscape change is also important in maintaining whitebark pine. Composition of landscapes before and after treatment should be compared to determine the impact of restoration. Detection of changes in landscape structure and composition across an entire landscape will provide insight into the rate and direction of succession by substituting space for succession time (Kratz et al. 1994). For example, if 10 percent of the high-elevation stands capable of supporting both whitebark pine and subalpine fir change from a whitebark pine cover type to a subalpine fir cover type in ten years, then an inference might be drawn that all high-elevation forests will be in subalpine fir cover type in 100 years. Change may be detected through remote sensing techniques, such as photointerpretation (Hessburg et al.1999a) and satellite imagery (Craighead et al. 1982; Sachs et al. 1998). Selection of the remote sensing media and subsequent analysis will depend on local image processing experience, funding, and monitoring objectives.

Concluding Comments

The seven steps that comprise the restoration process, are meant more as guidelines than actual procedures. The complex issues and lack of resources inherent in fire and resource management may require eliminating or abbreviating some steps. We feel it is more important to initiate and implement restoration projects than it is to adhere to the detailed process presented here. The primary emphasis on these restoration steps is to return native ecosystem processes to declining forests. These projects can be planned across many time and space scales to ensure their continued success, but they can also be planned for small demonstration areas when historical and spatial data are limited.

The restoration principles and technology presented in this chapter are applicable to any fire-dependent ecosystem. Anthropogenic fire exclusion is a repeated theme in many ecosystems of the world (Ferry et al. 1995), leading to the successional replacement of an early-seral, fire-tolerant species by a late-seral, shade-tolerant species. For example, open-grown ponderosa pine (*Pinus ponderosa*) forests are being successionally replaced by dense Douglas-fir (*Pseudotsuga menziesii*) thickets in many parts of the Rocky Mountains and east slope of the Cascade Range. Without fire, longleaf pine (*Pinus palustris*) stands succeed to a southern mixed-hardwood community (Wright and Bailey 1982). *Eucalyptus* spp. trees are replaced by *Acacia* spp. in tall, open Australian forests (Gill et al. 1981).

The majority of whitebark pine forests occur in remote settings. Keane (2000) estimated that more than 49 percent of whitebark pine's range occurs in wilderness and roadless areas, and 98 percent on public lands. Thus, many of the conventional silvicultural and prescribed fire treatments mentioned in this chapter are applicable only to the portion of whitebark pine forests that are accessible. Consequently, the most important restoration tool for whitebark pine may be a comprehensive prescribed natural fire program (i.e., wildland fire use for resource benefit). For a whitebark pine restoration program to succeed across an entire region, landscape-level treatments and primarily consistent prescribed natural fire programs are crucial for the conservation of the species.

Restoring whitebark pine ecosystems may seem a daunting task, but many land management agencies have successfully developed large-scale prescribed fire programs (Parsons and Landres 1996). Some management organizations may wish to start on a smaller scale by implementing prescribed fire restoration projects in small accessible stands to build up expertise and confidence. And, of course, not all whitebark pine communities are in need of restoration. Throughout the range of whitebark pine, high-elevation sites where whitebark pine is the indicated climax species have not experienced significant rust mortality. Some areas in the northern part of the range, where whitebark pine is a seral species, have not yet experienced the adverse effects of fire exclusion. However, the rust range is expanding, infection is intensifying, and succession is a continual process (Keane and Arno 1993, Chapter 11). One thing is certain—whitebark pine communities will continue to decline if we do not change current management practices.

Acknowledgments

Our thanks to James Agee, University of Washington, for reviewing this chapter.

Literature Cited

Ager, A. A., and R. J. McGaughey. 1997. UTOOLS: Microcomputer software for spatial analysis and landscape visualization. USDA Forest Service, Pacific Northwest Research Station, General Technical Report PNW-GTR0-397, Seattle, Washington.

Alexander, R. R., R. C. Shearer, and W. D. Shepperd. 1990. *Abies lasiocarpa* (Hook.) Nutt. Subalpine fir. Pages 60–70 *in* R. M. Burns and B. H. Honkala, technical coordinators. Silvics of North America, Volume 1, Conifers. USDA Agriculture Handbook 654, Washington, D.C.

Andrews, P. L. 1986. BEHAVE: Fire behavior prediction and fuel modeling system—BURN subsystem. USDA Forest Service, Intermountain Research Station, General Technical Report INT-194, Ogden, Utah.

Apfelbaum, S. I., and K. A. Chapman. 1997. Ecological restoration: A practical approach. Pages 301–322 *in* Ecosystem management: Applications for sustainable forest and wildlife resources. Yale University Press, New Haven, Connecticut.

Arno, S. F. 1980. Forest fire history of the northern Rockies. Journal of Forestry 78:460–465.

———. 1986. Whitebark pine cone crops—A diminishing source of wildlife food? Western Journal of Applied Forestry 1:92–94.

Arno, S. F., and R. Hoff. 1990. *Pinus albicaulis* Engelm. Whitebark pine. Pages 268–279 *in* R. M. Burns and B. H. Honkala, technical coordinators. Silvics of North America, Volume 1, Conifers. USDA Forest Service Agricultural Handbook 654, Washington, D.C.

Arno, S. F., and K. M. Sneck. 1977. A method for determining fire history in coniferous forests of the mountain West. USDA Forest Service, Intermountain Research Station, General Technical Report INT-42, Ogden, Utah.

Arno, S. F., E. Reinhardt, and J. Scott. 1993. Forest structure and landscape patterns in the subalpine lodgepole pine type: A procedure for quantifying past and present conditions. USDA Forest Service, Intermountain Research Station, General Technical Report INT-294, Ogden, Utah.

Ayres, H. B. 1901. Lewis and Clark forest reserve, Montana. 21st Annual Report, U.S. Geological Survey, Part 5:27–80.

Baker, R. G. 1990. Late Quaternary history of whitebark pine in the Rocky Mountains. Pages 40–49 *in* W. C. Schmidt and K. J. McDonald, compilers. Proceedings—Symposium on whitebark pine ecosystems: Ecology and management of a high-mountain resource. USDA Forest Service, Intermountain Research Station, General Technical Report INT-270, Ogden, Utah.

Baker, W. L., and Y. Cai. 1990. The R.LE programs for multiscale analysis of landscape structure using the GRASS geographical information system. Landscape Ecology 7:291–302.

Barrett, S. W., and S. F. Arno. 1988. Increment borer methods for determining fire his-

tory in coniferous forests. USDA Forest Service, Intermountain Research Station, General Technical Report INT-244, Ogden, Utah.
———. 1992. Classifying fire regimes and defining their topographic controls in the Selway-Bitterroot Wilderness. Pages 299–307 *in* P. L. Andrews and D. F. Potts, editors. Proceedings of the 11th Conference on Fire and Forest Meteorology. Society of American Foresters, Bethesda, Maryland.
Bartlein, P. J., C. Whitlock, and S. L. Shafer. 1997. Future climate in the Yellowstone National Park and its potential impact on vegetation. Conservation Biology 11:782–792.
Bell, S. S., M. S. Fonseca, and L. B. Motten. 1997. Linking restoration and landscape ecology. Restoration Ecology 5:318–323.
Beukema, S. J., and W. A. Kurz. 1998. Vegetation dynamics development tool—Users Guide Version 3.0. ESSA Technologies, #300-1765 West 8th Avenue, Vancouver, BC V6J 5C6, Canada.
Bonnicksen, T. M., and E. C. Stone. 1985. Restoring naturalness to national parks. Environmental Management 9:479–486.
Bradshaw, A. D. 1997. What do we mean by restoration. Pages 8–14 *in* K. M. Urbanska, N. R. Webb, and P. J. Edwards, editors. Restoration ecology and sustainable development. Cambridge University Press, Cambridge, UK.
Brown, D. H. 1969. Aerial application of antibiotic solutions to whitebark pine infected with *Cronartium ribicola*. Plant Disease Reporter 53:487–489.
Brown, J. K. 1970. A method for inventorying downed woody fuel. USDA Forest Service, Intermountain Research Station, General Technical Report INT-16, Ogden, Utah.
Brown, J. K., S. F. Arno, S. W. Barrett, and J. P. Menakis. 1994. Comparing the prescribed natural fire program with presettlement fires in the Selway-Bitterroot Wilderness. International Journal of Wildland Fire 4:157–168.
Carlson, C. E. 1978. Noneffectiveness of *Ribes* eradication as a control of white pine blister rust in Yellowstone National Park. Forest Insect and Disease Management Report 78-18. USDA Forest Service, Northern Region, Missoula, Montana.
Chambers, J. C. 1997. Restoring alpine ecosystems in the western United States: Environmental constraints, disturbance characteristics, and restoration success. Pages 161–187 *in* K. M. Urbanska, N. R. Webb, and P. J. Edwards, editors. Restoration ecology and sustainable development. Cambridge University Press, Cambridge, UK.
Chew, J. D. 1990. Timber management and target stands in the whitebark pine zone. Pages 310–315 *in* W. C. Schmidt and K. J. McDonald, compilers. Proceedings—Symposium on whitebark pine ecosystems: Ecology and management of a high-mountain resource. USDA Forest Service, Intermountain Research Station, General Technical Report INT-270, Ogden, Utah.
———. 1997. Simulating vegetative patterns and processes at landscape scales. Pages 300–309 *in* Conference Proceedings—GIS 97, 11th Annual Symposium on Geographic Information Systems—Integrating spatial information technologies for tomorrow. GIS World, Inc. Vancouver, British Columbia, Canada.
Craighead, J. J., J. S. Summer, and G. B. Scaggs. 1982. A definitive system for analysis of grizzly bear habitat and other wilderness resources utilizing LANDSAT multispectral imagery and computer technology. Wildlife-Wildlands Institute Monograph No. 1. Missoula, Montana.
Crow, T. R., and E. J. Gustafson. 1997. Ecosystem management: Managing natural resources in time and space. Pages 215–229 *in* K. A. Kohm and J. F. Franklin, editors. Creating forestry for the 21st century. Island Press, Washington, D.C.
DeBell, D. S., R. O. Curtis, C. A. Harrington, and J. C. Tappeiner. 1997. Shaping stand development through silvicultural practices. Pages 141–148 *in* K. A. Kohm and J. F.

Franklin, editors. Creating forestry for the 21st century. Island Press, Washington, D.C.

Dooling, O. J. 1974. Evaluation of pruning to reduce impact of white pine blister rust on selected areas in Yellowstone National Park. USDA Forest Service Northern Region, State and Private Forestry Report 74–19, Missoula, Montana.

Eggers, D. E. 1986. Management of whitebark pine as potential grizzly bear habitat. Pages 170–175 *in* G. P. Contreras and K. E. Evans, editors. Proceedings of a Grizzly Bear Habitat Symposium. USDA Forest Service, Intermountain Research Station, General Technical Report INT-207, Ogden, Utah.

———. D. E. 1990. Silvicultural management alternatives for whitebark pine. Pages 324–328 *in* W. C. Schmidt and K. J. McDonald, compilers. Proceedings—Symposium on whitebark pine ecosystems: Ecology and management of a high-mountain resource. USDA Forest Service, Intermountain Research Station, General Technical Report INT-270, Ogden, Utah.

Eyre, F. H. (Editor). 1980. Forest cover types of the United States and Canada. Society of American Foresters, Washington, D.C.

Ferguson, S. A. 1997. A climate-change scenario for the Interior Columbia River Basin. USDA Forest Service, Pacific Northwest Research Station, General Technical Report PNW-GTR-499, Seattle, Washington.

Ferry, G. W., R. G. Clark, R. E. Montgomery, R. W. Mutch, W. P. Leenhouts, and G. T. Zimmerman. 1995. Altered fire regimes within fire-adapted ecosystems. Pages 222–224 *in* Our living resources: A report to the nation on the distribution, abundance, and health of U.S. plants, animals, and ecosystems. USDI National Biological Survey, Washington, D.C.

Fischer, W. C., M. Miller, C. M. Johnston, J. K. Smith, D. G. Simmerman, and J. K. Brown. 1996. Fire effects information system: User's guide. USDA Forest Service, Intermountain Research Station, General Technical Report INT-GTR-327, Ogden, Utah.

Gabriel, H. W. 1976. Wilderness ecology: The Danaher Creek drainage, Bob Marshall Wilderness, Montana. Master's thesis. University of Montana, Missoula.

Gill, A. M., R. H. Groves, and I. R. Noble. 1981. Fire and the Australian biota. Australian Academy of Science, Canberra, Australia.

Hann, W. J. 1990. Landscape and ecosystem-level management in whitebark pine ecosystems. Pages 335–340 *in* W. C. Schmidt and K. J. McDonald, compilers. Proceedings—Symposium on whitebark pine ecosystems: Ecology and management of a high-mountain resource. USDA Forest Service, Intermountain Research Station, General Technical Report INT-270, Ogden, Utah.

Hartwell, M. 1997. Comparing historic and present conifer species compositions and structures on forested landscapes of the Bitterroot Front. Contract completion report for RJVA-94928 on file at the Fire Sciences Laboratory, P.O. Box 8089, Missoula, Montana 59807.

Hessburg, P. F., B. G. Smith, S. D. Kreiter, C. A. Miller, R. B. Salter, C. H. McNicoll, and W. J. Hann. 1999a. Historical and current forest and range landscapes in the Interior Columbia River Basin and portions of the Klamath and Great Basins. Part II: Linking vegetation patterns and landscape vulnerability to potential insect and pathogen disturbances. USDA Forest Service, Pacific Northwest Research Station, General Technical Report PNW-GTR-458, Seattle, Washington.

Hessburg, P. F., B. G. Smith, and R. B. Salter. 1999b. A method for detecting ecologically significant change in forest spatial patterns. Ecological Applications 9:1252–1272.

Hoff, R., and S. Hagle. 1990. Diseases of whitebark pine with special emphasis on white pine blister rust. Pages 335–340 *in* W. C. Schmidt and K. J. McDonald, compilers. Proceedings—Symposium on whitebark pine ecosystems: Ecology and management of a high-mountain resource. USDA Forest Service, Intermountain Research Station, General Technical Report INT-270, Ogden, Utah.

Hunt, R. S. 1998. Pruning western white pine in British Columbia to reduce white pine blister rust losses: 10-year results. Western Journal of Applied Forestry 13:60–64.

Hutchins, H. E., and Lanner, R. M. 1982. The central role of Clark's nutcrackers in the dispersal and establishment of whitebark pine. Oecologia 55:192–201.

Jensen, M. E., W. Hann, R. E. Keane, J. Caratti, and P. S. Bourgeron. 1993. ECODATA—A multiresource database and analysis system for ecosystem description and evaluation. Pages 249–265 *in* M. E. Jensen and P. S. Bourgeron, editors. Eastside forest ecosystem health assessment. Volume II: Ecosystem management: Principles and applications. USDA Forest Service, National Forest System Information Report, Washington, D.C.

Keane, R. E. 2000. The importance of wilderness to whitebark pine research and management. Pages 22–33 *in* D. N. Cole and S. F. McCool, editors. Proceedings—Wilderness science in a time of change. USDA Forest Service, Rocky Mountain Research Station, Proceedings RMRS-P-2, Fort Collins, Colorado.

Keane, R. E., and S. F. Arno. 1993. Rapid decline of whitebark pine in western Montana: Evidence from 20-year remeasurements. Western Journal of Applied Forestry 8:44–47.

———. 1996. Whitebark Pine (*Pinus albicaulis*) ecosystem restoration in western Montana. Pages 51–54 *in* S. F. Arno and C. C. Hardy, editors. The use of fire in forest restoration. Society of Ecosystem Restoration. USDA Forest Service, Intermountain Research Station, General Technical Report INT-GTR-341, Ogden, Utah.

Keane, R. E., S. F. Arno, J. Brown, and D. Tomback. 1990. Modeling disturbances and conifer succession in whitebark pine forests. Pages 274–288 *in* W. C. Schmidt and K. J. McDonald, compilers. Proceedings—Symposium on whitebark pine ecosystems: Ecology and management of a high-mountain resource. USDA Forest Service, Intermountain Research Station, General Technical Report INT-270, Ogden, Utah.

Keane, R. E., P. Morgan, and J. P. Menakis. 1994. Landscape assessment of the decline of whitebark pine (*Pinus albicaulis*) in the Bob Marshall Wilderness Complex, Montana, USA. Northwest Science 68:213–229.

Keane, R. E., S. F. Arno, and C. Stewart. 1996a. Restoration of upper subalpine whitebark pine ecosystems in western Montana. Pages 31–26 *in* R. L. Mathiasen, editor. Proceedings of the 43d Annual Western Internation Forest Disease Work Conference, Whitefish, Montana, USA. Society of American Foresters, Bethesda, Maryland.

Keane, R. E., J. P. Menakis, and W. J. Hann. 1996b. Coarse-scale restoration planning and design in Interior Columbia River Basin Ecosystems—An example using whitebark pine (*Pinus albicaulis*) forests. Pages 14–20 *in* S. F. Arno and C. C. Hardy, editors. The use of fire in forest restoration. Society of Ecosystem Restoration. USDA Forest Service, Intermountain Research Station, General Technical Report INT-GTR-341, Ogden, Utah.

Keane, R. E., P. Morgan, S. W. Running. 1996c. Fire-BGC—A mechanistic ecological process model for simulating fire succession on coniferous forest landscapes of the northern Rocky Mountains. USDA Forest Service, Intermountain Research Station, Research Paper INT-484, Ogden, Utah.

Keane, R. E., J. P. Menakis, D. G. Long, W. J. Hann, and C. Bevins. 1996d. Simulating coarse-scale vegetation dynamics using the Columbia River Basin Succession Model—CRBSUM. USDA Forest Service, Intermountain Research Station, General Technical Report INT-GTR-340, Ogden, Utah.

Keane, R. E., D. G. Long, D. Basford, and B. A. Levesque. 1997a. Simulating vegetation dynamics across multiple scales to assess alternative management strategies. Pages 310–315 *in* Conference Proceedings—GIS 97, 11th Annual Symposium on Geographic Information Systems—Integrating spatial information technologies for tomorrow. GIS World, Inc., Vancouver, British Columbia, Canada.

Keane, R. E., C. H. McNicoll, K. Schmidt, and J. L. Garner. 1997b. Spatially explicit ecological inventories for ecosystem management planning using gradient modeling and remote sensing. Pages 135–146 *in* J. D. Greer, editor. Proceedings of the sixth Forest Service remote sensing applications conference—Remote Sensing: People in partnership with technology. American Society for Photogrammetry and Remote Sensing, Bethesda, Maryland.

Keane, R. E., J. L. Garner, K. M. Schmidt, D. G. Long, J. P. Menakis, and M. A. Finney. 1998. Development of the input data layers for the FARSITE Fire Growth Model for the Selway-Bitterroot Wilderness Complex, USA. USDA Forest Service, Rocky Mountain Research Station, General Technical Report RMRS-GTR-3, Fort Collins, Colorado.

Keane, R. E., P. Morgan, and J. White. 1999. Temporal pattern of ecosystem processes on simulated landscapes of Glacier National Park, USA. Landscape Ecology 14:311–329.

Keenan, F. J., R. R. Glavicic, P. W. Swindle, and P. A. Cooper. 1970. Mechanical properties of whitebark pine. Forestry Chronicle 46:322–325.

Kendall, K. C., and S. F. Arno. 1990. Whitebark pine—An important but endangered wildlife resource. Pages 264–274 *in* W. C. Schmidt and K. J. McDonald, compilers. Proceedings—Symposium on whitebark pine ecosystems: Ecology and management of a high-mountain resource. USDA Forest Service, Intermountain Research Station, General Technical Report INT-270, Ogden, Utah.

Kipfer, T. R. 1992. Post-logging stand characteristics and crown development of whitebark pine (*Pinus albicaulis*). Master's thesis, Montana State University, Bozeman.

Kratz, T. K., J. J. Magnuson, T. M. Frost, B. J. Benson, and S. R. Carpenter. 1994. Landscape position, scaling, and the spatial and temporal variablility of ecological parameters: Considerations for biological monitoring. Pages 217–231 *in* Biological monitoring on aquatic ecosystems. Lewis Publishers, Boca Raton, Florida.

Krebill, R. G., and R. J. Hoff. 1995. Update on *Cronartium ribicola* in *Pinus albicaulis* in the Rocky Mountains, USA. Pages 119–126 *in* Proceedings: 4th IUFRO Rusts of Pines Working Party Conference, Tsukuba, Japan.

Lasko, R. J. 1990. Fire behavior characteristics and management implications in whitebark pine ecosystems. Pages 319–324 *in* W. C. Schmidt and K. J. McDonald, compilers. Proceedings—Symposium on whitebark pine ecosystems: Ecology and management of a high-mountain resource. USDA Forest Service, Intermountain Research Station, General Technical Report INT-270, Ogden, Utah.

Lueck, D. 1980. Ecology of *Pinus albicaulis* on Bachelor Butte, Oregon. Master's thesis, Oregon State University, Corvallis.

McCaughey, W. W., and W. C. Schmidt. 1990. Autecology of whitebark pine. Pages 85–95 *in* W. C. Schmidt and K. J. McDonald, compilers. Proceedings—Symposium on whitebark pine ecosystems: Ecology and management of a high-mountain resource. USDA Forest Service, Intermountain Research Station, General Technical Report INT-270, Ogden, Utah.

McCaughey, W. W., and T. Weaver. 1990a. Biotic and microsite factors affecting whitebark pine establishment. Pages 140–151 *in* W. C. Schmidt and K. J. McDonald, compilers. Proceedings—Symposium on whitebark pine ecosystems: Ecology and management of a high-mountain resource. USDA Forest Service, Intermountain Research Station, General Technical Report INT-270, Ogden, Utah.

———. 1990b. Reference guide to whitebark pine. Pages 140–151 *in* W. C. Schmidt and K. J. McDonald, compilers. Proceedings—Symposium on whitebark pine ecosystems: Ecology and management of a high-mountain resource. USDA Forest Service, Intermountain Research Station, General Technical Report INT-270, Ogden, Utah.

McCaughey, W. W., W. C. Schmidt, and R. C. Shearer. 1985. Seed dispersal characteristics of conifers of the Inland Mountain West. Pages 50–62 *in* R. C. Shearer, compiler. Proceedings—Conifer tree seed in the Inland Mountain West symposium. USDA Forest Service, Intermountain Research Station, General Technical Report INT-203, Ogden, Utah.

McGarigal, K., and B. J. Marks. 1995. FRAGSTATS: Spatial pattern analysis program for quantifying landscape structure. USDA Forest Service, Pacific Northwest Research Station, General Technical Report PNW-GTR-351, Seattle, Washington.

Mehringer, P. J., S. F. Arno, and K. L. Peterson. 1977. Postglacial history of Lost Trail Pass Bog, Bitterroot Mountains, Montana. Arctic and Alpine Research 9:345–368.

Michener, W. K. 1997. Quantitatively evaluating restoration experiments: Research design, statistical analysis, and data management considerations. Restoration Ecology 5:324–337.

Morgan, P., S. C. Bunting, R. E. Keane, and S. F. Arno. 1994. Fire ecology of whitebark pine (*Pinus albicaulis*) forests in the Rocky Mountains, USA. Pages 136–142 *in* W.C. Schmidt and F. K. Holtmeier, compilers. Proceedings—International workshop on subalpine stone pines and their environment: The status of our knowledge. USDA Forest Service, Intermountain Research Station, General Technical Report INT-GTR-309, Ogden, Utah.

Mueller-Dombois, D., and H. Ellenberg. 1974. Aims and methods of vegetation ecology. John Wiley and Sons, New York.

Norment, C. J. 1991. Bird use of forest patches in the subalpine forest–alpine tundra ecotone of the Beartooth Mountains, Wyoming. Northwest Science 65:1–10.

O'Hara, K., P. Latham, P. Hessburg, and B. Smith. 1996. Development of a forest stand structural stage classification for the Interior Columbia River Basin. Western Journal of Applied Forestry 11:97–102.

Parker, V. T. 1997. The scale of succession models and restoration objectives. Restoration Ecology 5:301–306.

Parker, V. T., and S. T. A. Pickett. 1997. Restoration as an ecosystem process: Implications of the modern ecological paradigm. Pages 17–32 *in* K. M. Urbanska, N. R. Webb, and P. J. Edwards, editors. Restoration ecology and sustainable development. Cambridge University Press, Cambridge, UK.

Parsons, D. J., D. M. Graber, J. K. Agee, and J. W. van Wagtendonk. 1986. Natural fire management in national parks. Environmental Managment 10:21–24.

Parsons, D. J., and P. B. Landres. 1998. Restoring natural fire to wilderness: How are we doing? Tall Timbers Fire Ecology Conference Proceedings 20:366–373.

Pickett, S. T. A., and P. S. White. 1985. The ecology of natural disturbance and patch dynamics. Academic Press, San Diego, California.

Pfister, R. D., B. L. Kovalchik, S. F. Arno, and R. C. Presby. 1977. Forest habitat types of Montana. Gen. Tech. Rep. INT-34. Ogden, Utah: USDA Forest Service, Intermountain Forest and Range Experiment Station.

Quigley, T. M., and S. J. Arbelbide (Editors). 1997. An assessment of ecosystem components in the Interior Columbia Basin and portions of the Klamath and Great Basins. USDA Forest Service, Pacific Northwest Research Station, General Technical Report PNW-GTR-405, Portland, Oregon.

Quigley, T. M., R. T. Graham, and R. W. Haynes. 1996. An integrated scientific assessment for ecosystem management in the Interior Columbia River Basin and portions of the Klamath and Great Basins. USDA Forest Service, Pacific Northwest Research Station, General Technical Report PNW-GTR-382, Portland, Oregon.

Rastetter, E. B. 1996. Validating models of ecosystem response to global change. Bioscience 46:190–197.

Redmond, R. L., and M. L. Prather. 1996. Mapping existing vegetation and land cover across western Montana and north Idaho. Report on file at USDA Forest Service, Northern Region, Ecosystem Management, P.O. Box 7669, Missoula, MT. Contract 53-0343-4-000012. Missoula, Montana.

Reinhardt, E., R. E. Keane, and J. K. Brown. 1997. First Order Fire Effects Model: FOFEM 4.0 User's Guide. USDA Forest Service, Intermountain Research Station, General Technical Report INT-GTR-344, Ogden, Utah.

Sachs, D. L., P. Sollins, and W. B. Cohen. 1998. Detecting landscape changes in the interior of British Columbia from 1975 to 1992 using satellite imagery. Canadian Journal of Forest Research 28:23–36.

Sund, S. K., D. F. Tomback, and L. A. Hoffmann. 1991. Post-fire regeneration of *Pinus albicaulis* in western Montana: Patterns of occurrence and site charactertistics. Unpublished report on file at U.S. Department of Agriculture, Forest Service, Intermountain Research Station, Intermountain Fire Sciences Laboratory, Missoula, Montana.

Swanson, F. J., J. A. Jones, and G. E. Grant. 1990. The physical environment as a basis for managing ecosystems. Pages 229–235 *in* K. A. Kohm and J. F. Franklin, editors. Creating forestry for the 21st century. Island Press, Washington, D.C.

Swanson, F. J., J. F. Franklin, and J. R. Sedell. 1997. Landscape patterns, disturbance, and management in the Pacific Northwest, USA. Pages 191–213 *in* I. S. Zonnneveld and R. T. Forman, editors. Changing landscapes: An ecological perspective. Springer-Verlag, New York.

Tomback, D. F. 1989. The broken circle: Fire, birds and whitebark pine. Pages 14–17 *in* T. Walsh, editor. Wilderness and wildfire. University of Montana, School of Forestry, Montana Forest and Range Experiment Station, Miscellaneous Publication 50, Missoula, Montana.

———. 1998. Clark's nutcracker (*Nucifraga columbiana*), No. 331. *In* A. Poole and F. Gill, editors. The birds of North America. The Birds of North America, Inc., Philadelphia.

Tomback, D. F., L. A. Hoffmann, and S. K. Sund. 1990. Coevolution of whitebark pine and nutcrackers: Implications for forest regeneration. Pages 118–130 *in* W. C. Schmidt and K. J. McDonald, compilers. Proceedings—Symposium on whitebark pine ecosystems: Ecology and management of a high-mountain resource. USDA Forest Service, Intermountain Research Station, General Technical Report INT-270, Ogden, Utah.

Tomback, D. F., S. K. Sund, and L. A. Hoffmann. 1993. Post-fire regeneration of *Pinus albicaulis:* Height-age relationships, age structure, and microsite characteristics. Canadian Journal of Forest Research 23:113–119.

Tomback, D. F., J. K. Clary, J. Koehler, R. J. Hoff, and S. F. Arno. 1995. The effects of blister rust on post-fire regeneration of whitebark pine: The Sundance burn of northern Idaho. Conservation Biology 9:654–664.

Turner, M. G., and R. H. Gardner (Editors). 1991. Quantitative methods in landscape ecology. Springer-Verlag, New York.

Turner, M. G., W. W. Hargrove, R. H. Gardner, and W. H. Romme. 1994. Effects of fire on landscape heterogeneity in Yellowstone National Park, Wyoming. Journal of Vegetation Science 5:731–742.

U.S. Fish and Wildlife Service. 1997. Grizzly bear recovery in the Bitterroot ecosystems—Draft environmental impact statement. Bitterroot Grizzly Bear Environmental Impact Statement, P.O. Box 5127, Missoula, MT.

Weaver, T., and F. Forcella. 1986. Cone production in *Pinus albicaulis* forests. Pages 68–76 *in* R. C. Shearer, compiler. Proceedings—Conifer tree seed in the Inland Mountain West symposium. USDA Forest Service, Intermountain Research Station, General Technical Report INT-203, Ogden, Utah.

Wright, H. A., and A. W. Bailey. 1982. Fire ecology: United States and southern Canada. John Wiley and Sons, New York.

Wykoff, W. R., N. L. Crookston, and A. R. Stage. 1982. User's guide to the stand prognosis model. USDA Forest Service, Intermountain Research Station, General Technical Report INT-133, Ogden, Utah.

Ziemer, R. R. 1997. Temporal and spatial scales. Pages 80–95 *in* J. E. Williams, C. A. Wood, and M. P. Dombeck, editors. Watershed restoration: Principles and practices. American Fisheries Society, Bethesda, Maryland.

Chapter 19

Social and Environmental Challenges to Restoring White Pine Ecosystems

Hal Salwasser and Dan E. Huff

The purpose of this chapter is to describe some of the social (i.e., political, institutional, and economic) and environmental challenges to our capacity to sustain or restore ecosystems characterized by the presence of five-needled white pines of the western United States. These species include whitebark pine (*Pinus albicaulis*), western white pine (*P. monticola*), sugar pine (*P. lambertiana*), limber pine (*P. flexilis*), the bristlecone pines (*P. longaeva* and *P. aristata*), foxtail pine (*P. balfouriana*), and southwestern white pine (*P. strobiformis*). Of particular concern is whitebark pine. This species, along with limber pine, tends to be characterized as early-successional and fire-dependent, but both may also occur in climax communities on harsh sites (Chapter 18). These white pine communities also provide vital seasonal habitats for a variety of wildlife species, some highly dependent upon the seed production of the pines. The high-elevation white pines also influence local hydrology and reduce soil erosion. The preponderance of white pine ecosystems in the western United States is located on public lands, notably those managed by the U.S. Forest Service (USFS) and the National Park Service (NPS) (see Chapters 1 and 12).

Whitebark pine is declining range wide as a consequence of the spread of nonnative white pine blister rust (*Cronartium ribicola*). Declines are particularly severe in southwestern Canada and northwestern United States, where blister rust losses are additive with successional losses that are caused by decades of fire exclusion (Chapters 1, 9, and 11). Urgent management actions will be required to mitigate further catastrophic loss of whitebark pine forests. Fortunately, useful technology is available (Chapter 18), but many barriers remain to timely and effective intervention. The greatest, perhaps, is lack of public will and consensus. And just beyond that lurk formidable economic and environmental concerns.

The term "restoration" has several popular meanings among natural resource managers and conservation biologists. For the purpose of this discussion, restoration refers to the reestablishment of ecosystems to sustain desired, measurable characteristics and includes the anthropogenic processes required to sustain them. It does not imply recreation of static, prescribed "target" reference period conditions as one might seek to do in restoring an historic scene or building to the conditions of a prior time.

Managing Ecosystems to Sustain Uses and Values

Managing ecosystems to sustain or restore desired conditions, human uses, and ecological values involves substantial departures from midcentury views of the "regulated" forest (Smith 1962), where wood production was often the primary value (Clawson 1975). Under that paradigm, the sustainable ecosystem was to be managed to minimize or eliminate competition (e.g., insect infestation, wildfire, or even unmanaged ecological succession) to priority species or products. With today's human pressures and invasive species as pervasive influences on many ecosystems, management for desired ecological values may not be possible with the traditional hands-off approaches operating in national parks and wilderness areas (Huff 1999). Also, management to sustain ecological values in any kind of ecosystem requires a more holistic and adaptive perspective on resource planning and management than the notion implied by a regulated forest. And it requires substantially more involvement from forest stakeholders (i.e., constituencies) in the design of conservation strategies.

The forces of change causing current departures from traditional forest management have multiple dimensions. Our evolving scientific understanding of how ecosystems work and change over space and time is one of the more important dimensions (Botkin 1990). Changing technologies is another, along with changing social, political, and economic conditions, climate change, invasive species, and dynamics in our institutions of governance.

Ecosystem management is the term often given to the new approach to forest management that responds to these diverse forces. As a policy, it has been evolving with America's public lands managers for about a decade (Salwasser 1995). But the somewhat amorphous concept was discussed in scientific circles and journals well before it emerged as proposed explicit management policy in the early 1990s (Van Dyne 1969). Ecosystem management has yet to be consistently defined by all federal ecosystem managers, but some commonly held concepts have emerged in recent years. Though there is some disagreement among conservationists, we take the position that all of Earth's ecosystems include humans either as integral parts or as factors that contribute external effects. An example of the former is a landscape with human communities and human activities embedded. An example of the latter might be the effects of air pollution on wilderness areas.

Possibly the most significant deviation from traditional forestry values in ecosystem management is the concern for "ecological diversity" (Noss and Cooperrider 1994). This concept generally refers to three things: (1) the maintenance of all desired ecological components (e.g., biological diversity), (2) the maintenance of structural variety both vertically and horizontally from sites to landscapes (i.e., structural diversity), and (3) the maintenance of processes through which ecosystems function (e.g., functional diversity). The challenge to managers is, of course, identification of the desired ecological characteristics—that is, what is possible given current and future influences and target conditions established by human stakeholders. Today's ecosystem managers are less concerned about the establishment of target outputs and/or outcomes (e.g., annual harvest of "n" board-feet of timber) than were their predecessors. But output/outcome goals can indeed be compatible with an "ecosystem approach" to management of forests and other renewable resources (Grumbine 1994; Salwasser 1995; Huff 1996). The key to such an approach is the consideration of the effects of target output/outcome management on ecosystems. Output/outcome management that threatens the sustainability of desired ecosystem components or processes must be revised or mitigated.

The concept of sustainability has been articulated in national and international dialogues since the mid-1980s (World Commission on Environment and Development 1987). A variety of ecosystem management goals have been identified under the concept of sustainability. These include conservation of biological diversity, provision of human benefits, and maintenance of soil productivity, water quality, and carbon sequestration. As such, ecosystem management is a process used to achieve sustainability of desired resource values through a predictable range of stochastic biotic and abiotic events (see also Christensen et al. 1996).

Principles of Ecosystem Management with Implications for Restoration

Numerous authors have suggested principles for ecosystem management (Slocombe 1993; Grumbine 1994; Salwasser and Pfister 1994; Christensen et al. 1996). We suggest the following:

Humans are integral parts of ecosystems, and we must consider their needs, capabilities, and impacts in establishing management goals. This principle is a major point of departure for ecosystem managers and some "traditionalists" who view humans, and their works, at worst, as despoilers of nature and, at best, as itinerant visitors. We view the separation of humans from nature (in its broadest context) as an artificial and outdated concept. Without the insertion of human values, there are no inherent goals or purposes to guide ecosystem restorations (see Primack 1998).

Ecosystems are constantly changing as a result of both internal developmental processes of populations and communities, and external influences including the actions of humans. Ecosystem resilience, or the capacity of ecosystem biota to adapt to the forces of change, is influenced by biological diversity and productive capacity. Restoring (i.e., returning) an ecosystem to the precise conditions that characterized it at some time in the past is practicably impossible. Even if we re-created the exact mosaic of biota, the biotic and abiotic evolutionary conditions that accompanied those species have changed. Even if the exact species complement could be reestablished in the approximate proportions of the restoration target, they would immediately begin moving in different directions (ecologically) from the historic trends.

The "ecosystem" is basically a human construct created by scientists to place general reference limits on extant biota and their environment for analysis. Ecosystems can be defined at all scales and may be determined to be "nested." Small ecosystems, such as individual rotting logs, occur nested within larger ecosystems, such as watersheds, which exist within larger landscape-scale ecosystems, and so on. Ecosystems have even been defined based on the limits of specific landforms or human uses, and on the boundaries of administrative jurisdictions. However, some resource or environmental issues or concerns can reasonably be addressed only at a particular geographic scale. For example, conservation of anadromous fish requires consideration of whole watersheds from the ocean to river headwaters. Some rare plant conservation strategies might require attention only to small catchment basins. Migratory wildlife, such as ungulates, birds, and anadromous fish, could require scales that range from specific elevations of single mountain ranges to major portions of a hemisphere. And some species (e.g., some cranes and terns) may even require habitat preservation in more than one hemisphere. Thus, many concerns in ecosystem restoration will require strategies that cross multiple geographic scales and integrate actions at several of those scales. Regardless of the various criteria used to designate ecosystems for study, management, or administration, many ecosystem components and processes can be effectively addressed only at the scales in which they occur on the biosphere.

A corollary to the above is that actions at one point in time or at one scale may have off-site or delayed responses at scales other than intended. An example of this effect is the influence that hydropower and flood control dams placed on our major rivers in midcentury have had on anadromous fishes. Salmon no longer get past the dams to their high-mountain spawning areas, the dams have modified riparian areas, and the water flows have been altered. Ecosystem restoration should also include consideration for the potential transfer effects of unmitigated actions on other ecosystems, for example, downstream effects of dams or the shifting of timber and other resource demands from one forest to another.

It is impossible to completely understand everything about ecosystems or to predict with precision all the potential responses to a particular action. We can predict some future conditions for some time, and the near future more accurately than the distant future. But uncertainty, unknowns, and risks are inherent in all ecosystem management considerations. For example, the dense, multilayered structure of some old-growth forests that provides unique habitat values for some species of wildlife is at greater risk of losing those conditions and associated values from catastrophic wildfires than are younger communities. Prolonged periods of wildfire protection may increase longevity, extent, and age of successional communities beyond the desired range of variation.

The upshot of these principles is that ecosystem restoration and management cannot be considered an "exact science." The longevity of ecosystem management successes is directly proportional to the level of applicable science and the continuity of human values. A dearth of science on which to base management decisions almost guarantees successes will be limited to the near term. And changes in human values (e.g., intensifying public concern for old-growth forest communities in national forests) can quickly reverse any ecosystem accomplishment, regardless of the level of science brought to bear on the decisions.

Forces of Change on Ecosystem Sustainability

Given the above perspectives, and given that the particular goal under discussion here is to restore and/or sustain desired conditions, uses, and values of white pine ecosystems (and whitebark pine ecosystems, especially), we can identify the major forces of change impinging on that goal.

Population Change

At all scales, assuming constant per capita consumption rates, growth of the human population constitutes the preeminent force affecting the sustainability of all the managed ecosystems on the planet (Wilson 1998). Projections show continued worldwide growth, although some local growth rates (e.g., United States and western Europe) have leveled off for now. In the United States, the strong economy drives increasing per capita consumption, resulting in higher demands on natural resources.

The incredible variety of scenic landscapes, flora and fauna, and associated recreational opportunities in our western states and provinces are attracting ever more visitation each year (Flather and Cordell 1995). The combined effects of increased local, national, and international demands on western forest resources for recreation and for commodities will continue to make resource use and conservation decisions ever more complex in the years to come.

Further complicating public awareness is the fact that our diversifying American cultures view natural resources from many different perspectives. Dominant western viewpoints include belief in the "right" to utilize resources locally while they are still available. Many people still believe that technology will provide the "fix" for sustained, or increasing, consumption of a limited supply (Cohen 1995). And there is evidently little public support for governmental policies and regulations aimed at reducing resource consumption.

Diversity of natural resource values could result in a strong synergy for decision making in our democratic society. But in the extreme, irresolvable conflict and "win-lose" litigated decisions can prevent optimization of resource management in the greatest public interest (Parker 1995). Increased allocation for specific uses often reduces ecosystem capacity for other resources. Some examples of this phenomenon include: harvesting timber from old-growth forests may alter their biodiversity value and, if carried out widely across regional landscapes, can threaten the sustainability of late-successional communities (Wilson 1984). Consuming surface or groundwater, or reduction of natural in-stream flows, can compromise the aquatic and riparian biodiversity values downstream (U.S. Army Corps of Engineers 1999). Managing for maximum sustained-yield populations of harvestable species reduces space and forage for declining sympatric species (Wagner et al. 1995). Technology may well provide for greater "yields" of renewable natural resources, and more efficient utilization of those yields, but there are ultimate limits to the use of global natural resources (Gowdy and McDaniel 1995).

Institutional Change

Institutional change is a second major force affecting our future. Structural shifts in our economies; globalization of markets; distrust of governments, authority, and expertise; substitution of legal processes (e.g., litigation) for collaborative problem solving and scientific scrutiny; and diverging institutions and interest groups all add to the complexity of resource management.

We have seen a virtual *conflict industry* become empowered by our current laws and policies designed to ensure public participation in public resource decisions. Reflecting the U.S. legal system, the conflict industry has often sparked litigation used to drive win-lose decisions about resources rather than catalyzing compromises that would better serve the myriad constituents. This industry seems to be perpetuated by single-purpose, not-for-profit, public organizations that, in the current economy, are able to generate the funds needed to litigate almost anything, including views of extreme minorities. The growing polarization of public opinion, and resultant fractionalization of interest groups, will haunt those who would restore white pine ecosystems on public lands, because significant active intervention in national parks and designated wilderness areas would be required (see Chapter 13).

Successful ecosystem management requires successfully meshing two highly dynamic processes, science and the institutionalization of public values. Like the chicken and the egg, it is impossible to determine which comes first. Science provides the information necessary to identify workable alternatives for resource management, the actions required to accomplish them, and the effects of management on other ecosystem dynamics. Public values affect science dollar allocation and, hence, the amount of information that can be collected, analyzed, and corroborated prior to, and during, planning for decision making. They also affect management goals and, therefore, the selection of alternatives. On the scale of a human lifetime, both are highly changeable.

Science provides us with volumes of data, though somewhat less usable "information" every day. Relatively few "laws of science" are generated in the short term (Wilson 1998). Rather, scientifically acquired information builds gradually with incremental additions from many associated directions, like the proverbial snowball rolling down a hill. But many hypothesized "cutting edge" relationships are only loosely connected through what may prove to be false assumptions or incomplete analyses. These adjunct hypotheses must await the arrival of either detailed corroborating research to be "proven," or "breakthrough" information to be exposed—like the snowball dashed against a boulder. The net result is forward momentum, but science growth is neither unidirectional nor linear. It may even move backward for some significant period of time until corrected, as described above. Resource management decisions made on the cusp of new, uncorroborated science (rather than a long-term accumulation) have greater potential for yielding the wrong results. Decisions that fail to consider such information have a greater potential for litigated challenge.

Human values change as well. And, like science, they don't always move "forward." General ecosystem values once professed by a large segment of the American public will be replaced with diverse values reflecting the diversification of American society. As such, the "truth" (i.e., scientific fact) can never fully "set us free." It is absolutely necessary to develop and institutionalize meaningful and responsible management goals for public resources that will be sustained by a preponderance of public support for some significant period of time. But it is important to note that no amount of science can ever guarantee universal acceptance or perpetual sustainability of public resource management decisions.

We believe that the greater the information base, the better the chance of support from a broader constituency, and the greater the longevity of management decisions—because science-based management actions have a greater probability of accomplishing desired objectives. All governmental entities with authority for natural resource management are affected by this phenomenon. The policy- and decision-making environment we find ourselves in today is ripe for new institutional arrangements to emerge. Thus, we see the growing emphasis on both large-scale, science-based assessments, such as the Interior Columbia River Basin Ecosystem Management Project (Lee 1993), the Northwest Forest Plan (USGAO 1999a), the South Florida Restoration Project (U.S. Army Corps of Engineers 1999), and community-based planning, such as the Central Platte River Project (U.S. Fish and Wildlife Service 1997), with honest attempts to forge interagency, collaborative models for solving problems. These projects are providing new hope for productive, collaborative decision making in an ever more diverse and litigious world.

Environmental Change

Environmental change on a global scale is a third major force affecting our ability to restore white pine ecosystems. Global environmental change refers not only to change in climate but also to changes in air and water quality, biodiversity, invasive species, and toxics in the environment. It is pervasive, though not precisely predictable in its consequences. It is probably irreversible. The rate of change appears to be getting faster all the time (Wilson 1998). We believe that if, in fact, this trend is long-term, the scientific "proof" will eventually be viewed as "conclusive" by American society. Only then will it be possible to invoke a national consciousness in the establishment of natural resource management goals reflecting the inevitable opportunities and limitations accompanying global environmental changes.

Science as an Agent of Change

As discussed earlier, science alone will not unify all diverse cultural and philosophical differences among Americans. Many Americans are skeptical, or ignorant, of scientific "facts" related to their personal belief systems. In a time when science is opening doors to understanding the submolecular to the cosmic, many people prefer to believe in pseudoscience (e.g., astrology) or counterscientific litany based in "ideology." This condition may point to a deteriorating educational system, growing impatience for self-analysis of complex issues, or some unconscious rebellion against the worldview that science portrays. As we described earlier, science is not immutable. But when consistently replicated and corroborated by numerous workers in a variety of disciplines, science most closely equates to objective "truth," at least for practical purposes in our lifetimes. Failure to accept that fact will relegate us to a Stone Age technology—continuously responding to some subjective set of instinctive, intuitive, experiential, and culturally engendered "facts."

Science is critical in our ability to understand demands for natural resources, processes for filling those demands, and the impacts of resource utilization on affected ecosystems and even the biosphere. Through science, we may be able to unify *some* highly variant public opinions and values and instill common knowledge of possibilities and outcomes. As such, we may reduce dissension wrought of misunderstanding, leaving a somewhat more limited range of true diversity in human values. Through a thorough analysis of the impacts of various alternatives, and a widescale disclosure of those impacts, we may be able to develop majority opinions on prudent decisions, if not consensus, among constituents. Clearly, without science, there is no common basis for predicting possibilities and outcomes.

Forces of Change in Ecosystems Containing White Pines

The following is a discussion of changes in the physical, biotic, and social environments that affect white pine ecosystems in particular.

Climate. Climate change is a significant factor in all aspects of ecosystem management. We have abundant evidence that older forests (>400 years old) were established during a period of much different climate than we now experience, probably wetter and colder. Some white pines, e.g., Great Basin bristlecone pine (*Pinus longaeva*), approach 4,000 or more years old and thus could have been established during the Holocene warm period (Cohen 1998). We also know that climate, local and global, has always and will always change. This will affect the elevations and aspects where restoration of white pines can be successful. Climate change will affect both species distribution and the processes that govern ecosystem dynamics (e.g., fire, precipitation, storm disturbance, insect infestations, growing season length, erosion rates).

Mining. Mining has historically impacted white pines through both direct displacement and mine-related pollutants. The remnants of old mines often remain as watershed restoration challenges. New mines will occur in the range of white pines as humans continue to demand more mineral resources. These new mines could have development plans linked to restoration programs as part of the permitting process.

Livestock grazing. Livestock grazing, part of the western economy for 150 years, is likely to continue in white pine ecosystems, even in the higher elevations inhabited by whitebark pine. In addition to the effects of grazing and trampling of seedlings of other species, the most significant effects of livestock on white pine ecosystems occurs through changes in understory fuels and the effects on natural fire regimes, and the invasion of exotic plant species that livestock grazing induces.

Timber harvest. Timber harvest (i.e., logging) has had a substantial historical impact on the commercially valuable species such as western white pine and sugar pine. Logging removed many of the biggest, oldest trees decades ago. Even for subalpine species of lower commercial value, logging for fuelwood and mine timbers in the late nineteenth century resulted in range reduction (Williams 1989). In the future, logging in the white pines will have less impact, as forests are managed for a broader suite of values and uses. Nevertheless, attention to the genetic variation and distribution and abundance of white pine populations is warranted in future harvest decisions.

Recreation. Recreational use in western forests, particularly in high-elevation areas that include white pines, has grown tremendously in the past half century, both in total person-days use and in variety of uses (Flather and Cordell 1995). Recreation impacts include the effects of roads, trails, and disturbance on terrestrial and aquatic ecosystems, vegetation, and wildlife. Another set of impacts has resulted from the development of residences and winter sports facilities in high-mountain resort areas. Such impacts have proliferated throughout the mountainous areas of the western United States and Canada. The mixed impacts include the elimination of white pine habitat for roads, buildings, and ski runs, as well as the air and water pollution from human activities. But there is also the potential to use revenues from users to fund mitigation and restoration activities in high-use areas.

Water development. Water development has been an historical factor affecting forest ecosystems. We have probably seen the peak of structural approaches to water development in the West (i.e., dams and diversions), but vegetation management has also been discussed as a method to augment water supplies. This could involve the deforestation of some forested areas to provide more runoff and less water "loss" to evapotranspiration. Many water structures generate substantial revenues for owners or managers. As they are re-licensed, consideration should be given to tapping those revenues for watershed restoration work.

Fire exclusion and management. Fire exclusion and management have been major factors altering forests that contain white pines. Currently, catastrophic forest fires are increasing in frequency and intensity in the West, and nonlethal ground fires are decreasing (USGAO 1999b). There may also be greater social tolerance for returning fire to its pre-Columbian ecological role. How we plan, manage, and allow fire to affect future forests will greatly influence the structure, composition, and function of those forest ecosystems. Changing fire management strategies from universal suppression to using fire to recreate desired disturbance regimes and reduce fuel accumulation would help restore white pine communities.

Invasive species. Invasive species (exotics and natives), including white pine blister rust, will produce major changes in forest ecosystems (Chapters 1, 11, and 12). We cannot eradicate most of these species, but we can temper their effects through management actions. Managers must consider the multiple consequences of those actions. For example, planting white pine seedlings that are genetically resistant to blister rust could simultaneously eliminate unknown desirable genetic qualities and variation (Millar et al. 1996). We need to think beyond the immediate and "obvious" and ask a widening variety of questions about the consequences of our actions.

Airborne pollutants. Various oxides of nitrogen, sulfur, and phosphorous can either stimulate or degrade ecosystems depending on dose and timing (Bytnerowicz 1997). Even though small "doses" might enhance primary productivity, concentrations of these nutrient-toxins in the environment already far exceed early-twentieth-century levels in many forest ecosystems. Such treatments, inadvertent or deterministic, could have more influence on future ecosystem conditions than climate.

Values and Uses of White Pine Ecosystems: Achieving Sustainability

Many forests containing white pines, and many with whitebark pine in the higher elevations, constitute the headwaters of our major western river systems. This is especially true of the Rocky Mountains and Sierra Nevada, where rivers such as the Columbia, Missouri, San Joaquin, Colorado, Rio Grande, and Arkansas all flow from basins containing white pines. This is also true of many major European and Canadian white pine forest ecosystems. The condition of those forests and watersheds in general is key to continuing the delivery of desired water quality and quantity downstream.

White pine forests make strong contributions to ecosystem biodiversity. Grizzly bears (*Ursus arctos*) and Clark's nutcrackers (*Nucifraga columbiana*) are wildlife species that are almost obligate seasonal users of white pine ecosystems. Many elements of biological diversity are affected by the health of white pines (Chapter 12). The trees themselves shape the structure and function of many forest ecosystems in addition to providing the shelter, environment, and foods that support other species of plants and animals (Chapter 12).

Without question, some of the most dramatic forests of the western United States are dominated by the white pines (e.g., sugar pine, limber pine, Great Basin bristlecone pine, whitebark pine). Unfortunately, with the current epidemic of blister rust in the West, we are finding out what these forests will look like without white pines. Preserving white pines for aesthetic and ecological values, particularly in national parks and wilderness areas, may require management interventions that are counter to existing policies (Chapter 13). Ultimately, the manager must decide whether to use interventive management to restore white pine ecosystems to conditions that are self-sustaining.

Forests that contain white pines supply important resources that sustain human well-being. These range from wood products in some cases, hunting and wildlife-viewing in all cases, and other forest values (rainwater runoff to scenic) in most cases. White pine forests are significant recreation settings. This is especially true of the higher-elevation forests, where broad vistas of whitebark and limber pines *are* the setting. Native Americans have used the products of white pines for millennia and continue to sustain cultural traditions based on the health and availability of these species (Malouf 1969; Moerman 1998).

Considerations in Restoration and Sustainable Management

The following issues and processes should be considered integral to the planning, management, restoration, and resource development projects that will involve white pine ecosystems:

Ecosystem dynamics: space and time. Management decisions should consider (1) the basics of ecosystem stewardship: conservation of soil productivity, biological diversity, water and air quality, perpetuation of ecological processes, and mitigation of vulnerability to invasive species, and (2) the natural dynamism of ecosystems, with or without management. Change happens, and ecological disturbance and succession must be factored into management objectives and actions.

Populations: genetic variation, distribution, and abundance. Consideration should be given to the long-term and evolutionary consequences of proposed management actions on the genetic variation and demographic viability of the white pine species and its populations. The effects on factors such as abundance, distribution, and connectivity at landscape and regional scales all must be evaluated. Stand-level actions alone are not sufficient as long-term conservation strategies.

Human uses and values. Managers must foster societal values for sustaining whitebark pine ecosystems. The benefits of preserving these communities must be well researched, documented, and communicated during the planning process. The roles of species of special concern (e.g., charismatic, keystone, rare) should also be clearly laid out.

Climate change. Management planning should consider the effects of alternative actions on sustaining target species (e.g., whitebark pine) in a period of rapid climate change.

Ownership patterns and land-use mandates. Forests containing white pines occur across a variety of land ownerships and under various land-use mandates. For example, many of the sugar pines in the Sierra Nevada are in forests open to management for multiple products and uses, while most of the whitebark pines occur in national parks and wilderness areas with significant constraints on acceptable management practices. Among the implications are that (1) interagency and interowner coordination is necessary in conservation and restoration strategies, because conditions in adjoining ownerships can have substantial effects across the boundary lines, and (2) ability to employ active intervention versus passive management depends on owner goals and land-use mandates.

Fire management. Forest restoration and fire management are inseparable. Forests in the West evolved with fire playing several significant roles in a complete life cycle. Long-term health of these forests depends on fire as a recurring disturbance factor. But short-term health depends on preventing catastrophic fires at unacceptable frequencies or intensities.

Adaptive management. Given the need for action, our lack of complete knowledge, and the many unknowns, uncertainties, and risks involved in any course of management, adaptive management is the only prudent course. Adaptive management means accepting that resource management is always an experiment of sorts, then designing the treatments, monitoring, and evaluation processes together as integral packages (Barrett 1985). Hypotheses related to management results are established, and revisions to management can be made when monitoring results so indicate.

Social Barriers to Sustainable Management of White Pine Ecosystems

The barriers to public acceptance of sustainable management practices for white pine ecosystems have been woven throughout this paper. The following is a concise review of those constraints.

Finding the common ground. The growing diversity of public opinions on how, when, and where natural resources should be managed, combined with an increased propensity toward, and fiscal support for, litigating minority views, severely threaten the manager's ability to make decisions regarding the sustainable management of ecosystems in the public domain. The public must accept and adhere to a resource-planning/decision-making process to ensure non-litigious, cooperative management decisions. Maintaining current resource values in the face of all global changes will require active intervention. Consensus must be reached among the diverse constituencies to intervene actively in many previously "inviolate" areas in order to accomplish sustainability goals (Brunson and Kennedy 1995).

Competition for resources. The dollars still drive the agenda. Resource management agencies must focus on resources and services that provide high economic value. As agency resources continue to shrink (as evidenced by out-year federal budget proposals), competition for dollars and people to plan and manage ecosystems, which support only limited commodity production and passive recreation will get worse, regardless of the relative significance of biodiversity issues. Societal values will have to change in order to ensure sustainability of many white pine ecosystems.

Limitations of science. Information is necessary in order to understand resource potentials and management alternatives, develop site-specific management techniques to accomplish them, and predict and monitor the impacts of these actions on other ecosystem components and processes. But science is dynamic. Today's predictions of the impact of these actions will certainly be revised in the future with new basic science and ecosystem monitoring. We cannot always wait for conclusive science that rarely if ever arrives in time; hence, adaptive management is the most reasonable approach for utilizing available science and technology to accomplish future goals. This process accommodates and considers new knowledge along the way. Solid, long-term, well-corroborated science can best ensure the accomplishment of our resource management goals and the longevity of our decisions. Science reduces controversy resulting from misinterpretation of the "facts," but it cannot dictate human values. Ultimately, the search for common ground is an on going process that requires the continuous input of new scientific information for its success.

Acknowledgments

We thank Constance Millar and Bohun Kinloch of the USDA Forest Service, Pacific Southwest Research Station, for their thoughts on this chapter. Thanks also to Steve Arno and Bob Keane of the USDA Forest Service, Rocky Mountain Research Station, and Diana Tomback of the University of Colorado at Denver for their thoughtful edits and suggestions to improve the manuscript.

Literature Cited

Barrett, G. W. 1985. A problem-solving approach to resource management. Bioscience 35:423–427.

Botkin, D. B. 1990. Discordant harmonies: A new ecology for the twenty-first century. Oxford University Press, New York.

Brunson, M. W., and J. J. Kennedy. 1995. Redefining "multiple use": Agency responses to changing social values. Pages 143–158 *in* R. L. Knight and S. R. Bates, editors. A new century for natural resources management. Island Press, Washington, D.C.

Bytnerowicz, A. 1997. Atmospheric and biospheric interactions of gases and energy in the Pacific Region of the United States, Mexico, and Brazil. USDA Forest Service, Pacific Southwest Research Station, General Technical Report PSW-GTR–161, Berkeley, California.

Christensen, N. L., A. M. Bartuska, J. H. Brown, S. Carpenter, C. D'Antonio, R. Francis, J. F. Franklin, J. A. MacMahon, R. F. Noss, D. J. Parsons, C. H. Peterson, M. G. Turner, and R. G. Woodmansee. 1996. The report of the Ecological Society of America Committee on the Scientific Basis for Ecosystem Management. Ecological Applications 6:665–691.

Clawson, M. 1975. Forests for whom and for what. Johns Hopkins University Press, Baltimore, Maryland.

Cohen, J. E. 1995. How many people can the earth support? W.W. Norton, New York.

Cohen, M. P. 1998. A garden of bristle cones: Tales of change in the Great Basin. University of Nevada Press, Reno.

Flather, C. H., and H. K. Cordell. 1995. Outdoor recreation: historical and anticipated trends. Pages 3–16 *in* R. L. Knight and K. J. Gutzwiller, editors. Wildlife and recreationists: Coexistence through management and research. Island Press, Washington, D.C.

Gowdy, J. M., and C. N. McDaniel. 1995. One world, one experiment: Addressing the biodiversity-economics conflict. Ecological Economics 15:181–192.

Grumbine, R. E. (Editor). 1994. Environmental policy and biodiversity. Island Press, Washington, D.C.

Huff, D. E. 1996. Defining ecosystem health in national parks. Pages 448–453 *in* K. G. Wadsworth, editor. Transactions of the 62nd North American wildlife and natural resources conference. Wildlife Management Institute, Washington, D.C.

———. 1999. Restoring to what? Do we know enough? Paper presented at the Annual Meeting of the Society for Ecological Restoration, San Francisco, California.

Lee, K. N. 1993. Compass and gyroscope: Integrating science and politics for the environment. Island Press, Washington, D.C.

Malouf, C. 1969. The coniferous forests and their uses in the northern Rocky Mountains through 9,000 years of prehistory. Pages 271–290 *in* R. D. Taber, editor. Coniferous forests of the northern Rocky Mountains: Proceedings of the 1968 symposium. University of Montana, Missoula.

Millar, C. I., B. B. Kinloch Jr., and R. D. Westfall. 1996. Conservation of biodiversity in sugar pine: Effects of the blister rust epidemic on genetic diversity. Pages 190–198 *in* B. B. Kinloch, M. Marosy, and M E. Huddleston, editors. Sugar pine status, values, and roles in ecosystems. University of California Division of Agriculture, Natural Resources Publication 3362, Oakland.

Moerman, D. E. 1998. Native American ethnobotany. Timber Press, Portland, Oregon.

Moffat, A. S. 1998. Global nitrogen overload problem grows critical. Science 279:988–999.

Noss, R. F., and A. Y. Cooperrider. 1994. Saving nature's legacy: Protecting and restoring biodiversity. Island Press, Washington, D.C.

Parker, K. 1995. Natural resources management by litigation. Pages 209–220 *in* R. L. Knight and S. F. Bates, editors. A new century for natural resources management. Island Press, Washington, D.C.

Primack, R. B. 1998. Essentials of conservation biology, 2nd edition. Sinauer Associates, Sunderland, Massachusetts.

Salwasser, H. 1995. Factors influencing the context and principles of ecosystem management. Pages 5–16 *in* F. H. Wagner, editor. Ecosystem management of natural resources in the Intermountain West. Natural Resources and the Environment Issues, Volume V, Utah State University, Logan.

Salwasser, H., and R.D. Pfister. 1994. Ecosystem management: From theory to practice. Pages 150–161 *in* W. W. Covington, and L. F. DeBano, compilers. Sustainable ecological systems: Implementing an ecological approach to land management. USDA Forest Service, Rocky Mountain Research Station, General Technical Report RM–247, Fort Collins, Colorado.

Slocombe, D. S. 1993. Implementing ecosystem-based management: Development of theory, practice, and research for planning and managing a region. Bioscience 43:612–622.

Smith, D. M. 1962. The practice of silviculture, 7th edition. John Wiley and Sons, New York.

U.S. Army Corps of Engineers. 1999. Central and Southern Florida Project comprehensive review study: Integrated feasibility report and programmatic impact statement. Jacksonville, Florida.

U.S. General Accounting Office (USGAO). 1999a. Ecosystem planning: Northwest forest and Interior Columbia River Basin plans demonstrate improvements in level-use planning: Report to congressional requestors. GAO/RCED-99-64. Washington, D.C.

———. 1999b. Western national forests: A cohesive strategy is needed to address catastrophic wildfire threats. GAO/RCED-99-65. Washington, D.C.

U.S. Fish and Wildlife Service. 1997. Cooperative agreement for Platte River research and other efforts relating to endangered species habitats along the Central Platte River, Nebraska. Region 6. Denver, Colorado.

Van Dyne, G. M. (Editor). 1969. The ecosystem concept in natural resource management. Academic Press, New York.

Wagner, F. H., R. Foresta, R. B. Gill, D. R. McCullough, M. R. Pelton, W. F. Porter, and H. Salwasser. 1995. Wildlife policies in the U.S. national parks. Island Press, Washington, D.C.

Williams, M. 1989. Americans and their forests. Cambridge University Press, New York.

Wilson, E. O. 1984. Biophilia. Harvard University Press, Cambridge, Massachusetts.

———. 1998. Consilience: The unity of knowledge. Vintage Books, New York.

World Commission on Environment and Development. 1987. Our common future. Oxford University Press, New York.

Chapter 20

Whitebark Pine Restoration: A Model for Wildland Communities

Stephen F. Arno, Diana F. Tomback, and Robert E. Keane

The contributors to this book have addressed the ecology, problems, and restoration needs for whitebark pine (*Pinus albicaulis*). A recurring theme, implicit to many chapters, is the paradox that envelops the stewardship of whitebark pine communities: Although whitebark pine communities occupy remote areas and generally are protected from human activities, whitebark pine is threatened by anthropogenic alterations of ecological processes (fire suppression) and introduction of an exotic pathogen (blister rust, *Cronartium ribicola*).

The dilemma of whitebark pine ecosystems, as presented here, epitomizes the problems associated with preservation of many "wildland" ecosystems worldwide, which are made up of native plants and animals situated in seemingly unaltered environments. An observer's initial impression may be that restricting human use—for example, by creating large natural areas—will provide the necessary protection. Careful inspection, however, reveals that the "natural system" has already been altered by pervasive human influences that will not be eliminated despite a "protected" status or a remote location. Examples of these human influences include disruption of natural disturbance cycles, such as periodic fires or floods; introduction of invasive plants, animals, or pathogens; elimination of important predators; chemical pollution, such as pesticides; regional air or water pollution; or global climatic change.

The case for restoration of whitebark pine communities provides insight and guidance for approaching stewardship of other wildland communities. As this volume demonstrates, environmental specialists need to identify the important components and relationships within the ecosystem under study. This information should then be incorporated into simulation models, which, with testing and refinement, can be used to predict outcomes of various management alternatives. These alternatives include continuing present protection and management regimes compared with different restoration strategies. Ecologists teamed up with experienced land managers should be developing and testing possible restoration strategies. Monitoring and evaluation of the effectiveness of treatments then provides feedback into improving and expanding restoration to larger scales, and ultimately to a landscape scale.

Whitebark pine communities are declining because of two forces, fire suppression and the introduction of an exotic fungal disease. These problems—altered fire cycles and invasion by exotics—are common worldwide. Issues and perceptions related to the restoration of whitebark pine communities, as presented in this book, also apply to many other kinds of wildland ecosystems. One issue is the important influence of historic fire regimes and the ecological disruption caused by fire suppression. For example, efforts to preserve California redwood (*Sequoia sempervirens*) forests have been successful in establishing large reserves, but few of these areas have attempted to return the frequent fires that were characteristic of these ecosystems (Fritz 1931; Finney and Martin 1992; Brown and Swetnam 1994). Farther north, the coastal Douglas-fir (*Pseudotsuga menziesii* var. *menziesii*) forests, habitat for the endangered northern spotted owl (*Strix occidentalis*), were recognized a century ago as being fire-dependent (Pinchot 1899). Detailed studies have documented the critical role of mixed-severity and stand-replacement fire regimes in these forests (Morrison and Swanson 1990; Agee 1993), but few plans exist to reintroduce fire in the extensive reserves within this ecosystem (Means et al. 1996). In the southeastern United States, historically one of the most extensive ecosystems was dominated by longleaf pine (*Pinus palustris*) and wiregrass (*Aristida stricta* and *Sporobolus gracilis*); this ecosystem is now reduced to a tiny remnant largely as a result of disruption of the historic regime of frequent fires (Christiansen 1981). One of the world's largest forest types, stretching from Norway to eastern Siberia, is a Scots pine (*Pinus sylvestris*), spruce (*Picea* spp.), birch (*Betula* spp.) mixed forest, where a high level of biodiversity was maintained historically by fires at intervals of about 30 to 100 years; ecologists and foresters are now concerned about the compounding effects of fire suppression here (Granström 1996; Arno 1998).

Another issue is the introduction of invasive and disruptive nonnative organisms. In the whitebark pine ecosystem, it is blister rust fungus. In intermountain ponderosa pine (*Pinus ponderosa*) forests, the most widespread invader is spotted knapweed (*Centaurea maculosa*), and in intermountain steppes, it is cheatgrass (*Bromus tectorum*). These exotic species are so widespread and such good dispersers that it is not possible to remove these disruptive organisms or to erase their effects.

The harsh realities of such disruptions of wildland ecosystems elicits two fundamentally different responses among ecologists and conservationists. Some argue for a hands-off policy that minimizes additional human intervention and lets nature take its course, perhaps producing ecological conditions and processes very different from historic conditions. The conceptual basis of this philosophy is discussed in McCool and Freimund (Chapter 13, this volume).

The other response to major disruptions in wildland ecosystems is to attempt to restore the natural processes and structures through application of concepts of restoration ecology (Cairns and Heckman 1996; Dobson et al. 1997), a specialty that has emerged in the last thirty years and now involves large numbers of scientists, technicians, students, lay conservationists, and land managers.

Restoration ecologists weigh and evaluate the relative merits and drawbacks of various restoration strategies. For instance, in the case of whitebark pine ecosystems, we have twenty years of experience with allowing some lightning fires to burn in some large natural areas. However, even the most successful program for reintroducing natural fires has proven inadequate in returning the historical fire regime to whitebark pine communities, because, for one thing, it does not allow fires to spread from lower elevations, in lands that are now developed (Brown et al. 1994). To counteract this deficit of natural fires, we could examine different prescribed fire strategies. Perhaps using some manager-ignited fires to restore a semblance of historic fire would be more beneficial than continuing to restrict fire. A similar kind of evaluation would be used in conjunction with studying alternative approaches for reducing the impacts of blister rust on whitebark pine ecosystems.

Understandably, scientists like to have an abundance of time to study ecological processes and interactions, and thus to evaluate the relative merits of various restoration approaches as compared with nonintervention. In the case of whitebark pine ecosystems, the impact of fire exclusion and blister rust are relatively clear and can be evaluated for any given area. The effects of alternative restoration strategies need testing, but initial results are encouraging (Chapter 18). In areas where the disruptions have had major impacts, an abundance of time is not available. For any disturbed natural community, the sooner ecologists become engaged in helping to restore a semblance of natural processes, the more likely that restoration will succeed. If, over time, too many components of the ecosystem are disrupted, only small-scale reconstructions of the ecosystem will be possible—as, for example, is the case with the once expansive midwestern prairie ecosystem (Samson and Knopf 1996).

The case for pro-active involvement of ecologists in restoration of whitebark pine ecosystems challenges learned people to apply their knowledge to develop restoration strategies. In a broader perspective, this is a call for ecologists to help maintain natural ecosystems in coexistence with human influences worldwide. We argue that this is a more realistic approach for maintaining these ecosystems than merely attempting to keep human influences out, which is useful, but not fully sufficient.

Literature Cited

Agee, J. K. 1993. Fire ecology of Pacific northwest forests. Island Press, Washington, D.C.

Arno, S. F. 1998. Fire ecology in Scandinavian forests: Parallels to western North America. Journal of Forestry 96:20–23.

Brown, J. K., S. F. Arno, S. W. Barrett, and J. P. Menakis. 1994. Comparing the prescribed natural fire program with presettlement fires in the Selway-Bitterroot Wilderness. International Journal of Wildland Fire 4:157–168.

Brown, P. M., and T. W. Swetnam. 1994. A cross-dated fire history from coast redwood near Redwood National Park, California. Canadian Journal of Forest Reseach 24:21–31.

Cairns, J., and J. R. Heckman. 1996. Restoration ecology: The state of an emerging field. Annual Review of Energy and the Environment 21:167–189.

Christiansen, N. L. 1981. Fire regimes in southeastern ecosystems. Pages 112–136 *in* H. A. Mooney, T. M. Bonnicksen, N. L. Christensen, J. E. Lotan, and W. A. Reinsers, editors. Fire regimes and ecosystem properties. USDA Forest Service, General Technical Report WO-26, Washington, D.C.

Dobson, A. P., A. D. Bradshaw, and A. J. M. Baker. 1997. Hopes for the future: Restoration ecology and conservation biology. Science 277:515–522.

Finney, M. A., and R. E. Martin. 1992. Short fire intervals recorded by redwoods at Annadel State Park, California. Madrono 39:251–262.

Fritz, E. 1931. The role of fire in the redwood region. Journal of Forestry 29:939–950.

Granström, A. 1996. Fire ecology in Sweden and future use of fire for maintaining biodiversity. Pages 445–452 *in* J. G. Goldammer and V. V. Furytaev, editors. Fire in ecosystems of boreal Eurasia. Kluwer Academic Publishers, Boston.

Means, J. E., J. H. Cissel, and F. J. Swanson. 1996. Fire history and landscape restoration in Douglas-fir ecosystems of western Oregon. Pages 61–67 *in* C. C. Hardy and S. F. Arno, editors. The use of fire in forest restoration. USDA Forest Service, Intermountain Research Station, General Technical Report INT-341, Ogden, Utah.

Morrison, P. H., and F. J. Swanson. 1990. Fire history and pattern in a Cascade Range landscape. USDA Forest Service, Pacific Northwest Forest and Range Experiment Station, General Technical Report PNW-254, Portland, Oregon.

Pinchot, G. 1899. The relation of forests and forest fires. National Geographic 10:393–403.

Samson, F. B., and F. L. Knopf (Editors). 1996. Prairie conservation: Preserving America's most endangered ecosystem. Island Press, Washington, D.C.

About the Contributors

Stephen F. Arno Research Forester, Retired, USDA Forest Service, Rocky Mountain Research Station, Fire Sciences Laboratory, P.O. Box 8089, Missoula, Montana 59807.

Leo P. Bruederle Associate Professor of Biology, Department of Biology, Campus Box 171, University of Colorado at Denver, P.O. Box 173364, Denver, Colorado 80217.

Karen E. Burr Horticulturist, USDA Forest Service, Coeur d'Alene Tree Nursery, 3600 Nursery Road, Coeur d'Alene, Idaho 83815.

Donna Dekker-Robertson Research Plant Geneticist, USDA Forest Service, Rocky Mountain Research Station, Forestry Sciences Laboratory, Moscow, Idaho 83843.

Kent Eggleston Horticulturist, USDA Forest Service, Coeur d'Alene Tree Nursery, 3600 Nursery Road, Coeur d'Alene, Idaho 83815.

Aram Eramian Forester, USDA Forest Service, Coeur d'Alene Tree Nursery, 3600 Nursery Road, Coeur d'Alene, Idaho 83815.

Dennis E. Ferguson Research Silviculturist and Project Leader, USDA Forest Service, Rocky Mountain Research Station, Forestry Sciences Laboratory, Moscow, Idaho 83843.

Wayne A. Freimund Arkwright Associate Professor of Wilderness Studies, School of Forestry, University of Montana, Missoula, Montana 59812.

Raymond J. Hoff Research Plant Geneticist, Retired, USDA Forest Service, Rocky Mountain Research Station, Forestry Sciences Laboratory, Moscow, Idaho 83843.

Dan E. Huff Wildlife Ecologist and Project Leader, Jackson Bison-Elk Management Plan, Mountain-Prairie Region, USDI Fish and Wildlife Service, P.O. Box 25286, Denver, Colorado 80225.

Robert E. Keane Research Ecologist, USDA Forest Service, Rocky Mountain Research Station, Fire Sciences Laboratory, P.O. Box 8089, Missoula, Montana 59807.

Katherine C. Kendall Research Ecologist, USDI, U.S. Geological Survey, Biological Resources Division, Glacier Field Station, Glacier National Park, West Glacier, Montana 59936.

Konstantin V. Krutovskii Senior Research Scientist and Courtesy Associate Professor, Department of Forest Science, Oregon State University, FSL 020, Corvallis, Oregon 97331; and Senior Research Scientist, Laboratory of Population Genetics, N. I. Vavilov Institute of General Genetics, Russian Academy of Sciences, Gubkin Str. 3, GSP-1 Moscow B-333, 117809, Russia.

David J. Mattson Research Wildlife Biologist, USDI, U.S. Geological Survey, Forest and Rangeland Ecosystem Science Center, Colorado Plateau Field Station, Northern Arizona University, P.O. Box 5614, Bldg. 24, Flagstaff, Arizona 86011.

Ward W. McCaughey Research Forester, USDA Forest Service, Rocky Mountain Research Station, Forestry Sciences Laboratory, Montana State University, Bozeman, Montana 59717.

Stephen F. McCool Professor of Wildland Recreation Management, School of Forestry, University of Montana, Missoula, Montana 59812.

Geral I. McDonald Research Plant Pathologist, USDA Forest Service, Rocky Mountain Research Station, Forestry Sciences Laboratory, Moscow, Idaho 83843.

Penelope Morgan Professor of Forest Resources, Department of Forest Resources, University of Idaho, Moscow, Idaho 83844.

Michael P. Murray Ecologist, Oregon Natural Heritage Program, 821 S.E. 14th Avenue, Portland, Oregon 97214.

Dmitri V. Politov Senior Research Scientist, Laboratory of Population Genetics, N. I. Vavilov Institute of General Genetics, Russian Academy of Sciences, Gubkin Str. 3, GSP-1 Moscow B-333, 117809, Russia.

Daniel P. Reinhart Management Biologist, USDI, National Park Service, Lake Ranger Station, Yellowstone National Park, Wyoming 82190.

Deborah L. Rogers Research Geneticist and Genetic Resources Analyst, Genetic Resources Conservation Program, University of California at Davis, One Shields Avenue, Davis, California 95616.

Hal Salwasser Dean, College of Forestry, 150 Peavy Hall, Oregon State University, Corvallis, Oregon 97331.

Wyman C. Schmidt Research Scientist and Project Leader, Retired, USDA Forest Service, Rocky Mountain Research Station, Forestry Sciences Laboratory, Montana State University, Bozeman, Montana 59717.

Diana F. Tomback Professor of Biology, Department of Biology and Center for Environmental Sciences, Campus Box 171, University of Colorado at Denver, P.O. Box 173364, Denver, Colorado 80217.

T. Weaver Professor of Plant Ecology, Department of Biology, Montana State University, Bozeman, Montana 59717.

Index

Italic numbers refer to figures; boldface numbers refer to tables.

Island Press Board of Directors